This monograph presents a complete, up-to-date guide to the theory of modern spectroscopy of atoms. It describes the contemporary state of the theory of many-electron atoms and ions, the peculiarities of their structure and spectra, the processes of their interaction with radiation, and some of the applications of atomic spectroscopy.

Atomic spectroscopy continues to be one of the most important subjects of contemporary physics. Spectra are fundamental characteristics of atoms and ions, and are the main source of information on their structure and properties. Modern atomic spectroscopy studies the structure and properties of practically every atom of the Periodical Table as well as of ions of any ionization degree. The book contains a large number of new results, which have been mainly published in Russian and are therefore almost unknown to western scientists.

A graduate level text and research reference, this book is aimed primarily at atomic physicists, astronomers and physical chemists, but will also be of value to physicists, astronomers and chemists in other areas who use spectroscopy in their work.

CAMBRIDGE MONOGRAPHS ON ATOMIC, MOLECULAR AND CHEMICAL PHYSICS

General Editors: A. Dalgarno, P. L. Knight, F. H. Read, R. N. Zare

THEORETICAL ATOMIC SPECTROSCOPY

CAMBRIDGE MONOGRAPHS ON ATOMIC, MOLECULAR AND CHEMICAL PHYSICS

Theoretical Atomic Spectroscopy

ZENONAS RUDZIKAS
Lithuanian Academy of Sciences

CAMBRIDGE UNIVERSITY PRESS
Cambridge, New York, Melbourne, Madrid, Cape Town, Singapore, São Paulo

Cambridge University Press
The Edinburgh Building, Cambridge CB2 8RU, UK

Published in the United States of America by Cambridge University Press, New York

www.cambridge.org
Information on this title: www.cambridge.org/9780521444255

First published 1997
This digitally printed version (with corrections) 2007

A catalogue record for this publication is available from the British Library

Library of Congress Cataloguing in Publication data

Rudzikas, Z. B. (Zenonas Bronislovaitis)
Theoretical atomic spectroscopy / Zenonasa Rudzikas.
p.   cm. – (Cambridge monographs on atomic, molecular, and chemical physics ; 7)
Includes bibliographical references.
ISBN 0 521 44425 X
1. Atomic spectroscopy.   I. Title.   II. Series
QC454.A8R89   1997
539.7′028′7–dc20   96-18913   CIP

ISBN 978-0-521-44425-5 hardback
ISBN 978-0-521-02622-2 paperback

Contents

Preface

*Nothing exists except atoms
and empty space; everything
else is opinion.*
Democritus

It has taken a very long time for this book to appear. For many years I had in my mind the idea of publishing it in English, but practical implementation became possible only recently, after drastic changes in the international political situation. The book was started in the framework (a realistic one) of the former USSR, and finished soon after its collapse, after my motherland, the Republic of Lithuania, regained its independence.

Academician of the Lithuanian Academy of Sciences, Adolfas Jucys, initiated the creation of a group of scientists devoted to the theory of complex atoms and their spectra. Later it was named the Vilnius (or Lithuanian) school of atomic physicists, often called by his name. However, for many years the results of these studies were published largely only in Russian and, therefore as a rule they were not known among Western colleagues, particularly those in English-speaking countries. A large number of the papers were published in Russian in the main Lithuanian physical journal *Lietuvos fizikos rinkinys* – *Lithuanian Journal of Physics*, translated into English by Allerton Press, Inc. (New York) as *Soviet Physics* – *Collection* (since 1993 – *Lithuanian Physics Journal*).

Recently the situation has become incomparably better. There is no problem publishing the main ideas and results in prestigious international journals in English. However, it would be very useful to collect, to analyse and to summarize the main internationally recognized results on the theory of many-electron atoms and their spectra in one monograph, written in English. This book is the result of the long process of practical realization of that idea. Many scientists from an international community of atomic physicists as well as many scientific institutions in various countries in one way or another have helped, encouraged me to start, to go ahead, to accomplish it. It is impossible to overestimate their role.

Let us enumerate some of them: Lithuanian Academy of Sciences, State Institute of Theoretical Physics and Astronomy, Lithuanian Branch of the International Centre for Scientific Culture – World Laboratory, Nordisk Institut for Teoretisk Fysik (NORDITA) in Denmark, the University of Lund in Sweden, International Science Foundation.

I am extremely grateful to Alexander Dalgarno, who inspired and encouraged me to start writing the book. I wish to express my deep appreciation to M. Cornille, C. Froese Fischer, B. C. Fawcett, I. P. Grant, W. R. Johnson, B. Judd, J. Karwowski, E. Knystautas, I. Lindgren, I. Martinson, J. Migdalek, C. A. Nicolaides, G. zu Putlitz, P. Pyykko, B. G. Wybourne and to many others for providing me with their papers and for numerous discussions by electronic mail or during various meetings. Special thanks to my colleagues in Lithuania, with whom I was cooperating for many years and whose names are represented in the list of References.

It is my pleasure to acknowledge the valuable help of Richard Ennals (Kingston University, Kingston upon Thames, UK) in improving the grammar and style of the book and to Gediminas Vilutis for converting the manuscript to an electronic version. I also wish to thank the staff of Cambridge University Press, who were so friendly, kind and helpful. And last but not least, I am much indebted to my wife Marija, son Andrius and daughter Gražina, who saw much less of me during the several years I was absorbed in writing and polishing the manuscript.

Zenonas Rudzikas
State Institute of Theoretical Physics and Astronomy
A. Goštauto 12, Vilnius 2600, Lithuania

Foreword to the Paperback Edition

1 General remarks

Ten years have elapsed since first publication of this monograph. This may, at first glance, seem a very short time in what has been a very stable area of fundamental physics. But the last decade has seen unprecedented technological progress and socio-economic changes. Have there been radical changes in "Theoretical Atomic Spectroscopy" in that time?

The answer is very much "yes". Just a few examples driving current interest in this subject are:

- In the explosively growing nanoscience area we are approaching the molecular and atomic scales where quantum mechanical effects become significant. This is opening new horizons in the nanotechnologies, including quantum computing and quantum cryptography.

- In astrophysics, there has been a requirement for much improved spectroscopy of highly charged ions.

- The practical considerations necessary for building thermonuclear power plants (driven through ITER, DEMO and other projects) require deeper insight into the structure and properties of ions, atoms and their separate isotopes.

- Progress in fission research, particularly in the development of higher efficiency reactors, in radioactive waste management, in transmutation studies and transuranic elements have contributed substantially to a renewed interest in the structure and properties of complex many-electron atoms and ions of arbitrary ionization.

Even before publication of the first edition of this book it became obvious to me that there was more that needed to be said. A very important development had been the formulation and practical implementation of a new version of Racah algebra based on the second-quantization method in the coupled tensorial form, angular momentum theory in three spaces (orbital, spin and quasispin) and a generalized graphical technique. It requires neither coefficients of fractional parentage nor unit tensors. Both diagonal and non-diagonal matrix elements of electronic configurations of any part of the energy operator are treated uniformly in such an approach. In order not to delay publication, I decided to mention these new developments in an Epilogue (pages 404 to 406).

The last sentence of the Epilogue reads: "Who will write a monograph on theoretical atomic spectroscopy based on such an approach?". Ten years on, it is still not written. But the situation is not hopeless. This approach has been the subject of a number of articles (see references at the end of this Foreword). Moreover, it is implemented in a number of computer codes.

I have taken the opportunity in this Foreword to the paperback edition of my book, to briefly describe some of these aspects.

A small errata list, and new references, appear at the end of the new Foreword.

2   Progress in calculation of spin-angular parts of matrix elements

Any two-particle operator $\widehat{G}$ may be generally presented in the following irrreducible form [1, 2]:

$$\widehat{G}^{(\kappa_1\kappa_2k,\sigma_1\sigma_2k)} = \sum_\alpha \sum_{k_{12}\sigma_{12}k'_{12}\sigma'_{12}} \Theta(\Xi)\Big\{ A_{p,-p}^{(kk)}(n_\alpha\lambda_\alpha,\Xi)\delta(U,1)$$

$$+ \sum_\beta \Big[B^{(k_{12}\sigma_{12})}(n_\alpha\lambda_\alpha,\Xi) \times C^{(k'_{12}\sigma'_{12})}(n_\beta\lambda_\beta,\Xi) \Big]_{p,-p}^{(kk)} \delta(U,2)$$

$$+ \sum_{\beta\gamma} \Big[[D^{(l_\alpha s)} \times D^{(l_\beta s)}]^{(\kappa_{12}\sigma_{12})} \times E^{(\kappa'_{12}\sigma'_{12})}(n_\gamma\lambda_\gamma,\Xi) \Big]_{p,-p}^{(kk)} \delta(U,3)$$

$$+ \sum_{\beta\gamma\delta} \Big[[D^{(l_\alpha s)} \times D^{(l_\beta s)}]^{(\kappa_{12}\sigma_{12})} \times [D^{(l_\gamma s)} \times D^{(l_\delta s)}]^{(\kappa'_{12}\sigma'_{12})} \Big]_{p,-p}^{(kk)} \Big\}\delta(U,4). \quad (1)$$

Thus, it is expressed in terms of operators of the type $A_{p,-p}^{(kk)}(n_\alpha\lambda_\alpha,\Xi)$, $B^{(k_{12}\sigma_{12})}(n_\alpha\lambda_\alpha,\Xi)$, $C^{(k'_{12}\sigma'_{12})}(n_\beta\lambda_\beta,\Xi)$, $D^{(l_\alpha s)}$ and $E^{(\kappa'_{12}\sigma'_{12})}(n_\gamma\lambda_\gamma,\Xi)$. These quantities may be written as the tensorial products presented in the Epilogue, whereas the multiplier $\Theta(\Xi)$ is proportional to the general two-electron submatrix element $(n_i\lambda_i n_j\lambda_j||g||n_{i'}\lambda_{i'}n_{j'}\lambda_{j'})$. Here U refers to the number of shells upon which the relevant operator acts.

Concrete irreducible forms of physical operators follow from (1) by defining ranks k_1, k_2, σ_1, σ_2, k and specifying the two-electron submatrix element (see e. g. [3]). Ξ represents an array of intermediate coupling parameters of the tensorial form of the operator, including k_{12}, σ_{12}, k'_{12}, σ'_{12}.

In order to calculate the spin-angular parts of matrix elements of the two-particle operator (1) with an arbitrary number of open shells, it is necessary to consider all possible distributions of shells upon which the second quantization operators are acting. In [2] they are found to be grouped into 42 different distributions, subdivided into 4 different classes. This also explains why operator (1) is written as the sum of four complex terms. The first term represents the case when all second-quantization operators act upon the same shell (distribution 1 in [2]), the second describes the situation when these operators act upon the two different shells (distributions 2–10), third and fourth are in charge of the interactions upon three and four shells respectively (distributions 11–18 and 19–42). Such expression is particularly convenient to take into account correlation effects, because it describes all possible superpositions of configurations for the case of two-electron operator.

The general form of the submatrix element of the operator between functions with u open shells (1) may be presented as follows [4]:

$$(\psi_u(LS)\|G\|\psi_u(L'S'))$$

$$= \sum_{n_i\lambda_i,n_j\lambda_j,n'_i\lambda'_i,n'_j\lambda'_j} (\psi_u(LS)\|\widehat{G}(n_i\lambda_i, n_j\lambda_j, n'_i\lambda'_i, n'_j\lambda'_j)\|\psi_u(L'S'))$$

$$= \sum_{n_i\lambda_i,n_j\lambda_j,n'_i\lambda'_i,n'_j\lambda'_j} \sum_{\kappa_{12},\sigma_{12},\kappa'_{12},\sigma'_{12}} \sum_{K_l,K_s} (-1)^{\Delta}\, \Theta'(n_i\lambda_i, n_j\lambda_j, n'_i\lambda'_i, n'_j\lambda'_j, \Xi)$$

$$\times T(n_i\lambda_i, n_j\lambda_j, n'_i\lambda'_i, n'_j\lambda'_j, \Lambda^{bra}, \Lambda^{ket}, \Xi, \Gamma)\, R(n_i, n_j, n'_i, n'_j, \Lambda^{bra}, \Lambda^{ket}, \Gamma), \quad (2)$$

where $\lambda \equiv l, s$, $\Lambda^{bra} \equiv (L_i, L_j, L'_i, L'_j)^{bra}$, $\Lambda^{ket} \equiv (S_i, S_j, S'_i, S'_j)^{ket}$ and Γ refers to the array of coupling parameters connecting the recoupling matrix $R(n_i, n_j, n'_i, n'_j, \Lambda^{bra}, \Lambda^{ket}, \Gamma)$ to $T(n_i\lambda_i, n_j\lambda_j, n'_i\lambda'_i, n'_j\lambda'_j, \Lambda^{bra}, \Lambda^{ket}, \Xi, \Gamma)$ – the submatrix element.

Similar expressions are found for the case of relativistic wave functions.

This methodology is implemented in a number of published universal computer codes facilitating theoretical studies of complex many-electron atoms and ions. For example, a universal multiconfiguration Hartree–Fock atomic structure package for large-scale calculations based on such a methodology is described in [5]. Analogical relativistic configuration interaction calculations may be performed using [6]. Codes described in [7,8] allow transformation of wave functions in an intermediate coupling scheme based on LS coupling, to those based on jj coupling and vice versa. These transformations also cover the case of changes of coupling scheme in a shell of equivalent electrons l^N.

The second group of codes is written adopting Maple symbolic procedures and are intended for studying the mathematical expressions of the theory of an atom. For example, computer code [9] considers the spin-angular coefficients for single-shell configurations, whereas [10] deals with hyperfine structure parametrisation in Maple.

3   Accounting for correlation effects

Let us also briefly discuss some developments in accounting for correlation effects while performing calculations of spectral properties of complex many-electron atoms and ions. The method of superposition of configurations based on the transformed radial orbitals, briefly described in Chapter 29, has demonstrated its effectiveness particularly for complex electronic configurations. Code for transformed radial orbitals is described in [11]. There the following transformed radial orbitals are recommended:

$$P_{\mathrm{TRO}} = N\Big[f(k, m, B|r)\, P_{\mathrm{HF}}(n_0 l_0|r)$$

$$- \sum_{n' < n} P(n'l|r) \int P(n'l|r)\, f(k, m, A|r)\, P_{\mathrm{HF}}(n_0 l_0|r)\, \mathrm{d}r\Big], \quad (3)$$

where

$$f(k, m, B|r) = r^k \exp(-Br^m), \ k \geq l - l_0, \ k > 0, \ m > 0, \ B > 0. \quad (4)$$

The second term in equation (3) ensures the orthogonality of the transformed radial orbital obtained. The criterion for determining the optimal values of integer parameters k and m and the freely variable parameter B (or A) is the maximum of the averaged energy correction, expressed in the second order of perturbation theory.

Analysis of the properties of the transformed radial orbitals testify that these functions are a realistic practical alternative to the solutions of multi-configuration (Hartree–Fock–Jucys) equations. Moreover, the generation of solutions using the code is fairly simple and not computer time consuming even for heavy atoms and ions.

Efficient methods of selection of admixed configurations in this approach are described in [12]. The computer code for generating the reduced configuration state list is presented in [13]. A special diagonalisation method is developed to reduce the orders of matrices to be diagonalised [14]. A more detailed overview of the methods considered may be found in [15]. All these improvements considerably speed up the calculations and widen the domains of applicability of the methods developed.

4 Errata

The table below lists some misprints in the monograph.

Page	Line or formulae	Printed	Must be
43	(5.21) and (5.22)	$j'_2 j$	$j'_2 j'$
45	10 from below	V^{k+1}	V^{k1}
49	(6.11)	$\sqrt{2j+1}$	$1/\sqrt{2j+1}$
81	18 from below	46,	26,
121	(14.1)	$\sqrt{2}$	$1/\sqrt{2}$
138	(15.2)	$\sqrt{2}$	$1/\sqrt{2}$
166	8 from above	odd	even
200	(18.1)	$t^{(1)}_+$	$t^{(1)}_-$
200	(18.2)	$t^{(1)}_-$	$t^{(1)}_+$
210	(18.49)	$V^{(100)}_{\Gamma'00}$	$\widetilde{V}^{(100)}_{\Gamma 00}$
262	(22.2)	C^k	$C^{(k)}$

References

[1] G. Gaigalas, Z. Rudzikas, On the secondly quantized theory of the many-electron atom, *J. Phys. B: At. Mol. Opt. Phys.*, **29**, 3303–3318 (1996).

[2] G. Gaigalas, Z. Rudzikas, Ch. Froese Fischer, An efficient approach for spin-angular integrations in atomic structure calculations, *J. Phys. B: At. Mol. Opt. Phys.*, **30**, 3747–3771 (1997).

[3] G. Gaigalas, A. Bernotas, Z. Rudzikas, Ch. Froese Fischer, Spin–other-orbit operator in the tensorial form of second quantization, *Physica Scripta*, **57**, 207–212 (1998).

[4] G. Gaigalas, Integration over spin-angular variables in atomic physics, *Lithuanian J. Phys.*, **39**, 79–105 (1999).

[5] Ch. Froese Fischer, G. Tachiev, G. Gaigalas, M. Godefroid, An MCHF atomic-structure package for large-scale calculations, *Comput. Phys. Commun.*, **176**, 559–579 (2007).

[6] S. Fritzsche, Ch. Froese Fischer, G. Gaigalas, RELCI: A program for relativistic configuration interaction calculations, *Comput. Phys. Commun.*, **148**, 103–123 (2002).

[7] T. Žalandauskas, G. Gaigalas, Routines for transformation of atomic state functions from LS- to jj-coupled basis, *Lithuanian J. Phys.*, **43**, 15–26 (2003).

[8] G. Gaigalas, T. Žalandauskas, S. Fritzsche, Spectroscopic LSJ notation for atomic levels obtained from relativistic calculations, *Comput. Phys. Commun.*, **157**, 239–253 (2004).

[9] G. Gaigalas, O. Scharf, S. Fritzsche, Maple procedures for the coupling of angular momenta. VIII. Spin-angular coefficients for single-shell configurations, *Comput. Phys. Commun.*, **166**, 141–169 (2005).

[10] G. Gaigalas, O. Scharf, S. Fritzsche, Hyperfine structure parametrisation in Maple, *Comput. Phys. Commun.*, **174**, 202–221 (2006).

[11] P. Bogdanovich, R. Karpuškienė, Transformed radial orbitals with a variable parameter for the configuration interaction, *Lithuanian J. Phys.*, **39**, 193–208 (1999).

[12] P. Bogdanovich, R. Karpuškienė, A. Momkauskaitė, A program of generation and selection of configurations for the configuration interaction method in atomic calculations SELECTCONF, *Comput. Phys. Commun.*, **172**, 133–143 (2005).

[13] P. Bogdanovich, A. Momkauskaitė, A program for generating configuration state lists in many-electron atoms, *Comput. Phys. Commun.*, **157**, 217–225 (2004).

[14] P. Bogdanovich, R. Karpuškienė, A. Momkauskaitė, Some problems of calculation of energy spectra of complex atomic configurations, *Comput. Phys. Commun.*, **143**, 174–180 (2002).

[15] P. Bogdanovich, Modern methods of multiconfiguration studies of many-electron highly charged ions, *Nucl. Instrum. Meth. Phys. Research*, **B235**, 92–99 (2005).

Zenonas Rudzikas
Lithuanian Academy of Sciences and
Vilnius University Research Institute of Theoretical Physics and Astronomy
May 3, 2007

Introduction

Atomic spectroscopy continues to be one of the most important branches of contemporary physics, and has a wealth of practical applications. Many domains of modern physics and other fields of science and technology utilize atomic spectroscopy. Modern atomic spectroscopy studies the structure and properties of all atoms of the Periodical Table, and ions of any ionization degree.

Spectra are fundamental characteristics of atoms and ions, the main source of information on their structure and properties. Diagnostics of both laboratory and astrophysical plasma is carried out, as a rule, on the basis of these spectra. Nowadays the possibilities of theoretical spectroscopy are much enlarged thanks to wide usage of powerful computing devices. This enables one to investigate fairly complicated mathematical models of the system under consideration and to obtain in this way results which are in close agreement with experimental measurements. Interest in spectroscopy has increased particularly in connection with the rise of non-atmospheric astrophysics and laser physics. New important applied problems have arisen: diagnostics of thermonuclear plasmas, creation of lasers generating in the X-ray region, identification of solar and stellar spectra, studies of Rydberg, 'hollow' atoms, etc. Due to the use of laser-produced plasmas, powerful thermonuclear equipments (Tokamaks), low inductance vacuum sparks, exploding wires, beam-foil spectroscopy, and other advanced ion sources and ion traps (such as EBIS, EBIT, ECR, etc.) there was discovered a new, extremely interesting and original world of very highly ionized atoms, their radiation being, as a rule, in the far ultraviolet and even the X-ray wavelengths region. The spectra due to the transitions between highly excited (Rydberg) levels are studied intensively. Investigations of complex atoms and ions with two, three and even more open shells are needed. Identification of the spectra of all these above-mentioned systems is practically impossible without corresponding theoretical investigations.

The data of atomic spectroscopy are of extreme importance in revealing the nature of quantum-electrodynamical effects. For the investigation of many-electron atoms and ions, it is of great importance to combine theoretical and experimental methods. Therefore, the methods used must be universal and accurate. A number of physical characteristics of the many-electron atom (e.g., a complete set of quantum numbers) may be found only on the basis of theoretical considerations. In many cases the mathematical modelling of physical objects and processes using modern computers may successfully replace the corresponding experiments. In this book we shall describe the contemporary state of the theory of many-electron atoms and ions, the peculiarities of their structure and spectra as well as the processes of their interaction with radiation, and some applications.

From the theoretical point of view the atom is considered a many-body problem which, as is well known, cannot be solved exactly. Normally, we replace the Schrödinger equation, unsolvable for the many-electron atoms, by the simplified eigenvalue problem, which can be quite easily solved for any atom or ion. In this way, instead of considering the wave function of the whole atom, we find it for each electron moving in the central nuclear charge field and in the screening field of the remaining electrons. The wave function of this electron is represented as a product of radial and spin-angular parts. The radial part is usually found by solving various modifications of the Hartree–Fock equations and can be represented in numerical or analytical form, whereas the angular part is expressed in terms of spherical functions. Then, the wave function of the whole atom can be constructed in some standard way, starting with these one-electron functions, and may be used further on for the calculation of any matrix elements, representing physical quantities.

The starting point of the creation of the theory of the many-electron atom was the idea of Niels Bohr [1] to consider each electron of an atom as orbiting in a stationary state in the field, created by the charge of the nucleus and the rest of the electrons of an atom. This idea is several years older than quantum mechanics itself. It allows one to construct an approximate wave function of the whole atom with the help of one-electron wave functions. They may be found by accounting for the approximate states of the passive electrons, in other words, the states of all electrons must be consistent. This is the essence of the self-consistent field approximation (Hartree–Fock method), widely used in the theory of many-body systems, particularly of many-electron atoms and ions. There are many methods of accounting more or less accurately for this consistency, usually named by correlation effects, and of obtaining more accurate theoretical data on atomic structure.

Since that time the theory of the structure of complex atoms and ions, as well as of their spectra, has rapidly and continuously developed, and has been described in many monographs [2–18]. Nowadays, it changes quickly, too. So do the experimental studies. Many interesting and important achievements are not yet summarized in monographs. Therefore, it is worthwhile trying to analyse the latest results in order to formulate problems to be solved in future. Data on atomic structure and properties are necessary for many domains of physics, astrophysics and neighbouring sciences. So, for example, atomic data are required even when studying the first hundreds of seconds of our Universe after the Big Bang [19]. This book intends to contribute to these issues. Special attention is paid to complex atoms and ions with several open electronic shells. Let us mention a few milestones in this field.

The monograph of Condon and Shortley [2] was a major work of reference for a whole generation of spectroscopists [20]. It treats an atom in the central-field approximation and it does not require deep knowledge of the theory of groups.

The next milestone was a group of papers by Racah [21–24]. Racah innovated the theory of atomic spectra by introducing spherical tensorial operators, the components of which satisfy certain commutation relations with respect to the components of an angular momentum vector. Thus, he defined the components of a spherical tensor by means of their commutation relations. To calculate matrix elements of irreducible tensorial operators he applied the so-called Wigner–Eckart theorem, the importance of which to spectroscopy it is difficult to overestimate. He widely utilized the vectorial coupling of angular momenta, $3nj$-coefficients and the coefficients of fractional parentage. All this is often called the Racah algebra. It opened the way to developing a consistent theory of complex atomic spectra, using this algebra for the studies.

The third important step was made by Jucys and his Lithuanian school of theoretical physicists in Vilnius. To calculate matrix elements of various operators, including those corresponding to physical quantities, particularly for complex configurations with several open shells, one has to cope with numerous sums over quantum numbers of the momentum type and their projections, leading to the occurrence of the so-called $3nj$- and jm-coefficients. The most general way to achieve this is through the exploitation of graphical methods developed by Jucys and his co-workers [9, 11, 25]. Now they are widely used in many other domains of physics, including nuclei, elementary particles, etc. [13]. They were also applied to perturbation theory [17].

The main advantage of graphical methods is their general character. Using them, one can sum any product of any combination of $3nj$- and (or) jm-coefficients. The latest generalization of the graphical technique [26] allows one to calculate graphically the irreducible tensorial products of the

second-quantization operators and their commutators, to find graphically
the normal form of the complex tensorial products of the operators,
their vacuum averages, etc. This allows one to apply it easily to atomic
perturbation theory for complex atoms and ions and even to quantum
electrodynamics.

During the last two decades a number of new versions of the Racah
algebra or its improvements have been suggested [27]. So, the exploitation
of the community of transformation properties of irreducible tensors and
wave functions allows one to adopt the notion of irreducible tensorial sets,
to deduce new relationships between the quantities considered, to simplify
further on the operators already expressed in terms of irreducible tensors,
or, in general, to offer a new method of calculating the matrix elements,
as an alternative to the standard Racah way. It is based on the utilization
of tensorial products of the irreducible operators and wave functions, also
considered as irreducible tensors.

Another option is the exploitation of the so-called generalized spherical
functions $D_{m,m'}^{(j)}$ instead of the usual spherical functions (harmonics) $Y_{m_l}^{(l)}$. In
this approach one can express all operators in terms of these D-functions.
It turned out [28] that it is possible to present a new form of relativistic
atomic wave function, the angular part of which is of the form:

$$\psi_{lm}^{(j)} = \sqrt{\frac{2j+1}{8\pi}} \begin{pmatrix} D_{m,1/2}^{(j)} \\ (-1)^{l+1/2-j} & D_{m,-1/2}^{(j)} \end{pmatrix} \tag{1}$$

where l and j are the quantum numbers of orbital and total angular
momenta of an electron, respectively. An expression of this kind also
exists for the non-relativistic case [29]. Their exploitation considerably
simplifies the process of finding the corresponding matrix elements of
physical operators, and leads to more convenient and simpler expressions
involving only radial integrals and phase multipliers which depend on
orbital quantum numbers.

By accounting for the symmetry properties of a many-electron system
via the theory of groups and their representations, one is in a position
to simplify the problem to be solved, to reduce it to the problem of one
or two particles moving in some external field. Another advantage of the
utilization of group theoretical methods is use of the parameters labelling
the representations of groups as sets of quantum numbers to classify
energy levels, an exploitation of the corresponding technique to find new
important relationships between the quantities of the theory of complex
spectra. This allows one to simplify the corresponding expressions or to
find analytical (algebraic) formulas for many such quantities.

An important step was made by applying the second-quantization
method to theoretical atomic spectroscopy [12, 30, 31]. It turned out

that instead of the seniority quantum number v it was possible to define a quasispin quantum number by $Q = \frac{1}{2}(2l + 1 - v)$ as well as a corresponding operator of the spin angular momentum type and its projection $M_Q = -\frac{1}{2}(2l + 1 - N)$, where N is the number of electrons in the shell nl^N. All angular momentum theory may be generalized to account for the new quasispin space. For example, applying the Wigner–Eckart theorem in the quasispin space we arrive at the dependence of matrix elements of any tensor on N in the form of a standard Clebsch–Gordan coefficient. This considerably simplifies many calculations. Both diagonal and off-diagonal matrix elements with respect to the configurations are treated in this approach in a uniform way, because their expressions follow from one general formula.

Coupling vectorially the quasispin operators of each shell $\mathbf{Q} = \mathbf{Q}_1 + \mathbf{Q}_2$ and assuming $M_Q = -\frac{1}{2}\{2(l_1 + l_2 + 1) - N\}$, where $N = N_1 + N_2$, we generalize the quasispin concept to cover the case of complex electronic configurations. Then we can define the total quasispin quantum number for any configuration. For two shells of equivalent electrons we have, in such a case, the following wave function:

$$|(l_1 + l_2)^N \alpha_1 Q_1 L_1 S_1 \alpha_2 Q_2 L_2 S_2 QLS)$$

$$= \sum_{M_1 M_2} \begin{bmatrix} Q_1 & Q_2 & Q \\ M_1 & M_2 & M \end{bmatrix} |l_1^{N_1} l_2^{N_2} \alpha_1 Q_1 L_1 S_1 \alpha_2 Q_2 L_2 S_2 LS). \qquad (2)$$

Formula (2) actually presents a new version of the superposition-of-configurations method. Its main advantage is that the weights of the superposed wave functions are the usual Clebsch–Gordan coefficients, in contrast to the standard method, where these weights are obtained after diagonalization of the energy matrix. There are cases (e.g., some configurations with doubly excited electrons) for which total Q is a fairly exact quantum number.

The use of the tensorial properties of both the operators and wave functions in the three (orbital, spin and quasispin) spaces leads to a new very efficient version of the theory of the spectra of many-electron atoms and ions. It is also developed for the relativistic approach.

For configurations of the type $n_1 l^{N_1} n_2 l^{N_2}$, additional symmetry properties exist, described by the so-called isospin basis [32]. In this basis two electrons belonging to different shells with the same l are interpreted as two states of a particle, having an additional degree of freedom in the isospin space. The wave function of $N = N_1 + N_2$ electrons in this basis may be found with the help of vectorial coupling of isospin, angular and spin momenta of separate electrons and then, instead of the quantum numbers $\alpha_i L_i S_i$ of separate shells, we have new isospin quantum number T. The corresponding operator $T_q^{(1)}$ is the irreducible tensor in the additional isospin space. In this way we offer a new version of the Racah algebra for such electronic configurations, again covering both the non-relativistic and relativistic cases [31, 32].

The wave function of isospin basis may be expressed in terms of the usual functions of separate shells with the help of some transformation matrix:

$$
\begin{aligned}
&\psi(n_1 n_2 (ll)^N \alpha T L S M_T M_L M_S) \\
&= \sum_{\alpha_1 L_1 S_1 \alpha_2 L_2 S_2} \psi(n_1 n_2 l^{N_1} l^{N_2} \alpha_1 L_1 S_1 \alpha_2 L_2 S_2 L S M_L M_S) \\
&\quad \times (l^{N_1} l^{N_2} \alpha_1 L_1 S_1 \alpha_2 L_2 S_2 L S | (ll)^N \alpha T L S M_T), \quad N = N_1 + N_2. \quad (3)
\end{aligned}
$$

Here $T = N/2, N/2 - 1, \ldots 1/2$ (or 0) is the isospin quantum number, whereas its projection is defined by $M_T = (N_2 - N_1)/2$. Thus, in this basis two shells make a 'supershell', being completely filled for $N = 8l + 4$.

The use of isospin basis allows one to widen the domain of applicability of the Brillouin theorem for excited states. In this approach it is also possible to account for part of the correlation effects. There are cases (e.g., configurations of the type $n_1 l n_2 l^{N_2}, n_1 l^{4l+1} n_2 l^{N_2}$ $(n_2 = n_1 + 1)$ of multiply charged ions) where the isospin quantum number is fairly exact [31, 32].

Configuration l^N may be considered as the configuration consisting of $4l + 2 - N$ holes (here $0 \leq N \leq 4l + 2$). All versions of the Racah algebra may also be rewritten for holes. The wave functions in particle and hole representations differ from each other only by a phase multiplier. The utilization of the particle–hole formalism gives optimal phase relations between the quantities connecting partially and almost filled shells. It turned out that again there are cases (e.g., $np^5 nd^2 = n\tilde{p} nd^2$ $n = 3$ and 4 for K and Pb isoelectronic sequences) when the particle–hole basis $(n\tilde{p} nd)nd$ is closer to reality than the usual basis $n\tilde{p} nd^2$. This shows that in such cases there exist, even in a separate atom, particle–hole pairs $(n\tilde{p} nd)$. The main part of the exchange electrostatic interaction is diagonal in this basis and the energy levels of such configurations are concentrated in two groups, leading to the energetic gap between them. Does this show that a separate atom can be an embryo of solid state?

The main shortcoming of the classical Racah algebra is the necessity to use tables of numerical values of many quantities. Adopting both the second-quantization and quasispin methods accounting for the properties of atomic quantities in quasispin space, algebraic expressions were found for angular parts of the wave functions and coefficients of fractional parentage in the case of all p^N and d^N electrons, as well as for f^N configurations ($N = 1 - 4, 10 - 14$). This means also that in the relativistic case there are algebraic expressions for coefficients of fractional parentage of any relativistic electronic shell needed in practice. A similar approach is also developed for isospin basis [31]. This also allows one to find algebraic expressions for a number of physical quantities.

There have been many attempts in atomic spectroscopy to use methods based on other groups or chains of groups, however, without much success [20, 33, 34]. Fortunately, the utilization of the so-called orthogonal operators [20] in semi-empirical studies of atomic spectra looks fairly promising. Their main advantage is that in this approach a least-squares fitting guarantees practically no changes in a set of fitted parameters when a new one is added. Usually, utilization of this method leads to smaller root-mean-square deviations of the calculated results from experimental ones compared to a standard fitting procedure.

Recently Judd [35] suggested that instead of thinking of the 15 electronic configurations f^N $(0 \leq N \leq 14)$ as distinct entities, one should construct all 16 384 states of the f-shell by coupling together four independent quark-like objects and including two parity labels. Each quark-like object is eight-dimensional and has the angular-momentum structure $s + f$. This model was developed further in [36–40]. Such an approach enabled the discovery of certain selection rules for some three-electron operators [41], and unexpected relations and proportionalities between their matrix elements in the atomic f-shell [42–43].

At first sight, it seems that much less progress was achieved in calculations of radial wave functions and radial integrals. Fortunately, this is not true. Both analytical and numerical radial wave functions are widely used, and can be obtained fairly easily for both non-relativistic and relativistic approaches. The corresponding methods are described in monographs [16, 18, 44, 45]. There exists a wealth of efficient computer programs to calculate various atomic structures and processes of the interaction of atoms with fields and other particles in various approximations. Let us mention only the names of C. Froese Fischer, R. D. Cowan and I. P. Grant in this connection. Many computational aspects of the theory of atoms and molecules (computer codes, simulations, supercomputing, parallelism for numerical algorithms, graphics and animation, interconnected moduli, superstructure of computer programs, software for personal computers, workstations and supercomputers) are discussed in detail in [46]. The necessity to utilize non-orthogonal radial orbitals and their theory are reviewed in [47].

There exist a number of methods to account for correlation [17, 45, 48] and relativistic effects as corrections or in relativistic approximation [18]. There have been numerous attempts to account for leading radiative (quantum-electrodynamical) corrections, as well [49, 50]. However, as a rule, the methods developed are applicable only for light atoms with closed electronic shells plus or minus one electron, therefore, they are not sufficiently general.

Thus, making use of modern methods of theoretical atomic spectroscopy and available computer programs, one is in a position to fulfill more or less accurate purely theoretical (*ab initio*) or semi-empirical calculations of the energy spectra, transition probabilities and of the other spectroscopic characteristics, in principle, of any atom or ion of the Periodical Table, their isoelectronic sequences, revealing in this way their structure and properties, to model the processes in low- and high-temperature plasma. Such calculations could be done prior to the corresponding experimental measurements, instead of them, or after them to help to interpret the interesting phenomena found in experimental studies.

Many interesting features have been revealed during such studies: the necessity to use various (differing from the usual *LS*) coupling schemes, the role of forbidden transitions, occurrence of satellite lines, unresolved transition arrays, splitting of electronic shells into subshells, etc.

However, there are still many problems to be solved, including: the necessity of further development of new and the improvement of already existing methods and computer-based software for theoretical modelling of the structure and spectra of many-electron atoms and ions, including very highly ionized ones; evaluation of the accuracy and the regions of validity of the methods and programs used; generation of highly accurate parameters describing the structure of many-particle systems and processes of their interaction with electromagnetic fields; standardization and systematization of the data obtained, creation of the corresponding computer databases, critical data compilation and tabulation; application of the theoretical results obtained for the interpretation of the data and processes in the thermonuclear, laser, non-atmospheric astrophysical plasma as well as in the other plasma sources; carrying out mathematical modelling of the objects and processes mentioned above; contribution to the checking of the fundamental laws of physics (parity violation, quantum-electrodynamical effects, etc.). All this will enrich our knowledge of the nature of the objects and phenomena under investigation.

The majority of the above-mentioned problems will be discussed in more or less detail in this book. The book is based on the variational approach, which is the most universal and efficient method of theoretical study of the spectra of any atom or ion of the Periodical Table. However, generalization of the perturbation theory to cover the case of configurations with several open shells is also presented [26, 51, 52].

Particular attention is paid to recently worked out methods of theoretical description of highly ionized atoms. They allow us to take into account relativistic effects. Accounting for them both as corrections and in relativistic approximation, as well as the use of various (differing from traditional *LS*) coupling schemes is considered in detail.

As concerns possible overlaps of this book with other books on a similar subject, it is worthwhile underlining that they are fairly small. This book contains a large number of new original results in the theory of many-electron atoms and ions, published, as a rule, in Russian and, therefore, almost unknown to Western scientists and not yet included in other monographs. Examples include second-quantization in a coupled tensorial form, quasispin and isospin, exploitation of group-theoretical symmetry properties, a large number of analytical expressions for the coefficients of fractional parentage and various matrix elements, generalized graphical methods and expressions for electric multipole radiation operators, relativistic approach, new type of wave function in relativistic jj coupling, brief theory of X-ray and electronic spectra, statistical methods to describe unresolved spectra, a general method to transform the wave functions of complex configurations from one coupling scheme to another, allowing classification of the energy spectra of any atom and ion with the help of an optimal coupling scheme, etc.

In fact, it is the only book in which you can find successive general non-relativistic and relativistic descriptions of the theory of energy spectra and transition probabilities in complex many-electron atoms and ions. The formulas and tables presented give the possibility, at least in principle, of calculating the energy spectra and electronic transitions of any multipolarity for any atom or ion of the Periodical Table. This book contains the bulk of new achievements in the non-relativistic and relativistic theory of an atom, especially as concerns the many-particle aspects of the non-relativistic and relativistic problem. It therefore complements books already available.

Although the material contained in this book concerns the theory of many-electron atoms and ions, its many ideas and methods (e.g., graphical methods, quasispin and isospin techniques, particle–hole formalism, etc.) are fairly universal and may be easily applied (or already are) to other domains of physics (nuclear theory, elementary particles, molecular, solid state physics, etc.).

The monograph is dedicated to those who are interested in the theory of many-electron atoms and ions, including very highly ionized ones, in the fundamental and applied spectroscopy of both laboratory (laser produced, thermonuclear, etc.) and non-atmospheric astrophysical low- and high-temperature plasma. To some extent it may serve as a reference book and textbook for physicists and astrophysicists.

The main ideas of the book are described in seven Parts divided into 33 Chapters, which are subdivided into Sections. In Part 1 we present the initial formulas to calculate the energy spectrum of a many-electron atom in non-relativistic and relativistic approximations, accounting for the relativistic effects as corrections and use perturbation theory in order to describe the energy spectra of an atom. Radiative and autoionizing electronic transitions and generalized expressions for electric multipole transitions are also discussed.

Part 2 is devoted to the foundations of the mathematical apparatus of the angular momentum and graphical methods, which, as it has turned out, are very efficient in the theory of complex atoms. Part 3 considers the non-relativistic and relativistic cases of complex electronic configurations (one and several open shells of equivalent electrons, coefficients of fractional parentage and optimization of coupling schemes). Part 4 deals with the second-quantization in a coupled tensorial form, quasispin and isospin techniques in atomic spectroscopy, leading to new very efficient versions of the Racah algebra.

In Part 5 we present expressions for the matrix elements of the energy operator, describe semi-empirical methods to calculate energy spectra and briefly discuss hyperfine structure, isotope and Lamb shifts. Part 6 contains similar considerations of non-relativistic and relativistic electric and magnetic multipole transitions. Part 7 deals with practical calculations of energy spectra and electronic transitions in the general case of complex configurations, peculiarities of the spectra of highly ionized atoms, statistical methods in the theory of an atom, and some specific features of the configurations with vacancies in inner shells. Generally accepted units and the so-called standard phase system, defined for the Hermitian component of the operator, e.g., the spherical function $C_q^{(k)}$, by the equality

$$C_q^{(k)\dagger} = (-1)^{k-q} C_{-q}^{(k)}, \tag{4}$$

are used. Here the symbol † denotes complex conjugation and transposition. The ranks of the tensors in the standard phase system are put in parentheses. Unit tensors and their sums are defined in a pseudostandard phase system (ranks without brackets), determined for unit tensor t^k, as follows:

$$t_q^{k\dagger} = (-1)^q t_{-q}^k. \tag{5}$$

Notice that in earlier books [2, 9, 10, 15] as a rule the pseudostandard phase system was utilized, whereas in later books [5, 11, 14, 53] the standard one was exploited.

Of course, the list of references is far from complete. As a rule, those papers are quoted which were a starting point for the corresponding considerations or which contain more detailed information on the problems considered (deduction of formulas, tables of necessary quantities, etc.) as well as monographs and review papers.

The book is organized in a 'user-friendly' manner, so that its separate parts are fairly autonomous and may be read or omitted according to the level of training, purposes or intentions of the reader.

Part 1

Energy Spectrum of Many-electron Atom. Radiative and Autoionizing Transitions (Initial Formulas)

1

Non-relativistic atomic Hamiltonian and relativistic corrections

The word 'atom' introduced by Democritus more than 2000 years ago in Greek means 'unseparable'. Only in the 20th century was it shown by Rutherford that the atom possesses a complex structure. The discovery of the complex inner structure of an atom, in fact, has led to the emergence of the main branch of physics describing the structure of a microworld, i.e. quantum mechanics, which, in its turn, has stimulated the development of many other domains of physics, neighbouring sciences and technology. Quantum mechanics continues to be of great importance for their future progress.

However, it is very far from enough to know that any atom consists of a nucleus and of electrons orbiting around it like planets around the Sun. The inner structure of an atom and its main fundamental characteristic – spectra – hide many fundamental laws of nature, the discovery of which has been a challenge for many generations of scientists. Among these laws it is worth mentioning parity violation and the manifestation of a number of fine quantum electrodynamical effects. This is particularly the case for complex atoms with many electronic shells and for highly ionized atoms, whose shell structure and spectra differ considerably from those of neutral or just a few times ionized atoms. These differences and changes are caused by the interplay of the relative role of existing intra-atomic (electron–nucleus or electron–electron) interactions. Therefore, the ability to describe them precisely in order to take them into account, is very important.

For many reasons atomic spectroscopy continues to be one of the most rapidly developing branches of physics. This is primarily due to the creation of very stable and monochromatic lasers, allowing one to selectively excite various atomic states, to create very highly excited (Rydberg) atoms, and due to the occurrence of new possibilities, given by non-atmospheric astrophysics, which allow one to register the electromagnetic radiation of various cosmic objects over the whole wavelength region, including the X-ray region. The problem of controlled thermonuclear fusion, or to be more precise, the problem of spectroscopic methods of diagnostics of high-temperature plasma, and the control of impurities in it, also plays an important role here. Of course, wide possibilities of precise theoretical methods to study the structure and spectra of many-electron atoms and ions have contributed greatly to the general progress in this field. And last but not least is the occurrence of more and more powerful computers, utilization of which allows one to work out special mathematical software for atomic calculations to implement the mathematical modelling of real physical objects and processes. This has contributed greatly to the efficient interpretation of atomic spectra and their other spectroscopic character- istics, measured or observed, and to identification and classification of the energy spectra of atoms and ions. Below, we shall start with the de- scription of the main theoretical methods of studies of the structure and spectra of many-electron atoms based, as a rule, on a quantum mechanical approach. Therefore, we have to start with the Schrödinger equation.

1.1   Schrödinger equation

In the framework of quantum mechanics, for every physical quantity F there is ascribed the corresponding operator $\hat{F}(\mathbf{r}, t)$ depending upon spatial $\mathbf{r}$ and time t variables which, acting on the wave function of the system under consideration $\psi(\mathbf{r}, t)$, transforms it, generally speaking, to the wave function of another state $\varphi(\mathbf{r}, t)$, i.e.

$$\hat{F}(\mathbf{r}, t)\psi(\mathbf{r}, t) = \varphi(\mathbf{r}, t). \tag{1.1}$$

Then the numerical value of that physical quantity at the moment t may be found as the mean value

$$\bar{F} = \int \psi^*(\mathbf{r}, t)\hat{F}\psi(\mathbf{r}, t)d\mathbf{r}, \tag{1.2}$$

where a star * denotes the operation of complex conjugation. Our main goal in this book will be to find, as precisely as possible, wave functions of the systems considered and then to use them to find the most exact values of various necessary physical quantities. If

$$\hat{F}\psi = F\psi, \tag{1.3}$$

where F is a constant, then the wave function ψ is an eigenfunction of the operator $\hat{F}$, and F is its eigenvalue. Then, according to the normalization condition of the wave function, the mean value of the operator $\hat{F}$ equals its eigenvalue F, i.e.

$$\bar{F} = F \int \psi^* \psi \, d\tau = F. \tag{1.4}$$

Here $\tau = \{\mathbf{r}, \sigma\}$ stands for all spatial and spin coordinates. If two operators $\hat{A}$ and $\hat{B}$ have the same eigenfunctions, then this means that they commute with each other

$$\left[\hat{A}, \hat{B}\right] = \hat{A}\hat{B} - \hat{B}\hat{A} = 0. \tag{1.5}$$

Two quantities, represented by commuting operators, possess definite values at the same time. In atomic theory it is very important to find a full set of commuting operators and wave functions, because in this case we can unambiguously describe the system considered. Having defined a full set of wave functions ψ_i $(i = 1, 2, \ldots, N)$ we are in a position to expand the function of arbitrary state ψ in terms of linear combination of these functions of the system considered, i.e.

$$\psi = \sum_{i=1}^{N} c_i \psi_i, \tag{1.6}$$

where, according to the normalization of wave functions,

$$\sum_{i=1}^{N} |c_i|^2 = 1. \tag{1.7}$$

It is very convenient, in quantum mechanics, to consider the wave functions and operators as matrices (matrix form of quantum mechanics). Then the wave function $\psi_j(x)$ of a set of states of the system under consideration is represented by a one-column matrix

$$\psi_j(x) = \begin{pmatrix} \psi_1(x) \\ \psi_2(x) \\ \vdots \end{pmatrix}, \quad j = 1, 2, \ldots, \tag{1.8}$$

whereas the operators are

$$\|F_{kj}\| = \begin{pmatrix} F_{11} & F_{12} & \cdots \\ F_{21} & F_{22} & \cdots \\ \vdots & \vdots & \end{pmatrix}, \tag{1.9}$$

where the matrices may be rectangular or quadratic. In this representation any physical quantity may be calculated as a matrix element (1.2). Let us notice that the energy operator is described by a quadratic matrix.

If the wave functions are the eigenfunctions of the operator $\hat{F}$, i.e. they satisfy condition (1.3), then the matrix of the relevant eigenvalues will be diagonal

$$\|F_{ii}\| = \begin{pmatrix} F_{11} & 0 & \cdots \\ 0 & F_{22} & \cdots \\ \vdots & \vdots & \end{pmatrix}. \tag{1.10}$$

Therefore, the finding of the eigenvalues of any operator may be considered as the diagonalization of the appropriate matrix, calculated in a given basis of the wave functions. This procedure is of particular importance in theoretical spectroscopy of many-electron atoms and ions.

The basic formula of quantum mechanics is the so-called Schrödinger equation

$$i\hbar \frac{\partial \psi}{\partial t} = \hat{H}\psi, \tag{1.11}$$

where $\hat{H}$ is the energy operator (Hamiltonian) of the system under consideration, $\hbar$ stands for the Planck constant, and $i = \sqrt{-1}$. In the general case, both the Hamiltonian and wave function depend on time t. In atomic physics such an approach is usually utilized in which $\hat{H}$ does not depend on t, whereas the dependence of the wave function on t has a special exponential form, i.e.

$$\psi(\mathbf{r}, t) = e^{-i/\hbar Et}\psi(\mathbf{r}). \tag{1.12}$$

Then, the energy spectrum of the system is described by the stationary Schrödinger equation

$$\hat{H}(\mathbf{r})\psi(\mathbf{r}) = E\psi(\mathbf{r}), \tag{1.13}$$

formally coinciding with Eq. (1.3). Thus, if the system considered is in a stationary state, then it has a definite energy value, and its dependence on time is described by Eq. (1.12).

1.2 Non-relativistic atomic Hamiltonian and wave function

Unfortunately, the stationary Schrödinger equation (1.13) can be solved exactly only for a small number of quantum mechanical systems (hydrogen atom or hydrogen-like ions, etc.). For many-electron systems (which we shall be dealing with, as a rule, in this book) one has to utilize approximate methods, allowing one to find more or less accurate wave functions. Usually these methods are based on various versions of perturbation theory, which reduces the many-body problem to a single-particle one, in fact, to some effective one-electron atom.

In perturbation theory one needs a small parameter, in powers of which it would be possible to expand the operators and wave functions. Each electron in an atom moves (let us assume independently) in the electrostatic field of the nucleus $U(r)$ and in the screening field of the remaining electrons. Both these fields, as we shall see later on, are of the same order of magnitude, therefore, it seems that there is no small parameter. However, in spite of the fact that the Coulomb (electrostatic) interaction of the electrons in a many-electron atom is rather strong, it is possible for its main (spherically symmetric) part to add to the interaction of the electron with the nucleus, making in this way some effective central symmetric field. The remaining part of the interelectronic interaction, having a more complex symmetry, is usually considered as perturbation and is accounted for in first-order perturbation theory.

Thus, the state of each electron in a many-electron atom is conditioned by the Coulomb field of the nucleus and the screening field of the charges of the other electrons. The latter field depends essentially on the states of these electrons, therefore the problem of finding the form of this central field must be coordinated with the determination of the wave functions of these electrons. The most efficient way to achieve this goal is to make use of one of the modifications of the Hartree–Fock self-consistent field method. This problem is discussed in more detail in Chapter 28.

The solution of the stationary Schrödinger equation for an electron, moving in the central symmetric field $U(r)$ of an atom, having nuclear charge Ze, where Z is the number of protons in the atomic nucleus, and e is the absolute value of electronic charge, may be written as follows:

$$\psi_{nlsm_lm_s}(r, \vartheta, \varphi, \sigma) = R_{nl}(r)Y_{m_l}^{(l)}(\vartheta, \varphi)\chi_{m_s}^{(s)}(\sigma), \qquad (1.14)$$

where n, l, s denote the principal, orbital and spin quantum numbers of an electron, respectively, m_l and m_s are the projections of l and s; and r, ϑ, φ are the variables of a spherical coordinate system, extremely useful in the central field approximation. σ represents spin variable, $R_{nl}(r) = r^{-1}P_{nl}(r)$, $Y_{m_l}^{(l)}(\vartheta, \varphi)$ and $\chi_{m_s}^{(s)}(\sigma)$ are the radial, angular and spin parts of the wave function of an electron, respectively. It follows from Eq. (1.14) that in the central field approximation the variables may be separated and we obtain individual equations for determining the radial and spin-angular parts of the wave function.

Spherical functions $Y_{m_l}^{(l)}(\vartheta, \varphi)$ have comparatively simple algebraic expressions. However, we shall not present them here because actually we shall need only their orthogonality, addition and transformation properties, which will be described in Chapter 5. Let us recall that $n = 1, 2, 3, \ldots$, $l = 0, 1, 2, \ldots, n-1$, $m_l = 0, \pm 1, \pm 2, \ldots, \pm l$, $s = 1/2$ and $m_s = \pm 1/2$.

For the case of a central field, the energy of an atom does not depend on magnetic quantum number m_l. This means that the energy level, characterized by quantum numbers n and l, is degenerated $2l+1$ times. For a pure Coulomb field there exists additional (hydrogenic) degeneration: the energy of such an atom does not depend on l. Wave function (1.14) may be characterized by the parity quantum number, defined by its behaviour with respect to the inversion of the coordinates. After this operation the wave function gains the phase multiplier $(-1)^l$. The states with positive l values are called even, whereas those with negative l values are called odd.

In the approximation considered, the non-relativistic (zero-order) Hamiltonian H_0 of the N-electron atom may be written as follows:

$$\hat{H}_0 = T + P + Q = \frac{1}{2m} \sum_{i=1}^{N} \mathbf{p}_i^2 - Ze^2 \sum_{i=1}^{N} \frac{1}{r_i} + e^2 \sum_{i>j=1}^{N} \frac{1}{r_{ij}}. \tag{1.15}$$

The first two terms describe the kinetic and potential energy of the electrons with respect to the nucleus, and they only contribute to the total energy of the atomic configuration, whereas the third term represents the Coulomb interaction between the electrons and is in charge of splitting the total energy into terms. In (1.15) m is the electron mass, $\mathbf{p}_i$ is its momentum, r_i is the distance of the ith electron from the nucleus, and r_{ij} describes the distance between electrons i and j.

1.3 Relativistic corrections

Regardless of the complexity of the atom and ion considered, in order to find more accurate wave functions one has to account for two kinds of effect: correlation and relativistic. The first effect will be discussed in more detail in Chapter 29, whereas the second one, when such effects are small and can be accounted for as corrections, we shall discuss now.

As we shall see later on, for a large variety of atoms and ions the relativistic effects can be accounted for fairly precisely in the framework of the so-called Hartree–Fock–Pauli (HFP) approximation, as corrections of the order α^2 ($\alpha = e^2/\hbar c$ is the fine structure constant and c stands for the velocity of light). Then, energy operator H will have the form

$$H = H_0 + W. \tag{1.16}$$

Here H_0 is defined by Eq. (1.15), and

$$W = H_1 + H_2 + H_3 + H_4 + H_5, \tag{1.17}$$

where

$$H_1 = -\frac{1}{8m^3c^2} \sum_i \mathbf{p}_i^4 \tag{1.18}$$

is the relativistic correction due to the dependence of the electron mass on the velocity,

$$H_2 = -\frac{e^2}{2m^2c^2} \sum_{i>j} \frac{1}{r_{ij}} \left\{ (\mathbf{p}_i \cdot \mathbf{p}_j) + \frac{(\mathbf{r}_{ij} \cdot (\mathbf{r}_{ij} \cdot \mathbf{p}_i)\mathbf{p}_j)}{r_{ij}^2} \right\} \tag{1.19}$$

stands for the orbit–orbit interaction operator,

$$H_3 = H_3' + H_3'' = \frac{\pi e^2 \hbar^2 Z}{2m^2c^2} \sum_i \delta(\mathbf{r}_i) - \frac{\pi e^2 \hbar^2}{m^2c^2} \sum_{i>j} \delta(\mathbf{r}_{ij}) \tag{1.20}$$

describes the contact interactions,

$$H_4 = \frac{e^2 \hbar}{2m^2c^2} \left(\left\{ \sum_i \frac{Z}{r_i^3} [\mathbf{r}_i \times \mathbf{p}_i] - \sum_{i>j} \frac{1}{r_{ij}^3} [\mathbf{r}_{ij} \times \mathbf{p}_i] \right. \right.$$
$$\left. \left. + \sum_{i>j} \frac{2}{r_{ij}^3} [\mathbf{r}_{ij} \times \mathbf{p}_j] \right\} \cdot \mathbf{s}_i \right) \tag{1.21}$$

represents the spin–orbit interaction,

$$H_5 = H_5' + H_5'' = -\frac{8\pi e^2 \hbar^2}{3m^2c^2} \sum_{i>j} (\mathbf{s}_i \cdot \mathbf{s}_j)\delta(\mathbf{r}_{ij})$$
$$+ \frac{e^2 \hbar^2}{m^2c^2} \sum_{i>j} \frac{1}{r_{ij}^3} \left[(\mathbf{s}_i \cdot \mathbf{s}_j) - \frac{3(\mathbf{s}_i \cdot \mathbf{r}_{ij})(\mathbf{s}_j \cdot \mathbf{r}_{ij})}{r_{ij}^2} \right] \tag{1.22}$$

is in charge of spin–contact (H_5') and spin–spin (H_5'') interactions. In these formulas $\delta(\mathbf{r})$ is the Dirac δ-function of vectorial argument $\mathbf{r}$, the sums are from 1 up to N, where N denotes the number of electrons, symbol $(\mathbf{a} \cdot \mathbf{b})$ means the scalar product, and $[\mathbf{a} \times \mathbf{b}]$ is the vectorial product of $\mathbf{a}$ and $\mathbf{b}$ (see also (5.14)).

One-particle operators H_1 and H_3' cause relativistic corrections to the total energy. Two-particle operators H_2, H_3'' and H_5' define more precisely the energy of each term, whereas H_4 and H_5'' describe their splitting (fine structure), i.e. they cause a qualitatively new effect. These operators are also often called describing magnetic interactions.

All these expressions may be deduced, if we start with the relativistic wave functions and relativistic Hamiltonian and look for the non-relativistic limit, retaining in the corresponding expansions terms of the order α^2. Notice that all terms in Eq. (1.17) contain small parameter $1/c^2$, which is why they usually cause small corrections.

2
Relativistic atomic Hamiltonian.
New wave function

2.1 Role of relativistic effects

Calculations using the methods of non-relativistic quantum mechanics have now advanced to the point at which they can provide quantitative predictions of the structure and properties of atoms, their ions, molecules, and solids containing atoms from the first two rows of the Periodical Table. However, there is much evidence that relativistic effects grow in importance with the increase of atomic number, and the competition between relativistic and correlation effects dominates over the properties of materials from the first transition row onwards. This makes it obligatory to use methods based on relativistic quantum mechanics if one wishes to obtain even qualitatively realistic descriptions of the properties of systems containing heavy elements. Many of these dominate in materials being considered as new high-temperature superconductors.

It is apparent that progress in our understanding of the properties of neutral heavy elements and their ions, including very highly ionized ones, as well as their role as constituents of molecules and solids, will depend on the development of theoretical methods and computational techniques, which are based on relativistic quantum mechanics. Fairly efficient methods of this kind have already been elaborated and many versions of relativistic codes for work with isolated atoms and ions are already available and in daily use by internationally known theoretical and experimental physicists and chemists [18, 54–57].

It is very important to evaluate the accuracy and the regions of applicability of the methods developed for the effective study of the structure and properties of many-body systems, to demonstrate their utility on selected problems of genuine physical and chemical interest, to improve understanding of the fundamentals on which these methods are based, and to perform mutual checking and validation of results obtained by various methods on the selected systems. In this chapter we shall start to study this problem by considering a relativistic atomic Hamiltonian suitable for use for complex electronic configurations.

2.2 Relativistic atomic Hamiltonian

If the relativistic effects are sufficiently large and therefore cannot be accounted for as corrections, then as a rule one has to utilize relativistic wave functions and the relativistic Hamiltonian, usually in the form of the so-called relativistic Breit operator. In the case of an N-electron atom the latter may be written as follows (in atomic units, in which the absolute value of electron charge e, its mass m and Planck constant $\hbar$ are equal to one, whereas the unit of length is equal to the radius of the first Bohr orbit of the hydrogen atom):

$$H = \sum_{i=1}^{N}(H_i^1 + H_i^2 + H_i^3) + \sum_{i>j=1}^{N}(H_{ij}^4 + H_{ij}^5 + H_{ij}^6). \tag{2.1}$$

Here

$$H_i^1 = c(\alpha_i^{(1)} \cdot p_i^{(1)}), \tag{2.2}$$

$$H_i^2 = \beta_i c^2, \tag{2.3}$$

$$H_i^3 = V(r_i), \tag{2.4}$$

$$H_{ij}^4 = 1/r_{ij}, \tag{2.5}$$

$$H_{ij}^5 = -\frac{1}{2r_{ij}}(\alpha_i^{(1)} \cdot \alpha_j^{(1)}), \tag{2.6}$$

$$H_{ij}^6 = -\frac{1}{2r_{ij}^3}(\alpha_i^{(1)} \cdot r_{ij}^{(1)})(\alpha_j^{(1)} \cdot r_{ij}^{(1)}). \tag{2.7}$$

In these formulas the Dirac matrices $\alpha^{(1)}$ and β are defined in the usual way:

$$\alpha^{(1)} = \begin{pmatrix} 0 & \sigma^{(1)} \\ \sigma^{(1)} & 0 \end{pmatrix}, \qquad \beta = \begin{pmatrix} \hat{I} & 0 \\ 0 & \hat{I} \end{pmatrix}, \tag{2.8}$$

where $\sigma^{(1)}$ and $\hat{I}$ are Pauli and unit matrices of the second order, respectively. H^1, H^2 and H^3 are the one-electron operators. H^1 represents the electron kinetic energy and one-electron part of the spin–orbit interaction, H^2 describes the mass effect, and H^3 denotes the potential energy, in the Coulomb approximation being of the form $-Z/r_i$. Two-electron operator H^4 stands for the electrostatic (Coulomb) interaction, having the same form as in a non-relativistic approximation; H^5 corresponds to the operator of the magnetic and part of the retardation effects, whereas H^6 describes the remaining part of retardation interactions.

The terms $H^5 + H^6$ are often written in the form of the sum of magnetic (H^m) and retarding (H^r) interactions, sometimes also called the relativistic Breit operator (H_{Br}):

$$H_{Br} = H^5 + H^6 = H^m + H^r, \qquad (2.9)$$

where

$$H_{ij}^m = -(\alpha_i^{(1)} \cdot \alpha_j^{(1)})/r_{ij}, \qquad (2.10)$$

$$H_{ij}^r = -\frac{1}{2}(\alpha_i^{(1)} \cdot \nabla_i^{(1)})(\alpha_j^{(1)} \cdot \nabla_j^{(1)})r_{ij}$$
$$= \frac{1}{2r_{ij}}\left[(\alpha_i^{(1)} \cdot \alpha_j^{(1)}) - \frac{(\alpha_i^{(1)} \cdot r_{ij}^{(1)})(\alpha_j^{(1)} \cdot r_{ij}^{(1)})}{r_{ij}^2}\right]. \qquad (2.11)$$

Here

$$\nabla^{(1)} = C^{(1)}\frac{\partial}{\partial r} + \frac{i\sqrt{2}}{r}\left[C^{(1)} \times L^{(1)}\right]^{(1)}, \qquad (2.12)$$

where $i = \sqrt{-1}$, $L^{(1)}$ is the angular momentum operator, and $C^{(1)}$ is the special case $k = 1$ of the spherical function normalized to $4\pi/(2k + 1)$, i.e. in the following way, connected with the usual spherical function $Y_m^{(l)}$ from Eq. (1.14):

$$C_q^{(k)} = \sqrt{\frac{4\pi}{2k + 1}}Y_q^{(k)}. \qquad (2.13)$$

The energy operator considered above is an approximation, in which only the lowest terms of the correction for the retardation of the interaction are taken into account. More general is the formal quantum-electrodynamical interaction energy operator in the approximation of the exchange of one virtual photon [58]

$$H = \sum_{i>j}\frac{1 - (\alpha_i^{(1)} \cdot \alpha_j^{(1)})}{r_{ij}}e^{i\omega_{ij}r_{ij}}. \qquad (2.14)$$

Here ω_{ij} is the difference in eigenvalues of the energy operator. Only the real part of Eq. (2.14) corresponds to the energy, therefore, the exponent may be replaced by $\cos\omega_{ij}r_{ij}$. The lowest terms of the expansion of $\cos\omega_{ij}r_{ij}$ in powers of $1/c$ lead to the Breit approximation discussed above.

2.3   Relativistic wave functions. New type of angular parts of non-relativistic and relativistic wave functions

The relativistic analogue of a one-electron wave function (1.14) is

$$\psi_{nljm}(r,\vartheta,\varphi) = \begin{pmatrix} \varphi_{nljm}(r,\vartheta,\varphi) \\ \chi_{nl'jm}(r,\vartheta,\varphi) \end{pmatrix}$$

$$= \begin{pmatrix} |ljm|\vartheta,\varphi)f(nlj|r) \\ (-1)^{\beta} \; |l'jm|\vartheta,\varphi)g(nl'j|r) \end{pmatrix}. \qquad (2.15)$$

Here $\beta = \frac{1}{2}(1+l-l')$, $l' = 2j-l$, f and g are the so-called large and small components of the relativistic radial wave function, respectively, and the spherical spinor $|ljm|\vartheta,\varphi)$ is

$$|ljm|\vartheta,\varphi) = \sum_{\sigma=\pm 1/2} \begin{bmatrix} l & s & j \\ m-\sigma & \sigma & m \end{bmatrix} Y_{m-\sigma}^{(l)}(\vartheta,\varphi)\Phi_{\sigma}, \qquad (2.16)$$

where the quantity Φ_{σ} is a Pauli spin matrix defined by

$$\Phi_{1/2} = \begin{pmatrix} 1 \\ 0 \end{pmatrix}, \qquad \Phi_{-1/2} = \begin{pmatrix} 0 \\ 1 \end{pmatrix}$$

and

$$\begin{bmatrix} l & s & j \\ m-\sigma & \sigma & m \end{bmatrix}$$

is the Clebsch–Gordan coefficient. Angular parts of wave function (2.15) obey the usual orthogonality conditions with respect to all quantum numbers, whereas for the radial parts the following relation must be preserved:

$$\int_0^{\infty} [f(nlj|r)f(n'lj|r) + g(nl'j|r)g(n'l'j|r)] \, r^2 dr = \delta(n,n'). \qquad (2.17)$$

Thus, the non-relativistic wave function (1.14) of an electron is a two-component spinor (tensor having half-integer rank) whereas its relativistic counterpart is already, due to the presence of large (f) and small (g) components, a four-component spinor. The choice of β in the form $\frac{1}{2}(1 + l - l')$ is conditioned by the requirement of a standard phase system for the wave functions (see Introduction, Eq. (4)).

Non-relativistic (1.14) and relativistic (2.15) wave functions are widely used for theoretical studies of the structure and spectra of many-electron atoms and ions. However, it has turned out that such forms of wave functions in the case of the jj coupling scheme are not optimal. Their utilization, particularly in the relativistic approximation, is rather inconvenient and tedious.

In a series of papers [59–64] a mathematical technique was described which made it possible to find the general expressions for matrix elements of relativistic energy and relativistic radiative transition operators in the case of complex electronic configurations. After the straightforward but tedious use of the mathematical apparatus of angular momentum theory and original graphical methods [11], it was shown that only radial integrals and some phase multipliers of these matrix elements depend on orbital quantum numbers. Therefore, it was worthwhile to elaborate a practical mathematical method that would lead directly to the simplest form of the matrix elements under consideration. This goal was achieved in the papers [28, 29], based on the use of generalized spherical functions.

Indeed, as was shown in [28, 29], instead of (2.16) in jj coupling, we can use the following function:

$$|l\,jm) = \left[\frac{2j+1}{8\pi}\right]^{1/2} \begin{pmatrix} D^{(j)}_{m,1/2} \\ (-1)^{l+1/2-j} & D^{(j)}_{m,-1/2} \end{pmatrix}, \tag{2.18}$$

where the generalized spherical functions $D^{(j)}_{m,\mu}$ are defined by [11]

$$D^{(j)}_{m,\mu}(\Phi,\Theta,\Psi) = i^{\mu-m}\sum_t (-1)^t \frac{[(j+m)!(j-m)!(j+\mu)!(j-\mu)!]^{1/2}}{t!(\mu-m+t)!(j+m-t)!(j-\mu-t)!}$$

$$\times e^{i(m\Phi+\mu\Psi)}\left(\sin\frac{\Theta}{2}\right)^{2t-m+\mu}\left(\cos\frac{\Theta}{2}\right)^{2j-2t+m-\mu}. \tag{2.19}$$

In our case we can choose $\Psi = 0$, and then $\Theta = \vartheta$, $\Phi = \varphi + \pi/2$. The functions $D^{(j)}_{m,\mu}(\Phi,\Theta)$ obey the orthogonality conditions and are normalized according to

$$\int D^{(j)^*}_{m,\mu}D^{(j)}_{m',\mu'}d\tau = \frac{4\pi}{2j+1}\delta(jm\mu, j'm'\mu'), \tag{2.20}$$

where $d\tau = \sin\Theta d\Theta d\Phi$, and

$$D^{(j)^*}_{m,\mu}(\Phi,\Theta) = (-1)^{m-\mu}D^{(j)}_{-m,-\mu}(\Phi,\Theta). \tag{2.21}$$

Making use of Eq. (2.21), we can find the form of Hermitian conjugated wave function (2.18), i.e.,

$$|l\,jm)^\dagger = (l\,jm| = (-1)^{j+l+m}\left((-1)^{l-1/2-j}\,D^{(j)}_{-m,-1/2}\quad D^{(j)}_{-m,1/2}\right). \tag{2.22}$$

Thus, formula (2.18) represents a new form of the non-relativistic wave function of an atomic electron (to be more precise, its new angular part in jj coupling). It is an eigenfunction of the operators l^2, s^2, j^2 and j_z, and it satisfies the one-electron Schrödinger equation, written in j-representation. Only its phase multiplier depends on the orbital quantum number to ensure selection rules with respect to parity.

In order to utilize the mathematical apparatus of the angular momentum theory, we have to write all the operators in j-representation, i.e., to express them in terms of the quantities which transform themselves like the eigenfunctions of operators $\mathbf{j}^2$ and j_z. For example, the explicit form of the spherical components of the spin operator in j-representation, according to [28], is as follows:

$$s_q^{(1)} = \frac{i}{2} \begin{pmatrix} D_{q,0}^{(1)} & -2^{1/2} D_{q,1}^{(1)} \\ 2^{1/2} D_{q,-1}^{(1)} & -D_{q,0}^{(1)} \end{pmatrix}. \tag{2.23}$$

For the components of the orbital angular momentum operator we have

$$l_q^{(1)} = j_q^{(1)} - s_q^{(1)} \tag{2.24}$$

and

$$l^{(1)^2} = j^{(1)^2} - 2(j^{(1)} \cdot s^{(1)}) + s^{(1)^2}.$$

The radius vector is

$$r_q^{(1)} = ir D_{q,0}^{(1)}, \tag{2.25}$$

where, in general, for integer k

$$D_{q,0}^{(k)}(\Phi, \Theta) = i^{k-q} C_q^{(k)}(\Theta, \Phi) = i^{-k} C_q^{(k)}(\vartheta, \varphi). \tag{2.26}$$

Now, by use of formulas (2.23)–(2.26) we are in a position to present in j-representation all the operators needed. For example, the non-relativistic operator of electric multipole radiation will have the form

$$Q_q^{(k)} = -i^k C_q^{(k)} r^k = (-1)^{k+1} D_{q,0}^{(k)} r^k. \tag{2.27}$$

We shall obtain a new type of a relativistic wave function if we use in Eq. (2.15) expression (2.18) instead of (2.16). However, usually in this case β is chosen to equal $-1/2$. Such a form leads to simpler phase multipliers for some quantities of the corresponding mathematical apparatus, but does not affect the matrix elements of the operators corresponding to physical quantities.

Thus, as we shall see in Chapters 19, 20 and 28, having in mind that the angular parts of new wave functions depend on orbital quantum number l only in the form of a phase multiplier, we are in a position to obtain straightforwardly optimal expressions for all matrix elements needed.

3

Perturbation theory for the energy of an atom

3.1 Methods of accounting for correlation effects

In order to improve the theoretical description of a many-body system one has to take into consideration the so-called correlation effects, i.e. to deal with the problem of accounting for the departures from the simple independent particle model, in which the electrons are assumed to move independently of each other in an average field due to the atomic nucleus and the other electrons. Making an additional assumption that this average potential is spherically symmetric we arrive at the central field concept (Hartree–Fock model), which forms the basis of the atomic shell structure and the chemical regularity of the elements. Of course, relativistic effects must also be accounted for as corrections, if they are small, or already at the very beginning starting with the relativistic Hamiltonian and relativistic wave functions.

In fact, electrons do not move independently, they 'feel' each other, there is a certain correlation between the electrons in their mutual Coulomb field (many-body effects).

Many-body calculations which go beyond the Hartree–Fock model can be performed in two ways, i.e. using either a variational or a perturbational procedure. There are a number of variational methods which account for correlation effects: superposition-of-configurations (or configuration interaction (CI)), random phase approximation with exchange, method of incomplete separation of variables, multi-configuration Hartree–Fock (MCHF) approach, etc. However, to date only CI and MCHF methods and some simple versions of perturbation theory are practically exploited for theoretical studies of many-electron atoms and ions.

The CI method is often based on the use of analytic functions, which form the basis set. The parameters of a many-body wave function obtained in this way, as well as the mixing coefficients, are then found while

16

minimizing the expectation value of the energy. The success of such calculations depends essentially on the choice of this basis set, on the degree of its optimization. However, slow convergence of this method, the absence of exact criteria for the choice of the most important electronic configurations, technical difficulties, growing rapidly with the increase of the order of the matrices to be diagonalized, crucially restrict the possibilities of the CI method.

In the MCHF approach a number of superposed configurations are chosen and the mixing coefficients (weights of the configurations) and also the radial parts of the wave functions are varied. This method does not depend on choice of the basis set and both analytical and numerical wave functions may be used. However, MCHF calculations for complex electronic configurations would require variation of a large number of parameters, which needs powerful computers. Problems may also occur with the convergence of the procedure [45].

In the framework of perturbation theory (PT) the correlation effects can be accounted for fairly accurately. However, various versions of PT as a rule have been practically applied only to light atoms and multiply charged ions having a number of electrons $N \leq 10$ [65, 66].

In any version of PT the Hamiltonian H of the system considered is split into two parts, the model Hamiltonian H_0 and the perturbation V [17]

$$H = H_0 + V. \tag{3.1}$$

Model Hamiltonian H_0 in Eq. (3.1) should be as close as possible to the 'exact' Hamiltonian. For atoms, the central field approach is a natural choice of H_0. Then H_0 may be defined by Eq. (1.15), if we take into account only the spherically symmetric part of Coulomb interaction. One of the main difficulties in utilizing PT for the studies of many-electron atoms consists in the fact that in the theory of complex atoms, the wave functions defined in the central field approximation are used as a basis. Energy levels of separate electrons, moving in a central field, are degenerate with respect to the projections of the orbital and spin angular momenta. This leads, in the case of open electronic shells, to the very high degree of degeneracy of the many-electron problem in a zero-order approximation. High degeneracy essentially complicates the use of the PT.

Moreover, for atoms with open shells, the difficulties in calculating the angular parts of the PT expansion grow very rapidly with the increase in the order of expansion terms. Even the methods of their calculations are not developed sufficiently, unlike the usual mathematical apparatus of the theory of atomic spectra. Therefore, in order to successfully apply the PT to complex atoms, further development of the many-body theory is necessary, accounting for the pecularities of many-electron atoms and ions, including very highly ionized atoms.

There are two main perturbation schemes: the Rayleigh–Schrödinger and the Brillouin–Wigner expansions. Straightforwardly these methods are rather hard to employ beyond, let us say, the third order. An important breakthrough was made in PT after the development of the so-called partitioning technique [67, 68], in which the functional space for the wave functions is separated into two parts, a model space and an orthogonal space. The principal idea here is to find such effective operators, which act only within the limited model space but which generate the same results as do the original operators acting on the entire functional space [17]. An effective operator formalism for many-electron atoms has been developed in [12, 33, 69, 70].

The Brillouin–Wigner formalism is best suited to problems where only a single energy level is involved, i.e. for non-degenerate systems, such as closed-shell atoms. Following [17] we shall devote prime attention to the Rayleigh–Schrödinger formalism. This scheme has the basic advantage that the exact energy does not appear explicitly. Therefore, it can be applied to a group of states simultaneously, to study the whole energy spectrum of a given atom or ion. For that reason this method looks very promising for generalization to cover the case of complex atoms with several open shells. This method also leads quite naturally to the concept of effective operators, very suitable to describe the contributions of various existing intra-atomic interactions. Below we shall briefly sketch the basic principles of the Brillouin–Wigner and Rayleigh–Schrödinger formalisms, mainly following monograph [17]. A number of aspects of many-body perturbation theory are discussed in [71–74].

3.2   Brillouin–Wigner (BW) and Rayleigh–Schrödinger (RS) expansions

Let us assume that zero-order Hamiltonian H_0 is a Hermitian operator having a complete set of eigenfunctions

$$H_0\Phi^a = E_0^a\Phi^a. \tag{3.2}$$

Let us assume again that functions Φ^a are non-degenerate, in which case they are automatically orthogonal. If one of Φ^a is a reasonable approximation of the exact wave function ψ for the state investigated, then this zero-order approximation is called the model function:

$$\psi_0 = \Phi^a. \tag{3.3}$$

The purpose of PT is to find a scheme for generating a sequence of successive improvements to this zero-order approximation. Let us define a projection operator

$$P = |\psi_0)(\psi_0| = |\Phi^a)(\Phi^a| \tag{3.4}$$

and a corresponding projection operator for the complementary part of the functional space (orthogonal or Q space)

$$Q = \sum_{b \neq a} |\Phi^b)(\Phi^b|. \tag{3.5}$$

According to these definitions, the P operator projects out of any function describing the system, the part that is proportional to the model function, and the Q operator projects the part that is orthogonal to this function. The sum of $P + Q$ satisfies the condition

$$P + Q = 1. \tag{3.6}$$

Operating with P on the exact wave function ψ, we find

$$P\psi = \psi_0. \tag{3.7}$$

If ψ_0 is normalized to unity, then exact function ψ is not (intermediate normalization). Having in mind Eqs. (3.6) and (3.7), we get

$$\psi = (P + Q)\psi = \psi_0 + Q\psi. \tag{3.8}$$

As was mentioned earlier, model function ψ_0 corresponds to the zero-order approximation and the remaining part, $Q\psi$, can be considered as a 'correction'. If ψ_0 is generated in an independent particle model (e.g., Hartree–Fock), then $Q\psi$ is often referred to as the correlation function [75, 76].

If model function ψ_0 is known, then we can find a series expansion for the remaining part $Q\psi$ of the exact wave function (3.8). Starting with the Schrödinger equation in the form

$$(E - H_0)\psi = V\psi, \tag{3.9}$$

we are in a position to obtain the following exact expression for ψ, which may be used to generate a series expansion for the wave function:

$$\psi = \psi_0 + T_E V\psi, \tag{3.10}$$

where

$$T_E = Q/(E - H_0), \tag{3.11}$$

or

$$\psi = \left(1 + \frac{Q}{E - H_0}V + \frac{Q}{E - H_0}V\frac{Q}{E - H_0}V + \ldots\right)\psi_0. \tag{3.12}$$

Eq. (3.12) represents the Brillouin–Wigner (BW) perturbation expansion of the exact wave function, whereas the corresponding expression for the energy expansion is

$$E = E_0 + (\psi_0| \left(V + V\frac{Q}{E - H_0}V + V\frac{Q}{E - H_0}V\frac{Q}{E - H_0}V + \dots \right) |\psi_0),$$

$$(3.13)$$

or

$$E = E_0 + E^{(1)} + E^{(2)} + E^{(3)} + \dots, \qquad (3.14)$$

where $E^{(n)}$ is the general nth-order energy contribution, containing n interactions of the perturbation V,

$$E^{(n)} = (\psi_0|V \left(\frac{Q}{E - H_0}V \right)^{n-1} |\psi_0). \qquad (3.15)$$

The sum

$$E_0 + E^{(1)} = (\psi_0|(H_0 + V)|\psi_0) = (\psi_0|H|\psi_0) = \bar{E} \qquad (3.16)$$

is the expectation value of the energy in the state represented by the model function. This quantity is minimized in the Hartree–Fock procedure and is known as the Hartree–Fock energy, whereas the remaining part is called the correlation energy. Exact energy E is not known *a priori*, therefore we cannot evaluate the energy contributions order by order without solving a self-consistency problem, since each term of the perturbation expansion depends on the initially unknown exact energy.

The BW form of PT is formally very simple. However, the operators in it depend on the exact energy of the state studied. This requires a self-consistency procedure and limits its application to one energy level at a time. The Rayleigh–Schrödinger (RS) PT does not have these shortcomings, and is, therefore, a more suitable basis for many-body calculations of many-electron systems than the BW form of the theory, it is applicable to a group of levels simultaneously.

Again, let us assume that in RS PT the eigenfunctions Φ^a of model Hamiltonian H_0 are known and they obey Eq. (3.2). If there are several independent eigenfunctions corresponding to the same eigenvalue, then these functions are not automatically orthogonal. They can be orthogonalized, for instance, by means of the Schmidt procedure. We assume that this is already done, so that we have a complete orthonormal set of the eigenfunctions of H_0,

$$(\Phi^a|\Phi^b) = \delta(a, b). \qquad (3.17)$$

For a degenerate system we cannot always determine beforehand the zero-order or model functions. The mixing of the degenerate eigenfunctions in the zero-order depends on the perturbation. Therefore, generally speaking, we can only confine the model function to a certain subspace of the entire functional space. Let us call this subspace the model (or P)

space. The remaining part of the functional space is called the orthogonal (or Q) space. We shall always have in mind that eigenfunctions of H_0, corresponding to the same eigenvalue, belong to the same subspace. Thus, projection operators P and Q for the model and orthogonal spaces are defined correspondingly (compare with (3.4), (3.5)):

$$P = \sum_{a \in P} |\Phi^a)(\Phi^a|, \qquad Q = \sum_{b \notin P} |\Phi^b)(\Phi^b|. \qquad (3.18)$$

The P operator projects out of any function the component that lies in the model space, and the Q operator projects out the component in the orthogonal space. They also obey relation (3.6).

In the Hartree–Fock approach the model space represents all functions associated with one or several electronic configurations. The main advantage of including several configurations in the model space is that it becomes possible to account for strongly interacting configurations to all orders and consider the weakly interacting ones by means of a low-order expansion. A wider model space must lead to a better zero-order approximation and better convergence, but at the expense of higher complexity. The optimal choice of the model space may, therefore, be of great importance for the success of the perturbation procedure.

Let us consider now a general model space, spanned by the eigenfunctions of the model Hamiltonian corresponding to one or several eigenvalues, e.g., $1s^2 2s^2$ and $1s^2 2p^2$ configurations in the case of the beryllium atom. If the dimensionality of the model space is d, then there are normally d well-defined eigenstates ψ^a of the full Hamiltonian, which have their major part within the model space. The projections of these exact states onto the model space

$$\psi_0^a = P\psi^a \qquad (a = 1, 2, \ldots, d) \qquad (3.19)$$

represent the model functions. We shall assume that the model functions are linearly independent, so that they span the entire model space. This means that there is a one-to-one correspondence between the d exact solutions of the Schrödinger equation and the model functions. Then we can introduce a single wave operator Ω, which transforms all the model states back into the corresponding exact states

$$\psi^a = \Omega\psi_0^a \qquad (a = 1, 2, \ldots, d). \qquad (3.20)$$

Notice that in RS theory the wave operator is the same for all states under investigation. It is convenient to use also another (correlation) operator χ, defined by

$$\Omega = 1 + \chi. \qquad (3.21)$$

Acting on the model function, operator χ generates the projection of the full wave function in Q space, i.e.

$$\chi\psi_0^a = Q\psi^a. \tag{3.22}$$

For generating the RS expansion, it is very suitable to use the so-called generalized Bloch equation

$$[\Omega, H_0]P = V\Omega P - \Omega P V\Omega P, \tag{3.23}$$

or, in an alternative form,

$$[\Omega, H_0]P = QV\Omega P - \chi P V\Omega P, \tag{3.24}$$

with commutator

$$[a, b] = ab - ba. \tag{3.25}$$

Equations (3.23) and (3.24) are valid also for a model space containing several unperturbed energies, e.g. several atomic configurations. These equations will form the basis for our many-body treatment. The generalized Bloch equation is exact and completely equivalent to the Schrödinger equation for the states considered.

Operating with P from the left on the Schrödinger equation

$$H\Omega\psi_0^a = E^a\psi^a \tag{3.26}$$

we get

$$PH\Omega\psi_0^a = E^a\psi_0^a. \tag{3.27}$$

Let us define the operator of the effective Hamiltonian in Rayleigh–Schrödinger perturbation theory in the following way:

$$H_{eff} = PH\Omega P. \tag{3.28}$$

It operates entirely within the model space and satisfies the eigenvalue equation

$$H_{eff}\psi_0^a = E^a\psi_0^a. \tag{3.29}$$

It follows from Eq. (3.29) that the eigenvectors of the effective Hamiltonian represent the model functions, whereas the eigenvalues are the exact energies of the corresponding true states. This indicates the basic differences between BW and RS approaches. In the former case there is one effective Hamiltonian for each energy, while in the latter case a single operator yields all the model states and corresponding energies.

Our main goal is to construct the effective Hamiltonian and to solve the eigenvalue equation (3.29). There are various representations of effective Hamiltonian, Eq. (3.28) being only one of them.

Let us consider further the case where the model space is completely degenerate. We assume also that all the states of the model space correspond to the same eigenvalue E_0 of the model Hamiltonian

$$H_0\psi_0^a = E_0\psi_0^a \quad (a = 1, 2, \ldots, d). \tag{3.30}$$

Then Eq. (3.24) leads to

$$(E_0 - H_0)\Omega P = QV\Omega P - \chi PV\Omega P. \tag{3.31}$$

Using the expansion

$$\Omega = 1 + \chi = 1 + \Omega^{(1)} + \Omega^{(2)} + \cdots \tag{3.32}$$

we obtain

$$\Omega^{(1)}P = RVP,$$
$$\Omega^{(2)}P = RVRVP - R^2VPVP,$$
$$\Omega^{(3)}P = RVRVRVP - RVR^2VPVP - R^2VRVPVP$$
$$+ R^3VPVPVP - R^2VPVRVP, \tag{3.33}$$

where

$$R = Q/(E_0 - H_0). \tag{3.34}$$

Using Eq. (3.33) we find the following expressions for the first few terms of the effective Hamiltonian (3.28):

$$H_{eff}^0 = PH_0P,$$
$$H_{eff}^{(1)} = PVP,$$
$$H_{eff}^{(2)} = PV\Omega^{(1)}P = PVRVP,$$
$$H_{eff}^{(3)} = PV\Omega^{(2)}P = PVRVRVP - PVR^2VPVP. \tag{3.35}$$

In general, the corresponding energies are obtained by diagonalizing this effective Hamiltonian in the model space. If the model functions are already found, then the energies can be calculated directly starting with Eq. (3.29) and evaluating the matrix element, i.e.

$$E^a = (\psi_0^a|H_{eff}|\psi_0^a). \tag{3.36}$$

This leads to the well-known expressions for the first and the second terms of the energy expansion

$$E^{a(1)} = (\psi_0^a|H_{eff}^{(1)}|\psi_0^a) = (\psi_0^a|V|\psi_0^a), \tag{3.37}$$

$$E^{a(2)} = (\psi_0^a|H_{eff}^{(2)}|\psi_0^a) = \sum_{b \notin P} \frac{(\psi_0^a|V|\psi_0^b)(\psi_0^b|V|\psi_0^a)}{E_0 - E_0^b}. \tag{3.38}$$

Again, as in the case of Eq. (3.16) the sum of the zero- and first-order energy contributions is equal to the expectation value of the full Hamiltonian in the unperturbed model state, i.e.

$$E_0 + E^{(1)} = (\psi_0^a|(H_0 + V)|\psi_0^a) = (\psi_0^a|H|\psi_0^a). \tag{3.39}$$

In order to calculate the energy spectrum of an atom or ion in the framework of the few lowest orders of PT, we have to be able to find the expressions for the corresponding matrix elements. For complex electronic configurations, having several open shells, this is a task very far from trivial. For optimization of their expressions one has to combine the methods of second quantization, angular momentum theory, irreducible tensorial sets, tensorial products in coupled form, coefficients of fractional parentage with utilization of graphical (diagrammatic) methods, and also account for the symmetry properties of the system considered in the additional spaces (e.g. in quasispin space). This will be done in the following chapters of this book.

4

Radiative and autoionizing electronic transitions. Generalized expressions for electric multipole (Ek) transition operators

4.1 Interaction of atomic electrons with electromagnetic radiation

Until now we have studied the methods of theoretical description of the energy of separate, non-interacting atoms or ions. Interaction was present indirectly while speaking about many states or excited configurations, saying nothing about the mechanisms of their creation or their lifetimes.

The wavelengths and intensities of the electronic transitions from the higher to lower (or, finally, ground) states are measured experimentally. From these data the energy levels (energy spectra) are deduced. Therefore, it is very important to have efficient and fairly general methods of theoretical description of the main quantities of electronic transitions in many-electron atoms and ions, including very highly ionized ones.

An atom in a ground state can exist for an infinitely long time, whereas any excited state exists only for a finite time interval. This time interval (lifetime of the excited state), together with the energy spectrum of the excited configuration, is one of the basic spectral characters of the quantum mechanical system considered. There exist various channels of decay of the excited states, and the realization of each of them depends on concrete physical conditions.

The finite lifetime of each excited state is the reflection of a fundamental law of nature – tendency towards minimum total energy of a system. The quantum mechanical system tends to occupy the state in which its total energy would be minimal. However, the transition of an atom to the lowest (ground) state depends on many circumstances (first of all, on the sort of excited configuration, on the presence of external fields, on the character of the matter itself – density of gas, vapours or plasma, etc.). There are two main channels of decay of the excited states: radiative and radiationless. In the first case the electronic transition from the higher to the lower state is connected with the radiation of one or several quanta of

25

the electromagnetic field, each having strictly defined wavelength, whereas in the second case the system considered returns to its initial (ground) state transferring a part of its energy to another particle (e.g. in the process of a collision with another atom or ion, exciting or even ionizing it). This indicates that the latter channel is possible when there are other particles (e.g., dense plasma, gas, etc.). Excited states are produced in the reciprocal way: by absorbing one or a few photons, or colliding with other atoms, ions or even separate electrons and taking from them some energy, or via recombination capturing a free electron.

Excited states may decay also in the process of autoionization, with the detachment of an excited electron and its transition to the state of a continuous spectrum (free electron).

As in the case of the energy spectra of atoms and ions, there occurs the problem of accounting for the correlation and relativistic effects. The former are usually taken into consideration by making the wave functions more precise: the latter, if they are small, may be included in the corresponding operators as small corrections, i.e. in the framework of first-order perturbation theory. If the relativistic effects are large and, therefore, cannot be treated as perturbation, then the corresponding theory must be based on the relativistic operators of electronic transitions and relativistic wave functions. We shall briefly sketch both these approaches. As we shall see further on, practical calculations show that for the optical transitions in the overwhelming majority of neutral atoms as well as the ions of low ionization degree, the relativistic effects do not play a leading role. Therefore, it is enough to account for them while calculating the energy spectra.

The expressions for the operators describing electronic transitions may be found following the classical approach, namely by considering the connection between the propagation of electromagnetic waves and the classical sources of radiation, a system of currents and charges. The connection between them and the vectors of the strengths of magnetic $\mathbf{H}$ and electric $\mathbf{E}$ fields is given by the Maxwell equations. Expanding these vectors, which are plane waves in space without charges and currents, in a series of spherical harmonics we arrive at two kinds of expressions, corresponding to the operators of electric (Ek) and magnetic (Mk) multipole radiation (k denotes the multipolarity of an electronic transition). In this way we deduce the non-relativistic operators of electronic transitions.

It is convenient to classify the emitted radiation by the quantities of angular momentum l carried out by each quantum of energy ω and by the properties of the radiation with respect to the inversion operation (parity). In this representation each photon emitted is characterized by its energy and quantum numbers l and m. There are two kinds of solutions of the Maxwell equations, characterized by different parity. This causes the occurrence of two types of radiation – electric and magnetic; moreover, electric multipole radiation of multipolarity l has parity $(-1)^l$ whereas magnetic radiation of the same multipolarity l is characterized by parity $(-1)^{l+1}$. Thus, division of the radiation into electric and magnetic is just formal and is connected with the parity of the photon emitted.

Multipolarity l is the quantity of the angular momentum type. However, in electron transition theory it normally characterizes the rank of the tensor of the corresponding operator; that is why further on we shall denote it as k.

There exists another more consistent way of obtaining the electron transition operators. We can start with the quantum-electrodynamical description of the interaction of the electromagnetic field with an atom. In this case we find the relativistic operators of electronic transitions with respect to the relativistic wave functions. After that they may be transformed to the well-known non-relativistic ones, accounting for the relativistic effects, if necessary, as corrections to the usual non-relativistic operators. Here we shall consider the latter in more detail. It gives us a closed system of universal expressions for the operators of electronic transitions, suitable to describe practically the radiation in any atom or ion, including very highly ionized atoms as well as the transitions of any multipolarity and any type of radiation (electric or magnetic).

4.2 Generalized expressions for relativistic electric multipole (Ek) transitions

It is convenient to treat the interaction of atomic electrons with an electromagnetic field in the framework of perturbation theory.

This is due to the comparative weakness of the electromagnetic interaction, the theory of which contains a small dimensionless parameter (fine structure constant), by the powers of which the corresponding quantities can be expanded. The electron transition probability of the radiation of one photon, characterized by a definite value of angular momentum, in the first order of quantum-electrodynamical perturbation theory may be described as follows [53] (a.u.):

$$W_{1\to2} = 2\pi |V_{21}|^2, \tag{4.1}$$

where the quantity V_{21} is given by

$$V_{21} = -\langle\psi_2|\{(\boldsymbol{\alpha}\cdot\mathbf{A}^*) - \Phi^*\}|\psi_1\rangle. \tag{4.2}$$

Here $\{\Phi, \mathbf{A}\}$ denotes the components of the four-dimensional vector-potential of the electromagnetic field corresponding to a definite state of the photon, in the coordinate representation, ψ_1 and ψ_2 are the eigenfunctions of the Dirac equation describing the stationary states with energy E_1 and E_2, α stands for the Dirac matrices, defined by Eq. (2.8).

Let us underline that the matrix elements denoted $\langle \ | \ | \ \rangle$ will always correspond to relativistic operators and wave functions. For the photons of electric multipole type (Ek) the expression for the electromagnetic field potential may be chosen using its different gauge conditions; moreover, it may be written in different forms, equivalent for the exact wave functions. In order to reveal the pecularities of various gauge conditions and to deduce the most universal expressions for transition operators, we shall consider the electromagnetic field potential with the arbitrary gauge condition K, because if we specify it at the very beginning, then we arrive at one or another particular form of the quantity studied.

Starting with Eq. (4.1) after tedious mathematical transformations [18] we finally find the following two general forms of the relativistic expressions for the electric multipole (Ek) transition probability (in a.u.):

$$W_{1\to2}^{Ek} = \frac{2(k+1)\omega^3}{k(2k+1)c} \left| \left\langle \psi_2 \left| \left[{}_rQ_{-q}^{(k)} + K\sqrt{\frac{k}{k+1}} \left\{ \frac{1}{\omega}{}_vQ_{-q}^{(k)} - {}_rQ_{-q}^{(k)} \right\} \right] \right| \psi_1 \right\rangle \right|^2 , \tag{4.3}$$

$$W_{1\to2}^{Ek} = \frac{2(k+1)\omega}{k(2k+1)c} \left| \left\langle \psi_2 \left| \left[{}_vQ_{-q}^{(k)} + K\sqrt{\frac{k}{k+1}} \left\{ {}_vQ_{-q}^{(k)} - \omega{}_rQ_{-q}^{(k)} \right\} \right] \right| \psi_1 \right\rangle \right|^2 . \tag{4.4}$$

Here

$${}_rQ_{-q}^{(k)} = -\frac{2k+1}{\omega} \left\{ g_k(z)C_{-q}^{(k)} + i\sqrt{\frac{2k+3}{k+1}} g_{k+1}(z) \left[C^{(k+1)} \times \alpha^{(1)} \right]_{-q}^{(k)} \right\}, \tag{4.5}$$

$$\begin{aligned}
{}_vQ_{-q}^{(k)} = -i \Bigg\{ &k\sqrt{\frac{2k+3}{k+1}} g_{k+1}(z) \left[C^{(k+1)} \times \alpha^{(1)} \right]_{-q}^{(k)} \\
&+ \sqrt{k(2k-1)}\, g_{k-1}(z) \left[C^{(k-1)} \times \alpha^{(1)} \right]_{-q}^{(k)} \Bigg\},
\end{aligned} \tag{4.6}$$

where $g_k(z)$ denotes the spherical Bessel function of the argument $z = \omega r/c$, ω being the frequency of the photon emitted, and c the velocity of light. K is an arbitrary parameter describing the gauge condition of the electromagnetic field potential.

Quantities (4.5) and (4.6) are two equivalent forms of relativistic Ek-transition operators. They may be connected with each other through the relation

$$\left\langle \psi_2 \left| \left[H_D, {}_rQ_{-q}^{(k)} \right] \right| \psi_1 \right\rangle = \left\langle \psi_2 \left| {}_vQ_{-q}^{(k)} \right| \psi_1 \right\rangle, \tag{4.7}$$

where H_D denotes the Dirac Hamiltonian.

The corresponding expression for the relativistic transition probability of the magnetic multipole (Mk) radiation has the form

$$W_{1\rightarrow 2}^{Mk} = \frac{2(2k+1)\omega}{c} \left| \left\langle \psi_2 \left| {}_mQ_{-q}^{(k)} \right| \psi_1 \right\rangle \right|^2, \tag{4.8}$$

where

$$_mQ_{-q}^{(k)} = g_k(z) \left[C^{(k)} \times \alpha^{(1)} \right]_{-q}^{(k)}. \tag{4.9}$$

Unlike Ek-radiation, there is only one expression for Mk-transitions, because for them the potential does not depend on gauge K.

Relativistic quantities ${}_rQ_{-q}^{(k)}$ and ${}_vQ_{-q}^{(k)}$, in the non-relativistic limit, correspond to the Ek-transition operators, leading, for $k = 1$, to the well-known 'length' and 'velocity' forms. However, in the case of relativistic transition probabilities (4.3) and (4.4) this limit depends on the K-value chosen.

4.3 Non-relativistic operators of electronic transitions

Let us consider the non-relativistic limit of the relativistic operators describing radiation. Expressing the small components of the four-component wave functions (bispinors) in terms of the large ones and expanding the spherical Bessel functions in a power series in $\omega r/c$, we obtain, in the non-relativistic limit, the following two alternative expressions for the probability of electric multipole radiation:

$$W_{1\rightarrow 2}^{Ek} = \frac{2(k+1)(2k+1)\omega^{2k+1}}{k[(2k+1)!!]^2 c^{2k+1}}$$
$$\times \left| \left(\psi_2 \left| \left[Q_{-q}^{(k)} + K\sqrt{\frac{k}{k+1}} \left\{ \frac{1}{\omega} Q'^{(k)}_{-q} - Q_{-q}^{(k)} \right\} \right] \right| \psi_1 \right) \right|^2, \tag{4.10}$$

$$W_{1\rightarrow 2}^{Ek} = \frac{2(k+1)(2k+1)\omega^{2k-1}}{k[(2k+1)!!]^2 c^{2k+1}}$$
$$\times \left| \left(\psi_2 \left| \left[Q'^{(k)}_{-q} + K\sqrt{\frac{k}{k+1}} \left\{ Q'^{(k)}_{-q} - \omega Q_{-q}^{(k)} \right\} \right] \right| \psi_1 \right) \right|^2. \tag{4.11}$$

In these formulas

$$Q_{-q}^{(k)} = -r^k C_{-q}^{(k)} \tag{4.12}$$

denotes the non-relativistic form of the electric multipole transition operator corresponding, in the case of $k = 1$, to the length form of the electric dipole radiation operator. The quantity

$$Q'^{(k)}_{-q} = -r^{k-1}\left(kC^{(k)}_{-q}\frac{\partial}{\partial r} + \frac{i}{r}\sqrt{k(k+1)}\left[C^{(k)} \times L^{(1)}\right]^{(k)}_{-q}\right) \qquad (4.13)$$

is a new form of the electric multipole radiation operator [77], which, for $k = 1$, turns into the velocity form of the electric dipole transition operator. Operator $Q'^{(k)}_{-q}$ may also be obtained by considering the commutator of $Q^{(k)}_{-q}$ with the non-relativistic Hamiltonian of atom H_{NR}, i.e.

$$\left[H_{NR}, Q^{(k)}_{-q}\right] = -Q'^{(k)}_{-q}. \qquad (4.14)$$

By evaluating the commutator of $Q'^{(k)}_{-q}$ with H_{NR} the generalization of the acceleration form of the transition operator to cover the case of electric multipole transitions of any multipolarity was obtained [77]. Unfortunately, the resultant operator has a much more complex and cumbersome form than $Q^{(k)}_{-q}$ and $Q'^{(k)}_{-q}$, since it contains both one-electron and two-electron parts.

The structure of expressions (4.10), (4.11) is similar to that of the corresponding relativistic formulas (4.3), (4.4). For exact wave functions (both non-relativistic and relativistic) these operators will not depend on gauge condition K, because the multiplier in the curly brackets will be equal to zero due to the commutation relations. However, for many-electron atoms and ions we have to use approximate functions, the result being that the numerical values of the corresponding transition probabilities, calculated using different values of K, may differ significantly. This dependence has a parabolic form and it may be utilized as a criterion for the accuracy of the wave functions used. Because for the more accurate functions, the term in the matrix element that depends on K will be closer to zero, the dependence of the transition probability upon K will be weaker. The well-known non-equivalency of length and velocity operators of electric dipole transitions [78, 79] is a particular case of this gauge dependence. On the other hand, gauge K may be chosen as some semi-empirical parameter to obtain fairly accurate values of transition quantities, e.g. along the whole isoelectronic sequence. By choosing K so that the theoretical values of transition probability coincide with the experimental data for a particular transition, we may then partially account for correlation effects in the whole isoelectronic sequence.

Now we can establish the correspondence between relativistic expressions (4.3), (4.4) and the non-relativistic ones obtained above, as well as the particular cases when specifying gauge condition K. The relationship between the values of $K = 0$, $-\sqrt{(k+1)/k}$ and the forms of the electron transition operators was discussed by Grant [80, 81]. Formula (4.4) corresponds to equation (4.11) giving, when $k = 1$ and $K = 0$, the velocity form, whereas for $K = -\sqrt{(k+1)/k}$ we get the length form of the transition operator. This conclusion is in agreement with the results of Grant [80]. However, we can also conclude that the relationship between the gauge value and the form of transition operator does not have a simple one-to-one correspondence. If $K = 0$, then relativistic expression (4.3) leads in the non-relativistic limit, to the length form of the transition probability, whereas if $K = \sqrt{(k+1)/k}$ we obtain the velocity form (see Eq. (4.10)), i.e., for Eq. (4.3) we have the situation opposite to that in the case of Eq. (4.4).

The transformation of the relativistic expression for the operator of magnetic multipole radiation (4.8) may be done similarly to the case of electric transitions. As has already been mentioned, in this case the corresponding potential of electromagnetic field does not depend on the gauge condition, therefore, there is only the following expression for the non-relativistic operator of Mk-transitions (in a.u.):

$$W^{Mk}_{1 \to 2} = \frac{2(k+1)(2k+1)}{k[(2k+1)!!]^2} \left(\frac{\omega}{c}\right)^{2k+1} \left|\left(\psi_2|_mQ^{(k)}_{-q}|\psi_1\right)\right|^2. \tag{4.15}$$

Here the non-relativistic operator of Mk-transitions is as follows:

$$_mQ^{(k)}_{-q} = -i^k \frac{r^{k-1}}{c} \sqrt{k(2k-1)} \left\{ \frac{1}{k+1} \left[C^{(k-1)} \times L^{(1)}\right]^{(k)}_{-q} \right.$$
$$\left. + \left[C^{(k-1)} \times S^{(1)}\right]^{(k)}_{-q} \right\}. \tag{4.16}$$

In the special case when $k = 1$, Eq. (4.16) gives

$$M^{(1)}_{-q} = -\frac{i}{2c} \left(L^{(1)}_{-q} + 2S^{(1)}_{-q}\right) = -\frac{i}{2c} \left(J^{(1)}_{-q} + S^{(1)}_{-q}\right). \tag{4.17}$$

It differs by multiplier i from the normally used one, but this multiplier does not affect the transition quantities.

4.4 Relativistic corrections to various forms of the electric multipole (Ek) radiation operator

While calculating the energy spectra we shall see that for a large variety of atoms and their ionization degrees, the relativistic effects may be very efficiently accounted for as corrections of the order v^2/c^2. While studying the electronic transitions the relativistic effects are taken into consideration firstly by way of wavelengths of the corresponding transitions as well as via the wave functions of the so-called intermediate coupling (see Chapter 11).

However, one may expect that there are some relativistic corrections to the transition operators themselves.

Relativistic corrections of order v^2/c^2 to the non-relativistic transition operators may be found either by expanding the relativistic expression of the electron multipole radiation probability in powers of v/c, or semiclassically, by replacing $\mathbf{p}$ in the Dirac–Breit Hamiltonian by $\mathbf{p} - (1/c)\mathbf{A}$ (here $\mathbf{A}$ is the vector-potential of the radiation field) and retaining the terms linear in $\mathbf{A}$. Calculations show that in the general case the corresponding corrections have very complicated expressions, therefore we shall restrict ourselves to the particular case of electric dipole radiation and to the main corrections to the length and velocity forms of this operator.

Let us consider the intercombination transitions. Then, we shall retain only the corrections containing the spin operator in the expansion. To find the form of the operator describing the electric multipole intercombination transitions and absorbing the main relativistic corrections, one has to retain in the corresponding expansion the terms containing spin operator $\mathbf{S} = \frac{1}{2}\boldsymbol{\sigma}$ and to take into account, for the quantities of order v/c, the first retardation corrections, whereas, for the quantities of order v^2/c^2 one must neglect the retardation effects. Then the velocity form of the electric dipole transition probability may be written as follows:

$$W^{E1}_{1\to2} = \frac{4\omega}{3c^3}\left|\left(\psi_2\left|Q'^{(1)}_{-q} + \frac{i\omega^2}{c^2\sqrt{2}}\left[Q^{(1)} \times S^{(1)}\right]^{(1)}_{-q}\right|\psi_1\right)\right|^2, \qquad (4.18)$$

$$W^{E1}_{1\to2} = \frac{4\omega}{3c^3}\left|\left(\psi_2\left|Q'^{(1)}_{-q} + \frac{i\omega}{c^2\sqrt{2}}\left[Q'^{(1)} \times S^{(1)}\right]^{(1)}_{-q}\right|\psi_1\right)\right|^2, \qquad (4.19)$$

$$W^{E1}_{1\to2} = \frac{4\omega}{3c^3}\left|\left(\psi_2\left|Q'^{(1)}_{-q} - \frac{i}{c^2\sqrt{2}}\left[Q''^{(1)} \times S^{(1)}\right]^{(1)}_{-q}\right|\psi_1\right)\right|^2. \qquad (4.20)$$

Here $Q''^{(1)}_{-q}$ denotes the acceleration form of the electric dipole transition operator

$$Q''^{(1)}_{-q} = -C^{(1)}_{-q}\frac{\partial U}{\partial r}, \qquad (4.21)$$

where U is the potential of the external field. Eqs. (4.18)–(4.20) show that the relativistic corrections to the velocity form of the transition operator may be represented by three equivalent expressions, containing the length, velocity or acceleration forms of the transition operator. Of particular interest is Eq. (4.19), because in this case one has no need to calculate the additional radial integrals, and the relativistic corrections may be found starting with the non-relativistic velocity form of the transition operator.

Eq. (4.3) with $K = 0$ may serve as starting point to find the relativistic corrections to the length form of the transition operator. The expansion of the matrix element (4.3) up to terms of order v^2/c^2 for the case of the electric dipole intercombination transition was obtained in [82]. It is easy to show that such relativistic corrections to the radiation operator $Q'^{(k)}_{-q}$ may be found by considering the commutator of the atomic Hamiltonian, including the spin–own-orbit interaction, with the expressions for the relativistic corrections to operator $Q^{(k)}_{-q}$. In the special case, when $k = 1$, the non-relativistic limit of Eq. (4.3), taking into account the intercombination corrections of order v^2/c^2, is as follows:

$$W^{E1}_{1 \to 2} = \frac{4\omega^3}{3c^3} \left| \left(\psi_2 \left| Q^{(1)}_{-q} - \frac{i}{c^2\sqrt{2}} \left\{ \omega \left[Q^{(1)} \times S^{(1)} \right]^{(1)}_{-q} \right. \right. \right. \right.$$
$$\left. \left. \left. \left. - \left[Q'^{(1)} \times S^{(1)} \right]^{(1)}_{-q} \right\} \right| \psi_1 \right) \right|^2. \tag{4.22}$$

While using (4.14) and $\omega = E_1 - E_2$, the expression in curly brackets is zero for the case of the exact wave functions. This supports the conclusion of Drake [83] that for electric dipole transitions, by considering the commutator of $Q^{(1)}_q$ with the atomic Hamiltonian in the Pauli approximation, we obtain $Q'^{(1)}_{-q}$ with relativistic corrections of order v^2/c^2 (see (4.18)–(4.20)). However, for many-electron atoms and ions, one has to use approximate (e.g., Hartree–Fock) wave functions, and then this term gives non-zero contribution, conditioned by the inaccuracy of the model adopted.

Thus, the kind and quantity of relativistic corrections to the length and velocity forms of $E1$-radiation are different. From this point of view the concept of the equivalency of these forms must be improved: both forms will lead to coinciding transition values for the accurate (exact) wave functions only if we account for the relativistic corrections of order v^2/c^2 to the transition operators (in practice, only for the velocity form). The other conclusion: accounting for the relativistic effects leads to qualitatively new results, namely, to new operators, which allow not only improved values of permitted transitions, but also describe a number of lines, which earlier were forbidden. These relativistic corrections usually are very small, but they are very important for weak intercombination lines of light neutral atoms (see Chapter 30).

4.5 Autoionizing transitions

Let us discuss briefly the peculiarities of the excited states. Highly excited (Rydberg) states form a separate class of excited states. Experimentally, using, for example, selective laser radiation, it is possible to excite one electron to the states described by very high n values. Rydberg states of atoms are observed in interstellar space. Rydberg atoms possess a number of pecularities: their sizes may be very large, comparable to the size of bacteria, the excited electron of a Rydberg atom exhibits an essential influence of the other (surrounding) atoms, etc.

Another group of excited states is formed by atoms with several excited electrons (usually, two). Theoretical studies of these states present considerable difficulties. Among all excited states one has to distinguish the so-called autoionizing states, when an electron occupying such a state has energy exceeding the first threshold of the ionization of an atom, i.e., of its detachment from an atom or ion and transition into the state of a continuous spectrum. However, this definition is not rigorous. Sometimes, states are called autoionizing when they decay via an autoionization process. As a rule, autoionizing states are created when two electrons are excited. Such a system is rather unstable, it tends to autoionize, i.e. to lose one electron which goes to continuum. There are other channels of decay for such states, too. Therefore, when studying the decay of excited states, one has to consider not only radiative, but also radiationless electronic transitions.

Difficulties while studying excited states are conditioned by the abundance of energy levels, their high density for very highly excited states and the complexity in finding fairly accurate wave functions of such systems. It follows from physical considerations that the wave function of an excited configuration must be orthogonal to all lower-lying configurations. Ensuring these orthogonality conditions, when in some cases there is an infinite number of such configurations, presents considerable difficulties and the need to allow for some simplifications, e.g. to ensure orthogonality only with the wave function of the lowest configuration of the same symmetry, etc. Autoionizing states and transitions are of extreme importance when studying various elementary processes in low and high temperature plasma, both laboratory and astrophysical.

Of particular interest is the formation and study of 'hollow' atoms, the accumulation of electrons in shells with high values of principal quantum number n while inner shells remain sparingly populated or empty [84].

Part 2

Foundations of the Angular Momentum Theory. Graphical Methods

5

Angular momentum
and tensorial algebra

5.1 Central field approximation, angular momentum and spherical functions

We have already mentioned the central field approximation in Chapter 1 while discussing the self-consistent field method and the zero-order Hamiltonian of many-electron atoms. Let us recall briefly its main idea. The central field approximation means that any given electron in the N-electron atom moves independently in the electrostatic field of the nucleus, which is considered to be stationary, and of the other $N - 1$ electrons. This field is assumed to be time-averaged over the motion of these $N - 1$ electrons and, therefore, to be spherically symmetric. Then the wave function of this electron will be described by a formula of the type (1.14). In any such central field, the wave function (1.14) will be an eigenfunction of $\mathbf{L}^2$ and L_z, where $\mathbf{L}$ is angular momentum of an electron, and L_z its projection. Thus, the angular momentum of the electron is a constant of motion, and the wave function of the type (1.14) is an eigenfunction of the one-electron angular and spin momentum operators $\mathbf{L}^2$, L_z and $\mathbf{s}^2$, s_z with eigenvalues $l(l + 1)$, m_L and $s(s + 1) = 3/4$, m_s, correspondingly (in units of $\hbar^2$).

Thus, in the central field approximation the wave function of the stationary state of an electron in an atom will be the eigenfunction of the operators of total energy, angular and spin momenta squared and one of their projections. These operators will form the full set of commuting operators and the corresponding stationary state of an atomic electron will be characterized by total energy E, quantum numbers of orbital l and spin s momenta as well as by one of their projections.

Formally, orbital angular momentum operator $\mathbf{L}$ of a particle moving with linear momentum $\mathbf{p} = -i\hbar\nabla$ at a position $\mathbf{r}$ with respect to some

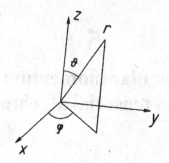

Fig. 5.1. Spherical coordinates.

given reference point (usually the origin of the coordinates) is

$$\mathbf{L} = -i\hbar[\mathbf{r} \times \nabla]. \tag{5.1}$$

A vectorial product will be defined below by (5.14), and ∇ as a tensor of first rank is defined by (2.12). Operator $\mathbf{L}$ may be defined also in a more general way by the commutation relations of its components. Such a definition is applicable to electron spin $\mathbf{s}$, as well. Therefore, we can write the following commutation relations between components of arbitrary angular momentum $\mathbf{j}$:

$$[j_x, j_y] = i\hbar j_z, \quad [j_y, j_z] = i\hbar j_x, \quad [j_z, j_x] = i\hbar j_y. \tag{5.2}$$

Here let us recall that

$$[j_i, j_k] = j_i j_k - j_k j_i. \tag{5.3}$$

Because of the spherical symmetry of an atom, it is very convenient to use spherical coordinates (Fig. 5.1), defined by

$$x = r \sin \vartheta \cos \varphi, \quad y = r \sin \vartheta \sin \varphi, \quad z = r \cos \vartheta. \tag{5.4}$$

Components of the angular momentum operator are connected with the infinitesimal operators of the group of rotations in three-dimensional space [11].

As we have seen from Eq. (1.14), the angular part of the wave function is defined by spherical functions (harmonics) $Y_m^{(l)}$ or $C_m^{(l)}$ connected with $Y_m^{(l)}$ according to Eq. (2.13). Therefore, we have to discuss briefly their properties. Normally, quantities $C_q^{(k)}$ are used when a spherical function plays the role of an operator. As we shall see, they are very important in theoretical atomic spectroscopy.

A set of spherical functions $C_m^{(l)}$ $(m = 0, \pm 1, \pm 2, \ldots, \pm l)$ composes the so-called irreducible tensor. In fact, we shall not need their explicit expressions, only their one-electron reduced matrix (submatrix) elements and the following addition theorem of spherical functions:

$$\left[C^{(k_1)} \times C^{(k_2)}\right]^{(K)} = (-1)^{\frac{k_1+k_2-K}{2}} \begin{bmatrix} k_1 & k_2 & K \\ 0 & 0 & 0 \end{bmatrix} C^{(K)}, \qquad (5.5)$$

will be utilized, in which the Clebsch–Gordan coefficient has projections equal to zero.

One-electron submatrix elements of the spherical functions operator occur in the expressions of any matrix element of a two-electron energy operator and the electron transition operators (except the magnetic dipole radiation), that is why we present in Table 5.1 their numerical values for the most practically needed cases $l, l' \leq 6$.

This submatrix element has the following fairly simple algebraic expression:

$$\left(l\|C^{(k)}\|l'\right) = \left(l'\|C^{(k)}\|l\right)$$
$$= g!\Delta(lkl')\sqrt{(2l+1)(2l'+1)}/\left[(g-l)!(g-k)!(g-l')!\right]!. \quad (5.6)$$

Here

$$g = (l + k + l')/2,$$

and

$$\Delta(lkl') = \left[(l+k-l')!(l-k+l')!(-l+k+l')!(l+k+l'+1)!\right]^{1/2}. \quad (5.7)$$

This submatrix element is always positive, and it is non-zero only when $l + k + l'$ is even. It is in the following way connected with a special case of the Clebsch–Gordan coefficient:

$$\left(l\|C^{(k)}\|l'\right) = (-1)^{g-l}\sqrt{2l'+1} \begin{bmatrix} l' & k & l \\ 0 & 0 & 0 \end{bmatrix}. \qquad (5.8)$$

This is described in more detail in Chapter 6. A relativistic analogue of this quantity will be considered in Chapter 7.

5.2 Irreducible tensorial sets

While finding the numerical values of any physical quantity one has to express the operator under consideration in terms of irreducible tensors. In the case of Racah algebra this means that we have to express any physical operator in terms of tensors which transform themselves like spherical functions $Y_{m_l}^{(l)}$. On the other hand, the wave functions (to be more exact, their spin–angular parts) may be considered as irreducible tensorial operators, as well. Having this in mind, we can apply to them all operations we carry out with tensors. As was already mentioned in the Introduction (formula (4)), spherical functions (harmonics) $Y_{m_l}^{(l)}$ are defined in the standard phase system.

Table 5.1.   Numerical values of $(l\|C^{(k)}\|l')$ for $l, l' \leq 6$.

$l\ l'\ k$	$(l\|C^{(k)}\|l')$	$l\ l'\ k$	$(l\|C^{(k)}\|l')$
0 k k	1	3 5 8	$2\cdot 7\sqrt{2}/\sqrt{13\cdot 17}$
1 1 0	$\sqrt{3}$	3 6 3	$2\cdot 5/\sqrt{3\cdot 11}$
1 1 2	$\sqrt{2\cdot 3}/\sqrt{5}$	3 6 5	$7/\sqrt{3\cdot 11}$
1 2 1	$\sqrt{2}$	3 6 7	$2\cdot 7\sqrt{2\cdot 3}/\sqrt{5\cdot 11\cdot 17}$
1 2 3	$3/\sqrt{7}$	3 6 9	$2\cdot 7\sqrt{3}/\sqrt{17\cdot 19}$
1 3 2	$3/\sqrt{5}$	4 4 0	3
1 3 4	$2/\sqrt{3}$	4 4 2	$2\cdot 3\sqrt{5}/\sqrt{7\cdot 11}$
1 4 3	$2\sqrt{3}/\sqrt{7}$	4 4 4	$3\cdot 9\sqrt{2}/\sqrt{7\cdot 11\cdot 13}$
1 4 5	$\sqrt{3\cdot 5}/\sqrt{11}$	4 4 6	$2\cdot 3\sqrt{5}/\sqrt{11\cdot 13}$
1 5 4	$\sqrt{5}/\sqrt{3}$	4 4 8	$3\cdot 7\sqrt{2\cdot 5}/\sqrt{11\cdot 13\cdot 17}$
1 5 6	$3\sqrt{2}/\sqrt{13}$	4 5 1	$\sqrt{5}$
1 6 5	$3\sqrt{2}/\sqrt{11}$	4 5 3	$2\cdot 3\sqrt{5}/\sqrt{7\cdot 13}$
1 6 7	$\sqrt{7}/\sqrt{5}$	4 5 5	$3\sqrt{2}/\sqrt{13}$
2 2 0	$\sqrt{5}$	4 5 7	$2\sqrt{2\cdot 5\cdot 7}/\sqrt{13\cdot 17}$
2 2 2	$\sqrt{2\cdot 5}/\sqrt{7}$	4 5 9	$9\cdot 7\sqrt{2}/\sqrt{13\cdot 17\cdot 19}$
2 2 4	$\sqrt{2\cdot 5}/\sqrt{7}$	4 6 2	$3\sqrt{5}/\sqrt{11}$
2 3 1	$\sqrt{3}$	4 6 4	$2\sqrt{5}/\sqrt{11}$
2 3 3	$2/\sqrt{3}$	4 6 6	$2\cdot 3\sqrt{7}/\sqrt{11\cdot 17}$
2 3 5	$5\sqrt{2}/\sqrt{3\cdot 11}$	4 6 8	$2\cdot 9\sqrt{2\cdot 7}/\sqrt{11\cdot 17\cdot 19}$
2 4 2	$3\sqrt{2}/\sqrt{7}$	4 6 10	$3\sqrt{2\cdot 5\cdot 7}/\sqrt{17\cdot 19}$
2 4 4	$2\cdot 5/\sqrt{7\cdot 11}$	5 5 0	$\sqrt{11}$
2 4 6	$3\cdot 5/\sqrt{11\cdot 13}$	5 5 2	$\sqrt{2\cdot 5\cdot 11}/\sqrt{3\cdot 13}$
2 5 3	$5\sqrt{2}/\sqrt{3\cdot 7}$	5 5 4	$\sqrt{2\cdot 11}/\sqrt{13}$
2 5 5	$5\sqrt{2}/\sqrt{3\cdot 13}$	5 5 6	$4\sqrt{5\cdot 11}/\sqrt{3\cdot 13\cdot 17}$
2 5 7	$\sqrt{3\cdot 7}/\sqrt{13}$	5 5 8	$7\sqrt{2\cdot 5\cdot 11}/\sqrt{13\cdot 17\cdot 19}$
2 6 4	$5/\sqrt{11}$	5 5 10	$2\cdot 3\sqrt{3\cdot 7\cdot 11}/\sqrt{13\cdot 17\cdot 19}$
2 6 6	$\sqrt{2\cdot 7}/\sqrt{11}$	5 6 1	$\sqrt{2\cdot 3}$
2 6 8	$2\sqrt{7}/\sqrt{17}$	5 6 3	$\sqrt{7}/\sqrt{3}$
3 3 0	$3\sqrt{7}$	5 6 5	$4\sqrt{5}/\sqrt{3\cdot 17}$
3 3 2	$2\sqrt{7}/\sqrt{3\cdot 5}$	5 6 7	$2\sqrt{3\cdot 5\cdot 7}/\sqrt{17\cdot 19}$
3 3 4	$\sqrt{2\cdot 7}/\sqrt{11}$	5 6 9	$2\sqrt{3\cdot 5\cdot 7}/\sqrt{17\cdot 19}$
3 3 6	$2\cdot 5\sqrt{7}/\sqrt{3\cdot 11\cdot 13}$	5 6 11	$3\cdot 11\sqrt{2\cdot 7}/\sqrt{17\cdot 19\cdot 23}$
3 4 1	2	6 6 0	$\sqrt{13}$
3 4 3	$3\sqrt{2}/\sqrt{11}$	6 6 2	$\sqrt{2\cdot 7\cdot 13}/\sqrt{5\cdot 11}$
3 4 5	$2\cdot 3\sqrt{5}/\sqrt{11\cdot 13}$	6 6 4	$2\sqrt{7\cdot 13}/\sqrt{11\cdot 17}$
3 4 7	$7\sqrt{5}/\sqrt{11\cdot 13}$	6 6 6	$4\cdot 5\sqrt{13}/\sqrt{11\cdot 17\cdot 19}$
3 5 2	$\sqrt{2\cdot 5}/\sqrt{3}$	6 6 8	$5\sqrt{2\cdot 7\cdot 13}/\sqrt{11\cdot 17\cdot 19}$
3 5 4	$2\sqrt{5}/\sqrt{13}$	6 6 10	$2\cdot 3\sqrt{3\cdot 7\cdot 13}/\sqrt{17\cdot 19\cdot 23}$
3 5 6	$7/\sqrt{3\cdot 13}$	6 6 12	$2\cdot 3\cdot 11\sqrt{7\cdot 13}/5\sqrt{17\cdot 19\cdot 23}$

The generalization of this conjugation phase condition, to cover the case of tensors with a complex inner structure and half-integer ranks, is presented in [85].

Fano and Racah [5] defined the concept of the irreducible tensorial set as a set of $2k + 1$ quantities $T_q^{(k)}$ (k is an integer or half-integer) transforming through each other according to the irreducible tensorial representations of the rotation group

$$T_{q'}^{(k)} = \sum_q D_{q',q}^{(k)} T_q^{(k)}, \tag{5.9}$$

where $D_{q',q}^{(k)}$ is the generalized spherical function (Wigner D-function, see Eq. (2.19)). The quantities $T_q^{(k)}$ may have a completely different nature (tensorial operators, wave functions (to underline this, we shall write the wave functions in the form of tensor $\psi_m^{(l)}$), etc.). The only condition is that they obey transformation property (5.9) and the consistent conjugation rule (more details may be found in Section 3 of [31]).

The exploitation of the community of the transformation properties of irreducible tensors and wave functions gives us the opportunity to deduce new relationships between the quantities considered, to further simplify the operators, already expressed in terms of irreducible tensors, or, in general, to offer a new method of calculating the matrix elements. Indeed, it is possible to show that the action of angular momentum operator $L_q^{(1)}$ on the wave function, considered as irreducible tensor $\psi_m^{(l)}$, may be represented in the form [86]:

$$\left[L^{(1)} \times \psi^{(l)} \right]_Q^{(K)} = -i\delta(K, l)\sqrt{l(l+1)}\psi_Q^{(l)}. \tag{5.10}$$

The commutation relations, written in the form of a tensorial product, are compact and also very useful for practical applications.

Thus, utilizing the concept of irreducible tensorial sets, one is in a position to develop a new method of calculating matrix (submatrix) elements, alternative to the standard way described in many papers [9–11, 14, 18, 21–23]. Indeed, the submatrix element of the irreducible tensorial operator $T_q^{(k)}$ can be expressed in terms of a zero-rank double tensorial (scalar) product of the corresponding operators (for simplicity we omit additional quantum numbers α, α'):

$$\left(j \| T^{(k)} \| j' \right) = (-1)^{j'-k+j} 4\pi \left[\psi^{(j)} \times \left[T^{(k)} \times \psi^{(j')} \right]^{(j)} \right]^0$$

$$= (-1)^{j'-k+j} \frac{4\pi}{\sqrt{2j+1}} \left(\psi^{(j)} \cdot \left[T^{(k)} \times \psi^{(j')} \right]^{(j)} \right). \tag{5.11}$$

Expressions of this kind can be found also for complex tensorial products of tensorial operators. Thus, we see that, indeed, we arrive at a new alternative method for the calculation of matrix elements in theoretical atomic spectroscopy.

5.3   Tensorial products and their matrix elements

Let us present the main definitions of tensorial products and their matrix or reduced matrix (submatrix) elements, necessary to find the expressions for matrix elements of the operators, corresponding to physical quantities. The tensorial product of two irreducible tensors $T^{(k_1)}$ and $U^{(k_2)}$ is defined as follows:

$$\left[T^{(k_1)} \times U^{(k_2)}\right]_q^{(k)} = \sum_{q_1 q_2} T_{q_1}^{(k_1)} U_{q_2}^{(k_2)} \begin{bmatrix} k_1 & k_2 & k \\ q_1 & q_2 & q \end{bmatrix}. \tag{5.12}$$

For $k = 0$, we find the definition of a scalar product

$$\sqrt{2k+1} \left[T^{(k)} \times U^{(k)}\right]_0^{(0)} = \sum_q (-1)^{k-q} T_q^{(k)} U_{-q}^{(k)} = \left(T^{(k)} \cdot U^{(k)}\right). \tag{5.13}$$

For $k = 1$ we have $(T^{(1)} \cdot U^{(1)}) = (\mathbf{T} \cdot \mathbf{U})$. For $k_1 = k_2 = k = 1$, formula (5.12) turns into a tensorial product, which can be expressed as a vectorial product of $\mathbf{T}$ and $\mathbf{U}$, i.e.

$$[\mathbf{T} \times \mathbf{U}] = -\sqrt{2} \left[T^{(1)} \times U^{(1)}\right]^{(1)}. \tag{5.14}$$

A fundamental role is played in theoretical atomic spectroscopy by the Wigner–Eckart theorem, the utilization of which allows one to find the dependence of any matrix element of an arbitrary irreducible tensorial operator on projection parameters,

$$\left(\alpha j m | T_q^{(k)} | \alpha' j' m'\right) = \frac{(-1)^{2k}}{\sqrt{2k+1}} \begin{bmatrix} j' & k & j \\ m' & q & m \end{bmatrix} \left(\alpha j \| T^{(k)} \| \alpha' j'\right). \tag{5.15}$$

The quantity denoted ($\|\ \ \|$) is called a reduced matrix (submatrix) element of operator $T^{(k)}$. It does not depend on projection parameters m, m', q. Dependence of the matrix element considered on these projections is contained in one Clebsch–Gordan coefficient. Such dependence is one of the indicators of the exceptional role played by the Clebsch–Gordan coefficients in the theory of many-particle systems. Their definitions and main properties will be discussed in the next paragraph.

The submatrix element of the tensorial product of two operators, acting on one and the same coordinate, may be calculated applying the formula

$$\left(\alpha j \| \left[T^{(k_1)} \times U^{(k_2)}\right]^{(k)} \| \alpha' j'\right) = (-1)^{j+j'+k} \sqrt{2k+1}$$

$$\times \sum_{\alpha'' j''} \left(\alpha j \| T^{(k_1)} \| \alpha'' j''\right) \left(\alpha'' j'' \| U^{(k_2)} \| \alpha' j'\right) \begin{Bmatrix} k_1 & k_2 & k \\ j & j' & j'' \end{Bmatrix}. \tag{5.16}$$

where symbol { } denotes the $6j$-coefficient, defined in Chapter 6. For the special case of a scalar product, instead of (5.16) we obtain

$$\left(\alpha j\|\left(T^{(k)}\cdot U^{(k)}\right)\|\alpha' j'\right) = \frac{\delta(j,j')}{\sqrt{2j+1}}\sum_{\alpha'' j''}\left(\alpha'' j''\|T^{(k)}\|\alpha j\right)^*\left(\alpha'' j''\|U^{(k)}\|\alpha' j\right),$$

(5.17)

where the condition

$$\left(\alpha j\|F^{(k)}\|\alpha' j'\right) = (-1)^{j'-j+k}\left(\alpha' j'\|F^{(k)}\|\alpha j\right)^*$$

(5.18)

has been taken into consideration. If the tensorial operators act on different coordinates, then the following equalities can be exploited:

$$\left(\alpha_1 j_1\alpha_2 j_2 j\|\left[T^{(k_1)}\times U^{(k_2)}\right]^{(k)}\|\alpha'_1 j'_1\alpha'_2 j'_2 j'\right) = \sqrt{(2j+1)(2j'+1)(2k+1)}$$

$$\times\left(\alpha_1 j_1\|T^{(k_1)}\|\alpha'_1 j'_1\right)\left(\alpha_2 j_2\|U^{(k_2)}\|\alpha'_2 j'_2\right)\begin{Bmatrix} j'_1 & j'_2 & j' \\ k_1 & k_2 & k \\ j_1 & j_2 & j \end{Bmatrix},$$

(5.19)

$$\left(\alpha_1 j_1\alpha_2 j_2 j\|\left(T^{(k)}\cdot U^{(k)}\right)\|\alpha'_1 j'_1\alpha'_2 j'_2 j'\right) = (-1)^{j_1+j_2+j}\delta(j,j')\sqrt{2j+1}$$

$$\times\left(\alpha'_1 j'_1\|T^{(k)}\|\alpha_1 j_1\right)^*\left(\alpha_2 j_2\|U^{(k)}\|\alpha'_2 j'_2\right)\begin{Bmatrix} j_1 & j_2 & j \\ j'_2 & j'_1 & k \end{Bmatrix}.$$

(5.20)

The last multiplier on the right side of (5.19) is the $9j$-coefficient, defined in Chapter 6. Such expressions for submatrix elements directly follow from (5.19) in the case when the operator acts only on coordinates with index 1 or 2:

$$\left(\alpha_1 j_1\alpha_2 j_2 j\|T_1^{(k)}\|\alpha'_1 j'_1\alpha'_2 j'_2 j\right) = (-1)^{j_1+j_2+j'+k}\delta(\alpha_2 j_2,\alpha'_2 j'_2)$$

$$\times\sqrt{(2j+1)(2j'+1)}\left(\alpha_1 j_1\|T_1^{(k)}\|\alpha'_1 j'_1\right)\begin{Bmatrix} j_1 & j_2 & j \\ j' & k & j'_1 \end{Bmatrix},$$

(5.21)

$$\left(\alpha_1 j_1\alpha_2 j_2 j\|U_2^{(k)}\|\alpha'_1 j'_1\alpha'_2 j'_2 j\right) = (-1)^{j_1+j'_2+j+k}\delta(\alpha_1 j_1,\alpha'_1 j'_1)$$

$$\times\sqrt{(2j+1)(2j'+1)}\left(\alpha_2 j_2\|U_2^{(k)}\|\alpha'_2 j'_2\right)\begin{Bmatrix} j_1 & j_2 & j \\ k & j' & j'_2 \end{Bmatrix}.$$

(5.22)

The elements of the theory of angular momentum and irreducible tensors presented in this chapter make a minimal set of formulas necessary when calculating the matrix elements of the operators of physical quantities for many-electron atoms and ions. They are equally suitable for both non-relativistic and relativistic approximations. More details on this issue may be found in the monographs [3, 4, 9, 11, 12, 14, 17].

5.4   Unit tensors

Unit tensors play, together with spherical functions, a very important role in theoretical atomic spectroscopy, particularly when dealing with the many-electron aspect of this problem. Unit tensorial operator u^k is defined via its one-electron submatrix element [22]

$$(l\|u^k\|l') = \{lkl'\}\delta(l,l') \qquad (k = 1, 2, \ldots, 2l), \tag{5.23}$$

and

$$(l\|u^0\|l') = \delta(l,l')\sqrt{2l+1}. \tag{5.24}$$

Let us also recall that the one-electron submatrix element of spin angular momentum operator s^1 and of scalar quantity s^0 (in $\hbar$ units) is given by

$$(s\|s^1\|s) = \sqrt{3/2}, \qquad (s\|s^0\|s) = \sqrt{2}. \tag{5.25}$$

Tensors in (5.23) and (5.25) are defined in a pseudostandard phase system (see Introduction, Eq. (5)). The symbol $\{abc\}$ in (5.23) means that parameters a, b, c obey the so-called triangular condition with integer perimeter (any number from these three is not smaller than the difference and not larger than the sum of the other two). From the operators introduced we can compose the operator

$$v^{k1} = u^k s^1 \tag{5.26}$$

as well as the sum of operators u^k and (5.26), namely

$$U_q^k = \sum_{i=1}^{N} u_q^k(i), \tag{5.27}$$

$$V_{qq'}^{k1} = \sum_{i=1}^{N} u_q^k(i)s_{q'}^1(i) = \sum_{i=1}^{N} v_{qq'}^{k1}(i). \tag{5.28}$$

Summation in (5.27), (5.28) is over the coordinates of a shell of equivalent electrons; moreover, u^k is related to spatial (orbital) and s^1 to spin coordinates. Quantities U_q^k are the infinitesimal operators (generators) of the groups of unitary transformations of the wave function of N particles, whereas u_q^k is that of one particle. Special combinations of the scalar products of operators (5.27), (5.28) define the so-called Casimir operators (invariants) of the corresponding groups. Let us consider these operators and their eigenvalues for the groups, the parameters of which representations are utilized to classify the states of a shell of equivalent electrons. Unitary unimodular group SU_{2l+1}, some special cases of orthogonal group

R_{2l+1} ($l = 1, 2, 3$), exceptional group G_2 and symplectic group Sp_{4l+2} belong to the above-mentioned groups.

The Casimir operator of $(2l + 1)$-dimensional orthogonal group R_{2l+1} may be expressed in terms of the sum of scalar products of operators U^k

$$G(R_{2l+1}) = \sum_{k_{odd}=1}^{2l-1} (2k + 1)(U^k \cdot U^k). \tag{5.29}$$

For a particular case of the group of three-dimensional rotations, we have the equation

$$G(R_3) = 3(U^1 \cdot U^1) = \frac{3(L^1 \cdot L^1)}{l(l + 1)(2l + 1)}, \tag{5.30}$$

again illustrating the exceptional role of angular momentum in atomic theory.

To classify the states of f-shell in LS coupling (see Chapter 9) the parameters of the representations of groups R_7 and G_2 are used [24]. The corresponding Casimir operators have the form

$$G(R_7) = 3(U^1 \cdot U^1) + 7(U^3 \cdot U^3) + 11(U^5 \cdot U^5), \tag{5.31}$$
$$G(G_2) = 3(U^1 \cdot U^1) + 11(U^5 \cdot U^5). \tag{5.32}$$

Formula (5.29) is the special case (odd k values) of the Casimir operator of the more general so-called unitary unimodular group SU_{2l+1}, i.e.

$$G(SU_{2l+1}) = \sum_{k>0}^{2l}(2k + 1)(U^k \cdot U^k). \tag{5.33}$$

The relevant Casimir operator of unitary group U_{2l+1} follows from (5.33) if we include in it the term with $k = 0$. The Casimir operator of symplectic group Sp_{4l+2}, whose generators are tensors U^k and V^{k+1} with the odd sums of ranks, may be defined in the following way:

$$G(Sp_{4l+2}) = 2 \sum_{k_{even}=0}^{2l} \frac{2k + 1}{(l\|u^k\|l)^2}(V^{k1} \cdot V^{k1}) + \frac{1}{2} \sum_{k_{odd}=1}^{2l-1} (2k + 1)(U^k \cdot U^k)$$

$$= 2 \sum_{k_{even}=0}^{2l} \frac{2k + 1}{(l\|u^k\|l)^2}(V^{k1} \cdot V^{k1}) + \frac{1}{2}G(R_{2l+1}). \tag{5.34}$$

The expressions for eigenvalues of the Casimir operators are presented below. They are completely defined by the transformation properties of the wave functions with regard to the corresponding groups and depend only on indices characterizing the representations of these groups. For the cases discussed above they are, respectively

$$g(R_3) = 3L(L + 1)/l(l + 1)(2l + 1), \tag{5.35}$$

$$g(R_7) = \frac{1}{2}\left[w_1(w_1 + 5) + w_2(w_2 + 3) + w_3(w_3 + 1)\right], \tag{5.36}$$

$$g(G_2) = \frac{1}{3}\left[u_1^2 + u_1 u_2 + u_2^2 + 5u_1 + 4u_2\right], \tag{5.37}$$

$$g(Sp_{4l+2}) = \frac{v}{2}(4l + 4 - v), \tag{5.38}$$

$$g(SU_{2l+1}) = -2S(S + 1) + \frac{4Nl(l + 1) - N(N - 1)}{2l + 1} - \frac{N(N - 4)}{2}. \tag{5.39}$$

Here w_1, w_2, w_3 are parameters characterizing the representations of group R_7; u_1, u_2 stand for the corresponding quantities of group G_2; v is the seniority quantum number, defined in a simpler way in Chapter 9. On the other hand, the eigenvalues of the Casimir operator of group R_{2l+1} may be expressed in the following way by v and S quantum numbers

$$g(R_{2l+1}) = -S(S + 1) + \frac{v}{4}(4l + 4 - v). \tag{5.40}$$

Hence, the quantum numbers, corresponding to the parameters of the representations of group R_{2l+1}, are superfluous, because they are strictly defined by Eq. (5.40) and the condition that indices characterizing the representations of group R_{2l+1} were positive and arranged in decreasing order (e.g., for $l = 3$ in (5.36) the condition $2 \geq w_1 \geq w_2 \geq w_3 \geq 0$ must be preserved). Therefore, for the classification of the states of the f-shell there is no need to use quantum numbers w_1, w_2, w_3, for d-shell – w_1, w_2 and for p-shell – L (see Chapter 9). Let us notice that in (5.37) parameters u_1, u_2 obey conditions $u_1 \geq u_2 \geq 0$, $\quad 4 \geq u_1 + u_2 \geq 0$. On the other hand, the use of the representations of orthogonal group R_{2l+1} in theoretical atomic spectroscopy gives additional information on the symmetry properties of a shell of equivalent electrons, allowing one to establish new relationships between the matrix elements of tensorial operators, including the operators, corresponding to physical quantities.

To conclude this chapter, let us present the main formulas for sums of unit tensors, necessary for evaluation of matrix elements of the energy operator. They will be necessary in Part 5. The matrix element of any irreducible tensorial operator $t^{(k)}$ may be written as follows:

$$t_i^{(k)} \stackrel{*}{=} (l\|t^{(k)}\|l)u_i^k/(l\|u^k\|l). \tag{5.41}$$

Here i indicates the coordinates, on which the operator acts. Symbol $\stackrel{*}{=}$ is used to emphasize that the equality is valid in the sense of matrix elements for a shell of equivalent electrons. When using (5.41), the spin–angular part of the scalar product of the operators, acting on spatial coordinates *i* and *j*, may be presented in such a form:

$$\sum_{j>i=1}^{N} (t_i^{(k)} \cdot t_j^{\prime(k)}) \doteq \frac{(l\|t^{(k)}\|l)(l\|t^{\prime(k)}\|l)}{2(l\|u^k\|l)^2} \left[\left(U^k \cdot U^k \right) - \sum_{i=1}^{N} (u_i^k \cdot u_i^k) \right]. \quad (5.42)$$

It is worthwhile to emphasize that all dependences of matrix element (5.42) on the structure of a shell are contained in the items in square brackets, whereas all pecularities of the operators themselves are contained in one-electron submatrix elements. The matrix element of the second term in (5.42) is equal, for any k, to

$$\frac{1}{(l\|u^k\|l)^2} \left(l^N \alpha LS \left| \sum_{i=1}^{N} (u_i^k \cdot u_i^k) \right| l^N \alpha' LS \right) = \frac{\delta(\alpha, \alpha')N}{2l+1}. \quad (5.43)$$

In the general case, there is no simple algebraic expression for the first term of (5.42), therefore, for its evaluation one must utilize the tables of numerical values of the submatrix elements of operators U^k. The most complete tables, covering also the case of operators V^{k1}, may be found in [87]. For operators, also depending on spin variables, the analogue of formula (5.41) will have the form

$$t_i^{(k)} s_i^1 \doteq (l\|t^{(k)}\|l) v^{k1} / (l\|u^k\|l), \quad (5.44)$$

and a similar problem of finding the expressions for submatrix elements of scalar products $(V^{k1} \cdot V^{k1})$ and $\sum_{i=1}^{N}(v_i^{k1} \cdot v_i^{k1})$ occurs. For the last term, similar to (5.43), we obtain

$$\frac{1}{(l\|u^k\|l)^2} \left(l^N \alpha LS \left| \sum_{i=1}^{N}(v_i^{k1} \cdot v_i^{k1}) \right| l^N \alpha' LS \right) = \frac{\delta(\alpha, \alpha')3N}{4(2l+1)}, \quad (5.45)$$

whereas the first term must be calculated using tables [87] or the corresponding computer programs. Division by $(l\|u^k\|l)$ is introduced in (5.41) and (5.44) in order to ensure the condition that the expressions obtained cover the case $k = 0$.

6

Main quantities of angular momentum theory

6.1 Clebsch–Gordan coefficients

Clebsch–Gordan coefficients have already occurred several times in our considerations: in the Introduction (formula (2)) while generalizing the quasispin concept for complex electronic configurations, while defining a relativistic wave function (formulas (2.15) and (2.16)), in the addition theorem of spherical functions (5.5); and in the definition of tensorial product of two tensors (5.12). Let us discuss briefly their definition and properties. There are a number of algebraic expressions for the Clebsch–Gordan coefficients [9, 11], but here we shall present only one:

$$
\begin{bmatrix} j_1 & j_2 & j \\ m_1 & m_2 & m \end{bmatrix} = \delta(m_1 + m_2, m)\Delta(j_1 j_2 j)
$$

$$
\times \left[\frac{(j+m)!(j-m)!(2j+1)}{(j_1+m_1)!(j_1-m_1)!(j_2+m_2)!(j_2-m_2)!} \right]^{1/2} \tag{6.1}
$$

$$
\times \sum_z (-1)^{j_2+m_2+z} \frac{(j+j_2+m_1-z)!(j_1-m_1+z)!}{z!(j+m-z)!(j-j_1+j_2-z)!(j_1-j_2-m+z)!},
$$

where $\Delta(j_1 j_2 j)$ is defined by Eq. (5.7). The Clebsch–Gordan coefficients have the following symmetry properties:

$$
\begin{bmatrix} j_1 & j_2 & j \\ m_1 & m_2 & m \end{bmatrix} = (-1)^{j_1+j_2-j} \begin{bmatrix} j_2 & j_1 & j \\ m_2 & m_1 & m \end{bmatrix}
$$

$$
= (-1)^{j_1+j_2-j} \begin{bmatrix} j_1 & j_2 & j \\ -m_1 & -m_2 & -m \end{bmatrix}, \tag{6.2}
$$

$$
= (-1)^{j_1-m_1} \sqrt{\frac{2j+1}{2j_2+1}} \begin{bmatrix} j_1 & j & j_2 \\ m_1 & -m & -m_2 \end{bmatrix}, \tag{6.3}
$$

$$
= (-1)^{j_2+m_2} \sqrt{\frac{2j+1}{2j_1+1}} \begin{bmatrix} j & j_2 & j_1 \\ -m & m_2 & -m_1 \end{bmatrix}, \tag{6.4}
$$

$$= (-1)^{j_2-j-m_1} \sqrt{\frac{2j+1}{2j_2+1}} \begin{bmatrix} j & j_1 & j_2 \\ -m & m_1 & -m_2 \end{bmatrix}, \quad (6.5)$$

$$= (-1)^{j_1-j+m_2} \sqrt{\frac{2j+1}{2j_1+1}} \begin{bmatrix} j_2 & j & j_1 \\ m_2 & -m & -m_1 \end{bmatrix}. \quad (6.6)$$

The Clebsch–Gordan coefficient may be expressed in terms of the Wigner coefficient (symbol in round brackets):

$$\begin{bmatrix} j_1 & j_2 & j_3 \\ m_1 & m_2 & -m_3 \end{bmatrix} = (-1)^{j_1-j_2-m_3} \sqrt{2j_3+1} \begin{pmatrix} j_1 & j_2 & j_3 \\ m_1 & m_2 & m_3 \end{pmatrix}, \quad (6.7)$$

which is much more symmetric than the Clebsch–Gordan coefficient itself. Its symmetry properties, analogous to those of (6.2)–(6.6), are as follows:

$$\begin{pmatrix} j_1 & j_2 & j_3 \\ m_1 & m_2 & m_3 \end{pmatrix} = \epsilon \begin{pmatrix} j_i & j_k & j_l \\ m_i & m_k & m_l \end{pmatrix} \quad (6.8)$$

where $\epsilon = 1$ if the permutation $\begin{pmatrix} 1 & 2 & 3 \\ i & k & l \end{pmatrix}$ is even, and $\epsilon = (-1)^{j_1+j_2+j_3}$, otherwise. One more symmetry property

$$\begin{pmatrix} j_1 & j_2 & j_3 \\ m_1 & m_2 & m_3 \end{pmatrix} = (-1)^{j_1+j_2+j_3} \begin{pmatrix} j_1 & j_2 & j_3 \\ -m_1 & -m_2 & -m_3 \end{pmatrix}. \quad (6.9)$$

Many special cases of the Clebsch–Gordan coefficients have very simple algebraic expressions, usually without sums. Thus, a general formula for the Clebsch–Gordan coefficient with all projection parameters equal to zero may be easily found from Eqs. (5.8) and (5.6). If one of the parameters $j_i = 0$, then

$$\begin{bmatrix} j & 0 & j' \\ m & 0 & m' \end{bmatrix} = \begin{bmatrix} 0 & j & j' \\ 0 & m & m' \end{bmatrix} = \delta(jm, j'm'), \quad (6.10)$$

$$\begin{bmatrix} j & j' & 0 \\ m & m' & 0 \end{bmatrix} = \begin{bmatrix} j' & j & 0 \\ -m' & -m & 0 \end{bmatrix} = (-1)^{j-m}\sqrt{2j+1}\,\delta(jm, j'-m'). \quad (6.11)$$

If two parameters equal zero, then

$$\begin{bmatrix} 0 & 0 & j \\ 0 & 0 & m \end{bmatrix} = \begin{bmatrix} 0 & j & 0 \\ 0 & m & 0 \end{bmatrix} = \begin{bmatrix} j & 0 & 0 \\ m & 0 & 0 \end{bmatrix} = \delta(jm, 00). \quad (6.12)$$

If $m_2 = \pm j_2$, then

$$\begin{bmatrix} j_1 & j_2 & j \\ m \mp j_2 & \pm j_2 & m \end{bmatrix} = (\mp 1)^{j_1+j_2-j} \, [(j_1+j_2 \mp m)!(j \pm m)!(2j_2)!$$
$$\times (j_1-j_2+j)!(2j+1)]^{1/2} \, [(j_1-j_2 \pm m)!(j \mp m)!(j_1+j_2-j)!$$
$$\times (j-j_1+j_2)!(j_1+j_2+j+1)!]^{-1/2}. \quad (6.13)$$

For $j_1 = 1$, $m_1 = 0$ we get

$$\begin{bmatrix} 1 & j_2 & j_3 \\ 0 & m & m \end{bmatrix} = -\frac{1}{2}[j_2(j_2 + 1) - j_3(j_3 + 1) + 2m]$$

$$\times [(j_2 - m)!(j_3 + m)!(2j_3 + 1)]^{1/2} [(j_2 + m)!(j_3 - m)!(j_2 + 1 - j_3)!$$
$$\times (j_3 - j_2 + 1)!(j_2 + j_3 + 2)(j_2 + j_3 + 1)(j_2 + j_3)]^{-1/2}. \qquad (6.14)$$

As we can see from Eq. (6.8), the Wigner coefficients are much more symmetrical than the Clebsch–Gordan coefficients. Therefore, usually their tables are presented (see, for example, in [11] the tables for $j_1 + j_2 + j_3 \leq 16$).

6.2   3nj-coefficients

In angular momentum theory a very important role is played by the invariants obtained while summing the products of the Wigner (or Clebsch–Gordan) coefficients over all projection parameters. Such quantities are called j-coefficients or $3nj$-coefficients. They are invariant under rotations of the coordinate system. A j-coefficient has $3n$ parameters ($n = 1, 2, 3, \ldots$), that is why the notation $3nj$-coefficient is widely used. The value $n = 1$ leads to the trivial case of the triangular condition $\{abc\}$, defined in Chapter 5 after formula (5.25). For $n = 2, 3, 4, \ldots$ we have $6j$-, $9j$-, $12j$-, $\ldots$ coefficients, respectively. $3nj$-coefficients ($n > 2$) may be also defined as sums of $6j$-coefficients. There are also algebraic expressions for $3nj$-coefficients. Thus, $6j$-coefficient may be defined by the formula

$$\begin{Bmatrix} a & b & e \\ d & c & f \end{Bmatrix} = \sum_z (-1)^z$$

$$\times \frac{\Delta(abc)\Delta(acf)\Delta(cde)\Delta(bdf)(z+1)!}{(z-a-b-e)!(z-a-c-f)!(z-c-d-e)!(z-b-d-f)!}$$

$$\times [(a+b+c+d-z)!(a+d+e+f-z)!(b+c+e+f-z)!]^{-1}. \quad (6.15)$$

It may also be expressed in terms of a sum of the Clebsch–Gordan coefficients:

$$\begin{Bmatrix} j_1 & j_2 & j_{12} \\ j_3 & j & j_{32} \end{Bmatrix} = (-1)^{2j} [(2j_{12} + 1)(2j_{32} + 1)]^{-1/2} (2j + 1)^{-1}$$

$$\times \sum_{\substack{m_1\ m_2\ m_3 \\ m_{12}\ m_{32}\ m}} \begin{bmatrix} j_1 & j_2 & j_{12} \\ m_1 & m_2 & m_{12} \end{bmatrix} \begin{bmatrix} j_{12} & j_3 & j \\ m_{12} & m_3 & m \end{bmatrix}$$

$$\times \begin{bmatrix} j_3 & j_2 & j_{32} \\ m_3 & m_2 & m_{32} \end{bmatrix} \begin{bmatrix} j_{32} & j_1 & j \\ m_{32} & m_1 & m \end{bmatrix}. \qquad (6.16)$$

$6j$-coefficients have the following symmetry properties:

$$\begin{Bmatrix} j_1 & j_2 & j_3 \\ l_1 & l_2 & l_3 \end{Bmatrix} = \begin{Bmatrix} j_a & j_b & j_c \\ l_a & l_b & l_c \end{Bmatrix} = \begin{Bmatrix} j_a & l_b & l_c \\ l_a & j_b & j_c \end{Bmatrix}, \qquad (6.17)$$

from which it follows that we can arrange the columns in any order as well as replace any pair of upper line symbols by those of the lower one without changing the numerical values of the coefficient itself. These coefficients are non-zero if the triangular conditions $\{j_1 j_2 j_3\}$, $\{j_1 l_2 l_3\}$, $\{l_1 l_2 j_3\}$ and $\{l_1 j_2 l_3\}$ are preserved. Let us present the algebraic expressions for the particular cases $d = 0$, $1/2$ and 1 of the $6j$-coefficient (6.15):

$$\begin{Bmatrix} a & b & c \\ 0 & c & b \end{Bmatrix} = (-1)^s \left[(2b+1)(2c+1)\right]^{-1/2}, \tag{6.18}$$

$$\begin{Bmatrix} a & b & c \\ 1/2 & c-1/2 & b+1/2 \end{Bmatrix}$$
$$= (-1)^s \left[\frac{(a+c-b)(a+b-c+1)}{(2b+1)(2b+2)2c(2c+1)}\right]^{1/2}, \tag{6.19}$$

$$\begin{Bmatrix} a & b & c \\ 1/2 & c-1/2 & b-1/2 \end{Bmatrix}$$
$$= (-1)^s \left[\frac{(a+b+c+1)(b+c-a)}{2b(2b+1)2c(2c+1)}\right]^{1/2}, \tag{6.20}$$

$$\begin{Bmatrix} a & b & c \\ 1 & c-1 & b-1 \end{Bmatrix}$$
$$= (-1)^s \left[\frac{s(s+1)(s-2a-1)(s-2a)}{(2b-1)2b(2b+1)(2c-1)2c(2c+1)}\right]^{1/2}, \tag{6.21}$$

$$\begin{Bmatrix} a & b & c \\ 1 & c-1 & b \end{Bmatrix}$$
$$= (-1)^s \left[\frac{(s+1)(s-2a)(s-2b)(s-2c+1)}{b(2b+1)(2b+2)(2c-1)2c(2c+1)}\right]^{1/2}, \tag{6.22}$$

$$\begin{Bmatrix} a & b & c \\ 1 & c-1 & b+1 \end{Bmatrix}$$
$$= (-1)^s \left[\frac{(s-2b)(s-2b-1)(s-2c+1)(s-2c+2)}{(2b+1)(2b+2)(2b+3)(2c-1)2c(2c+1)}\right]^{1/2}, \tag{6.23}$$

$$\begin{Bmatrix} a & b & c \\ 1 & c & b \end{Bmatrix}$$
$$= (-1)^s \frac{2X}{[2b(2b+1)(2b+2)2c(2c+1)(2c+2)]^{1/2}}. \tag{6.24}$$

In these formulas $s = a + b + c$, $X = a(a+1) - b(b+1) - c(c+1)$. A more complete listing of the formulas, covering the case when $d \leq 4$, is published in [88]. Fairly extensive tables of their numerical values may be found in [11].

There are many formulas for the summation of products of $6j$-coefficients

[9, 11]. Let us present here two of them:

$$\sum_x (2x+1)\begin{Bmatrix} a & b & e \\ d & c & x \end{Bmatrix}\begin{Bmatrix} a & b & e' \\ d & c & x \end{Bmatrix} = (2e+1)^{-1}\delta(e,e'), \quad (6.25)$$

$$\sum_x (2x+1)(-1)^{x+e+e'}\begin{Bmatrix} a & b & e \\ d & c & x \end{Bmatrix}\begin{Bmatrix} a & d & e' \\ b & c & x \end{Bmatrix} = \begin{Bmatrix} a & b & e \\ c & d & e' \end{Bmatrix}. \quad (6.26)$$

The $9j$-coefficient may be expressed in terms of the sum of the product of three $6j$-coefficients, namely,

$$\begin{Bmatrix} j_1 & j_2 & j_3 \\ l_1 & l_2 & l_3 \\ k_1 & k_2 & k_3 \end{Bmatrix} = \sum_x (-1)^{2x}(2x+1)\begin{Bmatrix} j_1 & j_2 & j_3 \\ l_3 & k_3 & x \end{Bmatrix}$$

$$\times \begin{Bmatrix} l_1 & l_2 & l_3 \\ j_2 & x & k_2 \end{Bmatrix}\begin{Bmatrix} k_1 & k_2 & k_3 \\ x & j_1 & l_1 \end{Bmatrix}. \quad (6.27)$$

It has the following symmetry properties:

$$\begin{Bmatrix} j_1 & j_2 & j_3 \\ l_1 & l_2 & l_3 \\ k_1 & k_2 & k_3 \end{Bmatrix} = \begin{Bmatrix} j_1 & l_1 & k_1 \\ j_2 & l_2 & k_2 \\ j_3 & l_3 & k_3 \end{Bmatrix} = (-1)^R \begin{Bmatrix} j_2 & j_1 & j_3 \\ l_2 & l_1 & l_3 \\ k_2 & k_1 & k_3 \end{Bmatrix}$$

$$= \begin{Bmatrix} j_3 & j_1 & j_2 \\ l_3 & l_1 & l_2 \\ k_3 & k_1 & k_2 \end{Bmatrix}. \quad (6.28)$$

Here $R = j_1+j_2+j_3+l_1+l_2+l_3+k_1+k_2+k_3$. $9j$-coefficients do not vanish if six triangular conditions, formed by the sums of parameters in each line and column, are fulfilled. Many algebraic formulas for $9j$-coefficients may be found in [9, 11, 88].

$12j$-coefficients of the first and the second kind are defined by the following formulas (R_4 stands for the sum of all 12 parameters):

$$\begin{Bmatrix} j_1 & & j_2 & & j_3 & & j_4 \\ & l_1 & & l_2 & & l_3 & & l_4 \\ k_1 & & k_2 & & k_3 & & k_4 \end{Bmatrix} = \sum_x (-1)^{R_4-x}(2x+1)\begin{Bmatrix} j_1 & k_1 & x \\ k_2 & j_2 & l_1 \end{Bmatrix}$$

$$\times \begin{Bmatrix} j_2 & k_2 & x \\ k_3 & j_3 & l_2 \end{Bmatrix}\begin{Bmatrix} j_3 & k_3 & x \\ k_4 & j_4 & l_3 \end{Bmatrix}\begin{Bmatrix} j_4 & k_4 & x \\ j_1 & k_1 & l_4 \end{Bmatrix}, \quad (6.29)$$

$$\begin{bmatrix} j_1 & j_2 & j_3 & j_4 \\ l_1 & l_2 & l_3 & l_4 \\ k_1 & k_2 & k_3 & k_4 \end{bmatrix} = (-1)^{l_1-l_2-l_3+l_4}\sum_x (2x+1)\begin{Bmatrix} k_1 & k_2 & x \\ j_3 & j_1 & l_1 \end{Bmatrix}$$

$$\times \begin{Bmatrix} k_3 & k_4 & x \\ j_3 & j_1 & l_2 \end{Bmatrix}\begin{Bmatrix} k_1 & k_2 & x \\ j_4 & j_2 & l_3 \end{Bmatrix}\begin{Bmatrix} k_3 & k_4 & x \\ j_4 & j_2 & l_4 \end{Bmatrix}. \quad (6.30)$$

The general case of $3nj$-coefficients is considered in detail in [9, 11]. There are many summation formulas for various products of $3nj$-

coefficients, but the easiest way to obtain them is the graphical approach, which will be sketched in Chapter 8.

6.3 Transformation matrices

Formula (3) of the Introduction was the first example of how to use transformation matrices. The simplest case of transformation matrices is presented by formulas (10.4) and (10.5), while considering the relationships between the wave functions of coupled and uncoupled momenta. The Clebsch–Gordan coefficients served there as the corresponding transformation matrices. However, the theory of transformation matrices is most widely utilized for transformations of the wave functions and matrix elements from one coupling scheme to another (Chapters 11,12) as well as for their calculations.

The theory of transformation matrices is described in detail in [9]. In [11] universal and efficient graphical methods of operating with Clebsch–Gordan coefficients and transformation matrices are described. Their use allows one easily to find expressions for the most complex matrix elements of the operators, corresponding to physical quantities, in various coupling schemes, and to carry out their processing if necessary. Here we shall present only the minimal data on transformation matrices and methods of their evaluation.

The addition of two angular momenta (formula of the type (10.4)) may be directly generalized to cover the case of an arbitrary number of momenta. However, in such a case it is not enough to adopt the total momentum and its projection for the complete characterization of the wave function of coupled momenta. Normally, the quantum numbers of intermediate momenta must be exploited too. Moreover, these functions depend on the form (order) of the coupling between these momenta. The relationships between the functions, belonging to different forms of coupling of their momenta, may be found with the aid of transformation matrices.

Let us consider the coupling of the three momenta j_1, j_2, j_3, as an example. The wave function of uncoupled momenta

$$\psi(j_1 j_2 j_3 m_1 m_2 m_3) = \psi(j_1 m_1)\psi(j_2 m_2)\psi(j_3 m_3) \tag{6.31}$$

will be an eigenfunction of the set of commuting operators

$$\mathbf{j}_1^2, \ \mathbf{j}_2^2, \ \mathbf{j}_3^2, \ j_{1z}, \ j_{2z}, \ j_{3z}. \tag{6.32}$$

The corresponding function of coupled momenta

$$\psi(j_1 j_2 j_3 J_{ik} J M) = \sum_{m_1 m_2 m_3} \psi(j_1 m_1)\psi(j_2 m_2)\psi(j_3 m_3) \begin{bmatrix} j_1 & j_2 & j_3 & J \\ m_1 & m_2 & m_3 & M \end{bmatrix}_{J_{ik}}$$
(6.33)

will be an eigenfunction of the following operators:

$$\mathbf{j}_1^2, \ \mathbf{j}_2^2, \ \mathbf{j}_3^2, \ \mathbf{J}_{ik}^2, \ \mathbf{J}^2, \ J_z,$$
(6.34)

where $\mathbf{J}_{ik} = \mathbf{j}_i + \mathbf{j}_k$ is the sum of any two coupling momenta. The last multiplier in (6.33) is called the generalized Clebsch–Gordan coefficient; it may be expressed in terms of the sum of the usual Clebsch–Gordan coefficients [9, 11]. Generalization of formulas (6.31)–(6.34) for an arbitrary number of coupled momenta is straightforward. As the number of intermediate momenta increases (for n momenta there occur $n-2$ intermediate momenta), the number of different ways of their coupling increases as well. The indices i and k at the intermediate momenta J_{ik} point out, the algebraic sum of which momenta is the considered intermediate momentum. The total sum of the indices at all intermediate momenta and the resultant (in the latter case they are usually skipped) define a way of coupling the momenta. It is convenient to distinguish the coupling scheme, often described by the location of brackets or commas in the notation of the wave function, and the coupling order, defined by the location of the indices of the coupled momenta. Thus, the wave function of the four momenta $\mathbf{j}_1$, $\mathbf{j}_2$, $\mathbf{j}_3$, $\mathbf{j}_4$, coupled in the following manner:

$$\mathbf{j}_1 + \mathbf{j}_3 = \mathbf{J}_{13}, \quad \mathbf{j}_2 + \mathbf{j}_4 = \mathbf{J}_{24}, \quad \mathbf{J}_{13} + \mathbf{J}_{24} = \mathbf{J}_{1324} \quad \text{(or simply } \mathbf{J}) \quad (6.35)$$

may be written as

$$\psi(j_1 j_2 j_3 j_4 J_{13} J_{24} J_{1324} M_{1324}),$$
(6.36)

$$\psi(((j_1 j_3) J_{13}(j_2 j_4) J_{24}) J_{1324} M_{1324}),$$
(6.37)

$$\psi(j_1 j_3 (J_{13}), \ j_2 j_4 (J_{24}), \ J_{1324} M_{1324}).$$
(6.38)

In general, not specifying the coupling scheme A, the wave function of coupled momenta may be presented in the form

$$\psi((j_1 j_2 \dots j_n)^A \ a J M),$$
(6.39)

where $a = a_1, a_2, \dots, a_{n-2}$ is a set of intermediate momenta.

It is necessary to emphasize that the indices of the wave functions denote not only the numbers of momenta, but also their coordinates. One has to preserve the marking of coordinates while dealing with the coinciding numerical values of coupled momenta (e.g. one-electron spin momenta, orbital momenta of equivalent electrons etc., in such cases the momenta must be denoted as, for example, $l(i)$, $l(k),\dots$). Neglecting the coordinates leads to mistakes in phase multipliers of the corresponding mathematical expressions.

If we have two different coupling schemes A and B, then the transition from the wave functions of coupling scheme A to those of B is carried out with the help of the transformation matrix, namely,

$$\psi((j_1 j_2 \ldots j_n)^B \ bJM) = \sum_{aJ'M'} \psi((j_1 j_2 \ldots j_n)^A \ aJ'M')$$
$$\times \left((j_1 j_2 \ldots j_n)^A \ aJ'M' | (j_1 j_2 \ldots j_n)^B \ bJM \right). \quad (6.40)$$

Multiplying the formula (6.40) by the function $\psi^*((j_1 j_2 \ldots j_n)^A \ a''J''M'')$ and integrating the expression obtained over all variables, we conclude that the transformation matrix is diagonal with respect to J and M and does not depend on M.

The transformation matrix may be expressed in terms of the product of Clebsch–Gordan coefficients, summed over all projection parameters, in other words, in terms of $3nj$-coefficients. Such sums are real numbers, therefore the transformation matrices will be real as well, i.e.

$$\left((j_1 j_2 \ldots j_n)^A \ aJ | (j_1 j_2 \ldots j_n)^B \ bJ \right) = \left((j_1 j_2 \ldots j_n)^B \ bJ | (j_1 j_2 \ldots j_n)^A \ aJ \right).$$
$$(6.41)$$

Exploiting the relationship between the transformation matrices and $3nj$-coefficients, we are in a position to study, for example, the symmetry properties of the first quantities, knowing the second and vice versa. The easiest way to express the transformation matrices in terms of the $3nj$-coefficients is through use of the universal graphical methods, described in [9, 11] and briefly sketched in Chapter 8. They are applicable to transformation matrices of any complexity. For a number of special cases it may be done algebraically, representing the matrix considered as a sum of products of simpler matrices, their expressions in terms of $3nj$-coefficients being already known. Below we shall present the corresponding formulas for three and four coupled momenta. For the case of three momenta we have to consider only one matrix. It is expressed in terms of a $6j$-coefficient [23]

$$(((j_1 j_2)J_{12} j_3)J | ((j_1 j_3)J_{13} j_2)J)$$
$$= (-1)^{j_2 + j_3 + J_{12} + J_{13}} \sqrt{[J_{12}, J_{13}]} \begin{Bmatrix} j_1 & j_2 & J_{12} \\ J & j_3 & J_{13} \end{Bmatrix}. \quad (6.42)$$

Here and further on $[a, b, \ldots] = (2a + 1)(2b + 1) \ldots$. For four momenta we have two formulas, corresponding to the successive coupling of the momenta, and in pairs, namely,

$$((((j_1 j_2)J_{12} j_3)J_{123} j_4)J|(((j_1 j_4)J_{14} j_3)J_{143} j_2)J)$$

$$= (-1)^{J_{12}-J_{14}-J_{123}+J_{143}} \sqrt{[J_{12}, J_{14}, J_{123}, J_{143}]} \begin{Bmatrix} j_1 & J_{12} & j_2 \\ J_{14} & j_3 & J_{143} \\ j_4 & J_{123} & J \end{Bmatrix}, \quad (6.43)$$

$$(((j_1 j_2)J_{12}(j_3 j_4)J_{34})J|((j_1 j_3)J_{13}(j_2 j_4)J_{24})J)$$

$$= \sqrt{[J_{12}, J_{34}, J_{13}, J_{24}]} \begin{Bmatrix} j_1 & j_2 & J_{12} \\ j_3 & j_4 & J_{34} \\ J_{13} & J_{24} & J \end{Bmatrix}. \quad (6.44)$$

We shall not consider here the more complex cases; if necessary, they may be found in [9, 11] or deduced utilizing methods described there. The expressions for matrix elements of the majority of energy operators (1.16) in terms of radial integrals and transformation matrices or $3nj$-coefficients for complex electronic configurations may be found in [14].

7

Angular momentum theory
for relativistic case

7.1 Angular momentum

The main ideas of angular momentum theory (e.g. central field approximation, self-consistent field method, etc.) are fairly general and are equally applicable for both the non-relativistic and relativistic theories of many-electron atoms and ions in spite of the fact that the corresponding energy operators differ significantly in nature. This distinction manifests itself in different kinds of wave functions (non-relativistic (1.14) and relativistic (2.15) or (2.18)). Strong interaction of orbital $\mathbf{L}$ and spin $\mathbf{s}$ momenta of an electron in relativistic approximation leads, as we shall see in Chapter 9, to the splitting of a shell of equivalent electrons into subshells according to $\mathbf{j} = \mathbf{L} + \mathbf{s}$ values and to the occurrence of jj coupling inside these subshells and between them. In order to use the mathematical apparatus of the angular momentum theory, we have to write all the operators in j-representation, i.e. express them in terms of the quantities which transform like the eigenfunctions of the operators $\mathbf{j}^2$ and j_z. From a mathematical point of view the main difference between non-relativistic and relativistic momenta is that in a non-relativistic case we have one-electron angular and spin momenta, which can acquire integer and half-integer numerical values, correspondingly, whereas in a relativistic case we have only one-electron total momentum, acquiring half-integer numerical values. However, they all satisfy the same commutation conditions (5.2), and all formulas of Chapter 5 on tensorial algebra are equally applicable in both cases.

The analogue of formula (5.10) for $j^{(1)}$ will be the following:

$$\left[j^{(1)} \times \psi^{(j)} \right]_Q^{(K)} = -i\delta(K,j)\sqrt{j(j+1)}\psi_Q^{(j)}. \qquad (7.1)$$

While calculating matrix elements of various items of the energy operator or electron transition quantities in relativistic approximation we shall

57

see that the analogue of the quantity $(l\|C^{(k)}\|l')$ (formula (5.8)) will be the Clebsch–Gordan coefficient

$$\begin{bmatrix} k & j_1 & j_2 \\ 0 & 1/2 & 1/2 \end{bmatrix}.$$

It plays a fundamental role in relativistic theory, therefore, below we shall present its algebraic values for $j_1 = j' \leq 7/2$ (Table 7.1) and for $k \leq 9$, $j_1 = j_2 = j$ (Table 7.2). These values are sufficient for the calculation of the spectroscopic characteristics of any atom or ion in the Periodical Table. Table 7.1 contains these coefficients presented in the form

$$\begin{bmatrix} j+\beta & j' & j \\ 0 & 1/2 & 1/2 \end{bmatrix},$$

whereas in Table 7.2 they are written in the form

$$\begin{bmatrix} k & j & j \\ 0 & 1/2 & 1/2 \end{bmatrix}.$$

Such presentation is necessary, because both these forms of the Clebsch–Gordan coefficient occur in relativistic formulas. They appear also in non-relativistic expressions while adopting jj coupling.

As was mentioned in the previous paragraph, the Wigner–Eckart theorem (5.15) is fairly general, it is equally applicable for both approaches considered, for tensorial operators, acting in various spaces (see, for example, Chapters 15, 17 and 18, concerning quasispin and isospin in the theory of an atom).

7.2   Unit tensors in a relativistic approach

The analogue of unit tensor (5.23) in j-space may be defined as follows:

$$(j\|t^k\|j') = \delta(j, j'), \qquad 0 \leq k \leq 2j. \tag{7.2}$$

In order to find expressions for relativistic matrix elements of the energy operator we have to utilize the following formula for two-particle scalar operators:

$$T^{kk0} = \sum_{i>j=1}^{N} \left(t_i^k \cdot t_j^k\right) = \frac{1}{2}\left(T^k \cdot T^k\right) - \frac{1}{2}\sum_i \left(t_i^k \cdot t_i^k\right), \tag{7.3}$$

where instead of (5.27) and (5.28), we have

$$T^k = \sum_i t_i^k. \tag{7.4}$$

The need to have only one sort of unit tensor (7.2) and its sum (7.4) is conditioned by the fact that, unlike a non-relativistic case, where we

Table 7.1. Algebraic expressions of the Clebsch–Gordan coefficient $\begin{bmatrix} j+\beta & j' & j \\ 0 & 1/2 & 1/2 \end{bmatrix}$ for $j', \beta \leq 7/2$

j'	β	$\begin{bmatrix} j+\beta & j' & j \\ 0 & 1/2 & 1/2 \end{bmatrix}$
$\frac{1}{2}$	$\frac{1}{2}$	$-\frac{1}{2}\sqrt{\frac{2j+1}{j+1}}$
$\frac{3}{2}$	$\frac{1}{2}$	$-\frac{1}{2^2}\sqrt{\frac{(2j+1)(2j+3)}{2j(j+1)}}$
	$\frac{3}{2}$	$\frac{1}{2^2}\sqrt{\frac{3(2j+1)(2j+3)}{2(j+1)(j+2)}}$
$\frac{5}{2}$	$\frac{1}{2}$	$\frac{1}{2^4}\sqrt{\frac{(2j-1)(2j+1)(2j+3)}{j(j+1)(j+2)}}$
	$\frac{3}{2}$	$\frac{1}{2^3}\sqrt{\frac{(2j+1)(2j+3)(2j+5)}{2j(j+1)(j+2)}}$
	$\frac{5}{2}$	$-\frac{1}{2^3}\sqrt{\frac{(2j+1)(2j+3)(2j+5)}{2(j+1)(j+2)(j+3)}}$
$\frac{7}{2}$	$\frac{1}{2}$	$\frac{3}{2^5}\sqrt{\frac{(2j-1)(2j+1)(2j+3)(2j+5)}{2(j-1)j(j+1)(j+2)}}$
	$\frac{3}{2}$	$-\frac{1}{2^5}\sqrt{\frac{3\cdot5(2j-1)(2j+1)(2j+3)(2j+5)}{2j(j+1)(j+2)(j+3)}}$
	$\frac{5}{2}$	$-\frac{1}{2^5}\sqrt{\frac{5(2j+1)(2j+3)(2j+5)(2j+7)}{2j(j+1)(j+2)(j+3)}}$
	$\frac{7}{2}$	$-\frac{1}{2^5}\sqrt{\frac{5\cdot7(2j+1)(2j+3)(2j+5)(2j+7)}{2(j+1)(j+2)(j+3)(j+4)}}$

have separate orbital angular and spin momenta, here we need only one momentum, i.e. their vectorial sum **j**.

The submatrix element of standard operator (7.4) is as follows:

$$\left(j^N\alpha J\|T^k\|j^N\alpha'J'\right) = N\sqrt{(2J+1)(2J'+1)}\sum_{\alpha_1 J_1}(-1)^{J_1+j+J+k}$$

$$\times \left(j^N\alpha J\|j^{N-1}(\alpha_1 J_1)j\right)\left(j^{N-1}(\alpha_1 J_1)j\|j^N\alpha'J'\right)\begin{Bmatrix} j & J & J_1 \\ J' & j & k \end{Bmatrix}, \quad (7.5)$$

where the quantities ($\|$) are the coefficients of fractional parentage (CFP) for subshell j^N of equivalent electrons, defined in Chapter 9. Submatrix elements of tensor (7.4), connecting partially j^N, if $N \leq (2j+1)/2$, and

Table 7.2. Algebraic expressions of the Clebsch–Gordan coefficient $\begin{bmatrix} k & j & j \\ 0 & 1/2 & 1/2 \end{bmatrix}$ for $k \le 9$

k	$\begin{bmatrix} k & j & j \\ 0 & 1/2 & 1/2 \end{bmatrix}$
0	1
1	$-\frac{1}{2}\sqrt{\dfrac{1}{j(j+1)}}$
2	$-\frac{1}{2^2}\sqrt{\dfrac{(2j-1)(2j+3)}{j(j+1)}}$
3	$\frac{3}{2^3}\sqrt{\dfrac{(2j-1)(2j+3)}{(j-1)j(j+1)(j+2)}}$
4	$\frac{3}{2^5}\sqrt{\dfrac{(2j-3)(2j-1)(2j+3)(2j+5)}{(j-1)j(j+1)(j+2)}}$
5	$-\frac{3\cdot5}{2^6}\sqrt{\dfrac{(2j-3)(2j-1)(2j+3)(2j+5)}{(j-2)(j-1)j(j+1)(j+2)(j+3)}}$
6	$-\frac{5}{2^7}\sqrt{\dfrac{(2j-5)(2j-3)(2j-1)(2j+3)(2j+5)(2j+7)}{(j-2)(j-1)j(j+1)(j+2)(j+3)}}$
7	$\frac{5\cdot7}{2^8}\sqrt{\dfrac{(2j-5)(2j-3)(2j-1)(2j+3)(2j+5)(2j+7)}{(j-3)(j-2)(j-1)j(j+1)(j+2)(j+3)(j+4)}}$
8	$\frac{5\cdot7}{2^{11}}\sqrt{\dfrac{(2j-7)(2j-5)(2j-3)(2j-1)(2j+3)(2j+5)(2j+7)(2j+9)}{(j-3)(j-2)(j-1)j(j+1)(j+2)(j+3)(j+4)}}$
9	$-\frac{3^2\cdot5\cdot7}{2^{12}}\sqrt{\dfrac{(2j-7)(2j-5)(2j-3)(2j-1)(2j+3)(2j+5)(2j+7)(2j+9)}{(j-4)(j-3)(j-2)(j-1)j(j+1)(j+2)(j+3)(j+4)(j+5)}}$

almost filled j^{2j+1-N} subshells , satisfy the equation ($k > 0$)

$$\left(j^{2j+1-N}\alpha v J \| T^k \| j^{2j+1-N}\alpha'v'J'\right)$$
$$= (-1)^{k+1+(v-v')/2}\left(j^{N}\alpha v J \| T^k \| j^{N}\alpha'v'J'\right). \tag{7.6}$$

Table 7.3 presents numerical values of submatrix elements (7.5) for the partially filled subshells $\left[\frac{3}{2}\right]^N$ and $\left[\frac{5}{2}\right]^N$. The corresponding tables for $\left[\frac{7}{2}\right]^N$ may be taken from [63]. Submatrix elements, connected by the

Table 7.3. Numerical values of submatrix elements (7.5) for $\left[\frac{3}{2}\right]^N$ and $\left[\frac{5}{2}\right]^N$

$$\left(\left[\tfrac{3}{2}\right]^2 vJ \| T^k \| \left[\tfrac{3}{2}\right]^2 v'J'\right)$$

v	J	v'	J'	T^2	T^3
0	0	2	2	1	0
2	2	2	2	0	$-\sqrt{2}$

$$\left(\left[\tfrac{5}{2}\right]^2 vJ \| T^k \| \left[\tfrac{5}{2}\right]^2 v'J'\right)$$

v	J	v'	J'	T^2	T^3	T^4	T^5
0	0	2	2	$\frac{\sqrt{2}}{\sqrt{3}}$	0	0	0
		2	4	0	0	$\frac{\sqrt{2}}{\sqrt{3}}$	0
2	2	2	2	$-\frac{5\sqrt{2}}{7\sqrt{3}}$	$-\frac{9}{7\sqrt{2}}$	$\frac{3\sqrt{5}}{7\sqrt{2}}$	0
		2	4	$\frac{9}{7\sqrt{2}}$	$\frac{5\sqrt{3}}{7}$	$\frac{\sqrt{5\cdot11}}{7\sqrt{3}}$	$-\frac{\sqrt{3\cdot5}}{\sqrt{2\cdot7}}$
2	2	2	4	$\frac{\sqrt{3\cdot11}}{7}$	$-\frac{\sqrt{11}}{7\sqrt{2}}$	$-\frac{\sqrt{11\cdot13}}{7\sqrt{2}}$	$-\frac{\sqrt{13}}{\sqrt{7}}$

$$\left(\left[\tfrac{5}{2}\right]^3 vJ \| T^k \| \left[\tfrac{5}{2}\right]^3 v'J'\right)$$

v	J	v'	J'	T^2	T^3	T^4	T^5
3	$\frac{3}{2}$	3	$\frac{3}{2}$	0	$-\frac{8}{7}$	0	0
		1	$\frac{5}{2}$	$\frac{2\sqrt{3}}{\sqrt{7}}$	0	$-\frac{2\sqrt{2}}{\sqrt{3\cdot7}}$	0
		3	$\frac{9}{2}$	0	$-\frac{2\cdot5\sqrt{2}}{7\sqrt{3}}$	0	$-\frac{2\sqrt{5}}{\sqrt{3\cdot7}}$
1	$\frac{5}{2}$	1	$\frac{5}{2}$	0	1	0	1
		3	$\frac{9}{2}$	$\frac{3}{\sqrt{7}}$	0	$-\frac{\sqrt{5\cdot11}}{\sqrt{3\cdot7}}$	0
3	$\frac{9}{2}$	3	$\frac{9}{2}$	0	$-\frac{\sqrt{11\cdot13}}{7\sqrt{3}}$	0	$\frac{\sqrt{5\cdot13}}{\sqrt{3\cdot7}}$

relationship

$$\left(j^N v' J' \| T^k \| j^N v J\right) = (-1)^{J-J'} \left(j^N v J \| T^k \| j^N v' J'\right), \qquad (7.7)$$

are skipped.

It is interesting to emphasize that submatrix elements (7.5) are proportional to the CFPs with two detached electrons, defined by (9.15), namely

$$2\sqrt{2(2J_2+1)Q(N,v_1)} \left(j^N \alpha v J \| T^{J_2} \| \alpha_1 v_1 J_1\right) = \sqrt{N(N-1)(2J+1)}$$

$$\times (-1)^{\varphi(N)} \left[Q(N,v_1) - Q(N+2,v)\right] \left(j^N \alpha v J \| j^{N-2}(\alpha_1 v_1 J_1) j^2 (J_2)\right), (7.8)$$

where $Q(N,v)$ is given by (9.21),

$$\varphi(N) = \begin{cases} 1, & \text{if } N > (2j+1)/2, \\ 0, & \text{if } N \le (2j+1)/2, \end{cases} \qquad (7.9)$$

and

$$\left(j^{N-2}(v_2 J_2) j^2 (v_0 J_0) \| j^N v J\right) = (-1)^{J_2+J+1} \sum_{v_1 J_1} \sqrt{[J_0, J_1]}$$

$$\times \begin{Bmatrix} J_2 & j & J_1 \\ j & J & J_0 \end{Bmatrix} \left(j^{N-2}(v_2 J_2) j \| j^{N-1} v_1 J_1\right) \left(j^{N-1}(v_1 J_1) j \| j^N v J\right). (7.10)$$

Let us present in conclusion the expression for the submatrix element of scalar product (7.3), necessary while calculating relativistic matrix elements of the energy operator. It is as follows:

$$\left(j^N \alpha J \| T^{kk0} \| j^N \alpha' J\right) = \frac{1}{2} \left[\frac{1}{\sqrt{2J+1}} \sum_{\alpha'' J''} (-1)^{J-J''} \right.$$

$$\times \left(j^N \alpha J \| T^k \| j^N \alpha'' J''\right) \left(j^N \alpha'' J'' \| T^k \| j^N \alpha' J\right) - \delta(\alpha, \alpha') N \frac{\sqrt{2J+1}}{(2j+1)} \right]. \quad (7.11)$$

For the particular case when $k = 0$, (7.11) becomes

$$\left(j^N \alpha J \| T^{000} \| j^N \alpha' J\right) = \delta(\alpha, \alpha') N(N-1) \sqrt{2J+1}/2(2j+1). \qquad (7.12)$$

All the main ideas of Chapter 6 are equally applicable for both non-relativistic and relativistic approaches.

8

Graphical methods: their generalization for perturbation theory

8.1 The graphs of Jucys, Levinson and Vanagas (JLV graphs)

The role of graphical methods to represent angular momentum, $3nj$- and jm-coefficients was emphasized in the Introduction as one of the most important milestones in the theory of many-electron atoms and ions.

As was already mentioned, in theoretical atomic spectroscopy, while considering complex electronic configurations, one has to cope with many sums over quantum numbers of the angular momentum type and their projections ($3nj$- and jm-coefficients). There are collections of algebraic formulas for particular cases of such sums [9, 11, 88]. However, the most general way to solve problems of this kind is the exploitation of one or another versions of graphical methods [9, 11]. They are widely utilized not only in atomic spectroscopy, but also in many other domains of physics (nuclei, elementary particles, etc.) [13].

The main advantage of graphic methods, compared to algebraic ones, is their general character: in fact, you can easily exploit them in any case. They are extremely efficient if you account for the peculiarities and symmetry properties of the quantities considered. The first version of graphical technique (graphs of Jucys, Levinson and Vanagas) was based on the Wigner coefficients (symmetric part of the Clebsch–Gordan coefficient (6.7)) and their symmetry properties. The main element of this technique was the symbol, presented in Fig. 8.1, describing the Wigner coefficient (Eq. (6.8)).

The sign by the vertex (Fig. 8.1) indicates whether the arguments of the Wigner coefficient are ordered in a counter-clockwise direction (positive sign) or a clockwise direction (negative sign). A change of the cyclic order of lines around a vertex corresponds to interchanging two of the columns of the Wigner coefficient. The arrows on lines indicate the signs of the corresponding projections m_1, m_2, and m_3, therefore there is no

Fig. 8.1. Graphical representation of the Wigner coefficient $\begin{pmatrix} j_1 & j_2 & j_3 \\ m_1 & m_2 & m_3 \end{pmatrix}$.

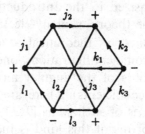

Fig. 8.2. 6j-coefficient.

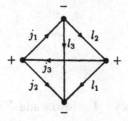

Fig. 8.3. 9j-coefficient.

need to indicate them on the graph. In- and out-going arrows correspond to the negative and positive m_i values, respectively [9]. In [9] certain rules of operations (summing, cutting, etc.) are formulated with these graphs, which are in [17] called the theorems of Jucys, Levinson and Vanagas. They are mostly used to sum up various products of jm- or jm- and $3nj$-coefficients. Thus, summing the following four Wigner coefficients over all projection parameters, we obtain the 6j-coefficient

$$\begin{Bmatrix} j_1 & j_2 & j_3 \\ l_1 & l_2 & l_3 \end{Bmatrix} = \sum_{m_i n_i} (-1)^{j_1-m_1+j_2-m_2+j_3-m_3+l_1-n_1+l_2-n_2+l_3-n_3}$$

$$\times \begin{pmatrix} j_1 & j_2 & j_3 \\ m_1 & m_2 & -m_3 \end{pmatrix} \begin{pmatrix} j_1 & l_2 & l_3 \\ -m_1 & n_2 & n_3 \end{pmatrix}$$

$$\times \begin{pmatrix} l_1 & l_2 & j_3 \\ n_1 & -n_2 & m_3 \end{pmatrix} \begin{pmatrix} l_1 & j_2 & l_3 \\ -n_1 & -m_2 & -n_3 \end{pmatrix}. \quad (8.1)$$

Its standard graphical form is presented in Fig. 8.2.

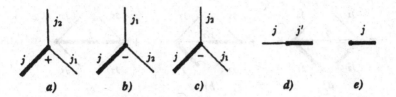

Fig. 8.4. Clebsch–Gordan coefficient.

Summing graphically the product of three $6j$-coefficients (Eq. (6.27)) we find the standard graphical form of the $9j$-coefficient (Fig. 8.3). In a similar way any $3nj$- or jm-coefficient may be defined graphically.

8.2 The graphs of Jucys and Bandzaitis (JB graphs)

However, it turned out that, while practically utilizing this method, one has many problems with multipliers of the kind $(2j+1)$ and phases. By choosing the Clebsch–Gordan coefficients as the main element of a graphical technique, the corresponding method becomes free of the shortcomings of previous techniques. Such a method, based on the Clebsch–Gordan coefficients and transformation matrices, was described in [11]. This version of graphical representation of $3nj$- and jm-coefficients will be termed Jucys–Bandzaitis graphs (JB graphs). In this technique the Clebsch–Gordan coefficient has the graphical symbol (a, b, c) presented in Fig. 8.4. The resulting momentum in the JB graphs corresponds to the heavy line. If there is a requirement to indicate the projections m_i, then they may be written together with j_i.

The sign $+$ or $-$ has the same meaning as in the case of the Wigner coefficient (Fig. 8.1). Thus, graphs 8.4a and 8.4b correspond to the same Clebsch–Gordan coefficient

$$\begin{bmatrix} j_1 & j_2 & j \\ m_1 & m_2 & m \end{bmatrix}$$

with positive and negative signs of the vertex, respectively. Graph 8.4c represents the Clebsch–Gordan coefficient

$$\begin{bmatrix} j_2 & j_1 & j \\ m_2 & m_1 & m \end{bmatrix}.$$

It follows from Eq. (6.2) that these coefficients are joined by the following symbolic equality:

$$8.4c = (-1)^{j_1 + j_2 - j} 8.4a$$

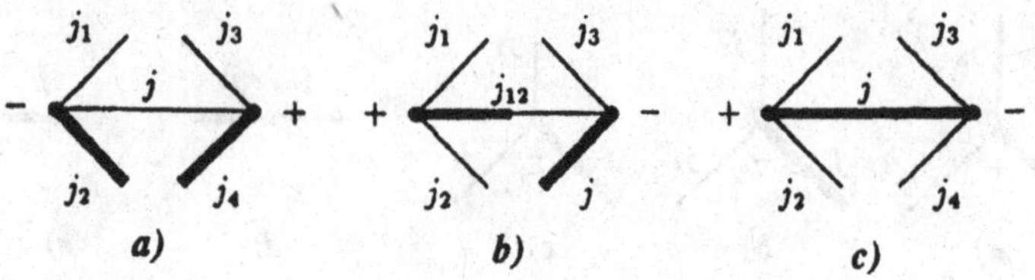

Fig. 8.5. Summation of two Clebsch–Gordan coefficients over one projection parameter.

If one of the parameters j_i equals zero, then

$$8.4d = \begin{bmatrix} j & 0 & j' \\ m & 0 & m' \end{bmatrix} = \begin{bmatrix} 0 & j & j' \\ 0 & m & m' \end{bmatrix} = \delta(jm, j'm'). \qquad (8.2)$$

Thus, such a coefficient is pictured in Fig. 8.4d. The case

$$8.4e = \begin{bmatrix} j & 0 & 0 \\ m & 0 & 0 \end{bmatrix} = \begin{bmatrix} 0 & j & 0 \\ 0 & m & 0 \end{bmatrix} = \begin{bmatrix} 0 & 0 & j \\ 0 & 0 & m \end{bmatrix} = \delta(jm, 00) \qquad (8.3)$$

is shown in Fig. 8.4e. As in the case of the Wigner coefficients [9], the summation over the projection parameter m_i in this graphical technique means connection of the corresponding lines of the graphs of two Clebsch–Gordan coefficients. Let us also take into account that the summation parameter in physical problems does not occur in the phase multiplier and repeats itself in two Clebsch–Gordan coefficients with the same sign. This means that the following equality holds:

$$8.5a = \sum_m \begin{bmatrix} j_1 & j & j_2 \\ m_1 & m & m_2 \end{bmatrix} \begin{bmatrix} j_3 & j & j_4 \\ m_3 & m & m_4 \end{bmatrix}$$

$$= \sum_m \begin{bmatrix} j_1 & j & j_2 \\ m_1 & -m & m_2 \end{bmatrix} \begin{bmatrix} j_3 & j & j_4 \\ m_3 & -m & m_4 \end{bmatrix}. \qquad (8.4)$$

Its graphical representation is given in Fig. 8.5a. Therefore, unlike the case of the Wigner coefficients, for the Clebsch–Gordan coefficients there is no necessity to indicate the arrows on the lines.

On graph 8.5a there are two connected thin lines. However, it is possible to connect a thin line with a heavy one (Fig. 8.5b) or two heavy lines (Fig. 8.5c). Thus, Fig. 8.5b corresponds to the sum

$$8.5b = \sum_{m_{12}} \begin{bmatrix} j_1 & j_2 & j_{12} \\ m_1 & m_2 & m_{12} \end{bmatrix} \begin{bmatrix} j_{12} & j_3 & j \\ m_{12} & m_3 & m \end{bmatrix}. \qquad (8.5)$$

Summation of the product of two Clebsch–Gordan coefficients over two

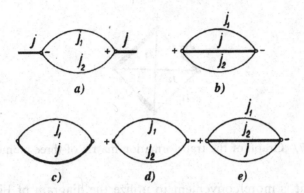

a)   b)

c)   d)   e)

Fig. 8.6. Summation of two Clebsch–Gordan coefficients over two (8.6a) and three (8.6b–e) projection parameters.

parameters (Fig. 8.6a) gives

$$8.6a = \sum_{m_1 m_2} \begin{bmatrix} j_1 & j_2 & j \\ m_1 & m_2 & m \end{bmatrix} \begin{bmatrix} j_1 & j_2 & j' \\ m_1 & m_2 & m' \end{bmatrix} = \delta(jm, j'm'). \qquad (8.6)$$

The further summation over m leads to the result

$$A = \sum_{m_1 m_2 m} \begin{bmatrix} j_1 & j_2 & j \\ m_1 & m_2 & m \end{bmatrix} \begin{bmatrix} j_1 & j_2 & j \\ m_1 & m_2 & m \end{bmatrix} = (2j+1)\delta(j_1 j_2 j). \qquad (8.7)$$

It is convenient to modify sum (8.7) in the way

$$B = (2j+1)^{-1} A = \{j_1 j_2 j\} \qquad (8.8)$$

and then to consider that a heavy line represents not only the summation over all permitted values of the corresponding projection parameter, but also the division by the number of its values. Thus, the graph on Fig. 8.6b corresponds to the expression

$$8.6b = (2j+1)^{-1} \sum_{m_1 m_2 m} \begin{bmatrix} j_1 & j_2 & j \\ m_1 & m_2 & m \end{bmatrix} \begin{bmatrix} j_1 & j_2 & j \\ m_1 & m_2 & m \end{bmatrix}. \qquad (8.9)$$

Thus, Fig. 8.6b pictures the triangular delta $\delta(j_1 j_2 j)$ or $\{j_1 j_2 j\}$, which equals zero if the parameters j_1, j_2 and j cannot compose the triangle having an integer perimeter. If $j_2 = 0$, then we have Fig. 8.6c, corresponding to the usual Kronecker delta condition, namely,

$$8.6c = \sum_{m_1 m} (2j+1)^{-1} \begin{bmatrix} j_1 & 0 & j \\ m_1 & 0 & m \end{bmatrix} \begin{bmatrix} j_1 & 0 & j \\ m_1 & 0 & m \end{bmatrix} = \delta(j_1, j). \qquad (8.10)$$

If in Fig. 8.6b $j = 0$, then we also have a Kronecker delta. Its symbol is Fig. 8.6d. In this case we have to indicate the signs of vertices, because the changing of a vertex sign leads to the occurrence of the phase $(-1)^{2j}$.

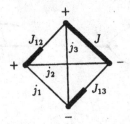

Fig. 8.7.   Graph of the transformation matrix of three momenta.

That is why it is more convenient to utilize the diagram of Fig. 8.6c, the value of which does not depend on the signs of vertices. If $j = 0$, then we have to transform the sum on Fig. 8.6b to the form

$$8.6e = (2j_2 + 1)^{-1} \sum_{m_1 m_2 m} \begin{bmatrix} j_1 & j & j_2 \\ m_1 & m & m_2 \end{bmatrix} \begin{bmatrix} j_1 & j & j_2 \\ m_1 & m & m_2 \end{bmatrix} = \delta(j_1 j_2 j), \quad (8.11)$$

graphically presented in Fig. 8.6e. Here m_2 and m have positive signs due to Eq. (8.4). Figure 8.6e illustrates the symmetry of the triangular delta graph. Indeed, Fig. 8.6c directly follows from Figs 8.6b and 8.6e for $j_2 = 0$ and $j = 0$, respectively. In a similar fashion, the more complex sums of products of the Clebsch–Gordan coefficients may be considered. Diagrams of such sums may be cut through a certain number of lines. This corresponds to the expression of the quantity considered in terms of another sum of products of the Clebsch–Gordan coefficients. However, we shall not discuss this problem here, nor the summation of diagrams over momentum quantum numbers. Interested readers will find descriptions of these procedures in [11].

Performing the summation of the products of Clebsch–Gordan coefficients over all projection parameters, we obtain the invariants, called $3nj$-coefficients. Their trivial cases were presented in Figs 8.6b–e (Kronecker and triangular deltas). The first non-trivial case is represented by formula (6.16) or the transformation matrix of three angular momenta, connected with $6j$-coefficient (see formula (6.42) and Fig. 8.7)).

Figure 8.7 represents the transformation matrix (6.42). One and the same transformation matrix may be represented by various graphs, which may differ from each other either by location of heavy lines but with identical geometrical figures, or by geometrical figures. This problem is discussed in more detail in [11].

The transformation matrices of four momenta lead to $9j$-coefficients. As in the case of three momenta, one and the same matrix can correspond to a number of graphs. Their common feature is that each graph has six vertices and nine lines. Here we shall present only two diagrams (Fig. 8.8a,b), corresponding to the transformation matrices (6.43) and

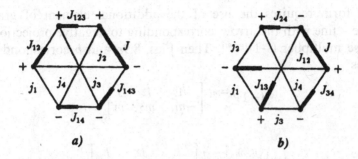

Fig. 8.8.   Graphs of the transformation matrix of four momenta.

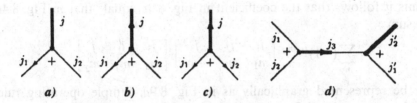

Fig. 8.9.   Another graphical representation of the Clebsch–Gordan coefficients.

(6.44), respectively. The general theory of transformation matrices and $3nj$-coefficients and of their graphical representation may be found in [11].

Both these techniques (graphs with arrows and heavy lines) in fact are equivalent and they were developed to sum up $3nj$- and jm-coefficients. There have been attempts to generalize them to cover the cases of summing up $3nj$-coefficients and coefficients of fractional parentage or even to calculate matrix elements, but without great success [8, 13, 17, 88]. Some refinements of graphical technique, particularly efficient for large n values of $3nj$-coefficients, are presented in [89].

During the last decades the second-quantization method has been increasingly used in atomic spectroscopy [12, 31]. Its application to non-trivial orders of perturbation theory [17] and even to the quantum-electrodynamic approach looks extremely promising. The next important step in improving the graphical technique was made in [26], where the generalization of the Jucys–Bandzaitis graphical technique was presented. It enabled one to simplify the operations with the second-quantization operators represented in the form of tensorial products of the electron creation and annihilation operators. The starting point is the Clebsch–Gordan coefficient with the arrows on one or two lines (Fig. 8.9(a–d)) and a number of simple rules to transform it.

It turned out that the method of second-quantization in the irreducible tensorial form requires the use of the additional element of graphical technique – line with the arrow, corresponding to negative projection $-m_1$ and phase multiplier $(-1)^{j_1-m_1}$. Then Figs. 8.9a and b correspond to the quantities

$$(-1)^{j_1-m_1} \begin{bmatrix} j_1 & j_2 & j \\ -m_1 & m_2 & m \end{bmatrix}$$

and

$$(-1)^{j_1-m_1+j-m} \begin{bmatrix} j_1 & j_2 & j \\ -m_1 & m_2 & -m \end{bmatrix},$$

respectively. From the symmetry properties of the Clebsch–Gordan coefficients it follows that the coefficient in Fig. 8.9c equals that in Fig. 8.4a. The sum

$$\sum_{m_3} (-1)^{j_3-m_3} \begin{bmatrix} j_1 & j_2 & j_3 \\ m_1 & m_2 & -m_3 \end{bmatrix} \begin{bmatrix} j_3 & j_1' & j_2' \\ m_3 & m_1' & m_2' \end{bmatrix}$$

may be represented graphically as in Fig. 8.9d. Simple operating rules have been formulated for graphs of this kind.

The latter method allows one to calculate graphically the irreducible tensorial products of the second quantization operators and their commutators, to find graphically the normal form of the complex tensorial products of the operators (the graphical analogue of the Wick's theorem) and their vacuum averages, etc. This approach gives one the possibility of finding in an efficient way the matrix elements of various second-quantized operators for complex electronic configurations; therefore, its application to atomic perturbation theory and even quantum electrodynamics seems very promising.

Part 3

Description of Complex Electronic Configurations

9

Non-relativistic and relativistic cases
of a shell of equivalent electrons

In previous chapters we considered the wave functions and matrix elements of some operators without specifying their explicit expressions. Now it is time to discuss this question in more detail. Having in mind that our goal is to consider as generally as possible the methods of theoretical studies of many-electron systems, covering, at least in principle, any atom or ion of the Periodical Table, we have to be able to describe the main features of the structure of electronic shells of atoms. In this chapter we restrict ourselves to a shell of equivalent electrons in non-relativistic and relativistic cases.

9.1 A shell of equivalent electrons

The non-relativistic wave function (1.14) or its relativistic analogue (2.15), corresponds to a one-electron system. Having in mind the elements of the angular momentum theory and of irreducible tensors, described in Part 2, we are ready to start constructing the wave functions of many-electron configurations. Let us consider a shell of equivalent electrons. As we shall see later on, the pecularities of the spectra of atoms and ions are conditioned by the structure of their electronic shells, and by the relative role of existing intra-atomic interactions.

N electrons with the same values of quantum numbers $n_i l_i$ (LS coupling) or $n_i l_i j_i$ (jj coupling) are called equivalent. The corresponding configurations will be denoted as nl^N (a shell) or $nl j^N$ (a subshell). A number of permitted states of a shell of equivalent electrons are restricted by the Pauli exclusion principle, which requires antisymmetry of the wave function with respect to permutation of the coordinates of the electrons.

The wave function for the particular case of two equivalent electrons may be constructed, using vectorial coupling of the angular momenta and antisymmetrization procedure. For LS and jj coupling, it will look as

73

follows:

$$\psi(nl^2 LSJ) = \frac{1}{2}\left[1 + (-1)^{L+S}\right]\psi(nlnlLSJ), \qquad (9.1)$$

$$\psi(nlj^2 J) = \frac{1}{2}\left[1 + (-1)^{J}\right]\psi(nljnljJ). \qquad (9.2)$$

Here $\psi(nlnlLSJ)$ and $\psi(nljnljJ)$ are the so-called wave functions of coupled momenta (see the case $N = 2$ of Eqs. (9.7) and (9.8)). It follows from Eqs. (9.1) and (9.2) that for two equivalent electrons only the states with even $L + S$ (for LS coupling) or even J (for jj coupling) are allowed, otherwise the corresponding wave functions equal zero. Thus, the Pauli exclusion principle allows only some of the states obtained by vectorial coupling of the quantum numbers of separate electrons. It follows also from the Pauli principle that for equivalent electrons only homogeneous (LS or jj) coupling schemes are possible, because only in such cases are the quantum numbers of all electrons participating in coupling symmetrically.

The existence of two coupling schemes for a shell of equivalent electrons is conditioned by the relative values of intra-atomic interactions. If the non-spherical part of electronic Coulomb interactions prevails over the spin–orbit, then LS coupling takes place, otherwise the jj coupling is valid. As we shall see later on, for the overwhelming majority of atoms and ions, including fairly highly ionized ones, LS coupling is valid in a shell of equivalent electrons, that is why we shall pay the main attention to it.

According to the vectorial model of an atom, the angular and spin momenta of equivalent electrons in LS coupling are coupled into total momenta $\mathbf{L}$ and $\mathbf{S}$, respectively. Their numerical values L and S define the term and its multiplicity (^{2S+1}L). The vectorial sum of $\mathbf{L}$ and $\mathbf{S}$, caused by a spin–orbit interaction, makes the quantum number of total momentum J, describing the levels of the given term $^{2S+1}L_J$. In jj coupling we straightforwardly obtain J.

Let us notice that in principle the shell nl^N may also be described by the so-called LL coupling, in which the electrons of a shell are divided into two groups in accordance with two possible projections of a spin of each electron. Then, the orbital momenta of these groups are coupled vectorially [34]. However, this coupling scheme is seldom used, because it contradicts some physical principles, therefore, we shall not discuss it any more.

If α denotes all additional quantum numbers needed for the one-to-one classification of the energy levels, then a complete set of quantum numbers of a shell (in LS coupling) or subshell (in jj coupling) will be written as $nl^N \alpha LSJ$ and $nlj^N \alpha J$, whereas that of a state will be written as $nl^N \alpha LSJM$ and $nlj^N \alpha JM$, respectively.

A number of levels of term LS (its statistical weight) is

$$g(LS) = (2L+1)(2S+1)$$

or

$$g(LS) = \sum_J g(J) = \sum_J (2J+1). \tag{9.3}$$

The statistical weight of a shell is the sum of the weights of all terms, i.e.,

$$g(l^N) = \sum_{LS} g(LS) = \binom{4l+2}{N} = \frac{(4l+2)!}{N!(4l+2-N)!}. \tag{9.4}$$

A shell is closed (filled) if $N = 4l + 2$, and then $g(l^{4l+2}) = 1$. For a closed shell $L = S = J = 0$ and then it has only one state $^1S_0$.

The parity π of shell l^N is defined by

$$\pi = (-1)^{\sum_i l_i} = (-1)^{Nl}. \tag{9.5}$$

It follows from Eq. (9.5) that the configuration itself defines the parity of the states obtained, therefore, we shall usually skip the parity quantum number π.

The shells of equivalent electrons according to the values of $l = 0, 1, 2, 3, 4, \ldots$ are usually denoted by the small Latin letters $s, p, d, f, g, \ldots$ respectively. Then we have the shells s^N, p^N, d^N, f^N, g^N. They are closed ($N = 4l + 2$) while having $2, 6, 10, 14, 18, \ldots$ electrons, respectively. Often the sets of shells with $n = 1, 2, 3, 4, 5, 6, \ldots$ are denoted by the capital letters $K, L, M, N, O, P, \ldots$. In this case, for example, the symbol K corresponds to $1s$, the symbol M to the sum of shells $3s + 3p + 3d$, etc.

9.2 Wave function of equivalent electrons and coefficients of fractional parentage

The efficient way of constructing the wave function of the states of equivalent electrons permitted by the Pauli exclusion principle is by utilization of the methods of the coefficients of fractional parentage (CFP). The antisymmetric wave function $\psi(l^N \alpha LS M_L M_S)$ of a shell nl^N is constructed in a recurrent way starting with the antisymmetric wave function of $N-1$ electrons $\psi(l^{N-1}\alpha_1 L_1 S_1 M_{L_1} M_{S_1})$. Let us construct the following wave function of coupled momenta:

$$\psi(l^{N-1}(\alpha_1 L_1 S_1)lLS M_L M_S) = \sum_{M_{L_1} M_{S_1} m_l m_s} \psi(l^{N-1}\alpha_1 L_1 S_1 M_{L_1} M_{S_1})$$

$$\times \psi(lsm_l m_s) \begin{bmatrix} L_1 & l & L \\ M_{L_1} & m_l & M_L \end{bmatrix} \begin{bmatrix} S_1 & s & S \\ M_{S_1} & m_s & M_S \end{bmatrix}. \tag{9.6}$$

The wave function, obeying the Pauli principle, will be a linear combination of functions (9.6)

$$
\begin{aligned}
&\psi(l^N \alpha L S M_L M_S) \\
&= \sum_{\alpha_1 L_1 S_1} \psi(l^{N-1}(\alpha_1 L_1 S_1) l L S M_L M_S)(l^{N-1}(\alpha_1 L_1 S_1) l \| l^N \alpha L S). \quad (9.7)
\end{aligned}
$$

The analogous formula for jj coupling will be

$$
\psi(l j^N \alpha J M) = \sum_{\alpha_1 J_1} \psi(l j^{N-1}(\alpha_1 J_1) j J M)(j^{N-1}(\alpha_1 J_1) j \| j^N \alpha J). \quad (9.8)
$$

The quantities $(\ \| \ )$ are called the coefficients of fractional parentage (CFP) with one detached electron. They ensure the antisymmetry of the wave function and the occurrence only of states permitted by the Pauli principle.

Coefficients of fractional parentage play a fundamental role in the theory of many-electron atoms. There are algebraic formulas for them (see Chapter 16), however, they are not very convenient for practical utilization, and normally tables of their numerical values are used. They can be generated in the recursive way, starting with the formula

$$
\begin{aligned}
&\sum_{\alpha_1 L_1 S_1} (-1)^{2 S_1} [L_0, S_0, L_1, S_1]^{\frac{1}{2}} \begin{Bmatrix} l & l & L_0 \\ L & L_2 & L_1 \end{Bmatrix} \begin{Bmatrix} s & s & S_0 \\ S & S_2 & S_1 \end{Bmatrix} \\
&\times \left(l^{N-2}(\alpha_2 L_2 S_2) l \| l^{N-1} \alpha_1 L_1 S_1 \right) \left(l^{N-1}(\alpha_1 L_1 S_1) l \| l^N \alpha' L S \right) = 0, \quad (9.9)
\end{aligned}
$$
$$
(L_0 + S_0 \text{ is an odd number}),
$$

the orthonormality condition

$$
\sum_{\alpha_1 L_1 S_1} \left(l^N \alpha L S \| l^{N-1}(\alpha_1 L_1 S_1) l \right) \left(l^{N-1}(\alpha_1 L_1 S_1) l \| l^N \alpha' L S \right) = \delta(\alpha, \alpha') \quad (9.10)
$$

and the corresponding values for one and two electrons, namely,

$$
(0(0) l \| l l s) = 1,
$$
$$
\left(l(l s) l \| l^2 L S \right) = \frac{1}{2} \left[1 + (-1)^{L+S} \right]. \quad (9.11)
$$

Numerical values of CFP with one detached electron for p^3 shell are presented in Table 9.1. Similar expressions for jj coupling are as follows:

$$
\begin{aligned}
&\sum_{\alpha_1 J_1} (-1)^{2 J_1} \sqrt{(2 J_0 + 1)(2 J_1 + 1)} \begin{Bmatrix} j & j & J_0 \\ J & J_2 & J_1 \end{Bmatrix} \\
&\times \left(j^{N-2}(\alpha_2 J_2) j \| j^{N-1} \alpha_1 J_1 \right) \left(j^{N-1}(\alpha_1 J_1) j \| j^N \alpha J \right) = 0 \quad (9.12)
\end{aligned}
$$
$$
(J_0 \text{ is an odd number}).
$$

$$
\sum_{\alpha_1 J_1} \left(j^N \alpha J \| j^{N-1}(\alpha_1 J_1) j \right) \left(j^{N-1}(\alpha_1 J_1) j \| j^N \alpha' J \right) = \delta(\alpha, \alpha'), \quad (9.13)
$$

Table 9.1. Coefficients of fractional parentage $(p^2(v_1L_1S_1)p\|p^3vLS)$

$p^2v_1L_1S_1$ \ p^3vLS	^4_3S	^2_1P	^2_3P
^1_0S		$\sqrt{2}/3$	
^3_2P	1	$-1/\sqrt{2}$	$1/\sqrt{2}$
^1_2D		$-\sqrt{5}/3\sqrt{2}$	$-1/\sqrt{2}$

$$(0(0)j\|jj) = 1, \quad \left(j(j)j\|j^2J\right) = \frac{1}{2}\left[1 + (-1)^J\right]. \qquad (9.14)$$

The quantities considered are the CFP with one detached electron, denoted in short form as $(N-1, 1\|N)$. Let us notice that $(N-1, 1\|N) = (N\|N-1, 1)$. In some physical problems CFP with more detached electrons $(N-\sigma, \sigma\|N)$ are used. They may be expressed in terms of $(N-1, 1\|N)$ [14].

Let us notice that a shell with $N < 2l+1$ is usually called partially filled, with $N = 2l+1$ – half-filled and with $N > 2l+1$ – almost filled. When $N \neq 4l+2$, then a shell is called unfilled or open. If a number of electrons in a partially filled shell is denoted as N, then the corresponding value in an almost filled shell will be $4l + 2 - N$. The configurations nl^N and l^{4l+2-N} are called supplementary shells. Their sum makes a closed (filled) shell with $L = S = 0$. It has turned out that the supplementary shells have the same set of terms. Indeed, the requirement for a filled (closed) shell $L = S = 0$ implements the condition

$$\mathbf{L}_{4l+2-N} = -\mathbf{L}_N, \qquad \mathbf{S}_{4l+2-N} = -\mathbf{S}_N.$$

On the other hand, the electrons in an almost filled shell may be considered as holes in a closed shell. A number of those holes will be equal to a number of electrons in a partially filled shell.

9.3 s^N, p^N, d^N and f^N shells. Seniority quantum number

s^N and p^N represent, in fact, trivial cases of a shell of equivalent electrons. For one-to-one classification of their terms the quantum numbers L and S are sufficient. For d^N and, especially, for f^N shells we need additional quantum numbers.

In atomic spectroscopy it is very important to have a complete set of quantum numbers for one-to-one classification of the energy levels. For s^N and p^N shells the quantum numbers L, S, J are sufficient for this purpose.

However, this is not the case for the d^N shell. For d^3 there are two of the same ^2D terms. This problem with the d^N shell and, partially, the f^N shell was solved by Racah in his paper [23] introducing the seniority quantum number. In accordance with formula (9.7) we can build the antisymmetric wave function of shell l^N with the help of the CFP with one detached electron. However, we could also use the CFP with two detached electrons

$$\psi(l^N \alpha v L S M_L M_S) = \sum_{\alpha_0 v_0 L_0 S_0 L_2 S_2} \psi(l^{N-2}(\alpha_0 v_0 L_0 S_0) l^2 (L_2 S_2) L S M_L M_S)$$

$$\times \left(l^{N-2}(\alpha_0 v_0 L_0 S_0) l^2 (L_2 S_2) \| l^N \alpha v L S \right). \qquad (9.15)$$

Here we have extracted one additional quantum number v_i from α_i and denoted it again as $\alpha_i v_i$. For the special case $L_2 = S_2 = 0$, the CFP in Eq. (9.15) equals

$$\left(l^{N-2}(\alpha_0 v_0 L_0 S_0) l^2 ({}_0^1 S) \| l^N \alpha v L S \right) = \delta(\alpha_0 v_0 L_0 S_0, \alpha v L S)$$

$$\times \sqrt{2Q(N, v) / [N(N-1)(2l+1)]}, \quad (9.16)$$

where

$$Q(N, v) = \frac{1}{4}(N - v)(4l + 4 - N - v). \qquad (9.17)$$

It follows from Eq. (9.16) that if we add to the wave function of the state $l^{N-2}\alpha_0 v_0 L_0 S_0 M_{L_0} M_{S_0}$ two electrons in the state ^1S, then we have the wave function of the configuration l^N with the same quantum numbers $LS = L_0 S_0$. Such a relationship of the terms ^{2S+1}L of the configurations l^N and l^{N-2}, $l^{N-4}, \ldots$ is adopted to distinguish the repeating terms. CFP (9.16) equals zero for $N = v$. This indicates that the term $\alpha v L S = {}_{\alpha v}^{2S+1}L$ has occurred for the first time in the configuration l^v and it cannot be obtained from the configuration l^{v-2} by adding two electrons $l^2 \, {}_0^1 S$. For $N > v$ ($N = v + 2, v + 4, \ldots$) CFP (9.16) is non-zero. This shows that the corresponding term can be obtained by adding $l^2 \, {}_0^1 S$ to the configuration l^{N-2}, and that it exists in the configuration l^{N-2}.

Thus, the repeating terms can be divided into two groups: those occurring for the first time in the configuration under consideration and those already existing in the configurations l^{N-2}, l^{N-4}, etc. Adopting this method we can find the minimal number of electrons, usually marked by v, for which the term LS occurs for the first time. It is called the seniority quantum number and is usually denoted as pre-subscript ${}_v^{2S+1}L$.

The simplest example of the use of this method is the case of two terms ^3D in d^3. Utilizing Eq. (9.16) we can see that the first ^2D term occurs for d^1 (for it $v = 1$) and the second one for d^3. Then we have two different terms ${}_1^2D$ and ${}_3^2D$.

Normally the terms of the supplementary shells are classified with the same v values (e.g. $d^3 \, {}_1^2 D$ and $d^7 \, {}_1^2 D$). However, having in mind that CFP (9.16) equals zero also for $v = 4l + 4 - N$, we could classify the terms of the almost filled shell by the quantum number $v' = 4l + 4 - v$. By the way, this has an interesting physical meaning: if v indicates the number of electrons for which a certain term occurs for the first time while filling up a shell with electrons, then v' will correspond to the number of electrons for which this term disappears. Thus, v and v' define the interval of the number of electrons in which the term with given v exists. The utilization of v' could simplify some mathematical relations, but traditionally, as a rule, only v is used.

Accounting for v, all terms of shell l^N ($l = 0, 1$ and 2) may be unambiguously classified in the following way: $s^1 - {}_1^2 S$; $s^2 - {}_0^1 S$; $p^1, p^5 - {}_1^2 P$; $p^2, p^4 - {}_0^1 S, {}_2^3 P, {}_2^1 D$; $p^3 - {}_3^4 S, {}_1^2 P, {}_3^2 D$; $d^1, d^9 - {}_1^2 D$; $d^2, d^8 - {}_0^1 S, {}_2^3 P, {}_2^1 D, {}_2^3 F, {}_2^1 G$; $d^3, d^7 - {}_3^4 P, {}_3^2 P, {}_1^2 D, {}_3^2 D, {}_3^4 F, {}_3^2 F, {}_3^2 G, {}_3^2 H$; $d^4, d^6 - {}_0^1 S, {}_4^1 S, {}_2^3 P, {}_4^3 P, {}_4^5 D, {}_2^3 D, {}_4^3 D, {}_2^1 D, {}_4^1 D, {}_2^3 F, {}_4^3 F, {}_4^1 F, {}_4^3 G, {}_2^1 G, {}_4^1 G, {}_4^3 H, {}_4^1 I$; $d^5 - {}_5^6 S, {}_5^2 S, {}_3^4 P, {}_3^2 P, {}_5^4 D, {}_1^2 D, {}_3^2 D, {}_5^2 D, {}_3^4 F, {}_3^2 F, {}_5^2 F, {}_5^4 G, {}_3^2 G, {}_5^2 G, {}_3^2 H, {}_5^2 I$.

There is no need to use v for s^N and p^N shells, however, often it is also indicated. There exists the equality for p^N shell [90]

$$L(L+1) = \frac{v}{2}(8 - v) - 2S(S+1), \tag{9.18}$$

from which, accounting for the positiveness of L, v and S, follows the possibility to classify the terms of shell p^N with the help of any pair of quantum numbers LS, vL or vS. However, traditionally the first pair is used.

Thus, a set of three quantum numbers vLS allows us to classify unambiguously all terms of the shells s^N, p^N and d^N. For them there is no need to indicate the additional quantum numbers α.

The presence of the repeating terms with the same L, S causes additional difficulties while orthogonalizing the CFP, finding their phase multipliers as well as the relationships between complementary shells. Also the problem of finding the algebraic expressions for the CFP becomes much more complicated (see Chapter 16).

Unfortunately, v is not enough to classify the terms of shell f^N. Racah [24] employed the parameters of the representations of the groups R_7 and G_2 for the additional classification of the terms of shell f^N, finding in this way new quantum numbers $w_1 w_2 w_3$ and $u_1 u_2$, respectively (see Chapter 5). Thus, the wave function of shell f^N may be written

$$\psi\left(f^N \alpha \begin{pmatrix} u_1 u_2 \\ w_1 w_2 w_3 \end{pmatrix} v L S M_L M_S\right). \tag{9.19}$$

However, the quantum numbers $w_1 w_2 w_3$ are superfluous, because they are connected with S and v [18].

Unfortunately, these quantum numbers cannot classify the terms of shell f^N $(10 > N > 4)$ unambiguously. This problem has not yet been solved.

9.4 A shell of equivalent electrons in jj coupling (a subshell)

As we have mentioned earlier, if spin–orbit interactions prevail over the non-spherical part of the electrostatic interaction in shell l^N, then jj coupling takes place, and shell l^N itself splits into subshells with $j = l \mp 1/2 = l_{\mp}$. Then, instead of the configuration $l^N \alpha LSJ$ we have $nl j_1^{N_1} j_2^{N_2} \alpha_1 J_1 \alpha_2 J_2 J$. In this case l is preserved only to show the parity of the configuration. Thus, the whole set of levels of shell l^N is grouped into a number of j-configurations with dependence on the number of allowed combinations of N_1 and N_2. For example, for p^3 we have in LS coupling the levels $^4S_{3/2}$, $^2P_{1/2,3/2}$ and $^2D_{3/2,5/2}$. In jj coupling we obtain the levels grouped in three subconfigurations $p_-^{N_1} p_+^{N_2} J_1 J_2 J$: p_+^3 0 3/2 3/2, $p_-^2 p_+^1$ 0 3/2 3/2 and $p_-^1 p_+^2$ 1/2 0 1/2, $p_-^1 p_+^2$ 1/2 2 3/2, $p_-^1 p_+^2$ 1/2 2 5/2.

Thus, in jj coupling we have, at the very beginning, to deal with the more complex structure of electronic configurations (the splitting of the usual shell l^N into subshells, the necessity to calculate the non-diagonal with respect to subconfigurations, matrix elements, etc.). However, these difficulties are compensated by the simplicity of the subshells themselves and by the ease of classification of their levels. Indeed, having in mind that the j shell is filled (closed) for $N = 2j+1$ we arrive at the conclusion that the analogues of the closed shells s^N, p^N, d^N and f^N, i.e. the subshells $[1/2]^N$, $[3/2]^N$, $[5/2]^N$ and $[7/2]^N$ will have 2, 4, 6 and 8 electrons, respectively. Half-filled $(2j+1)/2$ subshells will have 1, 2, 3 and 4 electrons. Let us notice that, for example, a subshell $[1/2]^N$ may originate from both p_+^N and d_-^N shells.

The seniority quantum number defined for $(N \leq (2j+1)/2)$ by

$$\left(j^{N-2}(vJ)j^2(0) \| j^N vJ \right) = \sqrt{2Q(N,v)/[N(N-1)(2j+1)]}, \qquad (9.20)$$

where

$$Q(N,v) = \frac{1}{2}(N-v)(2j+3-N-v), \qquad (9.21)$$

is sufficient for the one-to-one classification of the levels of subshell j^N ($j = 1/2, 3/2, 5/2, 7/2$). It is enough, in fact, for all elements of the Periodical Table. The permitted levels of a shell of j^N ($j = 1/2, 3/2, 5/2, 7/2$) are presented in Table 9.2. The relevant values of the CFP are listed in Table 9.3.

Table 9.2. Energy levels and statistical weights of a subshell j^N ($j \leq 7/2$)

j^N	Levels (vJ)	Number of levels	Stat. weight of a subshell j^N
$[1/2]^0$, $[1/2]^2$	0 0	1	1
$[1/2]^1$	1 1/2	1	2
$[3/2]^0$, $[3/2]^4$	0 0	1	1
$[3/2]^1$, $[3/2]^3$	1 3/2	1	4
$[3/2]^2$	0 0, 2 2	2	6
$[5/2]^0$, $[5/2]^6$	0 0	1	1
$[5/2]^1$, $[5/2]^5$	1 5/2	1	6
$[5/2]^2$, $[5/2]^4$	0 0, 2 2, 2 4	3	15
$[5/2]^3$	1 5/2, 3 3/2, 3 9/2	3	20
$[7/2]^0$, $[7/2]^8$	0 0	1	1
$[7/2]^1$, $[7/2]^7$	1 7/2	1	8
$[7/2]^2$, $[7/2]^6$	0 0, 2 2, 2 4, 2 6	4	28
$[7/2]^3$, $[7/2]^5$	1 7/2, 3 3/2, 3 5/2, 3 9/2, 3 11/2, 3 15/2	6	56
$[7/2]^4$	0 0, 2 2, 2 4, 4 2, 4 4, 4 5, 4 6, 4 8	8	70

9.5   Quasispin

Racah introduced the seniority quantum number rather formally. However, it also follows from group theory. Indeed, it turned out that the
seniority quantum number is the parameter characterizing the representations of the so-called symplectic group Sp_{4l+2}. Let us recall that the one-
electron quantum numbers of the orbital l and spin s momenta are also
connected with some symmetry groups. Orbital quantum number l characterizes the representations of the orthogonal R_3 group of the rotations
of a three-dimensional space, whereas spin s enumerates the representations of the unitary unimodular group SU_2. Thus, the above-mentioned
quantum numbers are actually the characteristics of basis functions of
the representations of the corresponding groups, describing the symmetry
properties of the quantities considered in the spaces of these groups.

The seniority quantum number can also have a group-theoretical interpretation. If we define

$$Q = \frac{1}{2}(2l + 1 - v),$$ (9.22)

then it is possible to show that the corresponding operator will have the

Table 9.3.　Numerical values of fractional parentage coefficients with one detached electron for $j \le 7/2$

$$\left([5/2]^2 (v_1J_1)5/2\| [5/2]^3 vJ\right)$$

v_1J_1 \ vJ	$3\frac{3}{2}$	$1\frac{5}{2}$	$3\frac{9}{2}$
0 0	0	$\frac{\sqrt{2}}{3}$	0
2 2	$-\frac{\sqrt{5}}{\sqrt{7}}$	$-\frac{\sqrt{5}}{3\sqrt{2}}$	$\frac{\sqrt{3}}{\sqrt{2\cdot7}}$
2 4	$\frac{\sqrt{2}}{\sqrt{7}}$	$-\frac{1}{\sqrt{2}}$	$-\frac{\sqrt{11}}{\sqrt{2\cdot7}}$

$$\left([7/2]^2 (v_1J_1)7/2\| [7/2]^3 vJ\right)$$

v_1J_1 \ vJ	$3\frac{3}{2}$	$3\frac{5}{2}$	$1\frac{7}{2}$	$3\frac{9}{2}$	$3\frac{11}{2}$	$3\frac{15}{2}$
0 0	0	0	$\frac{1}{2}$	0	0	0
2 2	$\frac{\sqrt{3}}{\sqrt{2\cdot7}}$	$\frac{\sqrt{11}}{3\sqrt{2}}$	$-\frac{\sqrt{5}}{2\cdot3}$	$\frac{\sqrt{13}}{3\sqrt{2\cdot7}}$	$-\frac{\sqrt{5}}{3\sqrt{2}}$	0
2 4	$-\frac{\sqrt{11}}{\sqrt{2\cdot7}}$	$\frac{\sqrt{2}}{\sqrt{3\cdot11}}$	$-\frac{1}{2}$	$-\frac{5\sqrt{2}}{\sqrt{7\cdot11}}$	$\frac{\sqrt{13}}{\sqrt{2\cdot3\cdot11}}$	$\frac{\sqrt{5}}{\sqrt{2\cdot11}}$
2 6	0	$-\frac{\sqrt{5\cdot13}}{3\sqrt{2\cdot11}}$	$-\frac{\sqrt{13}}{2\cdot3}$	$\frac{7}{3\sqrt{2\cdot11}}$	$\frac{2\sqrt{13}}{3\sqrt{11}}$	$-\frac{\sqrt{17}}{\sqrt{2\cdot11}}$

$$\left([7/2]^3 (v_1J_1)7/2\| [7/2]^4 vJ\right)$$

v_1J_1 \ vJ	$3\frac{3}{2}$	$3\frac{5}{2}$	$1\frac{7}{2}$	$3\frac{9}{2}$	$3\frac{11}{2}$	$3\frac{15}{2}$
0 0	0	0	1	0	0	0
2 2	$\frac{3}{2\sqrt{5\cdot7}}$	$-\frac{\sqrt{11}}{2\sqrt{2\cdot5}}$	$\frac{1}{\sqrt{3}}$	$-\frac{\sqrt{13}}{2\sqrt{2\cdot3\cdot7}}$	$-\frac{1}{2}$	0
4 2	$-\frac{\sqrt{11}}{\sqrt{5\cdot7}}$	$-\frac{1}{2\sqrt{2\cdot5}}$	0	$-\frac{3\sqrt{3\cdot13}}{2\sqrt{2\cdot7\cdot11}}$	$\frac{1}{\sqrt{11}}$	0
2 4	$-\frac{\sqrt{11}}{2\sqrt{3\cdot7}}$	$-\frac{1}{\sqrt{2\cdot3\cdot11}}$	$\frac{1}{\sqrt{3}}$	$\frac{5\sqrt{5}}{\sqrt{2\cdot3\cdot7\cdot11}}$	$\frac{\sqrt{13}}{2\sqrt{3\cdot11}}$	$\frac{\sqrt{5}}{\sqrt{3\cdot11}}$
4 4	$\frac{\sqrt{13}}{2\sqrt{3\cdot5\cdot7}}$	$\frac{\sqrt{13}}{\sqrt{2\cdot3\cdot5}}$	0	$\frac{\sqrt{3}}{\sqrt{2\cdot7\cdot13}}$	$\frac{\sqrt{5}}{2\sqrt{3}}$	$-\frac{2}{\sqrt{3\cdot13}}$
4 5	$\frac{\sqrt{3}}{2\sqrt{5}}$	$-\frac{7}{2\sqrt{2\cdot5\cdot11}}$	0	$\frac{3\sqrt{3}}{2\sqrt{2\cdot11}}$	$\frac{\sqrt{5\cdot13}}{2\sqrt{7\cdot11}}$	$\frac{\sqrt{17}}{\sqrt{7\cdot11}}$
2 6	0	$\frac{\sqrt{5}}{2\sqrt{2\cdot11}}$	$\frac{1}{\sqrt{3}}$	$-\frac{7\sqrt{5}}{2\sqrt{2\cdot3\cdot11\cdot13}}$	$\frac{\sqrt{2}}{\sqrt{11}}$	$-\frac{\sqrt{3\cdot17}}{\sqrt{11\cdot13}}$
4 8	0	0	0	$\frac{2\sqrt{5}}{\sqrt{11\cdot13}}$	$\frac{3\sqrt{2}}{\sqrt{7\cdot11}}$	$\frac{\sqrt{3\cdot19}}{\sqrt{7\cdot13}}$

transformation and commutation properties of the spin angular momentum. That is why Q is usually called the quasispin quantum number or simply quasispin. Its projection

$$M_Q = -\frac{1}{2}(2l + 1 - N) \qquad (9.23)$$

shows the interval of the number of electrons in a shell with a given l, in which the term LS, characterized by the quantum number v, exists. In this way the whole angular momentum theory may be extended to cover the quasispin space.

Usually the Wigner–Eckart theorem (5.15) is utilized to find the dependence of the matrix elements on the projections of angular and spin momenta. Its use in the quasispin space

$$\left(l^N \alpha QLSM_Q \| T_{M_1}^{(K_1 K_2 K_3)} \| l^N \alpha' Q'L'S'M_{Q'}\right) = \frac{(-1)^{2K_1}}{\sqrt{2Q+1}}$$

$$\times \begin{bmatrix} Q' & K_1 & Q \\ M_{Q'} & M_1 & M_Q \end{bmatrix} \left(l\alpha QLS \|\| T^{(K_1 K_2 K_3)} \|\| l\alpha' Q'L'S'\right) \qquad (9.24)$$

allows one to prove that the dependence of the submatrix element (9.24) (in this case – matrix element, to which the Wigner–Eckart theorem is already applied in orbital and spin spaces) on the number of electrons N in shell l^N is simply contained in one Clebsch–Gordan coefficient. The so-called completely reduced matrix (reduced submatrix) element ($\|\|$ $\|\|$) in (9.24) does not depend on N. The operator $T^{(K_1 K_2 K_3)}$ is a tensor of the ranks K_1, K_2 and K_3 in quasispin, orbital and spin spaces, respectively, and the following notation of the wave function is used, accounting for the concept of the quasispin Q and its projection M_Q,

$$|l^N \alpha vLS) = |l^N \alpha QLSM_Q). \qquad (9.25)$$

It would be more logical to denote the right side of Eq. (9.25) in the form $|\alpha QLSM_Q)$ because M_Q defines unambiguously the shell l^N, and, therefore, there is no need to indicate l^N. However, we retain this symbol as a traditional definition of the configuration.

We shall see in Part 4 that in the second-quantization representation the CFP are the matrix elements of electron creation or annihilation operators. Applying the Wigner–Eckart theorem (9.24) to them in a quasispin space we obtain the so-called subcoefficients of fractional parentage for the given v, not depending on N. They are much more useful than the standard ones [91]. Completely accounting for tensorial properties of the quantities considered in various spaces, the new version of the theory of many-electron atoms was developed [27], requiring much less starting numerical data.

Using the notation of a term as $^{2S+1}_Q L$ instead of $^{2S+1}_v L$ we can learn that the wave functions of the terms $^{2S+1}_{\ \ Q}L$ and $^{2Q+1}_{\ \ S}L$ are interrelated; we can transpose these quantum numbers. This property is considered in detail in [92]. Studies of the properties of the CFP and submatrix elements of the operators, including irreducible tensorial ones, with respect to permutation of the spin and quasispin quantum numbers lead to new interesting relations as well as to fairly simple algebraic expressions for some special cases.

10

Two and more shells of
equivalent electrons

10.1 Two non-equivalent electrons. Representation of coupled momenta

Equation (1.14) defines the wave function of uncoupled momenta. It obeys the following orthonormality condition:

$$\sum_\sigma \int_0^\infty \int_0^{2\pi} \int_0^\pi \psi_{nlsm_lm_s}^*(r,\vartheta,\varphi,\sigma)\psi_{n'l's'm_l'm_s'}(r,\vartheta,\varphi,\sigma)r^2dr\sin\vartheta d\vartheta d\varphi$$

$$= \delta(n,n')\delta(l,l')\delta(s,s')\delta(m_l,m_l')\delta(m_s,m_s'). \quad (10.1)$$

The condition $\delta(n,n')$ for many-electron atoms is normally ensured only approximately unlike the other $\delta(a,a)$, which are rigorous. The principle of the orthogonality of the wave functions reflects the fact that at one time only one state described by a given set of exact quantum numbers is realized, an electron cannot occupy simultaneously several physical states.

Spin–orbit interaction (1.21) couples the angular and spin momenta l and s. The main part of the spin–orbit interaction (1.21) is described by the operator

$$W = \xi(r)(l \cdot s), \quad (10.2)$$

i.e. its energy contribution is proportional to the numerical values of the corresponding momenta and depends on the angle between them, i.e. on their mutual orientation. The quantity $\xi(r)$ is

$$\xi(r) = -\frac{e\hbar}{2m^2c^2}\frac{1}{r}\frac{\partial U(r)}{\partial r}, \quad (10.3)$$

where $U(r)$ is the potential of the field in which the electron is orbiting.

It follows from the Eq. (10.2) that the spin–orbit interaction may be interpreted as the interaction of the appropriate momenta. If two momenta l and s interact with each other, then the corresponding quantum numbers become approximate. Only their vectorial sum $l + s = j$ and its projection

$m = m_l + m_s$ are exact quantum numbers. This leads to another representation, i.e. the representation of coupled momenta or *jm*-representation. Its wave function may be expressed in terms of the function of uncoupled momenta ($m_l m_s$-representation) and vice versa

$$\psi_{nlsjm} = R_{nl} \sum_{m_l m_s} \begin{bmatrix} l & s & j \\ m_l & m_s & m \end{bmatrix} Y_{m_l}^{(l)} \chi_{m_s}^{(s)} = \sum_{m_l m_s} \begin{bmatrix} l & s & j \\ m_l & m_s & m \end{bmatrix} \psi_{nlsm_l m_s},$$

(10.4)

$$\psi_{nlsm_l m_s} = \sum_{jm} \begin{bmatrix} l & s & j \\ m_l & m_s & m \end{bmatrix} \psi_{nlsjm}.$$

(10.5)

Each wave function forms a complete basis of functions, therefore, Eqs. (10.4) and (10.5) can be considered as the transformation from one representation to the other.

Let us consider now an atom having two non-equivalent electrons. The method of constructing wave functions may be easily generalized to cover the case of a many-electron atom having two or more open shells of equivalent electrons. However, a simple example reveals the majority of the problems one is facing while dealing with the many-electron system. For two-electron atoms the Schrödinger equation cannot be solved exactly and one has to adopt approximate methods. Usually, as has been mentioned in Chapter 1, the central field approximation is utilized and the many-body problem is reduced to the model of the motion of one electron in the effective field of the nuclear charge and the charge of the remaining electrons. One-electron wave functions (1.14) or (10.4) may be then used to construct the total wave function of the atom or ion considered. Thus, the angular and spin parts of the total wave function are constructed from the known one-electronic quantities, whereas in order to find the radial parts one has to solve (approximately) the corresponding equations, usually deduced from the variational principle.

The particles with half-integer spin values must obey the so-called Fermi–Dirac statistics. The requirement that the wave function of the system consisting of particles of this sort must be antisymmetric with respect to the permutation of the coordinates of any two particles is the mathematical expression of this fact. An electron has half-integer spin, therefore, the total wave function of any atom or ion must be antisymmetric with respect to the permutation of the coordinates (or quantum numbers) of each pair of electrons.

Antisymmetry of the atomic wave function is the mathematical expression of the requirements of the Pauli exclusion principle, demanding that two or more atomic electrons cannot simultaneously occupy the same state, i.e. cannot possess the same set of one-electron quantum numbers. In a particular case of two-electron atoms the wave function of an atom must change its sign under permutation of the sets of quantum numbers (coordinates) of the electrons. Determinants have such a feature, therefore, the atomic wave function may be expressed in terms of the determinant, composed of one-electron wave functions (orbitals). Then the antisymmetric wave function of the two-electron atom $\psi(x_1, x_2)$ will be

$$\psi(x_1, x_2) = \frac{1}{\sqrt{2}} \begin{vmatrix} \psi^1(x_1) & \psi^1(x_2) \\ \psi^2(x_1) & \psi^2(x_2) \end{vmatrix}$$

$$= \frac{1}{\sqrt{2}} \left[\psi^1(x_1)\psi^2(x_2) - \psi^2(x_1)\psi^1(x_2) \right], \qquad (10.6)$$

where $\psi^i(x_i)$ is the one-electron wave function of the ith electron and x_i stands for a set of all its coordinates. Multiplier $1/\sqrt{2}$ ensures normalization of the whole wave function. If $x_1 = x_2$, then two rows of the determinant are equal, and the determinant itself equals zero. Thus, the antisymmetric wave function of a two-electron atom may be composed of the antisymmetrized product of one-electronic functions. However, in this case it is written in the representation of the uncoupled momenta. Accounting for the interaction of the momenta, it is more convenient to deal with the representation of the coupled momenta (wave functions (10.4)). Therefore, we have to investigate the possible schemes of vectorial coupling of the momenta of two electrons and the methods of constructing their wave functions. This question will be considered in the following paragraph. Here we shall briefly discuss constructing the antisymmetric wave functions of two and more shells of equivalent electrons in the *LS* coupling.

10.2 Several shells of equivalent electrons

A set of pairs of quantum numbers $n_i l_i$ with the indicated number of electrons having these quantum numbers, is called an electronic configuration of the atom (ion). Thus, we have already discussed the cases of two non-equivalent electrons and a shell of equivalent electrons. If there is more than one electron with the same $n_i l_i$, then the configuration may look like this:

$$(n_1 l_1)^{N_1} (n_2 l_2)^{N_2} \dots (n_i l_i)^{N_i}. \qquad (10.7)$$

Usually the brackets in Eq. (10.7) are omitted. The configuration, having the lowest, for a given number of electrons, energy is called the ground configuration, the others are called the excited ones. The concept of configuration has a physical meaning, because the levels of the configuration studied usually make a compact group, separated from the levels belonging to the other configurations. This justifies in many cases the fairly high accuracy of the so-called single-configuration approximation. Usually real configurations consist of closed shells and a few open shells. While calculating the energy spectra, we have to account for the interactions in each shell and between all shells.

As was already mentioned, due to the Pauli exclusion principle, which states that no two electrons can have the same wave functions, a wave function of an atom must be antisymmetric upon interchange of any two electron coordinates. For a shell of equivalent electrons this requirement is satisfied with the help of the usual coefficients of fractional parentage. However, for non-equivalent electrons the antisymmetrization procedure is different. If we have N non-equivalent electrons, then a wave function that is antisymmetric upon interchange of any two electron coordinates can be formed by taking the following linear combination of products of one-electron functions [16]:

$$\Psi = (N!)^{-1/2} \sum_{P} (-1)^{p} \varphi_1(\mathbf{r}_{j_1}) \varphi_2(\mathbf{r}_{j_2}) \varphi_3(\mathbf{r}_{j_3}) \ldots \varphi_N(\mathbf{r}_{j_N}). \tag{10.8}$$

In each product function, the same set of one-electron quantum numbers is arranged in the same order (usually in the standard order $1, 2, \ldots, N$) but the electron coordinates $\mathbf{r}_1, \mathbf{r}_2, \mathbf{r}_3, \ldots$ have been rearranged into some new order $\mathbf{r}_{j_1}, \mathbf{r}_{j_2}, \mathbf{r}_{j_3}, \ldots$. The summation in (10.8) is over all $N!$ possible permutations $P = j_1 j_2 j_3 \ldots j_N$ of the normal coordinate ordering $1\ 2\ 3\ \ldots N$, and p is the parity of the permutation P ($p = 0$ if P is obtained from the normal ordering by an even number of interchanges, and $p = 1$ if an odd number of interchanges is involved).

Antisymmetrized function (10.8) has the property that if any two one-electron functions are identical, then ψ is identically zero (satisfying the Pauli exclusion principle). Its second very important property: if any two electrons lie at the same position, e.g., $\mathbf{r}_1 = \mathbf{r}_2$ (and they also have parallel spins $s_1 = s_2$), then $\Psi \equiv 0$. As the functions φ are continuous in the spatial variables (r, ϑ, φ), it follows that $|\Psi|$ must be unusually small whenever two electrons with parallel spin are close together. Thus, unlike the single product function, the antisymmetrized sum of product functions (10.8) shows a certain degree of electron correlation. This correlation is incomplete – it arises by virtue of the Pauli exclusion principle rather than as a result of electrostatic repulsion, and there is no correlation at all between two electrons with antiparallel spins [16].

Antisymmetrized wave function (10.8) may be written in the form of a determinant

$$\Psi = \frac{1}{N!^{1/2}} \begin{vmatrix} \varphi_1(\mathbf{r}_1) & \varphi_1(\mathbf{r}_2) & \varphi_1(\mathbf{r}_3) & \ldots \\ \varphi_2(\mathbf{r}_1) & \varphi_2(\mathbf{r}_2) & \varphi_2(\mathbf{r}_3) & \ldots \\ \varphi_3(\mathbf{r}_1) & \varphi_3(\mathbf{r}_2) & \varphi_3(\mathbf{r}_3) & \ldots \\ \vdots & \vdots & \vdots & \ddots \end{vmatrix}, \tag{10.9}$$

and is, therefore, referred to as a determinantal function or a Slater determinant. The antisymmetrization and Pauli exclusion principle then follow immediately from the well-known properties of determinants: (a) interchanging the coordinates of two electrons is equivalent to interchanging two columns of the determinant, which changes its sign; (b) if two one-electron functions have the same quantum numbers, then two rows of the determinant are identical, and the determinant is, therefore, zero; (c) if two electrons have the same coordinates, then two columns are identical, and the determinant is again zero.

Let us notice that antisymmetrization involves summations over permutations of electron coordinates, whereas the coupling of momenta and their projections involves summations over functions with different values of the magnetic quantum numbers. Since these are independent finite summations, the order in which they are performed is interchangeable: we can, in principle, either antisymmetrize, first, and couple, second, or vice versa. In practice, the Pauli principle restricts the possible values of the projections and as a result of these complications it proves to be more convenient to couple first, and antisymmetrize second, with the antisymmetrization not always being done by means of coordinate permutations.

Many configurations of practical interest, especially ground configurations, include only one open (non-filled) shell. More often, however, a configuration (particularly, the excited one) will contain two or more open shells. In such cases, the overall quantum numbers for, for example, LS coupled functions, may be found by applying the vector model to the vector sums of the quantum numbers of separate shells. Closed shells may be omitted since the corresponding values of their quantum numbers are zero. All intermediate and total quantum numbers predicted by the vector model are allowed because we are coupling the vectors associated with electrons having different values of $n_i l_i$, so that the Pauli exclusion principle imposes no limitations on the values of the magnetic quantum numbers.

For the general case of configurations consisting of q shells of equivalent electrons,

$$l_1^{N_1} \, l_2^{N_2} \ldots l_q^{N_q}, \qquad N_1 + N_2 + \ldots + N_q = N, \qquad (10.10)$$

we can write down the function Ψ_u consisting of the product of q functions $|l_i^{N_i} \alpha_i L_i S_i)$, namely,

$$\Psi_u = |l_1^{N_1} \alpha_1 L_1 S_1) |l_2^{N_2} \alpha_2 L_2 S_2) \ldots |l_q^{N_q} \alpha_q L_q S_q), \qquad (10.11)$$

where the first factor is a function of the coordinates $\mathbf{r}_1, \mathbf{r}_2, \ldots, \mathbf{r}_{N_1}$, the second factor is a function of the coordinates $\mathbf{r}_{N_1+1}, \ldots, \mathbf{r}_{N_1+N_2}$, etc., with the coordinates for the final factor being $r_{N-N_q+1}, \ldots, r_N$. We can then couple the various quantum numbers $L_1 S_1$, $L_2 S_2$,... according to some suitable coupling scheme (see Chapters 11 and 12).

Whether the quantum numbers $L_i S_i$ in (10.11) are coupled or not, this function Ψ_u is not antisymmetric with respect to exchange of pairs of electron coordinates between shells, e.g. the coordinates $\mathbf{r}_{N_1}$ and $\mathbf{r}_{N_1+i}$. The necessary additional antisymmetrization can be accomplished through use of the generalized coefficients of fractional parentage [14] or of modification of the coordinate permutation scheme employed with one-electron functions in (10.8).

In the wave function of a shell of equivalent electrons we already have antisymmetry with respect to electron coordinate interchanges within that shell. Additional antisymmetrization, done by summing over permutations of coordinates within the shell, leads to very awkward complications. Hence, we sum only over permutations of electron coordinates among different shells; within a given shell, the ordering of electron coordinates must be a single standard ordering, which we take to be that of the numerically increasing value of the subscript k of the coordinates $\mathbf{r}_k$.

Returning to general configuration (10.10), the number of different permutations to be included is the total number $N!$ of possible permutations of all N coordinates, divided by the product Π_j (over all shells) of the number of unallowed permutations within each shell:

$$N!/(N_1! N_2! \ldots N_q!) = N!/(\Pi_j N_j!). \tag{10.12}$$

Thus, the normalized, completely antisymmetrized wave function Ψ_a of N-electron atom is, similarly to (10.8),

$$\Psi_a = \left[\frac{\Pi_j N_j!}{N!} \right]^{1/2} \sum_P (-1)^p \Psi_u^{(P)}, \tag{10.13}$$

where $\Psi_u^{(P)}$ is a partially antisymmetrized coupled function formed from products like (10.11), except with electron coordinates permuted according to the permutation P with parity p, and the sum is over only those permutations ((10.12) is its number) that are of the type just described. The antisymmetrizing permutations of electron coordinates among different shells appreciably complicate the appearance of basis functions (10.13); however, these complications largely disappear in the evaluation of matrix elements of symmetric operators. The method sketched here is traditionally used to build the wave functions of complex electronic configurations. However, a more efficient way is the use of the second quantization technique, described in Part 4.

11
Classification of energy levels

11.1 Electronic interactions as interactions of momenta

From formula (10.2) we saw that the spin–orbit interactions may be interpreted as the interactions of angular and spin momenta. It is possible to show that in the central-field approximation the electrostatic (Coulomb) interaction may also be considered as the interaction of the corresponding momenta. This grounds the model of vectorial coupling of the momenta and allows one to introduce various schemes for vectorial coupling of the momenta (coupling schemes) and, thus, various ways of classifying electronic states. Indeed, as we shall see in Part 5, the direct part of the Coulomb interaction corresponds to the Coulomb interaction of two charges and it will depend only on the orbital part of the wave function, i.e. on the orientation of the angular momenta l_1 and l_2, whereas the exchange part of the corresponding matrix element will depend on the orientation of both orbital and spin momenta. The expression for the exchange part of this interaction contains dependence on the spins in the form of the multiplier $\delta(m_s^1, m_s^2)$ which causes the exchange interaction to equal zero if the electronic spins are oriented in the opposite directions, i.e. $m_s^1 = -m_s^2$. Thus, the exchange part of the electrostatic interaction depends on the angle between s_1 and s_2, which is defined by their scalar product. If we write

$$\delta(m_s^1, m_s^2) = \frac{1}{2}\left[1 + 4(s_1 \cdot s_2)\right], \qquad (11.1)$$

then we can easily see that the dependence of the above-mentioned part of the electrostatic interaction on spin quantum numbers may be considered as the interaction of the corresponding momenta. Thus, indeed, the main intra-atomic interactions may be represented as interactions of corresponding angular and (or) spin momenta.

91

11.2   Various coupling schemes

Zero-order energy of the central field approximation, described by the central symmetric part of the potential, does not contain interaction of the momenta. Therefore, in zero-order approximation all states of a given configuration differing from each other by quantum numbers m_l^i, m_s^i, i.e. by different orientation of orbital and spin momenta l_i and s_i, have the same energy, and the corresponding level is degenerated $(4l + 2)$ times.

Interaction of the momenta is contained in a non-spherical part of the Coulomb interaction and in the spin–orbit interaction. The value of the energy of the interaction of two momenta depends on the angle between them, therefore, in such a case the definite inter-orientation of all one-electronic momenta is settled. Differently oriented states have different energy, i.e. the zero-order level splits into sublevels and its degeneracy disappears.

Mutual orientation of the momenta occurs while adding (coupling) them vectorially, because the value of the resultant momentum is defined by the angle between the coupled momenta. This shows that the splitting of the initial zero level, i.e. the structure of the energy spectrum of the configuration considered, depends on the order of the coupling of the momenta, in other words, on the coupling scheme. Let us recall that if two momenta are interacting, then their values are already not exact quantum numbers, only total (resultant) momentum will be an exact quantum number.

The order of the vectorial coupling of momenta is caused by the decrease of the strength of their interaction. First, the most strongly interacting pair of momenta must be coupled. As a result, some splitting of the initial level and certain intermediate momenta, or what is the same, a certain mutual orientation of the momenta of this pair occurs. After this the next pair of momenta, being in second place according to the value of their interaction, must be coupled. This pair may be composed of the momenta of other electrons or of the intermediate momenta and the momentum of one of the remaining electrons. In this way we obtain further splitting of the sublevels. Continuing in such a way with this procedure, we finally obtain the total momentum J and all possible levels of the configuration considered.

Let us illustrate this method in the special case of two non-equivalent electrons $n_1l_1n_2l_2$, described by four momenta: two orbital l_1, l_2 and two spin s_1, s_2. Having in mind the commutativity of the addition as well as the fact that the interaction of the orbital momentum of a given electron with its own spin momentum is much stronger than that with the momentum of the other electron, we obtain the four main different coupling schemes:

$$l_1 + l_2 = L, \quad s_1 + s_2 = S, \quad L + S = J \quad (LS \text{ coupling}), \quad (11.2)$$
$$l_1 + l_2 = L, \quad L + s_1 = K, \quad K + s_2 = J \quad (LK \text{ coupling}), \quad (11.3)$$
$$l_1 + s_1 = j_1, \quad j_1 + l_2 = K, \quad K + s_2 = J \quad (jK \text{ coupling}), \quad (11.4)$$
$$l_1 + s_1 = j_1, \quad l_2 + s_2 = j_2, \quad j_1 + j_2 = J \quad (jj \text{ coupling}). \quad (11.5)$$

In all cases (11.2)–(11.5) we have the same resultant (total) momentum **J** and two intermediate ones, used to denote the coupling scheme obtained. These momenta may be considered as additional quantum numbers, allowing unambiguous classification of the energy levels.

These coupling schemes reflect the relative values of the interaction of the corresponding momenta. For LS coupling (11.2) the strongest interactions are between both angular and spin momenta separately, leading to the occurrence of the momenta **L** and **S**. After that they are coupled into total momentum **J**. In contrast, for jj coupling (11.5) the spin–orbit interactions of each electron are the strongest. In both these two cases the momenta of two electrons are coupled in pairs completely in a symmetric mode, that is why the LS and jj coupling schemes are called homogeneous.

The other two coupling schemes (LK and jK, formulas (11.3), (11.4)) are called inhomogeneous. They are characterized by successive coupling of the momenta. LK coupling is valid if one of the spin–orbit interactions becomes stronger than the exchange one, but remains smaller than the Coulomb non-spherical part of the electrostatic interaction of the two electrons. In this case the second spin–orbit interaction may be close to the exchange one. For jK coupling the spin–orbit interaction of one of the electrons prevails under both the exchange and Coulomb (direct) non-spherical part of the electrostatic interaction.

The total number of levels with given J is the same for all coupling schemes. If there is only one level with given J value, then all coupling schemes are equally valid for it. The level with maximal J value may serve as such an example. In order to obtain such a level we have all momenta directed parallel, i.e. we can couple them only in one fashion.

Unfortunately, as we shall see later, in the majority of cases none of the above-mentioned pure coupling schemes is accurate and, therefore, we are faced with the necessity to adopt the so-called intermediate coupling schemes, usually obtained by diagonalization of the energy matrices of the given configuration. However, in such cases, for unambiguous identification and classification of the energy levels, one has to start by coupling in a certain way the momenta and using quantum numbers of one of the coupling schemes ((11.2)–(11.5)) in spite of the fact that they are approximate.

11.3   Classification of the energy levels using various coupling schemes

In Chapter 9 we discussed the classification of the terms and energy levels of a shell of equivalent electrons using the LS coupling scheme. Here we shall consider the case of two non-equivalent electrons. As we shall see later on, generalization of the results for two non-equivalent electrons to the case of two or more shells of equivalent electrons is straightforward.

In LS coupling the energy levels of two-electron configuration will be characterized according to Eq. (11.2) by intermediate quantum numbers L, S and resultant momentum J. Usually, instead of LSJ, the notation $^{2S+1}L_J$ is utilized. Quantum numbers L and S denote the term of the multiplicity $2S + 1$, whereas J describes the so-called fine structure (levels) of the term. Each level in accordance with the values of $M = -J, -J + 1, \ldots, J$ consists of $2J + 1$ states. The number of levels of the term equals $2S + 1$, i.e. its multiplicity, if $L \geq S$, and equals $2L + 1$, if $L < S$. L acquire integer (whereas S and J may acquire also half-integer) positive values (including $L = S = J = 0$), however, usually the capital letters of the Latin alphabet are utilized to denote L (as well as small ones for l) in accordance with

$$L \text{ or } l : 0 \ 1 \ 2 \ 3 \ 4 \ 5 \ 6 \ 7 \ 8 \ 9 \ 10,$$
$$L : S \ P \ D \ F \ G \ H \ I \ K \ L \ M \ N, \tag{11.6}$$
$$l : s \ p \ d \ f \ g \ h \ i \ k \ l \ m \ n.$$

Starting with (11.2) it is easy to show that, for example, for the configuration $nsn'p$ we get levels $^1P_1$ and $^3P_{0,1,2}$, for $npn'p - {}^1S_0$, $^1P_1$, $^1D_2$, $^3S_1$, $^3P_{0,1,2}$, $^3D_{1,2,3}$, etc. For $2S + 1 = 1, 2, 3, 4, 5$, etc., the terms are correspondingly called singlet, doublet, triplet, quadruplet, quintuplet, etc.

In order to indicate the parity, defined here as $(-1)^{l_1+l_2}$, we have to add to the term a special symbol (e.g. for odd configurations the small letter o). Then, for example, the levels of the configuration $nsn'p$ will be $^1P_1^o$, $^3P_{0,1,2}^o$. Thus, the spectra of two non-equivalent electrons will consist of singlets and triplets.

Symbols of LS coupling are the most frequently used, partly due to the fact that for an overwhelming majority of neutral atoms and ions of moderate ionization degrees, it is closest to reality, partly traditionally.

The designations of the other coupling schemes are more cumbersome and less obvious. They can be easily obtained from formulas (11.3)–(11.5). For LK and jK coupling the levels may be denoted as XKJ or $X[K]_J$, where $X = L, j$. Quantum number L may be designated both using letters or numbers. For example, for the $npn'p$ configuration there are the following levels:

LK coupling:

$$S\left[\frac{1}{2}\right]_{0,1} \; ; \; P\left[\frac{1}{2}\right]_{0,1} \; ; \; P\left[\frac{3}{2}\right]_{1,2} \; ; \; D\left[\frac{3}{2}\right]_{1,2} \; ; \; D\left[\frac{5}{2}\right]_{2,3} , \tag{11.7}$$

jK coupling:

$$\frac{1}{2}\left[\frac{1}{2}\right]_{0,1} \; ; \; \frac{3}{2}\left[\frac{1}{2}\right]_{0,1} \; ; \; \frac{1}{2}\left[\frac{3}{2}\right]_{1,2} \; ; \; \frac{3}{2}\left[\frac{3}{2}\right]_{1,2} \; ; \; \frac{3}{2}\left[\frac{5}{2}\right]_{2,3} . \tag{11.8}$$

It is seen from Eqs. (11.7) and (11.8) that for LK and jK coupling schemes the energy spectrum consists of doublets. These doublets form groups according to the allowed L and j values, respectively.

For jj coupling the levels are denoted as $j_1 j_2 J$ or $[j_1 j_2]_J$. Then for the same $npn'p$ configuration we get:

$$\left[\frac{1}{2}\frac{1}{2}\right]_{0,1} \; ; \; \left[\frac{1}{2}\frac{3}{2}\right]_{1,2} \; ; \; \left[\frac{3}{2}\frac{1}{2}\right]_{1,2} \; ; \; \left[\frac{3}{2}\frac{3}{2}\right]_{0,1,2,3} . \tag{11.9}$$

Of course, in all cases (11.7)–(11.9) the writing of all symbols in one row is also used. Comparing the four coupling schemes it is easy to conclude that the total number of levels in the given configuration, as well as their number for a given J, does not depend on the type of coupling scheme.

Let us also recall that for a relativistic Hamiltonian and wave function only jj coupling is possible due to the prevailing spin–orbit interactions over the non-spherical part of the electrostatic interaction.

As is known, all levels of the energy spectrum are identified and classified with the help of sets of quantum numbers describing the coupling scheme used. However, these quantum numbers are exact only for the cases of pure coupling schemes, which is more the exception than the rule. For this reason one has to diagonalize the energy matrix, calculated in some pure coupling scheme supposed to be the closest to the real one. This means that we pass to another coupling scheme (usually called intermediate) where the wave function is a linear combination of the wave functions of pure (usually LS) coupling, i.e.

$$\psi(\beta J) = \sum_{\alpha_i L_i S_i} a(\alpha_i L_i S_i J)\psi(\alpha_i L_i S_i J), \tag{11.10}$$

where a are the weights of the wave functions of pure LS coupling, and β stands for the quantum numbers of intermediate coupling.

While calculating energy spectra one has to start with the coupling scheme closest to reality; then as a rule the intermediate coupling will differ insignificantly from the pure one. In this case there will be one weight much larger in comparison with the other weights in the wave function (11.10). The quantum numbers of this largest weight are then used to classify the corresponding energy level. In the case of an unfit coupling scheme there will be several almost equal weights in (11.10). This problem is described in more detail in [93]. Here let us mention only that pure coupling schemes are valid only in some special cases or limits as, for example, configurations with a vacancy or configurations with one highly excited electron (except the well-known limits of small and large Z). On the other hand, the presence of d-electrons in the configuration means, as a rule, the absence of a pure coupling scheme and strong mixing of the terms and their levels.

Configurations with three open shells represent a much more complicated case. In general, the number of possible coupling schemes is rather large in this case. However, it is possible to choose approximately 20 coupling schemes which are more likely from the physical point of view. The character of interactions is more complicated, the terms are mixed strongly and the probability of finding pure coupling is rather small. For example, the calculations for $5d^9 6s6p$ Hg^+ demonstrate that for $J = 1/2$ the coupling scheme $((((L_1 S_1)J_1, (L_2 L_3)L_{23})K_3, (S_2 S_3)S_{23})J)$ is optimal, whereas for $J = 7/2$ another scheme $(((L_1 S_1)J_1, ((L_2 L_3)L_{23}, (S_2 S_3)S_{23})J_{23})J)$ is valid, i.e. some combination of LS and JJ couplings takes place. In the case of $2s2p4p$ Fe^{21+} for all J, the coupling $(((((L_1 L_2)L_{12}, (S_1 S_2)S_{12})J_{12}, (L_3 S_3)J_3)J)$ is optimal, i.e. the first two electrons are coupled with LS coupling, whereas the third one is coupled in JJ coupling. For $2s2p^5 3s$ V^{14+} the coupling $(((((L_1 L_2)L_{12}, L_3)L, (S_1 S_2)S_{12})K, S_3)J)$ is preferable.

It is important to emphasize that there are cases where for the classification of the energy levels of a given configuration, one has to use different coupling schemes for separate groups of levels with the same value of J. For example, calculations show that in the case of the configuration $4d^9 5p$ of ion In^{3+} for $J = 2 - LS$, for $J = 3 - LK$ and for $J = 4 - JJ$, couplings are optimal.

Thus, for complex electronic configurations there is a large variety of possible coupling schemes. Therefore, the combination of theoretical and experimental investigations is of great importance for the interpretation of such spectra (see also Chapter 12).

12

Relations between various coupling schemes

In the previous chapter we discussed the problem of classification of the energy levels of many-electron atoms using various coupling schemes, including intermediate ones. For this purpose we start with the coupling scheme closest to reality and optimize it. In practice this may be achieved in two different ways, namely, utilizing the expressions for matrix elements of the energy operator in a different pure coupling scheme with subsequent diagonalization of the energy matrices, or starting with the energy matrices in one definite coupling scheme, not necessarily supposed to be the closest to reality, with further use of the special transformation matrices, converting the wave functions of a given coupling scheme (including an intermediate one) to the other. These transformation matrices have to cover the cases of complex electronic configurations, consisting of several open shells.

Usually, the first way is utilized in practice. This is due to the well developed mathematical technique necessary, by the presence of the expressions for both the matrix elements of the energy operator and of the electronic transitions in various coupling schemes. However, the second method is much more universal and easier to apply, provided that there are known corresponding transformation matrices. Now we shall briefly describe this method.

Let $|\psi_i\rangle$ and $|\varphi_j\rangle$ denote two full orthonormal sets of wave functions, corresponding to a given energy spectrum of a definite many-electron system, classified with the aid of quantum numbers of two different coupling schemes. Let us suppose that these functions are non-relativistic, then they will have one and the same electronic configuration. In a relativistic case this transformation will couple different electronic configurations, and this coupling will be approximate. Let us recall that in a relativistic approach, a shell of equivalent electrons splits into subshells, the number of electrons in which varies, depending on the value of total momentum J.

Two full orthonormal bases of wave functions $|\psi_j)$ and $|\varphi_i)$, describing the system under consideration, are related by a unitary transformation (moreover, if the corresponding transformation matrices are chosen to be real, then the transformation itself will be orthogonal)

$$|\psi_i) = \sum_j |\varphi_j)(\varphi_j|\psi_i), \tag{12.1}$$

$$|\varphi_j) = \sum_i |\psi_i)(\psi_i|\varphi_j). \tag{12.2}$$

Here the corresponding matrices satisfy the following condition:

$$(\psi_i|\varphi_j) = (\varphi_j|\psi_i)^\dagger. \tag{12.3}$$

Let us recall that the transformation matrix $D^\dagger$ is the Hermitian conjugate to matrix D and obtained from D by its transposition ($\tilde{D}$) and complex conjugation ($\tilde{D}^*$). The unitarity of the matrix implies that

$$D^\dagger D = 1, \qquad D^\dagger = D^{-1}, \tag{12.4}$$

where D^{-1} is the matrix inverse of D. For orthogonal matrices instead of (12.4) we get

$$\tilde{D}D = D\tilde{D} = 1, \qquad \tilde{D} = D^{-1}. \tag{12.5}$$

Formulas, similar to (12.1), (12.2), may be written for the wave functions of intermediate coupling (*ic*), too. Namely,

$$|\psi_i)_{ic} = \sum_j a_{ij}|\psi_j), \tag{12.6}$$

$$|\varphi_j)_{ic} = \sum_k c_{jk}|\varphi_k). \tag{12.7}$$

Here a_{ij} and c_{jk} are the weights of the wave functions for two pure coupling schemes. If wave functions (12.6) and (12.7) have the same eigenvalues (this is the case, if we consider in different coupling schemes the matrix elements of one and the same Hamiltonian, e.g., (1.16)), then we can find the unambiguous correspondence between them, i.e. with the accuracy of the phase multiplier (which we choose here to be equal to one) we can write

$$|\varphi_i)_{ic} = |\psi_i)_{ic}. \tag{12.8}$$

Then, substituting (12.6) and (12.7) into (12.8), having in mind the conditions of the orthonormality of the wave functions, we arrive at the following relationships between weights of the wave functions of two pure coupling schemes:

$$c_{jk} = \sum_r a_{jr}(\varphi_k|\psi_r), \tag{12.9}$$

$$a_{ij} = \sum_k c_{ik}(\psi_j | \varphi_k). \tag{12.10}$$

It follows from Eqs. (12.9) and (12.10) that, knowing the weights of the wave functions in a certain coupling scheme, for example, LS, we can easily find them for the case of any other coupling and choose the optimal one in this way, without diagonalizing the energy matrices in other coupling schemes. On the other hand, the above-mentioned equations may be adopted to check the correctness of the identification of the corresponding energy levels, because equalities of this type will not hold if, in at least one coupling scheme, the wrong set of quantum numbers is ascribed to the level.

Formulas (12.9) and (12.10) may also be utilized to transform the weights of the relativistic wave functions in jj coupling to the weights of other coupling schemes. However, in this case the equality of the type (12.8) becomes approximate due to the changes of the character of the wave function and the Hamiltonian. We lose a part of the relativistic effects as a result of such a transformation. Nevertheless, it gives another (compare with Hamiltonian (1.16)) way to account for the relativistic corrections to non-relativistic wave functions. However, in order to use this method of finding the optimal coupling scheme one has to have explicit expressions for the transformation matrices of the type $(\varphi_k | \psi_r)$. Therefore, let us briefly sketch their evaluation.

12.1 A shell of equivalent electrons l^N

In this case, we need only transformation matrices connecting the homogeneous LS and jj coupling schemes. In transforming the wave function of shell l^N from one coupling scheme to another, we have to account for its antisymmetry properties. This implies that the transformation matrices themselves must have multipliers ensuring this condition. As we shall see below, the coefficients of fractional parentage in LS and jj couplings will do this job. Transformations (12.1) and (12.2) for the wave functions $|l^N \alpha LSJ)$ (LS coupling) and $|lj_1^{N_1} j_2^{N_2} \beta_1 J_1 \beta_2 J_2 J)$ (jj coupling) will look like

$$\left| l^N \alpha LSJ \right) = \sum_{\beta_1 J_1 \beta_2 J_2 N_1} \left| lj_1^{N_1} j_2^{N_2} \beta_1 J_1 \beta_2 J_2 J \right) \left(lj_1^{N_1} j_2^{N_2} \beta_1 J_1 \beta_2 J_2 J | l^N \alpha LSJ \right),$$

$$\tag{12.11}$$

$$\left| lj_1^{N_1} j_2^{N_2} \beta_1 J_1 \beta_2 J_2 J \right) = \sum_{\alpha LS} \left| l^N \alpha LSJ \right) \left(l^N \alpha LSJ | lj_1^{N_1} j_2^{N_2} \beta_1 J_1 \beta_2 J_2 J \right). \tag{12.12}$$

We remind the reader that jj coupling inside a shell of equivalent electrons requires the use of relativistic wave functions, whereas for LS

coupling the non-relativistic ones must be adopted. However, formulas (12.11) and (12.12) imply equality of the eigenvalues of the corresponding wave functions. Therefore, one ought to have in mind the statement at the end of the previous section of this chapter about the accuracy of such an equality.

In theoretical atomic spectroscopy usually a phase system for the wave functions is chosen which ensures real values of the CFP. In this case the transformation matrices will acquire only real values, too. Let us notice that the transformation matrix in (12.12), according to (12.4), is reciprocal to that in (12.11). Due to the orthonormality of the sets of wave functions, these matrices obey the orthonormality conditions:

$$\sum_{\alpha LS} \left(l j_1^{N_1} j_2^{N_2} \beta_1 J_1 \beta_2 J_2 J | l^N \alpha LSJ \right) \left(l^N \alpha LSJ | l j_1^{N_1'} j_2^{N_2'} \beta_1' J_1' \beta_2' J_2' J \right)$$

$$= \delta(N_1 N_2 \beta_1 J_1 \beta_2 J_2, \ N_1' N_2' \beta_1' J_1' \beta_2' J_2'), \qquad (12.13)$$

$$\sum_{\beta_1 J_1 \beta_2 J_2 N_1} \left(l^N \alpha LSJ | l j_1^{N_1} j_2^{N_2} \beta_1 J_1 \beta_2 J_2 J \right) \left(l j_1^{N_1} j_2^{N_2} \beta_1 J_1 \beta_2 J_2 J | l^N \alpha LSJ \right)$$

$$= \delta(\alpha LS, \ \alpha' L' S'). \qquad (12.14)$$

Let us present the formula for these transformation matrices, convenient for practical utilization (the details of its deduction may be found in [18, 94]). It has the form of the following recurrence relation:

$$\left(l^N \alpha LSJ | l j_1^{N_1} j_2^{N_2} \beta_1 J_1 \beta_2 J_2 J \right) = \frac{1}{\sqrt{N}} \sum_{\alpha' L' S'} \left(l^N \alpha LS \| l^{N-1} (\alpha' L' S') l \right)$$

$$\times \sum_{J'} \left[\sqrt{N_1} (((L'l)L(S's)S)J|((L'S')J'(ls)j_1)J) \right.$$

$$\times \sum_{\beta_1' J_1'} (-1)^{2J_2} (((J_1' J_2)J' j_1)J|((J_1' j_1)J_1 J_2)J)$$

$$\times \left(j_1^{N_1-1}(\beta_1' J_1')j_1 \| j_1^{N_1} \beta_1 J_1 \right) \left(l^{N-1} \alpha' L' S' J' | l j_1^{N_1-1} j_2^{N_2} \beta_1' J_1' \beta_2 J_2 J' \right)$$

$$+ \sqrt{N_2} (((L'l)L(S's)S)J|((L'S')J'(ls)j_2)J)$$

$$\times \sum_{\beta_2' J_2'} (((J_1 J_2')J' j_2)J|(J_1(J_2' j_2)J_2)J)) \left(j_1^{N_2-1}(\beta_2' J_2')j_2 \| j_2^{N_2} \beta_2 J_2 \right)$$

$$\times \left. \left(l^{N-1} \alpha' L' S' J' | l j_1^{N_1} j_2^{N_2-1} \beta_1 J_1 \beta_2' J_2' J' \right) \right]. \qquad (12.15)$$

In (12.15) we have the two usual transformation matrices (see Chapter 6), describing the change of the coupling of three momenta, and two more for four momenta. Expressing them in terms of 6j- and 9j-coefficients according to formulas (6.42)–(6.44), we get the final form of the transformation matrix from LS to jj coupling

$$\left(l^N \alpha LSJ | l j_1^{N_1} j_2^{N_2} \beta_1 J_1 \beta_2 J_2 J\right)$$

$$= \sqrt{[L,S]/N} \sum_{\alpha' L' S'} \left(l^N \alpha LS \| l^{N-1}(\alpha' L' S') l\right)$$

$$\times \sum_{J'} [J'] \left[\sqrt{[N_1[j_1, J_1]]} \begin{Bmatrix} L' & l & L \\ S' & s & S \\ J' & j_1 & J \end{Bmatrix} \sum_{\beta_1' J_1'} (-1)^{j_1 + J_1 - J_2 + J'} \right.$$

$$\times \begin{Bmatrix} J_2 & J_1' & J' \\ j_1 & J & J_1 \end{Bmatrix} \left(j_1^{N_1-1}(\beta_1' J_1') j_1 \| j_1^{N_1} \beta_1 J_1\right)$$

$$\times \left(l^{N-1} \alpha' L' S' J' | l j_1^{N_1-1} j_2^{N_2} \beta_1' J_1' \beta_2 J_2 J'\right) + \sqrt{N_2[j_2, J_2]} \begin{Bmatrix} L' & l & L \\ S' & s & S \\ J' & j_2 & J \end{Bmatrix}$$

$$\times \sum_{\beta_2' J_2'} (-1)^{j_2 + J_1 + J_2' + J} \begin{Bmatrix} J_2' & J_1 & J' \\ J & j_2 & J_2 \end{Bmatrix} \left(j_2^{N_2-1}(\beta_2' J_2') j_2 \| j_2^{N_2} \beta_2 J_2\right)$$

$$\times \left. \left(l^{N-1} \alpha' L' S' J' | l j_1^{N_1} j_2^{N_2-1} \beta_1 J_1 \beta_2' J_2' J'\right) \right]. \tag{12.16}$$

Thus, using this equation, we are able in a recurrent way, starting with the two-electron configuration, to find the numerical values of this transformation matrix for all shells of equivalent electrons we are interested in. The following two-electron transformation matrices serve as the initial ones:

$$\left(l^2 LSJ | l j_1 j_2 J\right)$$

$$= \frac{1}{\sqrt{2}} \left(1 + (-1)^{L+S}\right) \sqrt{[j_1, j_2, L, S]} \begin{Bmatrix} l & l & L \\ s & s & S \\ j_1 & j_2 & J \end{Bmatrix}, \tag{12.17}$$

$$\left(l^2 LSJ | l j^2 J\right)$$

$$= \frac{1}{4} \left(1 + (-1)^{L+S}\right) \left(1 + (-1)^{J}\right) [j] \sqrt{[L,S]} \begin{Bmatrix} l & l & L \\ s & s & S \\ j & j & J \end{Bmatrix}. \tag{12.18}$$

In Eq. (12.18) the notation $j = j_1 = j_2$ is used, and the change of normalization of the wave function while passing to the equivalent electrons in the sense of jj coupling is taken into consideration.

Let us notice that even while calculating these matrices for the partially filled shells l^N we face the necessity to use in jj coupling the CFP of both partially and almost filled subshells. Therefore, in calculations of this sort it is very important to have a universal system of relations connecting complementary shells. The calculations show that this method gives the weights of the wave functions very close to those calculated using the diagonalization of the energy matrix [18, 93].

12.2　Two shells of equivalent electrons

In many physical problems we come across excited configurations consisting of several open shells or at least one electron above the open shell. Therefore, we have to be able to transform wave functions and matrix elements from one coupling scheme to the other for such complex configurations. If K denotes the configuration, and $\alpha\alpha'$, $\beta\beta'$ stand for the quantum numbers of two different coupling schemes, then for the corresponding wave functions formulas of the kind (12.1), (12.2) hold, whereas the matrix element of some scalar operator D transforms as:

$$(K\alpha J|D|K\alpha' J') = \sum_{\beta,\beta'} (K\alpha J|K\beta J)(K\beta J|D|K\beta' J')(K\beta' J'|K\alpha' J'). \quad (12.19)$$

For the case of several (let us say, two) open shells, there are two qualitatively different cases: when the transformation (1) does not affect, and (2) changes the coupling scheme inside a shell of equivalent electrons. The first case is much simpler than the second, because it does not require inclusion of the CFP in the corresponding transformation matrices. This case is identical to configurations consisting of two non-equivalent electrons, with possible coupling schemes (11.2)–(11.5), if we replace the momenta of separate electrons l_i, s_i by the total momenta of a shell L_i, S_i. Therefore, the transformation matrices, presented below, will be equally suitable for two non-equivalent electrons (when the momenta and, partially, coupling schemes are denoted by small Latin letters) as well as for two shells of equivalent electrons (momenta and coupling schemes are denoted by capital Latin letters). The latter designations will be used further on.

Formula (12.19) for the particular case of the energy operator and of the transformation from the LS coupling to another one, in the general case of two shells of equivalent electrons, inside which there is LS coupling and their total momenta are $L_1 S_1$ and $L_2 S_2$, respectively, will be of the form (for simplicity we skip the designations of the shells themselves and of the additional quantum numbers, distinguishing the terms of the shell with the same $L_i S_i$):

$$(L_1 S_1 L_2 S_2 T_1 T_2 J|H|L_1' S_1' L_2' S_2' T_1' T_2' J)$$
$$= \sum_{LSL'S'} (L_1 S_1 L_2 S_2 T_1 T_2 J|L_1 S_1 L_2 S_2 L S J)$$
$$\times (L_1 S_1 L_2 S_2 L S J|H|L_1' S_1' L_2' S_2' L' S' J)$$
$$\times (L_1' S_1' L_2' S_2' L' S' J|L_1' S_1' L_2' S_2' T_1' T_2' J). \quad (12.20)$$

Here $T_1 T_2$ and $T_1' T_2'$ stand for the intermediate momenta, the concrete symbols of which depend on the coupling scheme chosen. For the same reasons the order and the scheme of the addition of the momenta are not indicated. The explicit expressions of these matrices may be found while specifying the coupling scheme according to formulas of the kind (11.2)–(11.5). All these matrices correspond to the cases of changing the coupling of four or three (the fourth is left unchanged) momenta. Utilizing the ideas of Chapter 6, we obtain the following expressions for transformation matrices from LK, JK and JJ to LS coupling, and vice versa, in terms of $3nj$-coefficients:

$$((((L_1 L_2)L'S_1)KS_2)J|((L_1 L_2)L(S_1 S_2)S)J)$$

$$= (-1)^{L+S_1+S_2+J} \delta(L,L') \sqrt{[K,S]} \begin{Bmatrix} S_1 & S_2 & S \\ J & L & K \end{Bmatrix}, \tag{12.21}$$

$$((((L_1 S_1)J_1 L_2)KS_2)J|((L_1 L_2)L(S_1 S_2)S)J)$$

$$= (-1)^{L_2-S_2+J_1-J} \sqrt{[J_1,K,L,S]} \begin{Bmatrix} L_1 & L_2 & L \\ K & S_1 & J_1 \end{Bmatrix} \begin{Bmatrix} S_1 & S_2 & S \\ J & L & K \end{Bmatrix}, \tag{12.22}$$

$$(((L_1 S_1)J_1 (L_2 S_2)J_2)J|((L_1 L_2)L(S_1 S_2)S)J)$$

$$= \sqrt{[J_1,J_2,L,S]} \begin{Bmatrix} L_1 & S_1 & J_1 \\ L_2 & S_2 & J_2 \\ L & S & J \end{Bmatrix}. \tag{12.23}$$

In a similar way the remaining transformation matrices from JK to LK, and from JJ to LK and JK coupling schemes may be found

$$((((L_1 S_1)J_1 L_2)KS_2)J|(((L_1 L_2)LS_1)K'S_2)J)$$

$$= (-1)^{L_2+S_1+J_1+L} \delta(K,K') \sqrt{[J_1,L]} \begin{Bmatrix} L_1 & S_1 & J_1 \\ K & L_2 & L \end{Bmatrix}, \tag{12.24}$$

$$(((L_1 S_1)J_1 (L_2 S_2)J_2)J|(((L_1 L_2)LS_1)KS_2)J)$$

$$= (-1)^{L-S_1+S_2+J} \sqrt{[J_1,J_2,L,K]}$$

$$\times \begin{Bmatrix} L_1 & S_1 & J_1 \\ K & L_2 & L \end{Bmatrix} \begin{Bmatrix} L_2 & S_2 & J_2 \\ J & J_1 & K \end{Bmatrix}, \tag{12.25}$$

$$(((L_1 S_1)J_1 (L_2 S_2)J_2)J|(((L_1 S_1)J_1' L_2)KS_2)J)$$

$$= (-1)^{L_2+S_2+J_1+J} \delta(J_1,J_1') \sqrt{[J_2,K]} \begin{Bmatrix} L_2 & S_2 & J_2 \\ J & J_1 & K \end{Bmatrix}. \tag{12.26}$$

Formulas (12.21)–(12.26) are fairly simple, the corresponding transformation matrices are expressed only in terms of $6j$- and $9j$-coefficients, which can be easily calculated with the help of algebraic formulas (see [11]), tables of their numerical values or using standard computer codes. These formulas can be easily generalized to cover the case of a larger number of open shells.

Finally in this section let us discuss the case when the transformation changes the coupling scheme inside a shell of equivalent electrons. If inside a shell there is jj coupling, and if we are interested only in transformations preserving the jj coupling, then only transformations changing the order of coupling of the four momenta of the four subshells which occur from parallel to successive and vice versa, are important. The corresponding transformation matrix is very simple and is analogous to (12.21). Therefore, we ought only to consider the cases when the jj coupling changes to LS inside the shells of equivalent electrons of two-shell configuration.

The total momenta of two shells $l_1^{N_1} l_2^{N_2}$ may be coupled by four coupling schemes of the type (11.2)–(11.5), whereas the momenta of four subshells, obtained when jj coupling is valid in these shells, may be coupled according to above-mentioned schemes in parallel or successively. Let us consider first the case of parallel order coupling. Then, utilizing the same method as in the case of one shell, we find the following expression for the two-shell transformation matrix in terms of the one-shell matrices:

$$\left(l_1^{N_1} l_2^{N_2} LSJ | J_1 J_2 J \right)$$

$$= \sqrt{[J_1, J_2, L, S]} \begin{Bmatrix} K_1 & S_1 & J_1 \\ L_2 & S_2 & J_2 \\ L & S & J \end{Bmatrix} \left(N_1 | N_1 - N_{1+} \right) \left(N_2 | N_2 - N_{2+} \right), \quad (12.27)$$

where

$$\left(l_1^{N_1} l_2^{N_2} T_1 T_2 J | J_1 J_2 J \right) = \left(l_1^{N_1} l_2^{N_2} \alpha_1 L_1 S_1 \alpha_2 L_2 S_2 T_1 T_2 J \right|$$

$$\times l_1 j_{1-}^{N_{1-}} j_{1+}^{N_{1+}} l_2 j_{2-}^{N_{2-}} j_{2+}^{N_{2+}} \bar{\beta}_1 \bar{J}_1 \overset{+}{\beta}_1 \overset{+}{J}_1 (J_1), \bar{\beta}_2 \bar{J}_2 \overset{+}{\beta}_2 \overset{+}{J}_2 (J_2), J \Big), \quad (12.28)$$

$$\left(N_i | N_i - N_{i+} \right) = \left(l_i^{N_i} \alpha_i L_i S_i J_i | l_i j_{i-}^{N_{i-}} j_{i+}^{N_{i+}} \bar{\beta}_i \bar{J}_i \overset{+}{\beta}_i \overset{+}{J}_i J_i \right). \quad (12.29)$$

Passing in the left side of (12.27), with the help of the usual transformation matrices, to the coupling schemes desired, in the remaining cases of LK, JK and JJ coupling schemes between the shells we obtain

$$\left(l_1^{N_1} l_2^{N_2} LKJ | J_1 J_2 J \right) = (-1)^{L-J+S_1-S_2} \sqrt{[L, K, J_1, J_2]}$$

$$\times \begin{Bmatrix} L_1 & S_1 & J_1 \\ K & L_2 & L \end{Bmatrix} \begin{Bmatrix} L_2 & S_2 & J_2 \\ J & J_1 & K \end{Bmatrix} \left(N_1 | N_1 - N_{1+} \right) \left(N_2 | N_2 - N_{2+} \right), \quad (12.30)$$

$$\left(l_1^{N_1} l_2^{N_2} J_1' KJ | J_1 J_2 J \right) = (-1)^{J_1+L_2+S_2+J} \delta(J_1 J_1') \sqrt{[K, J_2]}$$

$$\times \begin{Bmatrix} L_2 & S_2 & J_2 \\ J & J_1 & K \end{Bmatrix} \left(N_1 | N_1 - N_{1+} \right) \left(N_2 | N_2 - N_{2+} \right), \quad (12.31)$$

$$\left(l_1^{N_1} l_2^{N_2} J_1' J_2' J | J_1 J_2 J \right) = \delta(J_1 J_2, J_1' J_2') \left(N_1 | N_1 - N_{1+} \right) \left(N_2 | N_2 - N_{2+} \right). \quad (12.32)$$

Let us denote in the following way the transformation matrix considered for the successive scheme of coupling of the momenta of subshells:

$$\left(l_1^{N_1} l_2^{N_2} T_1 T_2 J | J_1 J_{12} J\right) = \left(l_1^{N_1} l_2^{N_2} \alpha_1 L_1 S_1 \alpha_2 L_2 S_2 T_1 T_2 J \right|$$

$$\times l_1 j_{1-}^{N_1-} j_{1+}^{N_1+} l_2 j_{2-}^{N_2-} j_{2+}^{N_2+} \; \bar{\beta}_1 \bar{J}_1 \overset{+}{\beta}_1 \overset{+}{J}_1 (J_1), \bar{\beta}_2 \bar{J}_2 (J_{12}) \overset{+}{\beta}_2 \overset{+}{J}_2 J \Big). \tag{12.33}$$

Then in a similar way we find the following expressions for the corresponding transformation matrices:

$$\left(l_1^{N_1} l_2^{N_2} LSJ | J_1 J_{12} J\right) = (-1)^{\bar{J}_2 + \overset{+}{J}_2 + J_1 + J} \sqrt{[J_1, J_{12}, L, S]}$$

$$\times (N_1 | N_1 - N_{1+}) \sum_{J_2} (2J_2 + 1) (N_2 | N_2 - N_{2+}) \begin{Bmatrix} J_1 & \bar{J}_2 & J_{12} \\ \overset{+}{J}_2 & J & J_2 \end{Bmatrix}$$

$$\times \begin{Bmatrix} L_1 & S_1 & J_1 \\ L_2 & S_2 & J_2 \\ L & S & J \end{Bmatrix}, \tag{12.34}$$

$$\left(l_1^{N_1} l_2^{N_2} LKJ | J_1 J_{12} J\right) = (-1)^{L + J_1 + S_1 - S_2 + \bar{J}_2 + \overset{+}{J}_2} \sqrt{[J_{12}, J_1, L, K]}$$

$$\times (N_1 | N_1 - N_{1+}) \begin{Bmatrix} L_1 & S_1 & J_1 \\ K & L_2 & L \end{Bmatrix} \sum_{J_2} (2J_2 + 1) (N_2 | N_2 - N_{2+})$$

$$\times \begin{Bmatrix} J_1 & \bar{J}_2 & J_{12} \\ \overset{+}{J}_2 & J & J_2 \end{Bmatrix} \begin{Bmatrix} L_2 & S_2 & J_2 \\ J & J_1 & K \end{Bmatrix}, \tag{12.35}$$

$$\left(l_1^{N_1} l_2^{N_2} J_1' KJ | J_1 J_{12} J\right) = (-1)^{L_2 + S_2 - \bar{J}_2 - \overset{+}{J}_2} \delta(J_1, J_1') \sqrt{[J_{12}, K]}$$

$$\times (N_1 | N_1 - N_{1+}) \sum_{J_2} (2J_2 + 1) (N_2 | N_2 - N_{2+}) \begin{Bmatrix} J_1 & \bar{J}_2 & J_{12} \\ \overset{+}{J}_2 & J & J_2 \end{Bmatrix}$$

$$\times \begin{Bmatrix} L_2 & S_2 & J_2 \\ J & J_1 & K \end{Bmatrix}, \tag{12.36}$$

$$\left(l_1^{N_1} l_2^{N_2} J_1' J_2 J | J_1 J_{12} J\right) = (-1)^{\bar{J}_2 + \overset{+}{J}_2 + J_1 + J} \delta(J_1, J_1') \sqrt{2J_{12} + 1}$$

$$\times (N_1 | N_1 - N_{1+}) \sum_{J_2} \sqrt{2J_2 + 1} \, (N_2 | N_2 - N_{2+}) \begin{Bmatrix} J_1 & \bar{J}_2 & J_{12} \\ \overset{+}{J}_2 & J & J_2 \end{Bmatrix}. \tag{12.37}$$

All the presented transformation matrices obey orthogonality conditions of the type (12.13), (12.14). They may be utilized to transform the energy matrix, calculated in some coupling scheme, to other coupling schemes (trivial generalization of formula (12.20) for the case of a larger number of shells and other coupling schemes) with its successive diagonalization, or, more convenient and simple, for transformation of the weights of the wave functions from one coupling scheme to another (formula of the type (12.9)).

12.3   Optimization of the coupling scheme

The expressions for the transformation matrices presented above cover practically all necessary cases. However, their use is rather complicated. Therefore, for the most often considered configurations $l_1^{N_1} s$ and $l_1^{N_1} p$, one can start with somewhat simplified formulas for the corresponding matrices of the transformation to jj coupling presented in [94], or use general computer codes for the implementation of such transformations.

For the optimization of a coupling scheme, one may use the analysis of the relative contribution of separate interactions, the structure of the weights of the wave function of intermediate coupling (11.10), the qualitative picture of the spectrum (e.g., the grouping of the levels into multiplets) and the regularities in the isoelectronic sequences. However, the most efficient method is by use of the optimization procedure of the coupling scheme adopting the transformation matrices. It is worthwhile underlining the fact that for the correct and efficient classification of the measured energy spectra, collaboration of experimentalists and theoreticians is required, because only from theoretical calculations is it possible to find optimal sets of quantum numbers. This question is discussed in more detail in [93, 94].

Using the above formulas for the transformation matrices we can find, with the help of Eq. (12.19), expressions for the matrix elements in any coupling scheme and, after diagonalization of the total energy matrix, classify the energy levels with the aid of the coupling scheme closest to reality. However, the corresponding expressions, as a rule, are much more complicated than those in LS coupling, and, therefore, they are not very convenient for practical utilization (they contain $3nj$-coefficients of high orders, etc.). Thus the most efficient method is by calculations in one (usually LS) coupling scheme and the subsequent utilization of the transformation of the weights of the wave functions according to equation (12.9) or (12.10).

We conclude this chapter with the following remarks. The presence of JK coupling in ions having configurations $l^N l'$ shows that in shell l^N the relativistic effects dominate whereas the outer electron can be described fairly accurately with a non-relativistic wave function. Use of the transformation matrices from jj to other coupling schemes gives the possibility for studying many-electron atoms and ions (accounting for the main part of the relativistic effects), and for identifying and classifying the energy levels with the sets of quantum numbers of optimal coupling schemes.

Part 4

Second-quantization in the Theory of an Atom. Quasispin and Isospin

13

Second-quantization and irreducible tensorial sets

13.1 General remarks

As already mentioned in the Introduction, the exact solution of the main equation of quantum mechanics – the Schrödinger equation – lies beyond the potentialities of modern mathematics and computer technology. But a number of important inferences about the behaviour, structure and properties of a given quantum-mechanical many-particle system can be drawn without solving this equation, just by examining its symmetry properties.

Symmetry concepts come about in physics in two ways. First, since any physical process occurs in a real space, we have to make use of one or another coordinate system. The isotropy and homogeneity of space make physically meaningful only those mathematical relationships that remain unchanged under rotations of the axes of the coordinate system, and that impose fairly rigorous constraints on the possible physical laws. Second, every physical object and process features a symmetry which should be taken into account by the physical theory.

A common idea underlying particular forms of symmetry is the invariance of a system under a certain set (group) of transformations. The normally considered forms of symmetry are rotational symmetry, which is based on the equivalence of all directions in space, and permutation symmetry, which is caused by identical particles. The operations of the geometrical symmetry group are responsible for appropriate conservation laws. So, the rotational symmetry of a closed system gives rise to the law of conservation of angular momentum.

The group of rotations of a three-dimensional space stands apart in atomic spectroscopy. This is mostly due to the high accuracy of the central field approximation, on which the entire modern theory of complex atoms and ions is based.

In the central field approximation, the matrix elements of operators of physical quantities are generally found in terms of irreducible tensors of the group of three-dimensional rotations. There is one important factor contributing to the remarkable power and elegance of the mathematics of irreducible tensorial operators, and specifically of the quantum-mechanical theory of angular momentum. This factor is the commonness of the behaviour of the eigenfunctions of the squared operator of angular momentum, one of its components and irreducible tensorial operators with respect to the group of rotations of three-dimensional space. Therefore, they may be combined by introducing the concept of irreducible tensorial sets.

When constructing many-electron wave functions it is necessary to ensure their antisymmetry under permutation of any pair of coordinates. Having introduced the concepts of the CFP and unit tensors, Racah [22, 23] laid the foundations of the tensorial approach to the problem of constructing antisymmetric wave functions and finding matrix elements of operators corresponding to physical quantities.

Second-quantization formalism was introduced into the theory of many-electron atoms by Judd [12]. This formalism enables one to give a simple and elegant description of both the rotation symmetry of a system and its permutational symmetry: the tensorial properties of wave functions are translated to electron creation and annihilation operators, and the Pauli exclusion principle stems automatically from the anticommutation relations between these operators.

Moreover, the second-quantization technique is closely linked with the quasispin method, which turned out to be fruitful in atomic theory, since it enables physicists to improve significantly the theory of the spectra of many-electron atoms and ions, and to make it far more universal, and more simple and convenient for practical use.

The symmetry properties of the quantities used in the theory of complex atomic spectra made it possible to establish new important relationships and, in a number of cases, to simplify markedly the mathematical procedures and expressions, or, at least, to check the numerical results obtained. For one shell of equivalent electrons the best known property of this kind is the symmetry between the states belonging to partially and almost filled shells (complementary shells). Using the second-quantization and quasispin methods we can generalize these relationships and represent them as recurrence relations between respective quantities (CFP, matrix elements of irreducible tensors or operators of physical quantities) describing the configurations with different numbers of electrons but with the same sets of other quantum numbers. Another property of this kind is the symmetry of the quantities under transpositions of the quantum numbers of spin and quasispin.

The methods of theoretical description of many-electron atoms on the basis of tensorial properties of the orbital and spin angular momenta are well established [14, 18] and enable the spectral characteristics of these systems to be effectively found. The relation between the seniority quantum number and quasispin makes it possible to extend the mathematical tools to include the quasispin space and to work out new modifications of the mathematical techniques in the theory of spectra of many-electron atoms that take due account of the tensorial properties of the quasispin operator.

The account of the tensorial properties of the quasispin operator of one shell of equivalent electrons and the extension of that concept to several shells (by vectorial coupling of the quasispin momenta of each shell) provides us with valuable insights into the electron configurations using a method that is an alternative to the method of groups of higher ranks. In many respects this method is more convenient since it is based on the well-established mathematical technique of the theory of angular momentum and conserves essentially all of the quantum numbers available. Note that utilization of the concept of quasispin for several electron shells allows one to give a new treatment of configuration mixing in the atom.

The most effective way to find the matrix elements of the operators of physical quantities for many-electron configurations is the method of CFP. Their numerical values are generally tabulated. The methods of second-quantization and quasispin yield algebraic expressions for CFP, and hence for the matrix elements of the operators assigned to the physical quantities. These methods make it possible to establish the relationship between CFP and the submatrix elements of irreducible tensorial operators, and also to find new recurrence relations for each of the above-mentioned characteristics with respect to the seniority quantum number. The application of the Wigner–Eckart theorem in quasispin space enables new recurrence relations to be obtained for various quantities of the theory relative to the number of electrons in the configuration.

Thus, from these brief introductory remarks we can see that the second-quantization technique looks very promising in atomic spectroscopy. Therefore, let us discuss in this part all the above-mentioned questions in more detail. The discussion is mainly based on the monograph [31].

13.2 Second-quantization. Electron creation and annihilation operators

In the second-quantization method it is convenient to use the following designations of the N-electron function instead of (10.9), often omitting the subscripts of coordinates, if this does not lead to any confusion:

$$\left| \begin{matrix} \alpha_1, \alpha_2, \ldots, \alpha_N \\ (x_1, x_2, \ldots, x_N) \end{matrix} \right) = \left| \begin{matrix} \alpha_1, \alpha_2, \ldots, \alpha_N \\ (1, \ 2, \ldots, N) \end{matrix} \right) = |\alpha_1, \alpha_2, \ldots, \alpha_N). \tag{13.1}$$

Wave functions (13.1) form an orthonormal set, but their normalization factors are defined only up to a sign. The fact is that the wave function (13.1) is antisymmetric not only under coordinate permutations, but also under permutations of one-electron quantum numbers. Thus, to fix the sign of a wave function requires a way of ordering the set of quantum numbers $(\alpha) = \alpha_1, \alpha_2, \dots, \alpha_N$. There exists, however, a convenient formalism that allows us to include the constraints imposed by the requirement that the wave functions be antisymmetric in a simple operator form. This formalism became known as the second-quantization method. This chapter gives a detailed description of the fundamentals of the second-quantization method.

Let us define electron creation operator a_β as follows:

$$a_\beta \left| \begin{matrix} \alpha_1, \alpha_2, \dots, \alpha_N \\ (1, \ 2, \dots, N) \end{matrix} \right) = \left| \begin{matrix} \beta, \alpha_1, \dots, \alpha_N \\ 1, 2, \dots, N+1 \end{matrix} \right), \tag{13.2}$$

i.e. this is an operator that transforms the N-electron one-determinant wave function that does not contain the one-electron function φ_β, into the $(N+1)$-electron function that already includes function φ_β. If among quantum numbers $(\alpha) = \alpha_1, \dots, \alpha_N$ there is some number $\alpha_k = \beta$, then acting with a_β on wave function (10.9) gives, by (13.2), a determinant in which two lines will be the same and which then will be identically zero. Since any antisymmetric wave function can be expanded in terms of one-determinant wave functions, the creation operator is defined in the space of antisymmetric wave functions at all N.

The electron creation operator is introduced in such a way that the one-electron wave function produced by it would appear in the first line of the determinant with the same sign. Thereby the sign of the one-determinant wave function is uniquely specified also in the case where the resultant one-particle wave function is in any line of the determinant

$$a_{\alpha_k} |\alpha_1, \dots, \alpha_{k-1}, \alpha_{k+1}, \dots, \alpha_N) = (-1)^{k-1} |\alpha_1, \alpha_2, \dots, \alpha_N). \tag{13.3}$$

Let us look at operator $a^\dagger$, which is the Hermitian conjugate of the creation operator. Using the definition of the Hermitian conjugated operator

$$(\varphi_1 | A^\dagger | \varphi_2) = (\varphi_2 | A | \varphi_1)^* \tag{13.4}$$

(where the asterisk denotes the complex conjugation operation) and relationship (13.3), we shall transform the normalization integral as follows:

$$(\alpha_1, \dots, \alpha_N | \alpha_1, \dots, \alpha_N)$$
$$= (-1)^{k-1} (\alpha_1, \dots, \alpha_{k-1}, \alpha_{k+1}, \dots, \alpha_N | a^\dagger_{\alpha_k} | \alpha_1, \dots, \alpha_N). \tag{13.5}$$

Then we obtain

$$a^\dagger_{\alpha_k} |\alpha_1, \dots, \alpha_N) = (-1)^{k-1} |\alpha_1, \dots, \alpha_{k-1}, \alpha_{k+1}, \dots, \alpha_N), \tag{13.6}$$

i.e. $a^\dagger$ is the electron annihilation operator – it transforms an N-electron one-determinant wave function that contains one-electron wave function φ_α into an $(N-1)$-electron one-determinant function that no longer contains φ_α. If, however, the N-electron wave function does not have the appropriate state, then by definition

$$a^\dagger_\beta |\alpha_1, \ldots, \alpha_N) = 0 \quad \text{at} \quad \beta \neq \alpha_k \quad (k = 1, \ldots, N). \tag{13.7}$$

It should be stressed that in the literature one can come across a wide variety of notations for creation and annihilation operators. In this book we follow the authors [14, 95] who attach the sign of Hermitian conjugation to the electron annihilation operator, but not to the electron creation operator. Although the opposite notation is currently in wide use, it is inconvenient in the theory of the atom, since it is at variance with the common definitions of irreducible tensorial quantities.

13.3 Commutation relations. Occupation numbers

Let us now turn to the commutation relations between second-quantization operators. Acting in succession with two different creation operators on a one-determinant wave function, from (13.2), we get

$$a_{\beta_1} a_{\beta_2} |\alpha_1, \ldots, \alpha_N) = |\beta_1, \beta_2, \alpha_1, \ldots, \alpha_N),$$

where $\beta_1, \beta_2 \neq \alpha_k$ $(k = 1, \ldots, N)$. Acting with the same operators but now in the opposite sequence gives

$$a_{\beta_2} a_{\beta_1} |\alpha_1, \ldots, \alpha_N) = |\beta_2, \beta_1, \alpha_1, \ldots, \alpha_N).$$

Since these two equations hold for any one-determinant wave function, and the functions on the right side of these equations only differ in sign, we arrive at the following anticommutation relation for the creation operators:

$$\{a_{\beta_1}, a_{\beta_2}\} = a_{\beta_1} a_{\beta_2} + a_{\beta_2} a_{\beta_1} = 0. \tag{13.8}$$

Subjecting this to Hermitian conjugation, we find that the same anticommutation relation is also obeyed by the annihilation operators

$$\{a^\dagger_{\beta_1}, a^\dagger_{\beta_2}\} = a^\dagger_{\beta_1} a^\dagger_{\beta_2} + a^\dagger_{\beta_2} a^\dagger_{\beta_1} = 0. \tag{13.9}$$

From these anticommutation relations, specifically, we obtain the relation

$$a_\beta a_\beta = a^\dagger_\beta a^\dagger_\beta = 0, \tag{13.10}$$

which states that any expression is zero if it contains two identical annihilation or creation operators that stand side by side. In a similar way, we can obtain the anticommutation relations between the creation and annihilation operators. When the subscripts of these operators do not coincide, we have the anticommutation relations of the form (13.8)

$$\left\{a^\dagger_{\beta_1}, a_{\beta_2}\right\} = a^\dagger_{\beta_1} a_{\beta_2} + a_{\beta_2} a^\dagger_{\beta_1} = 0 \quad (\beta_1 \neq \beta_2). \qquad (13.11)$$

When the subscripts of creation and annihilation operators coincide, from (13.3) and (13.6), we find

$$a^\dagger_{\alpha_k} a_{\alpha_k} |\alpha_1, \ldots, \alpha_k, \alpha_{k+1}, \ldots, \alpha_N) = 0;$$

$$a^\dagger_\beta a_\beta |\alpha_1, \ldots, \alpha_N) = |\alpha_1, \ldots, \alpha_N); \quad \text{at } \beta \neq \alpha_k \quad (k = 1, \ldots, N).$$

These two equations can be merged by introducing occupation numbers n_β of one-particle states β; $n_\beta = 1$, if the appropriate state in the one-determinant wave function is occupied; and $n_\beta = 0$, if it is vacant. For any β then we shall obtain

$$a^\dagger_\beta a_\beta |\alpha_1, \ldots, \alpha_N) = (1 - n_\beta)|\alpha_1, \ldots, \alpha_N). \qquad (13.12)$$

Acting in the opposite direction with the same operators gives

$$a_\beta a^\dagger_\beta |\alpha_1, \ldots, \alpha_N) = n_\beta |\alpha_1, \ldots, \alpha_N). \qquad (13.13)$$

It follows from (13.12) and (13.13) that the following operator relation holds:

$$\left\{a_\beta, a^\dagger_\beta\right\} = a_\beta a^\dagger_\beta + a^\dagger_\beta a_\beta = 1. \qquad (13.14)$$

The above anticommutation relations for second-quantization operators have been derived using the symmetry properties of one-determinant wave functions with relation to the permutation of the coordinates of particles. Since the second-quantization operators are only defined in the space of antisymmetric wave functions, the reverse statement is true – in second-quantization formalism the permutative symmetry properties of wave functions automatically follow from the anticommutation relations for creation and annihilation operators. We shall write these relations together in the form

$$\{a_\alpha, a_\beta\} = \left\{a^\dagger_\alpha, a^\dagger_\beta\right\} = 0; \quad \left\{a_\alpha, a^\dagger_\beta\right\} = \delta(\alpha, \beta). \qquad (13.15)$$

13.4 Operators and wave functions in second-quantization representation

Many-electron wave functions in second-quantization form can conveniently be represented in an operator form. To this end, we shall introduce the vacuum state $|0)$, i.e. the state in which there are no particles. We shall define it by

$$|0) = a^\dagger_\alpha |\alpha) \qquad (13.16)$$

for any one-electron function $|\alpha\rangle$. Anticommutation relations (13.15) imply the simple properties of the vacuum

$$a_\alpha^\dagger|0\rangle = 0; \qquad a_\alpha|0\rangle = |\alpha\rangle; \qquad (0|0) = 1. \tag{13.17}$$

In principle, an N-electron wave function can always be represented as a certain combination of creation operators acting on the vacuum state. Specifically, for the one-determinant wave function (13.1) we have

$$|\alpha_1, \alpha_2, \ldots, \alpha_N\rangle = \hat{\varphi}(\alpha_1, \alpha_2, \ldots, \alpha_N)|0\rangle, \tag{13.18}$$

where

$$\hat{\varphi}(\alpha_1, \alpha_2, \ldots, \alpha_N) = a_{\alpha_1} a_{\alpha_2} \ldots a_{\alpha_N}. \tag{13.19}$$

We shall use operators with the sign '^' only when the same symbols are used both for operator and non-operator quantities.

Operators corresponding to physical quantities, in second-quantization representation, are written in a very simple form. In the quantum mechanics of identical particles we normally have to deal with two types of operators symmetric in the coordinates of all particles. The first type includes N-particle operators that are the sum of one-particle operators. An example of such an operator is the Hamiltonian of a system of non-interacting electrons (e.g. the first two terms in (1.15)). The second type are N-particle operators that are the sum of two-particle operators (e.g. the energy operator for the electrostatic interaction of electrons – the last term in (1.15)). In conventional representations these operators are

$$F = \sum_j f(x_j) \qquad (j = 1, \ldots, N); \tag{13.20}$$

$$G = \sum_{i<j} g(x_i, x_j) \qquad (i, j = 1, \ldots, N). \tag{13.21}$$

In second-quantization formalism, operators in coordinate space are replaced by operators defined in the space of occupation numbers

$$F = \sum_{(\alpha)} (\alpha_j|f|\alpha_k) a_{\alpha_j} a_{\alpha_k}^\dagger; \tag{13.22}$$

$$G = \frac{1}{2} \sum_{(\alpha)} (\alpha_j\alpha_k|g|\alpha_l\alpha_m) a_{\alpha_j} a_{\alpha_k} a_{\alpha_m}^\dagger a_{\alpha_l}^\dagger, \tag{13.23}$$

where

$$(\alpha_j|f|\alpha_k) = \int \varphi_{\alpha_j}^*(x) f(x) \varphi_{\alpha_k}(x) dx; \tag{13.24}$$

$$(\alpha_j\alpha_k|g|\alpha_l\alpha_m) = \iint \varphi_{\alpha_j}^*(x_1) \varphi_{\alpha_k}^*(x_2) g(x_1, x_2) \varphi_{\alpha_l}(x_1)$$
$$\times \varphi_{\alpha_m}(x_2) dx_1 dx_2 \tag{13.25}$$

are respectively the one-electron and two-electron matrix elements of the operators in question. The equivalence of the conventional (coordinate) and second-quantization forms of operators F and G is readily verified by comparing their matrix elements defined in the basis of one-determinant wave functions.

Equations (13.22) and (13.23) define the second-quantization form of an operator corresponding to a physical quantity, if its matrix elements are known in coordinate representations ((13.24) and (13.25)). Specifically, the operator of the total number of particles in a system will be

$$\hat{N} = \sum_\alpha a_\alpha a_\alpha^\dagger = \sum_\alpha \hat{n}_\alpha, \tag{13.26}$$

where

$$\hat{n}_\alpha = a_\alpha a_\alpha^\dagger \tag{13.27}$$

is the operator of the occupation number for one-electron state α. It follows from (13.13) that its eigenvalues are occupation numbers of one-electron states in the one-determinant wave function, i.e. zero or unity. Its square is

$$\hat{n}_\alpha^2 = \hat{n}_\alpha \hat{n}_\alpha = \hat{n}_\alpha. \tag{13.28}$$

The commutation relations are

$$[\hat{n}_\alpha, a_\alpha] = \hat{n}_\alpha a_\alpha - a_\alpha \hat{n}_\alpha = a_\alpha, \tag{13.29}$$

$$\left[\hat{n}_\alpha, a_\alpha^\dagger\right] = \hat{n}_\alpha a_\alpha^\dagger - a_\alpha^\dagger \hat{n}_\alpha = -a_\alpha^\dagger. \tag{13.30}$$

Linear combinations of one-determinant wave functions, in the general case, are no longer eigenfunctions of operators of one-particle occupation numbers. The concept of an exact one-electron state for such wave functions becomes pointless, since in the general case we can only indicate the probability of occupation ω $(0 \le \omega \le 1)$ of a one-electron state.

Generally speaking, the representation in terms of occupation numbers is considered to be an independent quantum-mechanical representation, distinct from the coordinate (or momentum) one. In that case, the occupation numbers for one-particle states are dynamic variables, and operators are the quantities that act on functions of these variables. In this section, second-quantization representation is directly related to coordinate representation in order that in what follows we may have a one-to-one correspondence between quantities derived in each of these representations.

13.5 Particle–hole representation

Apart from the occupation number operators for one-particle states we can introduce the operator

$$_h\hat{n}_\alpha = a_\alpha^\dagger a_\alpha. \tag{13.31}$$

According to (13.12), the eigenvalues of this operator will also be zero and unity, however, unity is now an eigenvalue of the one-determinant wave function in which this one-electron state α is vacant; and zero is an eigenvalue of a function for which this state is filled. Thus, quantity $_h\hat{n}_\alpha$ is the occupation number operator for the hole state α.

For commutation relations of this operator with electron creation and annihilation operators, instead of (13.29) and (13.30), we obtain

$$\left[_h\hat{n}_\alpha, a_\alpha\right] = -a_\alpha; \tag{13.32}$$

$$\left[_h\hat{n}_\alpha, a_\alpha^\dagger\right] = a_\alpha^\dagger. \tag{13.33}$$

For some one-determinant state $|\alpha_1, \ldots, \alpha_N)$ we can completely change over from particle description to hole description if, instead of electron creation and annihilation operators, we introduce, respectively, annihilation and creation operators for holes

$$b_\alpha = a_\alpha^\dagger; \tag{13.34}$$

$$b_\alpha^\dagger = a_\alpha. \tag{13.35}$$

State $|\alpha_1, \alpha_2, \ldots, \alpha_N)$ will be the vacuum state for holes

$$b_\alpha^\dagger |\alpha_1, \alpha_2, \ldots, \alpha_N) = b_{\alpha_i}^\dagger |_h0) = 0 \qquad (i = 1, 2, \ldots, N). \tag{13.36}$$

Wave function $|_h0)$ is also a vacuum state for the particles, which are not represented in one-determinant wave function $|\alpha_1, \alpha_2, \ldots, \alpha_N)$, i.e.

$$a_\beta^\dagger |\alpha_1, \alpha_2, \ldots, \alpha_N) = a_\beta^\dagger |_h0) = 0 \qquad (\beta \neq \alpha_k). \tag{13.37}$$

The description of a system of identical particles in terms of particles and holes has come to be known as the particle–hole representation. This representation appears to be very convenient when the ground state of a system of N non-interacting electrons can be described in a zero approximation by the one-determinant wave function. This description is used, in particular, in the theory of solid state and Fermi-gas. The maximal energy ϵ_F of one-particle levels occupied in the ground state is called the Fermi energy, and so vacuum $|_h0)$ describes the ground state in which all the levels with energy $\epsilon_\alpha \leq \epsilon_F$ are occupied, and the levels with energy $\epsilon_\beta > \epsilon_F$ are vacant. An excitation act will now correspond to the production of one or more particle–hole pairs; whereas the reverse process of transition to a lower state in terms of energy – to the 'annihilation' of these pairs.

If we confine ourselves to this method of describing the behaviour of a many-particle system, the problem of N interacting particles comes down to the model of several quasiparticles – particle–hole pairs. Then the detailed properties of individual particles occupying one-particle states with energy $\epsilon \leq \epsilon_F$ and forming a vacuum are ignored. Of course, this treatment is approximate. The true ground state of an N-electron system cannot be described by a one-determinant wave function, since it differs in principle from the ground state of a system of non-interacting electrons. Put another way, as a result of the interaction the vacuum state is rearranged, and this effect of vacuum rearrangement must be considered when determining the physical behaviour of the system in more exact models.

In the theory of many-electron atoms, the particle–hole representation is normally used to describe atoms with filled shells. To the ground state of such systems there corresponds a single determinant, composed of one-electron wave functions defined in a certain approximation. This determinant is now defined as the vacuum state. In the case of atoms with unfilled shells, this representation can be used for the atomic 'core' consisting only of filled shells. Then, the excitation of electrons from these shells will be described as the creation of particle–hole pairs.

The utilization of the particle–hole representation to describe the wave functions of unfilled shells yields no substantial simplifications. As has been noted above, these wave functions can be expressed in terms of linear combinations of one-determinant wave functions. Selection of one of these determinants as the vacuum state yields a linear combination of one-determinant wave functions, now described using the particle–hole representation.

Sometimes for unfilled atomic shells, too, vacuum is defined in terms of the wave function of completely filled shells. This turns out to be convenient when the number of holes that describe the wave function of electrons in an unfilled shell in the new representation is smaller than the number of electrons for this wave function.

13.6 Phase system

We were talking about phase systems in the conclusion of the Introduction (formulas (4) and (5)). Let us discuss this question here in more detail. Conjugate wave function $(jm|$ is transformed as follows:

$$(jm'| = \sum_m D^{(j)^*}_{m',m} (jm|,$$

i.e. its components do not form an irreducible tensor. If, however, we take into account the condition of the complex conjugation of the generalized spherical function (2.21), then we can introduce the quantities

$$\tilde{\Psi}_m^{(j)} = (j\tilde{m}| = (-1)^{\eta(j)-m}(j-m|, \tag{13.38}$$

which, under rotations, are transformed by (5.9). Phase factor $\eta(j)$ may be chosen in an arbitrary way. Let $\eta(j) = j$, then

$$(j\tilde{m}| = (-1)^{j-m}(j-m|. \tag{13.39}$$

Note, that this phase factor differs by $(-1)^{2j}$ from the phase factor that defines the standard phase system (Eq. (4) in the Introduction). This difference manifests itself only at half-integer values of momentum j. This change in the definition of the standard phase system is caused by the presence of commonly accepted phase relations in the second-quantization method [12, 96].

The set of $(2k+1)$ Hermitian conjugated operators $T_p^{(k)}$ does not form an irreducible tensor, but from (2.21) we can easily see that the operators

$$\tilde{T}_p^{(k)} = (-1)^{\eta(k)-p} T_{-p}^{(k)\dagger}$$

are transformed according to (5.9). Selecting, as in the case of wave functions, $\eta(k) = k$, we arrive at the irreducible tensorial operators

$$\tilde{T}_p^{(k)} = (-1)^{k-p} T_{-p}^{(k)\dagger}. \tag{13.40}$$

If $\tilde{T}_{-p}^{(k)}$ is an operator transposed in relation to $T_p^{(k)}$, i.e. if

$$\int \psi^* T_p^{(k)} \varphi d\tau = \int \varphi \tilde{T}_{-p}^{(k)} \psi^* d\tau,$$

then (13.40) defines the operation of complex conjugation of operators, since then (see also (4) of the Introduction)

$$T_p^{(k)\dagger} = (\tilde{T}_p^{(k)})^* = (-1)^{k+p} \tilde{T}_{-p}^{(k)}. \tag{13.41}$$

According to [11], in a standard phase system operators are defined that obey

$$T_p^{(k)\dagger} = (-1)^{k\pm p} T_{-p}^{(k)} \tag{13.42}$$

(the sign $\pm$ is meant to generalize this definition to tensors with half-integer ranks). It is to be noted that (13.42) does not always hold. It can hold, for example, for irreducible tensors, that have Hermitian Cartesian components. In general, a product of Hermitian operators is not a Hermitian operator. Therefore, operators with a complex internal structure and composed of tensors that are in agreement with phase system (13.42), will no longer be defined in the same phase system.

Apart from phase system (13.42) some of the tensorial operators (e.g. unit tensors and their sums (5.27), (5.28)) are defined in a pseudostandard phase system (see Introduction, Eq. (5)), for which

$$T_p^{k\dagger} = (-1)^p T_{-p}^k. \tag{13.43}$$

Ranks of such tensorial operators are not confined in parentheses, for the notation itself is used to indicate the phase system. Equation (13.43) does not hold in general, for the same reasons as (13.42).

In conclusion, it is only necessary that operators obey (13.40); if in addition in certain special cases expression $\tilde{T}_{-p}^{(k)} = (-1)^{p\mp p} T_{-p}^{(k)}$ is valid, then (13.42) is valid too, and if the expression $\tilde{T}_{-p}^{(k)} = (-1)^{-k} T_{-p}^k$ holds, then (13.43) also holds.

14
Operators and matrix elements in second-quantization representation

14.1 Irreducible tensors in the space of occupation numbers. One-particle states

The mathematical apparatus of the angular momentum theory can be applied to describe the tensorial properties of electron creation and annihilation operators in the space of occupation numbers of a certain definite one-particle state $|\alpha\rangle$. It follows from (13.29) and (13.30) that the operators

$$u_1^{(1)} = -i\sqrt{2}a_\alpha; \quad u_{-1}^{(1)} = i\sqrt{2}a_\alpha^\dagger; \quad u_0^{(1)} = i(\hat{n}_\alpha - 1/2) \tag{14.1}$$

meet the conventional commutation relations

$$\left[J_\rho^{(1)}, J_{\rho'}^{(1)} \right] = i\sqrt{2} \begin{bmatrix} 1 & 1 & 1 \\ \rho & \rho' & \rho+\rho' \end{bmatrix} J_{\rho+\rho'}^{(1)}. \tag{14.2}$$

The eigenfunctions of operator $u_z = \hat{n}_\alpha - 1/2$ are the one-particle and vacuum states

$$u_z|\alpha\rangle = \frac{1}{2}|\alpha\rangle; \quad u_z|0\rangle = -\frac{1}{2}|0\rangle, \tag{14.3}$$

i.e. these two wave functions form an irreducible tensor of rank $1/2$ in the space of occupation numbers of one-electron state α:

$$|\alpha\rangle = \begin{vmatrix} \frac{1}{2} & \frac{1}{2} \end{vmatrix}; \quad |0\rangle = \begin{vmatrix} \frac{1}{2} & -\frac{1}{2} \end{vmatrix}. \tag{14.4}$$

The expressions that define the action of second-quantization operators $a_\alpha^\dagger$ and a_α on wave functions $|\alpha\rangle$ and $|0\rangle$ can be presented in terms of the conventional relation

$$J_\rho^{(1)}|jm\rangle = \sum_{m'} |jm'\rangle\langle jm'|J_\rho^{(1)}|jm\rangle = i[j(j+1)]^{1/2} \begin{bmatrix} j & 1 & j \\ m & \rho & m+\rho \end{bmatrix} |jm+\rho\rangle, \tag{14.5}$$

or in irreducible tensorial form

$$\left[J^{(1)} \times |j)\right]_{\Pi}^{(K)} = -i[j(j+1)]^{1/2}\delta(K,j)|j\Pi).\qquad(14.6)$$

The finite rotations in the space of occupation numbers for the one-electron state define the transition to the quasiparticle wave functions

$$|\alpha') = D_{1/2,1/2}^{(1/2)}\left|\frac{1}{2}\ \ \frac{1}{2}\right) + D_{1/2,-1/2}^{(1/2)}\left|\frac{1}{2}\ -\frac{1}{2}\right),$$

$$|0') = D_{-1/2,1/2}^{(1/2)}\left|\frac{1}{2}\ \ \frac{1}{2}\right) + D_{-1/2,-1/2}^{(1/2)}\left|\frac{1}{2}\ -\frac{1}{2}\right),$$

that represent linear combinations of the one-particle and vacuum states. These functions are no longer eigenfunctions of occupation number operator $\hat{n}_\alpha$. However, if together with the above-mentioned transformation of the wave functions, we accomplish unitary transformation (5.9), which yields creation and annihilation operators $A_{\alpha'}$ and $A_{\alpha'}^\dagger$ for the quasiparticle, then the occupation number operator for the one-quasiparticle state will be

$$\hat{n}_{\alpha'} = A_{\alpha'}A_{\alpha'}^\dagger,\qquad(14.7)$$

whereas $|\alpha')$ and $|0')$ will be the eigenfunctions

$$\hat{n}_{\alpha'}|\alpha') = |\alpha');\qquad \hat{n}_{\alpha'}|0') = 0.\qquad(14.8)$$

Again, the change-over from the particle to the hole treatment can be represented in this case by a special kind of the aforementioned transformations. Indeed, by selecting in the D-matrix the values of parameters defining the rotation through angle π about the Oy-axis, we obtain

$$|\alpha') = |0);\qquad |0') = -|\alpha).\qquad(14.9)$$

The pertinent transformation of the second-quantization operators now yields creation and annihilation operators for holes (13.34) and (13.35).

14.2 Second-quantization operators as irreducible tensors. Tensorial properties of electron creation and annihilation operators

In the second-quantization representation the atomic interaction operators are given by relations (13.22) and (13.23), which do not include the operators themselves in coordinate representations, but rather their one-electron and two-electron matrix elements. Therefore, in terms of irreducible tensors in orbital and spin spaces, we must expand the products of creation and annihilation operators that enter (13.22) and (13.23). In this approach, the tensorial properties of one-electron wave functions are translated to second-quantization operators.

We shall confine ourselves to the case of one shell of equivalent electrons. In the central field approximation, for LS coupling the states of the l^N configuration are characterized by orbital and spin momenta. In fact, using (13.15), (5.15) and the condition

$$(j\|J^{(1)}\|j) = i[j(j+1)(2j+1)]^{1/2}, \qquad (14.10)$$

we can show that the commutation relations for electron creation operators with irreducible components of the operators of the orbital

$$L_\rho^{(1)} = \sum_{mm'\mu\mu'} \left(nlm\mu|L_\rho^{(1)}|nlm'\mu'\right) a_{m\mu}^{(ls)} a_{m'\mu'}^{(ls)\dagger} \qquad (14.11)$$

and spin

$$S_\rho^{(1)} = \sum_{m\mu m'\mu'} \left(nlm\mu|S_\rho^{(1)}|nlm'\mu'\right) a_{m\mu}^{(ls)} a_{m'\mu'}^{(ls)\dagger} \qquad (14.12)$$

momenta of one shell can be given by

$$\left[J_{\pm 1}^{(1)}, T_p^{(k)}\right] = \mp i[(k \pm p + 1)(k \mp p)/2]^{1/2} T_{p\pm 1}^{(k)};$$

$$\left[J_0^{(1)}, T_p^{(k)}\right] = ip T_p^{(k)}. \qquad (14.13)$$

Unlike general formula (13.22), the summation in (14.11) and (14.12) is only over the momentum projections, since in this section all the operators under consideration are defined in the subspace of the states of one shell.

Electron annihilation operators, as Hermitian conjugates of creation operators, are no longer the components of an irreducible tensor. According to (13.40), such a tensor is formed by $(2l+1)(2s+1)$ components of the operator

$$\tilde{a}_{m\mu}^{(ls)} = (-1)^{l+s-m-\mu} a_{-m-\mu}^{(ls)\dagger}. \qquad (14.14)$$

Utilization of the tensorial properties of the electron creation and annihilation operators allows us to obtain expansions in terms of irreducible tensors of any operators in the second-quantization representation. So, using the Wigner–Eckart theorem (5.15) in (14.11) and (14.12), then coupling ranks of second-quantization operators by (5.12) and utilizing (14.10), we can represent one-shell operators of angular momentum in the irreducible tensor form

$$L_\rho^{(1)} = -i\sqrt{l(l+1)[l,s]/3} \left[a^{(ls)} \times \tilde{a}^{(ls)}\right]_{\rho 0}^{(10)}; \qquad (14.15)$$

$$S_\rho^{(1)} = -i\sqrt{s(s+1)[l,s]/3} \left[a^{(ls)} \times \tilde{a}^{(ls)}\right]_{0\rho}^{(01)}. \qquad (14.16)$$

The particle number operator for a shell will be

$$\hat{N} = -(4l+2)^{1/2} \left[a^{(ls)} \times \tilde{a}^{(ls)}\right]^{(00)}, \qquad (14.17)$$

and hole number operator $_h\hat{N}$

$$_h\hat{N} = (4l + 2)^{1/2} \left[\tilde{a}^{(ls)} \times a^{(ls)} \right]^{(00)}. \tag{14.18}$$

The anticommutation relations between the electron creation and annihilation operators, accounting for (14.14), become

$$\left\{ a_{m\mu}^{(ls)}, \tilde{a}_{m'\mu'}^{(ls)} \right\} = (-1)^{l+s-m'-\mu'} \delta(m\mu, -m' - \mu'). \tag{14.19}$$

Writing (14.19) as

$$\left\{ a_{m\mu}^{(ls)}, \tilde{a}_{m'\mu'}^{(ls)} \right\} = -\sqrt{4l+2} \begin{bmatrix} l & l & 0 \\ m & m' & 0 \end{bmatrix} \begin{bmatrix} s & s & 0 \\ \mu & \mu' & 0 \end{bmatrix}, \tag{14.20}$$

then multiplying both sides by the Clebsch–Gordan coefficients and summing over the projections using (8.6) and (5.12), we arrive at the commutation rule for operators $a^{(ls)}$ and $\tilde{a}^{(ls)}$ in the irreducible tensorial product

$$\left[a^{(ls)} \times \tilde{a}^{(ls)} \right]^{(kK)} + (-1)^{2l+2s-k-K} \left[\tilde{a}^{(ls)} \times a^{(ls)} \right]^{(kK)}$$
$$= -\sqrt{4l + 2}\delta(kK, 00). \tag{14.21}$$

With $k = K = 0$ (14.21) transforms into the obvious relationship

$$\hat{N} + {}_h\hat{N} = 4l + 2.$$

Using second-quantization, it is often necessary to transform complicated tensorial products of creation and annihilation operators. If, to this end, conventional anticommutation relations (14.19) are used, then one proceeds as follows: write the irreducible tensorial products in explicit form in terms of the sum over the projection parameters of conventional products of creation and annihilation operators, then place these operators in the required order, and finally sum the resultant expression again over the projection parameters. On the other hand, the use of (14.21) enables the irreducible tensorial products of second-quantization operators to be transformed directly.

14.3   Unit tensors

If we take into account the tensorial properties of creation and annihilation operators, using (13.22) we shall be able to expand in terms of the irreducible tensors

$$\left[a^{(ls)} \times \tilde{a}^{(ls)} \right]^{(kK)} \tag{14.22}$$

any one-particle operator F given by (13.22). The theory of atomic spectra draws heavily on unit tensorial operators. Different authors define these operators in different ways. So, according to Racah [22], unit tensorial operator u^k is defined in terms of its one-electron reduced matrix element (5.23), (5.24). Then for its sum U^k (5.27) in the second-quantization representation we shall have

$$U^k = -\left[\frac{2s+1}{2k+1}\right]^{1/2} \left[a^{(ls)} \times \tilde{a}^{(ls)}\right]^{(k0)} \qquad (k = 1, 2, \ldots, 2l); \quad (14.23)$$

$$U^0 = -(4l+2)^{1/2} \left[a^{(ls)} \times \tilde{a}^{(ls)}\right]^{(00)} = \hat{N}.$$

Comparison of (14.23) and (14.15) suggests that

$$L_\rho^{(1)} = i[l(l+1)(2l+1)]^{1/2} U_\rho^{(1)}. \tag{14.24}$$

For another unit tensorial operator V^{k1} (5.28) in the second-quantization representation we have

$$V^{k1} = -\frac{1}{\sqrt{2(2k+1)}} \left[a^{(ls)} \times \tilde{a}^{(ls)}\right]^{(k1)} \qquad (k = 1, 2, \ldots, 2l);$$

$$V^{01} = -\sqrt{(2l+1)/2} \left[a^{(ls)} \times \tilde{a}^{(ls)}\right]^{(01)} = -iS^{(1)}. \tag{14.25}$$

If, according to [14], we introduce unit tensorial operator w'^{kK}, the submatrix element of which is

$$(ls\|w'^{kK}\|ls) = [l, s]^{1/2}, \tag{14.26}$$

then

$$W'^{kK} = -\sqrt{[l, s]/[k, K]} \left[a^{(ls)} \times \tilde{a}^{(ls)}\right]^{(kK)}. \tag{14.27}$$

The ranks of all the above operators are not put in brackets in order to indicate explicitly that for these operators relationships (13.43) hold.

Equations of the kind (14.23) or (14.25) are inconsistent in the way that tensors in their left and right sides are defined in different (pseudostandard and standard) phase systems. This is done to underline that historically tensors composed of unit tensors were defined in pseudostandard phase systems whereas in this book main tensorial quantities obey standard phase systems.

Judd [10, 12] introduced tensorial operator $w^{(kK)}$, the submatrix element of which is

$$(ls\|w^{(kK)}\|ls) = [k, K]^{1/2}. \tag{14.28}$$

The second-quantization form of the operator

$$W^{(kK)} = \sum_{i=1}^{N} w_i^{(kK)} \tag{14.29}$$

will then be

$$W^{(kK)} = -\left[a^{(ls)} \times \tilde{a}^{(ls)}\right]^{(kK)}. \tag{14.30}$$

This differs only by sign from irreducible tensorial product (14.22) which makes especially convenient the use of tensor $W^{(kK)}$ in the second-quantization method. By its definition, and by (14.15)–(14.17), (14.23) and (14.25), we can always go from (14.30) to other unit tensors defined above:

$$W^{(00)} = (4l+2)^{-1/2}\hat{N}; \tag{14.31}$$

$$W^{(10)}_{\rho 0} = -i\sqrt{\frac{3}{l(l+1)[l,s]}}L^{(1)}_\rho; \tag{14.32}$$

$$W^{(01)}_{0\rho} = -i\sqrt{\frac{2}{2l+1}}S^{(1)}_\rho; \tag{14.33}$$

$$W^{(k0)}_{\rho 0} = \sqrt{(2k+1)/2}U^k_\rho \qquad (k=1,2,\dots,2l); \tag{14.34}$$

$$W^{(k1)}_{\rho \pi} = \sqrt{2(2k+1)}V^{k1}_{\rho \pi} \qquad (k=1,2,\dots,2l). \tag{14.35}$$

This can conveniently be done after the final expressions for the operators of various physical quantities have been derived, since it is for operators U^k and V^{k1} that detailed tables [87, 97] of their submatrix elements are available.

The number of tensorial operators $W^{(kK)}_{\rho\pi}$ with different possible values of ranks k, K and appropriate values of the projections is predetermined by the number of possible projections of the second-quantization operators that enter into the tensorial product. This number is equal to $(4l+2)^2$.

Unit tensors are especially important for group-theoretical methods of studying the l^N configuration. We can express the infinitesimal operators of the groups [10, 24, 98], the parameters of irreducible representations of which are applied to achieve an additional classification of states of a shell of equivalent electrons, in terms of them.

14.4   Group-theoretical methods of classification of the states of a shell of equivalent electrons. Casimir operators

For the p-shell, quantum numbers LSM_LM_S completely classify the possible antisymmetrical N-electron states. Beginning with the d-shell, these quantum numbers are no longer sufficient. Thus, we are faced with the problem of classification of degenerate terms. The simplest example is the d^3 configuration where there are two ^2D terms (see Chapter 9).

The treatment in this book relies on the techniques of angular momentum theory. The author has sought, as far as possible, to do without higher-rank groups. Therefore, we shall only give here a glimpse of those methods, so as to be able to show later on how and to what degree they can be replaced by the techniques of angular momentum theory in other spaces.

Rearranging the electron creation and annihilation operators, we can obtain for tensors (14.30) the following commutation relation:

$$
\left[W^{(kK)}_{\rho\pi}, W^{(k'K')}_{\rho'\pi'} \right] = - \sum_{k''K''\rho''\pi''} [k,k',K,K']^{1/2}
$$

$$
\times \left\{ (-1)^{k+k'+K+K'} - (-1)^{k''+K''} \right\}
\begin{bmatrix} k & k' & k'' \\ \rho & \rho' & \rho'' \end{bmatrix}
\begin{bmatrix} K & K' & K'' \\ \pi & \pi' & \pi'' \end{bmatrix}
$$

$$
\times \begin{Bmatrix} k & k' & k'' \\ l & l & l \end{Bmatrix}
\begin{Bmatrix} K & K' & K'' \\ s & s & s \end{Bmatrix} W^{(k''K'')}_{\rho''\pi''},
\tag{14.36}
$$

which shows that the set of $(4l + 2)^2$ operators $W^{(kK)}_{\rho\pi}$ is complete in relation to the commutation operation. Then [10, 98] these operators are the generators of unitary group U_{4l+2}, a group of unitary matrices of order $(4l + 2)$. Operator (14.31) commutes with all other tensors (14.30). If we eliminate this operator from consideration, then the remaining $(4l + 3)(4l + 1)$ operators will, as is clearly seen from (14.36), also form a set that is complete in relation to the commutation operation and will, thus, represent generators of SU_{4l+2} group (the group of unitary unimodular matrices of order $4l + 2$).

Commutation relation (14.36) can be represented as

$$
\left[W^{(kK)} \times W^{(k'K')} \right]^{(k''K'')} - (-1)^{k+k'-k''+K+K'-K''}
$$

$$
\times \left[W^{(k'K')} \times W^{(kK)} \right]^{(k''K'')}
$$

$$
= - \left[(-1)^{k+k'+K+K'} - (-1)^{k''+K''} \right] [k,k',K,K']^{1/2}
$$

$$
\times \begin{Bmatrix} k & k' & k'' \\ l & l & l \end{Bmatrix}
\begin{Bmatrix} K & K' & K'' \\ s & s & s \end{Bmatrix} W^{(k''K'')}.
\tag{14.37}
$$

The characteristics of the irreducible representations of the U_{4l+2} group cannot be used for additional classification of degenerate terms, since all the possible wave functions of a shell with a fixed number of equivalent electrons N belong [99] to a single representation of this group, denoted as $\{l^N\}$. To obtain other classifying symbols it is necessary in some way to select the chain of eigensubgroups of U_{4l+2} group so that the characteristics of the irreducible representations of these subgroups could be used to classify the states of the atomic shell. As it has turned out [100, 101] the only possible group chains that conserve quantum numbers L and S are the reduction schemes

$$
U_{4l+2}
\begin{array}{c} \nearrow \\ \searrow \end{array}
\begin{array}{c} Sp_{4l+2} \\ \\ SU^S_2 \times SU_{2l+1} \end{array}
\begin{array}{c} \searrow \\ \nearrow \end{array}
SU^S_2 \times (R_{2l+1} \to R^L_3).
\tag{14.38}
$$

In the special case of f-electrons we have additionally the group chain

$$R_7 \rightarrow G_2 \rightarrow R_3^L. \tag{14.39}$$

In (14.38) we have denoted by Sp_{4l+2} the symplectic group of order $(4l + 2)$; by R_{2l+1} and SU_{2l+1}, the orthogonal and unitary unimodular groups of order $(2l + 1)$, respectively. R_3^L and SU_2^S groups are generated by the operators of orbital (14.15) and spin (14.16) angular momenta of electrons in a shell. The generators of Sp_{4l+2}, SU_{2l+1} and R_{2l+1} groups are some subsets of operators $W^{(kK)}$ that are complete relative to the commutation operation. In the case of Sp_{4l+2} groups, the generators will be tensors $W^{(kK)}$ with an odd sum of ranks [10, 98]. Indeed, if in relationship (14.36) rank sums $k + K$ and $k' + K'$ are odd, then sum $k'' + K''$ must also be odd due to the phase factor on the right side of the same relationship. In a similar manner we can show that operators $W^{(k0)}$ $(1 \le k \le 2l)$ are the generators of SU_{2l+1} group, and their subset – tensors $W^{(k0)}$ with odd values of rank $k(k = 1, 3, \ldots, 2l + 1)$ – are the generators of R_{2l+1} group. With f-electrons the generators of G_2 group are the fourteen operators $W_{\rho 0}^{(50)}$ and $W_{\rho 0}^{(10)}$.

Irreducible representations of various groups are generally classified using the eigenvalues of the corresponding Casimir operators [10, 24, 98]. These operators, which commute with all the generators of the group, are the generalization of operator $\mathbf{J}^2$ in the angular momentum theory, and in the space of antisymmetric wave functions can be expressed in terms of the scalar products of operators (5.27) and (5.28), or, using the second quantization operators, in terms of (14.23) and (14.25). For the Casimir operators of Sp_{4l+2} group we have Eq. (5.34), for R_{2l+1}, Eq. (5.29), for SU_{2l+1}, Eq. (5.33) and, finally for G_2, Eq. (5.32).

14.5   Commutation relations for various tensors

Apart from irreducible tensors (14.30) we can also introduce other operators that are expressed in terms of irreducible tensorial products of second-quantization operators, and establish commutation relations for them. As was shown in [12, 102, 103], using relations of this kind, we can relate standard quantities of the theory, which at first sight seem totally different. Consider the operator

$$\left[a^{(ls)} \times a^{(ls)} \right]_{\rho\pi}^{(kK)} = \sum_{mm'\mu\mu'} \begin{bmatrix} l & l & k \\ m & m' & \rho \end{bmatrix} \begin{bmatrix} s & s & K \\ \mu & \mu' & \pi \end{bmatrix} a_{m\mu}^{(ls)} a_{m'\mu'}^{(ls)}. \tag{14.40}$$

Rearranging the electron creation operators in the right side of this expression and using the symmetry properties of Clebsch–Gordan coefficients

(6.2)–(6.6) we can obtain the relationship

$$\left\{1 - (-1)^{k+K}\right\} \left[a^{(ls)} \times a^{(ls)}\right]^{(kK)} = 0, \qquad (14.41)$$

that shows that tensorial product (14.40), just like

$$\left[\tilde{a}^{(ls)} \times \tilde{a}^{(ls)}\right]^{(kK)}_{\rho\pi} = \sum_{mm'\mu\mu'} \begin{bmatrix} l & l & k \\ m & m' & \rho \end{bmatrix} \begin{bmatrix} s & s & K \\ \mu & \mu' & \pi \end{bmatrix} \tilde{a}^{(ls)}_{m\mu} \tilde{a}^{(ls)}_{m'\mu'}, \qquad (14.42)$$

is zero, when the sum of ranks $k + K$ is odd.

The commutation relations between operators (14.40) and (14.42) can be represented in the irreducible tensor form [103, 104]

$$\left[\left[\tilde{a}^{(ls)} \times \tilde{a}^{(ls)}\right]^{(kK)} \times \left[a^{(ls)} \times a^{(ls)}\right]^{(k'K')}\right]^{(LS)}$$

$$-(-1)^{L+S} \left[\left[a^{(ls)} \times a^{(ls)}\right]^{(k'K')} \times \left[\tilde{a}^{(ls)} \times \tilde{a}^{(ls)}\right]^{(kK)}\right]^{(LS)}$$

$$= \left\{1 + (-1)^{k+K}\right\} \left(1 + (-1)^{k'+K'}\right) [k, k', K, K']^{1/2} \begin{Bmatrix} k & k' & L \\ l & l & l \end{Bmatrix}$$

$$\times \begin{Bmatrix} K & K' & S \\ s & s & s \end{Bmatrix} \left[a^{(ls)} \times a^{(ls)}\right]^{(LS)} - 2\delta(LS, 00)[k, K]^{1/2}. \qquad (14.43)$$

If in the irreducible tensorial product of operators (14.40) and (14.42) we interchange the second-quantization operators connected with an arrow,

$$\left[\left[a^{(ls)} \times a^{(ls)}\right]^{(k_1 K_1)} \times \left[\tilde{a}^{(ls)} \times \tilde{a}^{(ls)}\right]^{(k_2 K_2)}\right]^{(LS)}$$

then, recoupling the momenta and using anticommutation relation (14.21), we can express it in terms of the sums of tensorial products of operators $W^{(kK)}$:

$$\left[\left[a^{(ls)} \times a^{(ls)}\right]^{(k_1 K_1)} \times \left[\tilde{a}^{(ls)} \times \tilde{a}^{(ls)}\right]^{(k_2 K_2)}\right]^{(LS)}$$

$$= \sum_{k_1' k_2' K_1' K_2'} [k_1, k_2, k_1', k_2', K_1, K_2, K_1', K_2']^{1/2} \begin{Bmatrix} k_2' & L & k_1' \\ l & k_1 & l \\ l & k_2 & l \end{Bmatrix}$$

$$\times \begin{Bmatrix} K_2' & S & K_1' \\ s & K_1 & s \\ s & K_2 & s \end{Bmatrix} \left[W^{(k_1' K_1')} \times W^{(k_2' K_2')}\right]^{(LS)}$$

$$+ (-1)^{L+S} [k_1, k_2, K_1, K_2]^{1/2} \begin{Bmatrix} k_1 & k_2 & L \\ l & l & l \end{Bmatrix}$$

$$\times \begin{Bmatrix} K_1 & K_2 & S \\ s & s & s \end{Bmatrix} W^{(LS)}. \qquad (14.44)$$

Hence the scalar product of operators (14.40) and (14.42) is given by

$$\left(\left[a^{(ls)} \times a^{(ls)}\right]^{(kK)} \cdot \left[\tilde{a}^{(ls)} \times \tilde{a}^{(ls)}\right]^{(kK)}\right)$$

$$= \frac{1 + (-1)^{k+K}}{2}[k,K]\left\{\sum_{k'K'}(-1)^{k'+K'}\begin{Bmatrix} l & l & k' \\ l & l & k \end{Bmatrix}\begin{Bmatrix} s & s & K' \\ s & s & K \end{Bmatrix}\right.$$

$$\left. \times \left(W^{(k'K')} \cdot W^{(k'K')}\right) - (4l+2)^{-1}\hat{N}\right\}. \tag{14.45}$$

Using the anticommutation properties of electron creation and annihilation operators we can establish any necessary commutation relations for their tensorial products. For example, [103]

$$\left[\left[a^{(ls)} \times a^{(ls)}\right]^{(k_1K_1)}_{\rho_1\pi_1}, W^{(k_2K_2)}_{\rho_2\pi_2}\right]$$

$$= 2[k_1,k_2,K_1,K_2]^{1/2}\sum_{kK\rho\pi}\begin{bmatrix} k_2 & k_1 & k \\ \rho_2 & \rho_1 & \rho \end{bmatrix}\begin{bmatrix} K_2 & K_1 & K \\ \pi_2 & \pi_1 & \pi \end{bmatrix}$$

$$\times \begin{Bmatrix} k_1 & k_2 & k \\ l & l & l \end{Bmatrix}\begin{Bmatrix} K_1 & K_2 & K \\ s & s & s \end{Bmatrix}\left[a^{(ls)} \times a^{(ls)}\right]^{(kK)}_{\rho\pi}. \tag{14.46}$$

The expressions become more compact if represented in the irreducible tensor form

$$\left[\left[a^{(ls)} \times a^{(ls)}\right]^{(k_1K_1)} \times W^{(k_2K_2)}\right]^{(kK)}$$

$$- (-1)^{k_1+k_2-k+K_1+K_2-K}\left[W^{(k_2K_2)} \times \left[a^{(ls)} \times a^{(ls)}\right]^{(k_1K_1)}\right]^{(kK)}$$

$$= 2[k_1,k_2,K_1,K_2]^{1/2}\begin{Bmatrix} k_1 & k_2 & k \\ l & l & l \end{Bmatrix}\begin{Bmatrix} K_1 & K_2 & K \\ s & s & s \end{Bmatrix}$$

$$\times \left[a^{(ls)} \times a^{(ls)}\right]^{(kK)}. \tag{14.47}$$

Relations (14.46) and (14.47) remain valid if we substitute (14.42) for (14.40).

We shall also provide the commutation relations for the electron creation operator with tensor (14.40) in the irreducible tensor form

$$\left[a^{(ls)} \times \left[a^{(ls)} \times a^{(ls)}\right]^{(k_1K_1)}\right]^{(kK)}$$

$$- (-1)^{l+s+k_1+K_1-k-K}\left[\left[a^{(ls)} \times a^{(ls)}\right]^{(k_1K_1)} \times a^{(ls)}\right]^{(kK)}$$

$$= 2\delta(kK,ls)[k_1,K_1,l,s]^{1/2}a^{(ls)}. \tag{14.48}$$

A similar relationship for operator $\tilde{a}^{(ls)}$ and tensor (14.42) can be obtained from (14.48) in which all the operators without the tilde are replaced by operators with the tilde. The commutators of electron creation and annihilation operators with tensor $W^{(kK)}$ can also be represented in the irreducible form

$$\left[a^{(ls)} \times W^{(k_1 K_1)}\right]^{(kK)} - (-1)^{l+k_1-k+s+K_1-K} \left[W^{(k_1 K_1)} \times a^{(ls)}\right]^{(kK)}$$
$$= -\delta(kK, ls)[k_1, K_1]^{1/2}[l, s]^{-1/2} a^{(ls)}; \tag{14.49}$$
$$\left[\tilde{a}^{(ls)} \times W^{(k_1 K_1)}\right]^{(kK)} - (-1)^{l+k_1-k+s+K_1-K} \left[W^{(k_1 K_1)} \times \tilde{a}^{(ls)}\right]^{(kK)}$$
$$= (-1)^{k_1+K_1} \delta(kK, ls)[k_1, K_1]^{1/2}[l, s]^{-1/2} \tilde{a}^{(ls)}. \tag{14.50}$$

14.6   One-particle operators of physical quantities

In Chapter 19 a number of operators corresponding to physical quantities will be expressed in coordinate representation in terms of irreducible tensors. In the most general form, the tensorial structure of one-electron operator f can be written as follows:

$$f^{(kK\gamma)}_{\Gamma} = \sum_{\rho\pi} f^{(kK)}_{\rho\pi} \begin{bmatrix} k & K & \gamma \\ \rho & \pi & \Gamma \end{bmatrix}, \tag{14.51}$$

where k is the rank of an operator in an orbital space, K is its rank in a spin space, and γ is the rank in the space of total angular momentum for one electron, which is the vectorial sum of k and K.

Let us now look at one-particle operators in the second-quantization representation, defined by (13.22). Substituting into (13.22) the one-electron matrix element and applying the Wigner–Eckart theorem (5.15) in orbital and spin spaces, we obtain by summation over the projections

$$F^{(kK\gamma)}_{\Gamma} = [k, K]^{-1/2} (nl \| f^{(kK)} \| nl) \sum_{\rho\pi} \begin{bmatrix} k & K & \gamma \\ \rho & \pi & \Gamma \end{bmatrix} W^{(kK)}_{\rho\pi}. \tag{14.52}$$

Specifically, the operators of kinetic and potential energy of electrons (see (1.15)) are scalars in orbital and spin spaces. Therefore, their sum is at once found to be for a shell nl^N

$$T + U = \hat{N}(nlm\mu|(T + U)|nlm\mu), \tag{14.53}$$

where the last factor is given by (19.23). For one shell similar simple expressions are derived for relativistic corrections (1.18) and the first term H'_3 in (1.20)

$$\hat{H}_1 = \hat{N}(nlm\mu|H_1|nlm\mu); \qquad H'_3 = \hat{N}(nlm\mu|H'_3|nlm\mu). \tag{14.54}$$

A more complex tensorial structure is inherent in the operator of the spin–own-orbit interaction. Using (19.58) and (14.52) we get

$$H^{so} = \sum_{\rho}(-1)^{\rho} \left[\frac{l(l+1)(2l+1)}{2\cdot 3}\right]^{1/2} \eta(nl)W^{(11)}_{\rho\,-\rho}. \qquad (14.55)$$

This operator, too, is a scalar in the space of total angular momentum for an electron. Tensors in this space are, for example, the operators of electric and magnetic multipole transitions (4.12), (4.13), (4.16). So, the operator of electric multipole transition (4.12) in the second-quantization representation is

$$Q^{(k)}_{-\rho} \stackrel{*}{=} -(2k+1)^{-1/2}(nl\|Q^{(k)}\|nl)W^{(k0)}_{-\rho 0}. \qquad (14.56)$$

Exactly the same expressions can be established for the second form of this operator by substituting $Q^{(k)}_{-\rho}$ for $Q'^{(k)}_{-\rho}$. The relativistic one-electron operators may be considered in a similar way.

14.7   Two-particle operators of physical quantities

Since all the two-electron interaction operators are invariant under rotations in the space of total momentum J of two particles, we shall consider only the scalar two-electron operators

$$G^{(k_1k_2k,K_1K_2k)} = \sum_{j>i}g_{ij}^{(k_1k_2k,K_1K_2k)} \qquad (14.57)$$

Here

$$g_{ij}^{(k_1k_2k,K_1K_2k)} = \sum_{\rho}(-1)^{k-\rho}g(r_i,r_j)\left[g_i^{(k_1K_1)} \times g_j^{(k_2K_2)}\right]^{(kk)}_{\rho\,-\rho}, \qquad (14.58)$$

where $g(r_i,r_j)$ is the radial part of the operator, whereas i and j are the coordinates of the respective electrons. k_1 and k_2 are the ranks of operators in orbital spaces, and K_1 and K_2 are the ranks of operators in spin spaces of the electrons.

    If the specific tensorial structure (14.58) of a two-electron operator is known, then we can obtain its representation in terms of the product of operators (14.30) acting in the space of states of one shell. In fact, if we substitute into the two-electron matrix element which enters into (13.23), the operator (14.58) in the form

$$g_{12}^{(k_1k_2k,K_1K_2k)} = g(r_1,r_2)\sum_{\rho_1\rho_2\rho\pi_1\pi_2}(-1)^{k-\rho}\begin{bmatrix}k_1 & k_2 & k \\ \rho_1 & \rho_2 & \rho\end{bmatrix}$$

$$\times \begin{bmatrix}K_1 & K_2 & k \\ \pi_1 & \pi_2 & -\rho\end{bmatrix}g_{1\rho_1\pi_1}^{(k_1K_1)}g_{2\rho_2\pi_2}^{(k_2K_2)}$$

and then apply the Wigner–Eckart theorem (5.15) in the orbital and spin spaces of each electron separately, this matrix element will be

$$(nlm\mu nlm'\mu'|g_{12}|nlm''\mu''nlm'''\mu''')$$

$$= (4l+2)^{-1} \sum_{\rho_1\rho_2\rho\pi_1\pi_2} (-1)^{k-\rho} \begin{bmatrix} k_1 & k_2 & k \\ \rho_1 & \rho_2 & \rho \end{bmatrix} \begin{bmatrix} K_1 & K_2 & k \\ \pi_1 & \pi_2 & -\rho \end{bmatrix}$$

$$\times \begin{bmatrix} l & k_1 & l \\ m'' & \rho_1 & m \end{bmatrix} \begin{bmatrix} l & k_2 & l \\ m''' & \rho_2 & m' \end{bmatrix} \begin{bmatrix} s & K_1 & s \\ \mu'' & \pi_1 & \mu \end{bmatrix}$$

$$\times \begin{bmatrix} s & K_2 & s \\ \mu''' & \pi_2 & \mu' \end{bmatrix} (nlnl\|g_{12}\|nlnl), \tag{14.59}$$

where

$$(nlnl\|g_{12}\|nlnl)$$
$$= (ls\|g^{(k_1K_1)}\|ls)(ls\|g^{(k_2K_2)}\|ls)(nlnl\|g(r_1,r_2)\|nlnl). \tag{14.60}$$

Let us now put (14.59) into the general formula (13.23) and rearrange the creation and annihilation operators so that when the second-quantization operators are placed side by side the rank projections enter into the same Clebsch–Gordan coefficient. A summation over the projections gives

$$G^{(k_1k_2k,K_1K_2k)} = \frac{1}{2}(nlnl\|g_{12}\|nlnl)\sum_{\rho}(-1)^{k-\rho}\left\{\frac{1}{\sqrt{[k_1,k_2,K_1,K_2]}}\right.$$

$$\times \left[W^{(k_1K_1)} \times W^{(k_2K_2)}\right]_{\rho\,-\rho}^{(kk)} + \begin{Bmatrix} k_1 & k_2 & k \\ l & l & l \end{Bmatrix}$$

$$\times \left. \begin{Bmatrix} K_1 & K_2 & k \\ s & s & s \end{Bmatrix} W_{\rho\,-\rho}^{(kk)}\right\}. \tag{14.61}$$

This expression is the second-quantization form of an arbitrary two-electron operator with tensorial structure of the kind (14.57). Examination of this formula enables us to work out expressions for operators that correspond to specific physical quantities. Many of these operators possess a complex tensorial structure, and their two-electron submatrix elements have a rather cumbersome form [14]. Therefore, by way of example, we shall consider here only the most important of the two-electron operators – the operator of the energy of electrostatic interaction of electrons (the last term in (1.15)). If we take into account the tensorial structure of that operator and put its submatrix element into (14.61), we arrive at

$$V = \frac{1}{2}\sum_k (nlnl\|V^{(kk)}\|nlnl)\left\{\frac{2}{2k+1}(W^{(k0)}\cdot W^{(k0)}) - (2l+1)^{-1}\hat{N}\right\}. \tag{14.62}$$

Two-electron operators can also be represented in terms of tensorial products of operators (14.40) and (14.42). To this end, we expand the bilinear combinations of creation and annihilation operators in (13.23) separately in terms of irreducible tensors. Using (14.14) and the analogue of formula (10.5) for two tensorial components gives

$$G = -\frac{1}{2} \sum_{\substack{(m,\mu) \\ L_{12}S_{12}L'_{12}S'_{12}}} (nlm\mu nlm'\mu'|g_{12}|nlm''\mu'' nlm'''\mu''') \begin{bmatrix} l & l & L_{12} \\ m & m' & m_{12} \end{bmatrix}$$

$$\times \begin{bmatrix} l & l & L'_{12} \\ m'' & m''' & m'_{12} \end{bmatrix} \begin{bmatrix} s & s & S_{12} \\ \mu & \mu' & \mu_{12} \end{bmatrix} \begin{bmatrix} s & s & S'_{12} \\ \mu'' & \mu''' & \mu'_{12} \end{bmatrix}$$

$$\times \left[a^{(ls)} \times a^{(ls)} \right]^{(L_{12}S_{12})}_{m_{12}\mu_{12}} \left[\tilde{a}^{(ls)} \times \tilde{a}^{(ls)} \right]^{(L'_{12}S'_{12})}_{m'_{12}\mu'_{12}}.$$

Here (m, μ) denotes the summation over all projections. Summing the right side of this equation over all projections but $m_{12}, \mu_{12}, m'_{12}, \mu'_{12}$, we obtain

$$G = -\frac{1}{2} \sum_{\substack{m_{12}m'_{12}\mu_{12}\mu'_{12} \\ L_{12}S_{12}L'_{12}S'_{12}}} (nlnlL_{12}S_{12}m_{12}\mu_{12}|g_{12}|nlnlL'_{12}S'_{12}m'_{12}\mu'_{12})$$

$$\times \left[a^{(ls)} \times a^{(ls)} \right]^{(L_{12}S_{12})}_{m_{12}\mu_{12}} \left[\tilde{a}^{(ls)} \times \tilde{a}^{(ls)} \right]^{(L'_{12}S'_{12})}_{m'_{12}\mu'_{12}}.$$

The matrix element is defined relative to two-electron wave functions of coupled momenta. If now we take into account the tensorial structure of operator (14.57), apply the Wigner–Eckart theorem to this matrix element and sum up over the appropriate projections, we have

$$G = -\frac{1}{2} \sum_{L_{12}S_{12}L'_{12}S'_{12}k\rho} (nlnlL_{12}S_{12}\|g_{12}\|nlnlL_{12}S_{12}) (-1)^{k-\rho}$$

$$\times \left[\left[a^{(ls)} \times a^{(ls)} \right]^{(L_{12}S_{12})} \times \left[\tilde{a}^{(ls)} \times \tilde{a}^{(ls)} \right]^{(L'_{12}S'_{12})} \right]^{(kk)}_{\rho\,-\rho}. \tag{14.63}$$

The submatrix element that enters into this expression is, by (5.19), related to submatrix element (14.60) by

$$(nlnlL_{12}S_{12}\|g_{12}\|nlnlL'_{12}S'_{12}) = [L_{12}, S_{12}, L'_{12}, S'_{12}]^{1/2}$$

$$\times (2k+1)(nlnl\|g_{12}\|nlnl) \begin{Bmatrix} l & l & L'_{12} \\ k_1 & k_2 & k \\ l & l & L_{12} \end{Bmatrix} \begin{Bmatrix} s & s & S'_{12} \\ K_1 & K_2 & k \\ s & s & S_{12} \end{Bmatrix}. \tag{14.64}$$

For the operator of electrostatic interaction, from general expressions (14.63) and (14.64), we get

$$V = -\frac{1}{2} \sum_{L_{12}S_{12}k} \left(\left[a^{(ls)} \times a^{(ls)} \right]^{(L_{12}S_{12})} \cdot \left[\tilde{a}^{(ls)} \times \tilde{a}^{(ls)} \right]^{(L_{12}S_{12})} \right)$$

$$\times \left(nlnlL_{12}S_{12} \| V^{(kk)} \| nlnlL_{12}S_{12} \right); \tag{14.65}$$

where

$$\left(nlnlL_{12}S_{12}\|V^{(kk)}\|nlnlL_{12}S_{12}\right)$$
$$= (-1)^L \begin{Bmatrix} l & l & L_{12} \\ l & l & k \end{Bmatrix} (l\|C^{(k)}\|l)^2 F_k(nl, nl). \tag{14.66}$$

It is to be stressed that, although the two-electron submatrix elements in (14.63) and (14.65) are defined relative to non-antisymmetric wave functions, some constraints on the possible values of orbital and spin momenta of the two particles are imposed in an implicit form by second-quantization operators. Really, tensorial products (14.40) and (14.42), when the sum of ranks is odd, are zero. Thus, the appropriate terms in (14.63) and (14.65) then also vanish.

The operator of the energy of electrostatic interaction of electrons in (14.65) is represented as a sum of second-quantization operators, and the appropriate submatrix element of each term is proportional to the energy of electrostatic interaction of a pair of equivalent electrons with orbital L_{12} and spin S_{12} angular momenta. The values of these submatrix elements are different for different pairing states, since, as follows from (14.66), the two-electron submatrix elements concerned are explicitly dependent on L_{12}, and, hence, implicitly – on S_{12} (sum $L_{12} + S_{12}$ is even). It is in this way that, in the second-quantization representation for the l^N configuration, the dependence of the energy of electrostatic interaction on the angles between the particles shows up. This dependence violates the central field approximation.

14.8 Operator of averaged electrostatic interaction

In constructing the energy functional of an atom using the Hartree–Fock method, the so-called averaging of the energy of electrostatic interaction over all the states of the configuration has gained wide acceptance [6]. This simplification makes it possible to use the same radial orbitals for all the states of the configuration under consideration.

The second-quantization counterpart of this approach is the replacement (for the l^N configuration) of operator (14.65) by some effective operator, whose two-particle submatrix elements are independent of characteristics L_{12}, S_{12} of the pairing state of electrons. To this end, we introduce the submatrix element averaged over the number of various antisymmetric pairing states in shell, equal to $(4l + 2)(4l + 1)/2$:

$$\left(nlnl\|V_{av}^{(kk)}\|nlnl\right) = \sum_{L_{12}S_{12}} \left[1 + (-1)^{L_{12}+S_{12}}\right] [L_{12}, S_{12}]$$
$$\times \left(nlnlL_{12}S_{12}\|V_{12}^{(kk)}\|nlnlL_{12}S_{12}\right) / [(4l + 2)(4l + 1)]. \tag{14.67}$$

The first factor under the summation sign takes care of the antisymmetry of the two-electron states of the shell. The submatrix elements are summed in accordance with the statistical weights of these states. Using (14.66), (6.25), (6.26) and (6.18), we sum the right side of (14.67) in the explicit form

$$\left(nlnl\|V_{av}^{(kk)}\|nlnl\right) = \frac{2\left(nlnl\|V^{(kk)}\|nlnl\right)}{(4l+1)}\left[\delta(k,0) - \frac{1}{4l+2}\right]. \quad (14.68)$$

Since the right side of this expression is independent of the values of parameters L_{12}, S_{12}, the summation over these parameters in the effective operator

$$V_{av} = -\frac{1}{2}\sum_{kL_{12}S_{12}}\left(\left[a^{(ls)} \times a^{(ls)}\right]^{(L_{12}S_{12})} \cdot \left[\tilde{a}^{(ls)} \times \tilde{a}^{(ls)}\right]^{(L_{12}S_{12})}\right)$$

$$\times \left(nlnl\|V^{(kk)}\|nlnl\right)$$

yields the final result

$$V_{av} = \frac{1}{2}\sum_{k}\hat{N}(\hat{N}-1)(nlnl\|V_{av}^{(kk)}\|nlnl). \quad (14.69)$$

We can also introduce another effective operator, one that explicitly includes the exchange interaction of electrons. For this purpose, we shall have, in averaging, to take into account the dependence of the electrostatic interaction of two particles on their total spin. For a pair of equivalent electrons the number of various singlet states ($S_{12} = 0$) is given by

$$\sum_{L_{12}}\{llL_{12}\}\left\{1 + (-1)^{L_{12}}\right\}(2L_{12}+1)/2 = (l+1)(2l+1), \quad (14.70)$$

and the number of triplet states ($S_{12} = 1$) is given by

$$\sum_{L_{12}}3\{llL_{12}\}\left\{1 - (-1)^{L_{12}}\right\}(2L_{12}+1)/2 = 3l(2l+1). \quad (14.71)$$

In these relationships the summation is carried out using (6.25), (6.26) and (6.18). Let us now introduce averaged submatrix elements of two-particle interactions separately for singlet states

$$\left(nlnlS_{12} = 0\|V_{av}^{(kk)}\|nlnlS_{12} = 0\right) = \frac{1}{2}\sum_{L_{12}}\left\{1 + (-1)^{L_{12}}\right\}(2L_{12}+1)$$

$$\times \left(nlnlL_{12}S_{12} = 0\|V^{(kk)}\|nlnlL_{12}S_{12} = 0\right)/[(l+1)(2l+1)]$$

$$= (nlnl\|V^{(kk)}\|nlnl)[1 + (2l+1)\delta(k,0)]/[2(l+1)(2l+1)], \quad (14.72)$$

and for triplet states

$$\left(nlnlS_{12} = 1 \| V_{av}^{(kk)} \| nlnlS_{12} = 1\right) = \frac{3}{2} \sum_{L_{12}} \left\{(-1)^{L_{12}} - 1\right\}(2L_{12} + 1)$$

$$\times \left(nlnlL_{12}S_{12} = 1 \| V^{(kk)} \| nlnlL_{12}S_{12} = 1\right) / [3l(2l+1)]$$

$$= (nlnl \| V^{(kk)} \| nlnl)[(2l+1)\delta(k,0) - 1]/[2l(2l+1)]. \quad (14.73)$$

The relevant effective operator of electrostatic interaction of electrons

$$V'_{av} = -\frac{1}{2} \sum_{kL_{12}} \left\{ \left(\left[a^{(ls)} \times a^{(ls)}\right]^{(L_{12}0)} \cdot \left[\tilde{a}^{(ls)} \times \tilde{a}^{(ls)}\right]^{(L_{12}0)}\right)\right\}$$

$$\times \left(nlnlS_{12} = 0 \| V_{av}^{(kk)} \| nlnlS_{12} = 0\right)$$

$$+ \left(\left[a^{(ls)} \times a^{(ls)}\right]^{(L_{12}1)} \cdot \left[\tilde{a}^{(ls)} \times \tilde{a}^{(ls)}\right]^{(L_{12}1)}\right)$$

$$\times \left(nlnlS_{12} = 1 \| V_{av}^{(kk)} \| nlnlS_{12} = 1\right) \quad (14.74)$$

after summation over the projections will read [6]

$$V'_{av} = \frac{1}{2}\hat{N}(\hat{N} - 1)F_0(nl, nl) - \sum_{k>0} \frac{(nlnl \| V^{(kk)} \| nlnl)}{4l(l+1)(2l+1)}$$

$$\times \left[\frac{2l+3}{4}\hat{N}^2 - \frac{4l+3}{2}\hat{N} + (2l+1)(S^{(1)} \cdot S^{(1)})\right], \quad (14.75)$$

i.e. the explicit form of the operator contains the dependence on the spin angular momentum. The matrix element of operator V'_{av} defines the average value of the energy of terms for a shell of equivalent electrons with a fixed multiplicity.

In this chapter we have found the relationship between the various operators in the second-quantization representation and irreducible tensors of the orbital and spin spaces of a shell of equivalent electrons. In subsequent chapters we shall be looking at the techniques of finding the matrix elements of these operators.

15

Quasispin for a shell of equivalent electrons

15.1 Wave functions in second-quantization representation

The traditional description of the wave function of a shell of equivalent electrons was presented in Chapter 9. Here we shall utilize the second-quantization method for this purpose. In fact, the one-electron wave function is

$$|lsm\mu) = a_{m\mu}^{(ls)}|0)$$ (15.1)

and the normalized two-electron function $|l^2LS)$ with given L and S is obtained by acting with

$$\hat{\varphi}(l^2)^{(LS)} = \sqrt{2}\left[a^{(ls)} \times a^{(ls)}\right]^{(LS)}$$ (15.2)

on the vacuum state, i.e.

$$|l^2LS) = \hat{\varphi}(l^2)^{(LS)}|0).$$ (15.3)

For N electrons the generalization of (15.3) is

$$|l^N\alpha LS) = \hat{\varphi}(l^N)^{\alpha(LS)}|0).$$ (15.4)

Operator $\hat{\varphi}$ in (15.4) can be represented as one irreducible tensorial product of N creation operators only if there are no degenerate terms. So, at $N = 3$, we have

$$\hat{\varphi}(l^3)^{(L_3S_3)} = N_3\left[a^{(ls)} \times \left[a^{(ls)} \times a^{(ls)}\right]^{(L_2S_2)}\right]^{(L_3S_3)},$$ (15.5)

where N_3 is the factor that takes care of normalization of the wave function produced by this operator. It is defined by

$$\left(0|\hat{\varphi}^{\dagger}(l^N)_{M_LM_S}^{\alpha(LS)}\hat{\varphi}(l^N)_{M_LM_S}^{\alpha(LS)}|0\right) = 1.$$ (15.6)

For this vacuum average to be computed, all the annihilation operators will have to be placed to the right of the creation operators. The result can be expressed in terms of $3nj$-coefficients.

If an l^3 configuration $^{2S_3+1}L_3$ term does not recur, then at different allowed values of L_2 and S_2 at which (15.5) does not vanish, we shall obtain, up to the phase, the same antisymmetric wave function. Otherwise, operator (15.5) may no longer be used, since the different values of L_2S_2 do not separate the degenerate terms. To see that this is so, it is sufficient to consider the vacuum average

$$\left(0 \middle| \left\{ \left[a^{(ls)} \times \left[a^{(ls)} \times a^{(ls)} \right]^{(L_2'S_2')} \right]^{(LS)}_{M_L M_S} \right\}^\dagger \right.$$
$$\left. \times \left[a^{(ls)} \times \left[a^{(ls)} \times a^{(ls)} \right]^{(L_2 S_2)} \right]^{(LS)}_{M_L M_S} \middle| 0 \right), \quad (15.7)$$

which at $L_2'S_2' \neq L_2 S_2$ is non-zero. Therefore, the functions produced by irreducible tensorial operators with different $L_2 S_2$ are not orthogonal with one another and in the general case are linearly dependent on one another. For the same reason, no additional classification of terms is obtained using the operators producing the four-electron functions

$$\hat{\varphi}(l^4)^{(L_4 S_4)} = N_4 \left[\left[a^{(ls)} \times a^{(ls)} \right]^{(L_2'S_2')} \times \left[a^{(ls)} \times a^{(ls)} \right]^{(L_2 S_2)} \right]^{(L_4 S_4)}, \quad (15.8)$$

$$\hat{\varphi}'(l^4)^{(L_4 S_4)} = N_4' \left[a^{(ls)} \times \left[a^{(ls)} \times [a^{(ls)} \times a^{(ls)}]^{(L_2 S_2)} \right]^{(L_3 S_3)} \right]^{(L_4 S_4)}, \quad (15.9)$$

where N_4, N_4' are the normalization factors, and the intermediate momenta are so selected that (15.8) and (15.9) do not vanish identically.

In the general case, the second-quantized operator $\hat{\varphi}(l^N)^{\alpha(LS)}$ must be some linear combination of irreducible tensorial products of electron creation operators. The combination must be selected so that a classification of states according to additional quantum numbers be provided for. Without loss of generality, all the numerical coefficients in the linear combinations can be considered real. Then, from (14.14), we can introduce the operators

$$\hat{\tilde{\varphi}}(l^N)^{\alpha(LS)}_{M_L M_S} = (-1)^{L+S-M_L-M_S+N(N-1)/2} \left[\hat{\varphi}(l^N)^{\alpha(LS)}_{-M_L-M_S} \right]^\dagger. \quad (15.10)$$

Comparison of (13.40) and (15.10) shows that on the right side of the latter there is an additional phase factor $(-1)^{N(N-1)/2}$. This factor has been introduced so that operator $\hat{\tilde{\varphi}}(l^N)^{\alpha(LS)}$ has the same internal tensorial structure as operator $\hat{\varphi}(l^N)^{\alpha(LS)}$ (i.e. in the tensorial products all the creation operators are replaced by (14.14)). For example,

$$\hat{\bar{\varphi}}(l^2)^{(L_2 S_2)} = 2^{-1/2} \left[\tilde{a}^{(ls)} \times \tilde{a}^{(ls)} \right]^{(L_2 S_2)} ; \tag{15.11}$$

$$\hat{\bar{\varphi}}(l^3)^{(L_3 S_3)} = N_3 \left[\tilde{a}^{(ls)} \times \left[\tilde{a}^{(ls)} \times \tilde{a}^{(ls)} \right]^{(L_2 S_2)} \right]^{(L_3 S_3)} . \tag{15.12}$$

15.2   Submatrix elements of creation and annihilation operators. Coefficients of fractional parentage

Submatrix elements of an arbitrary tensor of the form (14.52), $F_\Gamma^{(k\kappa\gamma)}$, are related by an equation similar to (5.19)

$$(l^N \alpha L S J \| F^{(k\kappa\gamma)} \| l^{N'} \alpha' L' S' J')$$

$$= [J, J', \gamma]^{1/2} \begin{Bmatrix} L' & S' & J' \\ k & \kappa & \gamma \\ L & S & J \end{Bmatrix} (l^N \alpha L S \| F^{(k\kappa)} \| l^{N'} \alpha' L' S'). \tag{15.13}$$

The submatrix elements of the creation operator, according to (5.15), can be given by the integral not depending on M_L, M_S, namely

$$(l^N \alpha L S \| a^{(ls)} \| l^{N-1} \alpha' L' S') = (-1)^{L+S+l+s-L'-S'} [L, S]^{1/2}$$

$$\times (l^N \alpha L S M_L M_S | \left[a^{(ls)} \times \| l^{N-1} \alpha' L' S') \right]_{M_L M_S}^{(LS)} . \tag{15.14}$$

Assuming these to be real, we obtain the transposition condition

$$(l^{N-1} \alpha' L' S' \| \tilde{a}^{(ls)} \| l^N \alpha L S)$$

$$= (-1)^{L+S+l+s-L'-S'} (l^N \alpha L S \| a^{(ls)} \| l^{N-1} \alpha' L' S'), \tag{15.15}$$

which relates the submatrix elements of creation and annihilation operators to one another.

These submatrix elements are exceedingly important for atomic theory, since they are proportional to the usual coefficients of fractional parentage $(l^{N-1}(\alpha_1 L_1 S_1) l \| l^N \alpha L S)$. In order to establish the relation between the coefficients of fractional parentage and the submatrix elements of creation operators we shall consider the irreducible tensorial product

$$\left[a^{(ls)} \times \| l^{N-1} \alpha_1 L_1 S_1) \right]_{M_L M_S}^{(LS)} = \sum_{M_{L_1} M_{S_1} m\mu} \begin{bmatrix} l & L_1 & L \\ m & M_{L_1} & M_L \end{bmatrix}$$

$$\times \begin{bmatrix} s & S_1 & S \\ \mu & M_{S_1} & M_S \end{bmatrix} a_{m\mu}^{(ls)} | l^{N-1} \alpha_1 L_1 S_1 M_{L_1} M_{S_1}). \tag{15.16}$$

Using the Laplace expansion in terms of the first row of the determinant and changing over the right side of (15.16) again to the wave function of coupled momenta, we have

$$\left[a^{(ls)} \times \|l^{N-1}\alpha_1 L_1 S_1)\right]_{M_L M_S}^{(LS)} = N^{-1/2} \sum_i (-1)^{i+1+L_1+S_1+l+s-L-S}$$

$$\times \left|\begin{array}{c} l^{N_1-1}(\alpha_1 L_1 S_1)lLSM_L M_S \\ (x_1, x_2, \ldots, x_{i-1}, x_{i+1}, \ldots, x_N)x_i \end{array}\right). \quad (15.17)$$

Here the wave function under the sign of summation is defined according to (9.6), however, the coordinates are specified in the order as indicated in the lower row. Multiplying (15.17) by the conjugate, completely antisymmetric wave function for N electrons and integrating gives

$$\left(l^N \alpha L S M_L M_S\right| \left[a^{(ls)} \times \|l^{N-1}\alpha_1 L_1 S_1)\right]_{M_L M_S}^{(LS)} = (-1)^{L_1+S_1+l+s-L-S}$$

$$\times N^{-1/2} \sum_i (-1)^{i+1} \left(\begin{array}{c|c} l^N \alpha L S M_L M_S & l^{N-1}(\alpha_1 L_1 S_1)lLSM_L M_S \\ 1, 2, \ldots, N & 1, 2, \ldots, i-1, i+1, \ldots, N, i \end{array}\right). \quad (15.18)$$

To sum the right side of this expression over i, we apply the relationship

$$\left|\begin{array}{c} l^N \alpha L S M_L M_S \\ 1, 2, \ldots, N \end{array}\right) = (-1)^{N-i} \left|\begin{array}{c} l^N \alpha L S M_L M_S \\ 1, 2, \ldots, i-1, i+1, \ldots, N, i \end{array}\right) \quad (15.19)$$

and then renumber the coordinates of each summand in the order of the natural series of numbers. Then,

$$\left(l^N \alpha L S M_L M_S\right| \left[a^{(ls)} \times \|l^{N-1}\alpha_1 L_1 S_1)\right]_{M_L M_S}^{(LS)} = (-1)^{N+1+L_1+S_1+l-s-L-S}$$

$$\times N^{1/2} \left(l^N \alpha L S M_L M_S | l^{N-1}(\alpha_1 L_1 S_1)lLSM_L M_S\right). \quad (15.20)$$

It then follows that

$$\left(l^N \alpha L S \| a^{(ls)} \| l^{N-1}\alpha_1 L_1 S_1\right) = (-1)^N (N[L, S])^{1/2}$$

$$\times \left(l^N \alpha L S \| l^{N-1}(\alpha_1 L_1 S_1)l\right). \quad (15.21)$$

This formula relates the submatrix element of the creation operator to the CFP. The last expression fully corresponds to similar relations in [12, 96]. The only exception is the monograph [14], where formula (16.4), according to relation (2.8) in the same work, differs from (15.21) by the phase factor $(-1)^N$. This difference is explained by the fact that in [14] the wave function $|lsm\mu)$ that corresponds to creation operator $a^{(ls)}$ appears without phase in the last row of the determinant, and not in its first row, as defined earlier by (2.6). As a consequence, although in the second-quantization representation the explicit form of one-determinant functions is not used, one should have in mind the phase convention for these functions (2.6) [14]. The relation we have used yielded the simplest phase factors.

The sums of products of CFP obey some additional relations. In fact, the operators of particle number N, of orbital L and spin S momenta are expressed in terms of tensorial products of electron creation and annihilation operators – relationships (14.17), (14.15) and (14.16), respectively. We can expand the submatrix elements of such tensorial products using (5.16) and then go over, using (15.21) and (15.15), to the CFP. On the other hand, these submatrix elements are given by the quantum numbers of the states of the l^N configuration. Then we obtain

$$
\sum_{\alpha_1 L_1 S_1} (-1)^{l-L-L_1+1} N \left\{ \frac{l(l+1)[l,s,L]}{L(L+1)} \right\}^{1/2} \left(l^N \alpha L S \| l^{N-1}(\alpha_1 L_1 S_1) l \right)
$$

$$
\times \left(l^{N-1}(\alpha_1 L_1 S_1) l \| l^N \alpha' L S \right) = \delta(\alpha, \alpha'); \quad (15.22)
$$

$$
\sum_{\alpha_1 L_1 S_1} (-1)^{s-S-S_1} N \left\{ \frac{s(s+1)[l,s,S]}{S(S+1)} \right\}^{1/2} \left(l^N \alpha L S \| l^{N-1}(\alpha_1 L_1 S_1) l \right)
$$

$$
\times \left(l^{N-1}(\alpha_1 L_1 S_1) l \| l^N \alpha' L S \right) = \delta(\alpha, \alpha') \quad (15.23)
$$

as well as the condition of orthonormalization of the CFP with respect to the quantum numbers of the N-electron state (9.10). To find this condition for the quantum numbers of the $(N-1)$-electron state, it is sufficient to work out the submatrix elements of the hole number operator (14.18), using the technique just described. Then

$$
\sum_{\alpha' L' S'} [L', S'] (l^{N-1}(\alpha L S) l \| l^N \alpha' L' S')(l^N \alpha' L' S' \| l^{N-1}(\alpha'' L S) l)
$$

$$
= [L, S] \delta(\alpha'', \alpha)(4l + 3 - N)/N. \quad (15.24)
$$

In order to obtain the recurrence relations between the CFP we shall consider the submatrix element of operator (14.40). This operator couples the states of l^N and l^{N-2} configurations, however, it vanishes at odd $k+K$. Applying relationship (5.16) gives straightforwardly the relation (9.9) we are looking for.

15.3 Many-particle coefficients of fractional parentage

A generalization of the above CFP for σ detached electrons is

$$
|l^N \alpha L S M_L M_S) = \sum_{\substack{\alpha_1 L_1 S_1 \\ \alpha_2 L_2 S_2}} |l^{N-\sigma}(\alpha_1 L_1 S_1) l^\sigma(\alpha_2 L_2 S_2) L S M_L M_S)
$$

$$
\times \left(l^{N-\sigma}(\alpha_1 L_1 S_1) l^\sigma(\alpha_2 L_2 S_2) \| l^N \alpha L S \right) \quad (15.25)
$$

or the inverse relation

$$\left(l^{N-\sigma}(\alpha_1 L_1 S_1)l^\sigma(\alpha_2 L_2 S_2)\|l^N \alpha LS\right)$$

$$= \sum_{M_L M_S} \left(l^{N-\sigma}(\alpha_1 L_1 S_1)l^\sigma(\alpha_2 L_2 S_2)LSM_L M_S|l^N \alpha LS M_L M_S\right). \quad (15.26)$$

The relation between the CFP with σ detached electrons and the reduced matrix elements of operator $\hat{\varphi}(l^N)^{\alpha(LS)}$ generating [see (15.4)] the σ-electron wave function is established in exactly the same way as in the derivation of (15.21). Only now in the appropriate determinants we have to apply the Laplace expansion in terms of σ rows. The final expression takes the form

$$\left(l^{N-\sigma}(\alpha_1 L_1 S_1)l^\sigma(\alpha_2 L_2 S_2)\|l^N \alpha LS\right) = (-1)^{\sigma(N-\sigma)+L_1+S_1+L_2+S_2-L-S}$$

$$\times \left(\frac{\sigma!(N-\sigma)!}{N!}\right)^{1/2} \left(0|\left\{\hat{\varphi}(l^N)^{\alpha(LS)}_{M_L M_S}\right\}^\dagger \left[\hat{\varphi}(l^\sigma)^{\alpha_2(L_2 S_2)}\right.\right.$$

$$\left.\left.\times \hat{\varphi}(l^{N-\sigma})^{\alpha_1(L_1 S_1)}\right]^{(LS)}_{M_L M_S}|0\right) \quad (15.27)$$

or, by (5.15),

$$\left(l^{N-\sigma}(\alpha_1 L_1 S_1)l^\sigma(\alpha_2 L_2 S_2)\|l^N \alpha LS\right) = (-1)^{N\sigma}\left(\frac{\sigma!(N-\sigma)!}{N!}\right)^{1/2}$$

$$\times \left(l^N \alpha LS\|\hat{\varphi}(l^\sigma)^{\alpha_2(L_2 S_2)}\|l^{N-\sigma}\alpha_1 L_1 S_1\right). \quad (15.28)$$

For the phase factor, we have used the fact that σ and $2S_2$ have the same parity.

By interchanging operators $\hat{\varphi}(l^\sigma)$ and $\hat{\varphi}(l^{N-\sigma})$ on the right side of (15.27), we arrive at

$$\left(l^{N-\sigma}(\alpha_1 L_1 S_1)l^\sigma(\alpha_2 L_2 S_2)\|l^N \alpha LS\right) = (-1)^{\sigma(N-\sigma)+L_1+S_1+L_2+S_2-L-S}$$

$$\times \left(l^\sigma(\alpha_2 L_2 S_2)l^{N-\sigma}(\alpha_1 L_1 S_1)\|l^N \alpha LS\right). \quad (15.29)$$

Here, the dependence of the phase factor on the number of electrons is determined by the number of commutations of appropriate creation operators.

The CFP with σ detached electrons can also be expressed in terms of the submatrix element of the irreducible tensorial product of σ annihilation operators (15.10), for which we have

$$\left(l^{N-\sigma}\alpha_1 L_1 S_1\|\hat{\tilde{\varphi}}(l^\sigma)^{\alpha_2(L_2 S_2)}\|l^N \alpha LS\right) = (-1)^{L_2+S_2-L_1-S_1+L+S+\sigma(\sigma-1)/2}$$

$$\times \left(l^N \alpha LS\|\hat{\varphi}(l^\sigma)^{\alpha_2(L_2 S_2)}\|l^{N-\sigma}\alpha_1 L_1 S_1\right). \quad (15.30)$$

Then

$$\left(l^{N-\sigma}(\alpha_1 L_1 S_1)l^\sigma(\alpha_2 L_2 S_2)\|l^N \alpha LS\right) = (-1)^{L_2+S_2-L_1-S_1+L+S+\sigma N+\sigma(\sigma-1)/2}$$
$$\times \left(\frac{\sigma!(N-\sigma)!}{N![L,S]}\right)^{1/2} \left(l^{N-\sigma}\alpha_1 L_1 S_1 \|\hat{\hat{\varphi}}(l^\sigma)^{\alpha_2(L_2 S_2)}\|l^N \alpha LS\right). \quad (15.31)$$

Of great importance is a special case of the above coefficients – the CFP with two detached electrons:

$$\left(l^{N-2}(\alpha_1 L_1 S_1)l^2(L_2 S_2)\|l^N \alpha LS\right) = \{N(N-1)[L,S]\}^{-1/2}$$
$$\times \left(l^N \alpha LS \| \left[a^{(ls)} \times a^{(ls)}\right]^{(L_2 S_2)} \|l^{N-2}\alpha_1 L_1 S_1\right). \quad (15.32)$$

We now expand the submatrix element of the irreducible tensorial product of creation operators on the right side of this equation by using (5.16), and then go over to CFP according to (15.21)

$$\left(l^{N-2}(\alpha_1 L_1 S_1)l^2(L_2 S_2)\|l^N \alpha LS\right) = \sum_{\alpha' L' S'} (-1)^{L_1+S_1+L+S+1}$$
$$\times [L_2, S_2, L', S']^{1/2} \begin{Bmatrix} l & l & L_2 \\ L & L_1 & L' \end{Bmatrix} \begin{Bmatrix} s & s & S_2 \\ S & S_1 & S' \end{Bmatrix}$$
$$\times \left(l^{N-2}(\alpha_1 L_1 S_1)l\|l^{N-1}\alpha' L' S'\right) \left(l^{N-1}(\alpha' L' S')l\|l^N \alpha LS\right), \quad (15.33)$$

i.e. the CFP with two detached electrons can readily be expressed in terms of conventional coefficients.

Tables of numerical values of the CFP with two detached electrons are available, for example, in [14]. But, as we shall see later, the quasispin method allows us to do without tabulation of these quantities, since they are simply related to other standard quantities of the theory – the submatrix elements of tensors $W^{(k\kappa)}$, i.e.

$$\left(l^N \alpha LS \| W^{(k\kappa)}\|l^N \alpha' L' S'\right) = (-1)^{L+S+L'+S'+k+\kappa+1}[k,\kappa]^{1/2}$$
$$\times \sum_{\alpha_1 L_1 S_1} \left(l^N \alpha LS\|a^{(ls)}\|l^{N-1}\alpha_1 L_1 S_1\right) \left(l^{N-1}\alpha_1 L_1 S_1\|\tilde{a}^{(ls)}\|l^N \alpha' L' S'\right)$$
$$\times \begin{Bmatrix} l & l & k \\ L' & L & L_1 \end{Bmatrix} \begin{Bmatrix} s & s & \kappa \\ S' & S & S_1 \end{Bmatrix}. \quad (15.34)$$

Considering (15.15) and (15.21), the right side of (15.34) can be rewritten in terms of CFP.

The relationships between various CFP and submatrix elements of unit tensors have been derived in a number of works [103, 105]. Their derivations drew specifically on commutation relations (14.45)–(14.50). An explanation of these results and their generalization, can be obtained within the framework of the quasispin method.

15.4 Quasispin

The concept of quasispin quantum number was discussed in the Introduction and Chapter 9 (see formulas (9.22) and (9.23)). Now let us consider it in the framework of the second-quantization technique. We can introduce the following bilinear combinations of creation and annihilation operators obeying commutation relations (14.2) – the quasispin operator:

$$Q_1^{(1)} = -i2^{-1} \left[a^{(\tau)} \times a^{(\tau)} \right]^{(0)} ;$$

$$Q_{-1}^{(1)} = -i2^{-1} \left[\tilde{a}^{(\tau)} \times \tilde{a}^{(\tau)} \right]^{(0)} ; \tag{15.35}$$

$$Q_0^{(1)} = -i2^{-3/2} \left\{ \left[a^{(\tau)} \times \tilde{a}^{(\tau)} \right]^{(0)} + \left[\tilde{a}^{(\tau)} \times a^{(\tau)} \right]^{(0)} \right\}.$$

Here τ may represent ls or j, whereas z-projection of the quasispin operator is determined by the difference between the particle number operator and the hole number operator in the pairing state (α, β)

$$Q_z(\alpha, \beta) = -iQ_0^{(1)} = \left[\hat{N}(\alpha, \beta) - {}_h\hat{N}(\alpha, \beta) \right] /4. \tag{15.36}$$

Operator (15.36), thus, has eigenvalue $\pm 1/2$ depending on whether or not the pairing state is occupied. If only one particle is in that state, then the eigenvalue of operator Q_z is zero. It is worthwhile mentioning the fact that vacuum states, both for holes and for particles, are the components of a tensor in the quasispin space

$$|\alpha\beta) = |_h 0) = \left| Q = \frac{1}{2}; Q_z = \frac{1}{2} \right),$$

$$|0) = \left| Q = \frac{1}{2}; Q_z = -\frac{1}{2} \right). \tag{15.37}$$

By computing the commutators of the components of the quasispin operator with electron creation and annihilation operators, we can directly see that the latter behave as the components of a tensor of rank $q = 1/2$ in quasispin space and obey the relationship of the type (14.2)

$$\left[Q_\rho^{(1)}, a_{vm}^{(q\tau)} \right] = i\sqrt{q(q+1)} \begin{bmatrix} q & 1 & q \\ v & \rho & v+\rho \end{bmatrix} a_{v+\rho m}^{(q\tau)},$$

where

$$a_{vm}^{(q\tau)} = \begin{cases} a_m^{(\tau)} & \text{at } v = 1/2, \\ \tilde{a}_m^{(\tau)} & \text{at } v = -1/2. \end{cases} \tag{15.38}$$

Quasispin formalism can be used in describing the properties of the occupation number space for an arbitrary pairing state. For one shell of equivalent electrons, these pairing states can be chosen to be two one-particle states with the opposite values of angular momentum projections.

The operator of total quasispin angular momentum of the shell can be obtained by the vectorial coupling of quasispin momenta of all the pairing states. For a shell of equivalent electrons, instead of (15.35) we have

$$Q_1^{(1)} = -\frac{i}{2}(2l+1)^{1/2} \left[a^{(ls)} \times a^{(ls)}\right]^{(00)} ; \tag{15.39}$$

$$Q_{-1}^{(1)} = -\frac{i}{2}(2l+1)^{1/2} \left[\tilde{a}^{(ls)} \times \tilde{a}^{(ls)}\right]^{(00)} ; \tag{15.40}$$

$$Q_0^{(1)} = -\frac{i}{4}(4l+2)^{1/2} \left(\left[a^{(ls)} \times \tilde{a}^{(ls)}\right]^{(00)} + \left[\tilde{a}^{(ls)} \times a^{(ls)}\right]^{(00)} \right). \tag{15.41}$$

Tensors (15.39)–(15.41) meet commutation relations (14.2) for irreducible components of the momentum operator, and, in addition, they commute with the operators of orbital (14.15) and spin (14.16) angular momenta for the l^N configuration, since they are scalars in their respective spaces. Accordingly, the states of the l^N configuration can be characterized by the eigenvalues of operators $\mathbf{L}^2$, L_z, $\mathbf{S}^2$, S_z, $\mathbf{Q}^2$, Q_z.

According to (14.17), the z-projection of the quasispin operator is related to the operator of the particle number in the shell

$$Q_z = -iQ_0^{(1)} = (\hat{N} - 2l - 1)/2, \tag{15.42}$$

therefore, the states of a shell of equivalent electrons can be classified by using the eigenvalues of the operator

$$\mathbf{Q}^2 = 2Q_1^{(1)}Q_{-1}^{(1)} + Q_z(Q_z - 1). \tag{15.43}$$

Let us recall that the seniority quantum number, by definition, is the number of unpaired particles in a given state, and two electrons are called paired when their orbital and spin momenta are zero. Since it follows from definitions (15.39) and (15.40) that operators $Q_1^{(1)}$ and $Q_{-1}^{(1)}$ operating on wave function $|l^N \alpha LS)$ respectively, create and annihilate two paired electrons, the seniority and quasispin quantum numbers v and Q must be somehow related. Let a certain N-particle state have absent the paired electrons:

$$|l^N \alpha v LS) = |l^v \alpha v LS). \tag{15.44}$$

Out of the set of additional quantum numbers α, the seniority number v is separated in an explicit form.

Then, by definition,

$$Q_{-1}^{(1)}|l^v \alpha v LS) = 0.$$

Further, from (15.43),

$$\mathbf{Q}^2|l^v \alpha v LS) = Q_z(Q_z - 1)|l^v \alpha v LS)$$

$$= \frac{1}{4}(v - 2l - 1)(v - 2l - 2)|l^v \alpha v LS) = Q(Q+1)|l^v \alpha v LS). \tag{15.45}$$

Consequently, the quantum numbers of quasispin Q and seniority v are related by expression (9.22) which is also valid for the wave function in the general case at $N \neq v$. Operator $Q_1^{(1)}$, acting on wave function (15.44), increases the number of particles by two paired electrons, leaving, by definition, number v unchanged.

In summary, there exists a one-to-one correspondence between quantum numbers Q, M_Q and v, N that is described by (9.22) and (15.42). The wave function will now be (9.25) or

$$|l^N \alpha v LS\rangle \equiv |l\alpha QLSM_Q\rangle. \tag{15.46}$$

15.5 Tensors in quasispin space

Using (14.18) we can write the operator (15.42) as

$$Q_z = (\hat{N} - {}_h\hat{N})/4. \tag{15.47}$$

This suggests that in the particle–hole representation each occupied one-particle state in the l^N configuration can be assigned a value of the z-projection of the quasispin angular momentum $1/4$ and each unoccupied (hole) state $-1/4$. When acting on an N-electron wave function the operator $a^{(ls)}$ produces an electron and, simultaneously, annihilates a hole. Therefore, the projection of the quasispin angular momentum of the wave function on the z-axis increases by $1/2$ when the number of electrons increases by unity. Likewise, the annihilation operator reduces this projection by $1/2$. Accordingly, the electron creation and annihilation operators must possess some tensorial properties in quasispin space. Examination of the commutation relations between quasispin operators, and creation and annihilation operators

$$\left[Q_1^{(1)}, a_{m\mu}^{(ls)}\right] = 0; \qquad \left[Q_1^{(1)}, \tilde{a}_{m\mu}^{(ls)}\right] = -i2^{-1/2} a_{m\mu}^{(ls)};$$

$$\left[Q_{-1}^{(1)}, a_{m\mu}^{(ls)}\right] = i2^{-1/2} \tilde{a}_{m\mu}^{(ls)}; \quad \left[Q_{-1}^{(1)}, \tilde{a}_{m\mu}^{(ls)}\right] = 0; \tag{15.48}$$

$$\left[Q_0^{(1)}, a_{m\mu}^{(ls)}\right] = \tfrac{i}{2} a_{m\mu}^{(ls)}; \qquad \left[Q_0^{(1)}, \tilde{a}_{m\mu}^{(ls)}\right] = -\tfrac{i}{2} \tilde{a}_{m\mu}^{(ls)}$$

and comparison with (14.13) shows that $a^{(ls)}$ and $\tilde{a}^{(ls)}$ are the components of an irreducible tensor of rank $q = 1/2$ in a quasispin space

$$a_{\rho m\mu}^{(qls)} = \begin{cases} a_{m\mu}^{(ls)} & \text{at } \rho = 1/2; \\ \tilde{a}_{m\mu}^{(ls)} & \text{at } \rho = -1/2, \end{cases} \tag{15.49}$$

i.e. we thus combine the electron creation and annihilation operators into a triple irreducible tensor operating in the spaces of quasispin, orbital and spin angular momenta.

It follows from (15.49) that the tensors $a^{(qls)}$ have the following property with respect to the operation of Hermitian conjugation:

$$a_{\rho m \mu}^{(qls)\dagger} = (-1)^{q+\rho+l-m+s-\mu} a_{-\rho-m-\mu}^{(qls)}, \qquad (15.50)$$

which agrees with (13.42).

Anticommutation relations (14.19) can be rewritten for the components of triple tensors:

$$\left\{ a_{\rho m \mu}^{(qls)}, a_{\rho' m' \mu'}^{(qls)} \right\} = (-1)^{q+l+s+\rho+m+\mu+1} \delta(\rho m \mu, -\rho' - m' - \mu')$$

$$= -2(2l+1)^{1/2} \begin{bmatrix} q & q & 0 \\ \rho & -\rho & 0 \end{bmatrix} \begin{bmatrix} l & l & 0 \\ m & -m & 0 \end{bmatrix} \begin{bmatrix} s & s & 0 \\ \mu & -\mu & 0 \end{bmatrix}. \quad (15.51)$$

Since $a_{\rho m \mu}^{(qls)}$ are irreducible tensors, we can define their irreducible tensorial product, i.e. the operator

$$W_{\Pi p \pi}^{(Kk\kappa)} = \left[a^{(qls)} \times a^{(qls)} \right]_{\Pi p \pi}^{(Kk\kappa)}$$

$$= \sum_{m \mu \rho m' \rho' \mu'} a_{\rho m \mu}^{(qls)} a_{\rho' m' \mu'}^{(qls)} \begin{bmatrix} q & q & K \\ \rho & \rho' & \Pi \end{bmatrix} \begin{bmatrix} l & l & k \\ m & m' & p \end{bmatrix} \begin{bmatrix} s & s & \kappa \\ \mu & \mu' & \pi \end{bmatrix}. \quad (15.52)$$

Anticommutation relations (15.51) impose some constraints on the values of ranks of operator $W^{(Kk\kappa)}$. Multiplying (15.51) by

$$\begin{bmatrix} q & q & K \\ \rho & \rho' & \Pi \end{bmatrix} \begin{bmatrix} l & l & k \\ m & m' & p \end{bmatrix} \begin{bmatrix} s & s & \kappa \\ \mu & \mu' & \pi \end{bmatrix}$$

and summing over the projections using the corresponding formulas or graphical methods, we arrive at the expression

$$\left[1 + (-1)^{K+k+\kappa} \right] \left[a^{(qls)} \times a^{(qls)} \right]^{(Kk\kappa)} = -\delta(Kk\kappa, 000)[q, l, s]^{1/2}, \quad (15.53)$$

from which it follows that $W^{(Kk\kappa)}$ at certain values of ranks becomes simply a number

$$W_{\Pi p \pi}^{(Kk\kappa)} = \begin{cases} 0, & \text{if } K + k + \kappa \neq 0 \text{ is even,} \\ -(2l+1)^{1/2}, & \text{if } K = k = \kappa = 0. \end{cases} \quad (15.54)$$

Tensors (15.52) in the quasispin method are the main operators. Their irreducible tensorial products are used to expand operators corresponding to physical quantities. That is why we shall take a closer look at their properties. Examining the internal structure of tensor $W^{(Kk\kappa)}$ yields

$$W_{\Pi \rho \pi}^{(Kk\kappa)\dagger} = (-1)^{\Pi + \rho + \pi} W_{-\Pi - \rho - \pi}^{(Kk\kappa)}. \quad (15.55)$$

This expression has a form similar to (13.43). However, for reasons laid down in the conclusion of Chapter 13, in the case of the tensors obeying (13.43) we also enclose ranks into parentheses.

For triple tensors $a^{(qls)}$ we shall express the irreducible components of quasispin operator (15.39)–(15.41) as follows:

$$Q_\rho^{(1)} = -\frac{i}{2}(2l+1)^{1/2} W_{\rho 00}^{(100)}. \qquad (15.56)$$

Similar expressions for irreducible components of the operators of orbital (14.11) and spin (14.12) momenta have the form

$$L_\rho^{(1)} = -i(l(l+1)(2l+1)/3)^{1/2} W_{0\rho 0}^{(010)}; \qquad (15.57)$$

$$S_\rho^{(1)} = -\frac{i}{2}(2l+1)^{1/2} W_{00\rho}^{(001)}. \qquad (15.58)$$

In general, the double tensor $W^{(k\kappa)}$ [see (14.30)] is related to the triple tensor

$$W_{\rho\pi}^{(k\kappa)} = -\frac{1}{\sqrt{2}} W_{0\rho\pi}^{(Kk\kappa)}. \qquad (15.59)$$

Here the required value of rank K (zero or unity) is determined by the condition that sum $K + k + \kappa$ is an odd number. Hence, by (14.34) and (14.35),

$$
\begin{aligned}
W_{0\rho 0}^{(0k0)} &= -(2k+1)^{1/2} U_\rho^k \qquad (k \text{ is odd}); \\[4pt]
W_{0\rho 0}^{(1k0)} &= -(2k+1)^{1/2} U_\rho^k \qquad (k \text{ is even}); \\[4pt]
W_{0\rho\pi}^{(1k1)} &= -2(2k+1)^{1/2} V_{\rho\pi}^{k1} \qquad (k \text{ is odd}); \\[4pt]
W_{0\rho\pi}^{(0k1)} &= -2(2k+1)^{1/2} V_{\rho\pi}^{k1} \qquad (k \text{ is even}).
\end{aligned}
\qquad (15.60)
$$

In addition, other equations are valid:

$$
\begin{aligned}
W_{1\rho\pi}^{(1k\kappa)} &= \left[a^{(ls)} \times a^{(ls)} \right]_{\rho\pi}^{(k\kappa)}; \\[4pt]
W_{-1\rho\pi}^{(1k\kappa)} &= \left[\tilde{a}^{(ls)} \times \tilde{a}^{(ls)} \right]_{\rho\pi}^{(k\kappa)}.
\end{aligned}
\qquad (15.61)
$$

Using (15.51) or (15.53), we can derive the commutation relations between any operators composed of tensors $W_{\Pi\rho\pi}^{(Kk\kappa)}$ and $a_{\rho m\mu}^{(qls)}$. For example,

$$
\left[W_{\Pi\rho\pi}^{(Kk\kappa)}, a_{\rho m\mu}^{(qls)} \right] = -\left[1 - (-1)^{K+k+\kappa} \right] \begin{bmatrix} K, k, \kappa \\ q, l, s \end{bmatrix}
$$
$$
\times \begin{bmatrix} q & K & q \\ \rho' & \Pi & \rho \end{bmatrix} \begin{bmatrix} l & k & l \\ m' & \rho & m \end{bmatrix} \begin{bmatrix} s & k & s \\ \mu' & \pi & \mu \end{bmatrix} a_{\rho'm'\mu'}^{(qls)}. \qquad (15.62)
$$

It is more convenient to write this formula in the irreducible tensorial form

$$\left[W^{(Kk\kappa)} \times a^{(qls)}\right]^{(K'k'\kappa')} - (-1)^{K+k+\kappa-K'-k'-\kappa'+l+1}$$

$$\times \left[a^{(qls)} \times W^{(Kk\kappa)}\right]^{(K'k'\kappa')} = \delta(K'k'\kappa', qls)$$

$$\times \left[1 - (-1)^{K+k+\kappa}\right] \begin{bmatrix} K, k, \kappa \\ q, l, s \end{bmatrix} a^{(qls)}. \tag{15.63}$$

This commutator is a generalization of (14.48)–(14.50). The commutation relations between the components of two triple tensors $W^{(Kk\kappa)}$ have the form

$$\left[W^{(Kk\kappa)}_{\Pi\rho\pi}, W^{(K'k'\kappa')}_{\Pi'\rho'\pi'}\right] = \sum_{K''k''\kappa''} \left[(-1)^{K+k+\kappa} - 1\right]$$

$$\times \left[(-1)^{K''+k''+\kappa''} - (-1)^{K+k+\kappa+K'+k'+\kappa'}\right] [K, K', k, k', \kappa, \kappa']^{1/2}$$

$$\times \begin{Bmatrix} q & q & K'' \\ K & K' & q \end{Bmatrix} \begin{Bmatrix} l & l & k'' \\ k & k' & l \end{Bmatrix} \begin{Bmatrix} s & s & \kappa'' \\ \kappa & \kappa' & s \end{Bmatrix}$$

$$\times \begin{bmatrix} K & K' & K'' \\ \Pi & \Pi' & \Pi'' \end{bmatrix} \begin{bmatrix} k & k' & k'' \\ \rho & \rho' & \rho'' \end{bmatrix} \begin{bmatrix} \kappa & \kappa' & \kappa'' \\ \pi & \pi' & \pi'' \end{bmatrix} W^{(K''k''\kappa'')}_{\Pi''\rho''\pi''} \tag{15.64}$$

or for the coupled momenta

$$\left[W^{(Kk\kappa)} \times W^{(K'k'\kappa')}\right]^{(K''k''\kappa'')} - (-1)^{K+k+\kappa+K'+k'+\kappa'+K''+k''+\kappa''}$$

$$\times \left[W^{(K'k'\kappa')} \times W^{(Kk\kappa)}\right]^{(K''k''\kappa'')} = \left[(-1)^{K+k+\kappa} - 1\right]$$

$$\times \left[(-1)^{K''+k''+\kappa''} - (-1)^{K+k+\kappa+K'+k'+\kappa'}\right] [K, K', k, k', \kappa, \kappa']^{1/2}$$

$$\times \begin{Bmatrix} q & q & K'' \\ K & K' & q \end{Bmatrix} \begin{Bmatrix} l & l & k'' \\ k & k' & l \end{Bmatrix} \begin{Bmatrix} s & s & \kappa'' \\ \kappa & \kappa' & s \end{Bmatrix} W^{(K''k''\kappa'')}. \tag{15.65}$$

Expressions (15.64) and (15.65) imply that the set of operators $W^{(Kk\kappa)}$ at all possible values of ranks and their respective projections are complete in relation to the commutation operation.

15.6   Connection of quasispin method with other group-theoretical methods

We noted in Chapter 6 that the seniority quantum number v in reduction chain (14.38) unambiguously classifies the irreducible representations of Sp_{4l+2} group. Then one may well ask: how can the earlier group-theoretical schemes include a rotation group defined by the operators of quasispin angular momentum?

It has long been believed that there are no groups of higher rank than U_{4l+2} that might be used to classify degenerate terms in a shell of equivalent electrons. However, Racah has established that for states with different numbers of electrons in the shell, some standard quantities of the theory are interrelated in a fairly simple manner, and the relationships (apart from the number of electrons) include quantum numbers v [23]. Since the generators of U_{4l+2} group (and its subgroups (14.38)) are tensors (14.35), which acting on the wave functions leave the number of particles unchanged, then from the group-theoretical viewpoint an explanation of the above feature can only be obtained using some more complex group associated with the transformations that change the number of particles in the shell. The total number of states in the shell is determined by the sum of numbers (9.4) of the possible antisymmetric states of the configuration over all possible N

$$\sum_{N=0}^{4l+2} C_{4l+2}^N = 2^{4l+2}.$$

The transformations providing the orthogonality of individual wave functions comprise the unitary group of dimension 2^{4l+2}, and all 2^{4l+2} states belong to one irreducible representation $\{1\}$ of this group. This group is, of course, too complicated, but Judd [99, 104] was successful in finding its important subgroup. It turned out that the operator set

$$a_{m\mu}^{(ls)\dagger}; \quad a_{m\mu}^{(ls)}; \quad a_{m\mu}^{(ls)} a_{m'\mu'}^{(ls)}; \quad a_{m\mu}^{(ls)\dagger} a_{m'\mu'}^{(ls)\dagger}; \quad a_{m\mu}^{(ls)} a_{m'\mu'}^{(ls)\dagger}; \quad a_{m\mu}^{(ls)\dagger} a_{m'\mu'}^{(ls)}$$

is complete with relation to the commutation operation and specifies rotation group R_{8l+5} of $8l + 5$ dimensions. Representation $\{1\}$ of group $U_{2^{4l+2}}$ is reduced to one irreducible representation of the R_{8l+5} group. Further, keeping only the bilinear combinations of electron creation and annihilation operators, we again return to the complete set of operators with relation to the commutation operation. The group defined by these operators is the rotation group of dimensions $8l+4$. The single representation of the R_{8l+5} group that characterizes all the states of the shell of equivalent electrons appears to be reducible with respect to R_{8l+4} subgroup. Thus, there exist two representations of this group that classify the states by the parity of the number of electrons in the shell [99, 104].

Bilinear combinations composed of creation and annihilation operators can be expressed in terms of triple tensors $W^{(K k \kappa)}$ (see (15.59) and (15.61)). In this case, these tensors are generators of the R_{8l+4} group, and relationship (15.64) determines the completeness condition for the generator set of this group in relation to the commutation operation. Out of this set we shall single out two subsets of operators – $W_{\Pi 00}^{(100)}$ and $W_{0\rho\pi}^{(0k\kappa)}$. Each of these subsets is complete with respect to the commutation operation, and besides,

$$\left[W_{\Pi 00}^{(100)}, W_{0\rho\pi}^{(0k\kappa)} \right] = 0.$$

Operator $W_{\Pi 00}^{(100)}$, by (15.56), is proportional to the quasispin operator, and operator $W^{(0k\kappa)}$, by (15.59), to double tensor $W^{(k\kappa)}$ with an odd sum of ranks $k + \kappa$. These tensors, as shown in Chapter 14, are generators of the Sp_{4l+2} group, and so the above selection of generator subsets corresponds to the reduction of the R_{8l+4} group on the direct product of two subgroups:

$$R_{8l+4} \to SU_2^Q \times Sp_{4l+2}.$$

In this case, the two subgroups are complementary to each other: corresponding unambiguously to the characteristics of the irreducible representations of the Sp_{4l+2} group are the characteristics of the irreducible representations of a rotation group in quasispin space SU_2^Q. From the group-theoretical viewpoint, this accounts for the possibility of using for a shell of equivalent electrons the techniques of the angular momentum theory in quasispin space.

When using reduction scheme (14.38) the states of the l^N configuration are classified by eigenvalues of the Casimir operators of the groups concerned. If we agree to do without higher-rank groups, then we shall be able to characterize wave functions successfully by quantum numbers Q, L, S, M_Q, M_L, M_S. To establish a relationship between the characteristics of interest we shall formulate a method that will make it possible to form linear combinations of tensorial products of operators $W^{(Kk\kappa)}$, which are given in terms of quantities composed of operators of momentum type. To this end, in the tensorial product

$$\left[W^{(Kk\kappa)} \times W^{(K'k'\kappa')} \right]^{(K''k''\kappa'')}$$

$$= \left[\left[a^{(qls)} \times a^{(qls)} \right]^{(Kk\kappa)} \times \left[a^{(qls)} \times a^{(qls)} \right]^{(K'k'\kappa')} \right]^{(K''k''\kappa'')}$$

we shall interchange the operators connected by the arrows and change the scheme of moments coupling so that the ranks of the operators being interchanged will be directly coupled. Applying then relation (15.53) and passing to the initial scheme of coupling the momenta, we obtain [85]

$$\left[W^{(Kk\kappa)} \times W^{(K'k'\kappa')} \right]_{\Pi\rho\pi}^{(K''k''\kappa'')}$$

$$= -\sum_{\substack{K_1 k_1 \kappa_1 \\ K_2 k_2 \kappa_2}} \left[W^{(K_1 k_1 \kappa_1)} \times W^{(K_2 k_2 \kappa_2)} \right]_{\Pi\rho\pi}^{(K''k''\kappa'')}$$

$$\times [K, K', K_1, K_2, k, k', k_1, k_2, \kappa, \kappa', \kappa_1, \kappa_2]^{1/2} \begin{Bmatrix} q & q & K \\ q & q & K' \\ K_1 & K_2 & K'' \end{Bmatrix}$$

$$\times \begin{Bmatrix} l & l & k \\ l & l & k' \\ k_1 & k_2 & k'' \end{Bmatrix} \begin{Bmatrix} s & s & \kappa \\ s & s & \kappa' \\ \kappa_1 & \kappa_2 & \kappa'' \end{Bmatrix} - (-1)^{K''+k''+\kappa''},$$

$$\times [K,K',k,k',\kappa,\kappa']^{1/2} \begin{Bmatrix} K & K' & K'' \\ q & q & q \end{Bmatrix} \begin{Bmatrix} k & k' & k'' \\ l & l & l \end{Bmatrix}$$

$$\times \begin{Bmatrix} \kappa & \kappa' & \kappa'' \\ s & s & s \end{Bmatrix} W_{\Pi\rho\pi}^{(K''k''\kappa'')}. \tag{15.66}$$

The ranks of triple tensors in (15.66) can be selected so that the left side contains operators with known eigenvalues. Then in the right side, we have the desired linear combinations of irreducible products of triple tensors.

To begin with we consider the case of complete scalars $K'' = k'' = \kappa'' = 0$ and assume that the ranks of operators $W^{(Kk\kappa)}$ and $W^{(K'k'\kappa')}$ in (15.66) take on different values. So, letting $K = K' = k = k' = \kappa = \kappa' = 0$ we get

$$\sum_k (2k+1)^{1/2} \left\{ \left[W^{(0k0)} \times W^{(0k0)} \right]^{(000)} + 3^{1/2} \left[W^{(0k1)} \times W^{(0k1)} \right]^{(000)} \right.$$

$$+ 3^{1/2} \left[W^{(1k0)} \times W^{(1k0)} \right]^{(000)} + 3 \left[W^{(1k1)} \times W^{(1k1)} \right]^{(000)} \right\}$$

$$= -(2l+1)(2l+3). \tag{15.67}$$

At $K = K' = 1$ and $k = k' = \kappa = \kappa' = 0$,

$$\sum_k (2k+1)^{1/2} \left\{ 3^{1/2} \left[W^{(0k0)} \times W^{(0k0)} \right]^{(000)} + 3 \left[W^{(0k1)} \times W^{(0k1)} \right]^{(000)} \right.$$

$$- \left[W^{(1k0)} \times W^{(1k0)} \right]^{(000)} - 3^{1/2} \left[W^{(1k1)} \times W^{(1k1)} \right]^{(000)} \right\}$$

$$= 3^{-1/2} \left[16Q^2 - 9(2l+1) \right]. \tag{15.68}$$

The third relationship can be obtained by putting $K = K' = k = k' = 0$ and $\kappa = \kappa' = 1$

$$\sum_k 3(2k+1)^{1/2} \left\{ \left[W^{(0k0)} \times W^{(0k0)} \right]^{(000)} - 3^{1/2} \left[W^{(0k1)} \times W^{(0k1)} \right]^{(000)} \right.$$

$$+ 3^{3/2} \left[W^{(1k0)} \times W^{(1k0)} \right]^{(000)} - 3 \left[W^{(1k1)} \times W^{(1k1)} \right]^{(000)} \right\}$$

$$= 16S^2 - 9(2l+1), \tag{15.69}$$

and the fourth one is found when $K = K' = \kappa = \kappa' = 1$ and $k = k' = 0$:

$$\sum_k (2k+1) \left\{ 3 \left[W^{(0k0)} \times W^{(0k0)} \right]^{(000)} - 3^{1/2} \left[W^{(0k1)} \times W^{(0k1)} \right]^{(000)} \right.$$

$$+ 3^{1/2} \left[W^{(1k0)} \times W^{(1k0)} \right]^{(000)} + \left[W^{(1k1)} \times W^{(1k1)} \right]^{(000)} \right\}$$

$$= 3(2l+1). \tag{15.70}$$

Now we shall solve the resultant set of equations (15.67)–(15.70) for the sums of a certain parity of orbital ranks and a given structure of spin and quasispin ranks to obtain

$$\sum_k (2k+1)^{1/2} \left[W^{(0k0)} \times W^{(0k0)} \right]^{(000)}$$

$$= S^2 + Q^2 - (2l+1)(2l+3)/4, \tag{15.71}$$

$$\sum_k [3(2k+1)]^{1/2} \left[W^{(0k1)} \times W^{(0k1)} \right]^{(000)}$$

$$= 3Q^2 - S^2 - 3(2l+1)(2l+3)/4, \tag{15.72}$$

$$\sum_k [3(2k+1)]^{1/2} \left[W^{(1k0)} \times W^{(1k0)} \right]^{(000)}$$

$$= 3S^2 - Q^2 - 3(2l+1)(2l+3)/4, \tag{15.73}$$

$$\sum_k (2k+1)^{1/2} \left[W^{(1k1)} \times W^{(1k1)} \right]^{(000)}$$

$$= -S^2 - Q^2 - 3(2l-1)(2l+1)/4. \tag{15.74}$$

Eigenvalues of the operators on the right sides of these relationships are given in terms of quantum numbers of the states of the l^N configurations, thereby defining the values of the matrix elements of linear combinations of tensors on the left sides.

Let us now return to the Casimir operators for groups Sp_{4l+2}, SU_{2l+1}, R_{2l+1}, which can also be expressed in terms of linear combinations of irreducible tensorial products of triple tensors $W^{(K\kappa k)}$. To this end, we insert into the scalar products of operators U^k (or V^{k1}), their expressions in terms of triple tensors (15.60) and then expand the direct product in terms of irreducible components in quasispin space. As a result, we arrive at

$$\left(U^k \cdot U^k \right) = - [(2k+1)3]^{-1/2} \left\{ \left[W^{(1k0)} \times W^{(1k0)} \right]^{(000)}_{000} \right.$$
$$\left. - \sqrt{2} \left[W^{(1k0)} \times W^{(1k0)} \right]^{(200)}_{000} \right\};$$

$$\left(V^{k1} \cdot V^{k1} \right) = -\frac{1}{4} [3/(2k+1)]^{1/2} \left[W^{(0k1)} \times W^{(0k1)} \right]^{(000)}_{000} \tag{15.75}$$

at even k; and

$$\left(U^k \cdot U^k \right) = -(2k+1)^{-1/2} \left[W^{(0k0)} \times W^{(0k0)} \right]^{(000)}_{000};$$

$$\left(V^{k1} \cdot V^{k1}\right) = -\frac{1}{4}(2k+1)^{-1/2}\left\{\left[W^{(1k1)} \times W^{(1k1)}\right]_{000}^{(000)} \right.$$
$$\left. -\sqrt{2}\left[W^{(1k1)} \times W^{(1k1)}\right]_{000}^{(200)}\right\} \quad (15.76)$$

at odd k.

It is worth mentioning that expressions (15.75) and (15.76), apart from terms that are scalar with respect to quasispin, also include terms that contain a second-rank tensor part in quasispin space. Substituting (15.75) and (15.76) into the expressions for the Casimir operators of appropriate groups (5.34), (5.29) and (5.33) yields

$$\mathscr{G}[Sp_{4l+2}] = -\frac{1}{2}\sum_k \frac{[3(2k+1)]^{1/2}}{(l\|u^k\|l)^2}\left[W^{(0k1)} \times W^{(0k1)}\right]_{000}^{(000)}$$
$$-\frac{1}{2}\mathscr{G}[R_{2l+1}]; \quad (15.77)$$

$$\mathscr{G}[R_{2l+1}] = \sum_{k>1}(2k+1)^{1/2}\left[W^{(0k0)} \times W^{(0k0)}\right]_{000}^{(000)}; \quad (15.78)$$

$$\mathscr{G}[SU_{2l+1}] = \mathscr{G}[R_{2l+1}] + \sum_k(2k+1)^{1/2}$$
$$\times\left\{3\left[W^{(1k0)} \times W^{(1k0)}\right]_{000}^{(000)} - 2^{1/2}\left[W^{(1k0)} \times W^{(1k0)}\right]_{000}^{(200)}\right\}. \quad (15.79)$$

The right side of (15.77) can be transformed using (15.71) and (15.72). Then,

$$\mathscr{G}[Sp_{4l+2}] = [(2l+1)(2l+3) - 4\mathbf{Q}^2]/2. \quad (15.80)$$

Similarly, the Casimir operator of R_{2l+1} group can, from (15.71), be written as

$$\mathscr{G}[R_{2l+1}] = \mathbf{S}^2 + \mathbf{Q}^2 - (2l+1)(2l+3)/4. \quad (15.81)$$

In consequence, classification of antisymmetric wave functions of l^N configurations using the characteristics of irreducible representations of Sp_{4l+2} and R_{2l+1} groups is fully equivalent to classification by the eigenvalues of operators $\mathbf{S}^2$ and $\mathbf{Q}^2$. If now we take into account formula (9.22) relating the quasispin quantum number to the seniority quantum number, we can establish the equivalence of (15.80) and (5.38).

With the SU_{2l+1} group, application of (15.71) and (15.73) enables us to express in terms of Q, L, S only the part of the Casimir operator that is scalar in quasispin space. To arrive at similar expressions for the tensor part of this operator, we first put in (15.66) $k = k' = k'' = \kappa = \kappa' = \kappa'' = 0$; $K = K' = 1$; $K'' = 2$, and then $k = k' = k'' = \kappa'' = 0$;

$K = K' = \kappa = \kappa' = 1; K'' = 2$, and isolate terms of different parity. Then,

$$\sum_{k>0}(2k+1)^{1/2}\left[W^{(1k0)}\times W^{(1k0)}\right]^{(200)}_{\rho00} = \frac{2(2l+3)}{2l+1}Q^{(2)}_{\rho}; \qquad (15.82)$$

$$\sum_{k}(2k+1)^{1/2}\left[W^{(1k1)}\times W^{(1k1)}\right]^{(200)}_{\rho00} = 2\cdot 3^{1/2}Q^{(2)}_{\rho}. \qquad (15.83)$$

Using (15.75), (15.73) and (15.82), we can find the Casimir operator of the SU_{2l+1} group:

$$\sum_{k>0}(2k+1)\left(U^{k}\cdot U^{k}\right) = -2\mathbf{S}^2 - [2(2l+3)/(2l+1)]Q_z Q_z + (2l+1)(2l+3)/2.$$
$$(15.84)$$

It is seen from this equation that this operator depends on the number of electrons in a simpler way than, for example, in [98].

In deriving (15.84) we have used the relationship

$$Q^{(2)}_0 = -[3/2]^{1/2}Q_z Q_z + 6^{-1/2}\mathbf{Q}^2. \qquad (15.85)$$

In an analogous way, using (15.74), (15.76) and (15.83), we find a new relationship, which yields a much simpler expression for the appropriate matrix element than in [90]

$$16\sum_{k}(2k+1)(V^{k1}\cdot V^{k1}) = 4\mathbf{S}^2 + 12\mathbf{Q}^2 - 24Q_z Q_z + 3(2l-1)(2l+1). \qquad (15.86)$$

All the expressions derived so far are suitable for shells with any values of quantum number l. For specific l values, additional relations can be constructed using (15.66), which makes it possible in some cases to obtain formulas similar to those derived above, not for sums, but for their individual terms.

For the p-shell we do not need these additional relationships, since from the set of equations (15.71)–(15.74), (15.82) and (15.83) we can work out expressions for all the irreducible tensorial products. Directly from (15.71) we obtain (hereinafter a letter over the equality sign shows what shell this particular equation is valid for)

$$\mathbf{L}^2 \overset{p}{=} -2\mathbf{Q}^2 - 2\mathbf{S}^2 + 3\cdot 5/2. \qquad (15.87)$$

It follows from (15.87) that when the states of p^N configuration are classified, one of the three quantum numbers Q, L, S is extraneous, because these numbers are related [90] by

$$L(L+1) \overset{p}{=} -2Q(Q+1) - 2S(S+1) + 3\cdot 5/2$$

or by (9.18).

Further,

$$2 \cdot 3\sqrt{3 \cdot 5} \left[W^{(120)} \times W^{(120)} \right]^{(000)} \overset{p}{=} -\mathbf{L}^2 + 4 \cdot 4\mathbf{S}^2 - 3 \cdot 4 \cdot 5;$$

$$2\sqrt{2} \left[W^{(111)} \times W^{(111)} \right]^{(000)} \overset{p}{=} \mathbf{L}^2 - 3 \cdot 4;$$

$$2 \cdot 3\sqrt{3 \cdot 5} \left[W^{(021)} \times W^{(021)} \right]^{(000)} \overset{p}{=} \mathbf{L}^2 - 4 \cdot 4\mathbf{Q}^2 - 3 \cdot 4 \cdot 5;$$

$$3\sqrt{5} \left[W^{(120)} \times W^{(120)} \right]_{\rho 00}^{(200)} \overset{p}{=} 2 \cdot 5Q_\rho^{(2)};$$

$$\left[W^{(111)} \times W^{(111)} \right]_{\rho 00}^{(200)} \overset{p}{=} 2Q_\rho^{(2)}.$$

Using (15.87), we obtain by (15.86)

$$2 \cdot 3 \cdot 4(V^{11} \cdot V^{11}) \overset{p}{=} -4\mathbf{S}^2 - 3\mathbf{L}^2 - 3 \cdot 4Q_z Q_z + 3. \tag{15.88}$$

For the d-shell, the expressions for irreducible tensorial products in terms of operators whose eigenvalues are determined by the quantum numbers of the states of the shell, are available only for complete scalars in all three spaces [85].

In [90] the relationship between eigenvalues of the Casimir operators of higher-rank groups and quantum numbers v, N, L, S is taken into account to work out algebraic expressions for some of the reduced matrix elements of operators $(U^k \cdot U^k)$ and $(V^{k1} \cdot V^{k1})$. However, the above formulas directly relate the operators concerned, and some of these formulas are not defined by the Casimir operators of respective groups.

For repeating terms of the f^N configuration to be classified, the seniority quantum number is no longer sufficient; in this case (see (14.39)) the characteristics of irreducible representations of the G_2 group [24] are used. The generators of this group are the fourteen operators $W_{0\rho 0}^{(010)}$, $W_{0\rho 0}^{(050)}$, and Casimir operator (5.32) can be expressed in terms of their tensorial product of the zeroth rank

$$\mathscr{G}[G_2] = -\sqrt{3} \left[W^{(010)} \times W^{(010)} \right]^{(000)} - \sqrt{11} \left[W^{(050)} \times W^{(050)} \right]^{(000)}. \tag{15.89}$$

Using (15.57) at once gives

$$\sqrt{11} \left[W^{(050)} \times W^{(050)} \right]^{(000)} = -\mathscr{G}[G_2] - \mathbf{L}^2/28. \tag{15.90}$$

Solving the sets of equations that follow from (15.66) at special values of ranks k and κ we find other algebraic expressions of the same kind [85].

15.7   Expansion of operators of physical quantities in quasispin space

We have established above that one-electron operators are expressible in terms of tensors $W^{(k\kappa)}$ related to triple tensors $W^{(Kk\kappa)}$ by (15.59). Therefore, we shall find here the expansion in terms of irreducible tensors in quasispin space only for the two-particle operator that is a scalar in the total momentum.

In Chapter 14 we derived, in the second-quantization representation, two different forms of the expressions for that operator – (14.61) and (14.63). To begin with, we consider expression (14.63) in which we, by (15.49), go over to triple tensors. Then, after some transformations and coupling the momenta in quasispin space, we arrive at

$$G = -\frac{1}{2} \sum_{LSL'S'Kk\rho} (2k+1)^{-1/2} (nlnlLS \| g_{12} \| nlnlL'S')$$

$$\times (-1)^{k-\rho} \begin{bmatrix} 1 & 1 & K \\ 1 & -1 & 0 \end{bmatrix} \left[W^{(1LS)} \times W^{(1L'S')} \right]_{0\rho-\rho}^{(Kkk)}. \quad (15.91)$$

Likewise, for (14.61)

$$G = \frac{1}{4}(nlnl \| g_{12} \| nlnl) \left\{ [k_1, k_2, \kappa_1, \kappa_2]^{-1/2} \right.$$

$$\times \sum_{K_1K_2Kk\rho} (-1)^{k-\rho} \left(\begin{bmatrix} K_1 & K_2 & K \\ 0 & 0 & 0 \end{bmatrix} \left[W^{(K_1k_1\kappa_1)} \times W^{(K_2k_2\kappa_2)} \right]_{0\rho-\rho}^{(Kkk)} \right.$$

$$\left. - \sqrt{2} \begin{Bmatrix} k_1 & k_2 & k \\ l & l & l \end{Bmatrix} \begin{Bmatrix} \kappa_1 & \kappa_2 & k \\ s & s & s \end{Bmatrix} W_{0\rho-\rho}^{(Kkk)} \right) \right\}. \quad (15.92)$$

Let us take the example of the operator of electrostatic interaction between electrons. According to (15.92), it is

$$V = \frac{1}{2}\hat{N}(\hat{N}-1)F_0(nl, nl) + \frac{1}{2} \sum_{k>0;K_1} (l \| C^{(k)} \| l)^2$$

$$\times \left\{ \frac{1}{\sqrt{2k+1}} \begin{bmatrix} 1 & 1 & K_1 \\ 0 & 0 & 0 \end{bmatrix} \left[W^{(1k0)} \times W^{(1k0)} \right]_{000}^{(K_100)} - \frac{\hat{N}}{2l+1} \right\}$$

$$\times F_k(nl, nl). \quad (15.93)$$

For specific values of quantum numbers of the shell, the part of this expression that is scalar in all three spaces, can be represented in terms of operators, whose eigenvalues are determined by the quantum numbers of the states of a shell. For the p-shell we have [90]

$$V \stackrel{p}{=} \frac{1}{2}\hat{N}(\hat{N}-1)F_0(np, np) - \frac{1}{2 \cdot 5 \cdot 5} \left[3L^2 + 3 \cdot 4S^2 + 5N(N-4) \right] F_2(nl, nl). \quad (15.94)$$

For the *d*-shell [85]

$$V \stackrel{d}{=} \frac{1}{2}\hat{N}(\hat{N}-1)F_0(nd,nd) - \frac{1}{7^2}\left[\frac{1}{2}\mathbf{L}^2 + \frac{4}{3}\mathbf{Q}^2 + 4\mathbf{S}^2 - 2\cdot 7Q_z\right]$$

$$\times F_2(nd,nd) - \frac{5}{3\cdot 7^2}\left[\mathbf{S}^2 - \frac{1}{2\cdot 3}\mathbf{L}^2 - \frac{5^2}{3^2}\mathbf{Q}^2 + \frac{7^2}{4} + \frac{2\cdot 7^2}{3\cdot 5}Q_zQ_z\right.$$

$$\left. + \frac{2\cdot 3\cdot 7}{5}Q_z\right]F_4(nd,nd) + \frac{1}{7}\sqrt{\frac{2\cdot 5}{3}}\left[W^{(120)} \times W^{(120)}\right]^{(200)}_{000}$$

$$\times \left[F_2(nd,nd) - \frac{5}{9}F_4(nd,nd)\right], \tag{15.95}$$

i.e. only one term remains for which the coefficients of radial integrals cannot be expressed in terms of the quantum numbers of the shell. This operator determines all the non-diagonal matrix elements of the operator of the energy of electrostatic interaction for the *d*-shell.

In summary, the techniques described in this chapter allow us to derive expansions of the operators that correspond to physical quantities, in terms of irreducible tensors in the spaces of orbital, spin and quasispin momenta, and also to separate terms that can be expressed by operators whose eigenvalues have simple analytical forms. Since the operators of physical quantities also contain terms for which this separation is impossible, the following chapter will be devoted to the general technique of finding the matrix elements of quantities under consideration.

16

Algebraic expressions for coefficients of fractional parentage (CFP)

16.1 Wave functions in quasispin space

It follows from (14.5) that when $Q_{\pm 1}^{(1)}$ operates on wave function (15.44), this changes only the z-projection of the quasispin momentum, and so we can obtain from the wave function of N electrons appropriate values for $N + 2$ electrons with the same quantum numbers

$$|l\alpha QLS(M_Q + 1)) = i\sqrt{2}\,[(Q + M_Q + 1)(Q - M_Q)]^{-1/2}\,\hat{Q}_1^{(1)}|l\alpha QLSM_Q).$$

(16.1)

Since the wave functions with $N > v$ can be found from the wave functions with $N = v$ using (16.1), in the second-quantization representation it is necessary to construct in an explicit form only wave functions with the number of electrons minimal for given v, i.e. $|l\alpha QLSM_Q = -Q)$. But even such wave functions cannot be found by generalizing directly relation (15.4): if operator $\hat{\varphi}$ is still defined so that it would be an irreducible tensor in quasispin space, then the wave function it produces in the general case will not be characterized by some value of quantum number $Q(v)$. This is because the vacuum state $|0)$ in quasispin space of one shell is not a scalar, but a component of a tensor of rank $Q = l + 1/2$

$$|0) = \left|lQ = l + \frac{1}{2};\ L = 0;\ S = 0;\ M_Q = -l - \frac{1}{2}\right).$$

All the other components of this tensor – the wave functions of $_0^1 S$ with different numbers of electrons – can be obtained from (16.1).

Operator $\hat{\varphi}(l^N)^{\alpha v(LS)}$ in the general case consists of a linear combination of creation operators that provides a classification of states according to quantum numbers α, v. Since each term of this expansion contains N creation operators then from (15.49) the quasispin rank of operator $\hat{\varphi}(l^N)^{\alpha v(LS)}$ and its projection are equal to $N/2$ (regardless of the values of quantum numbers α, v). Accordingly, for a function with a certain

160

quantum number $Q(v)$ the counterpart of (15.4) will be

$$|l^N \alpha Q L S M_L M_S\rangle = \hat{\varphi}(l^N)^{\alpha v(\frac{N}{2} L S)}_{\frac{N}{2} M_L M_S}|0\rangle. \qquad (16.2)$$

The rank of the operator $\hat{\varphi}$ in quasispin space will only be determined by the number of electrons N and can be ignored as a characteristic. In particular, operator $Q_1^{(1)}$ – a tensor of rank one in quasispin space – acting on the vacuum state gives rise to the two-electron wave function with $v = 0$

$$\hat{\varphi}^{0(100)}_{100}|0\rangle = i[2/(2l+1)]^{1/2}Q_1^{(1)}|0\rangle = |l^2{}^2_0 S\rangle. \qquad (16.3)$$

When there is sufficient to have quantum number $Q(v)$ to achieve an unambiguous classification of terms, then the wave functions can conveniently be constructed in a recurrence way – by forming the irreducible tensorial product

$$|lQLSM_Q M_L M_S\rangle = \mathcal{N}\left[T^{(K k \kappa)} \times |lQ'L'S'\rangle\right]^{(Q\ L\ S)}_{M_Q M_L M_S} \qquad (16.4)$$

of a tensor $T^{(K k \kappa)}$ and known wave functions $|lQ'L'S'M_{Q'}M_{L'}M_{S'}\rangle$ ($\mathcal{N}$ is the normalization factor). In (16.4) we substitute operator $W^{(1 k \kappa)}$, $k+\kappa \neq 0$, defined by (15.52), for tensor $T^{(K k \kappa)}$, and the wave function $|lQ+1L'S'\rangle$ with $v' = v - 2$ for function $|lQ'L'S'\rangle$. Then, using (16.1) and commutation relations (14.13) we have after some algebra

$$|lQLS(M_Q = -Q)M_L M_S\rangle = \mathcal{N}\left[\frac{2Q+3}{2Q+1}\right]^{1/2}\sum_{\rho \pi M_{L'} M_{S'}}\begin{bmatrix} k & L' & L \\ \rho & M_{L'} & M_L \end{bmatrix}$$

$$\times \begin{bmatrix} \kappa & S' & S \\ \pi & M_{S'} & M_S \end{bmatrix}\left\{W^{(1 k \kappa)}_{1\rho\pi} - \frac{(2l+1)^{1/2}}{2(Q+1)}W^{(100)}_{100}W^{(1 k \kappa)}_{0\rho\pi}\right.$$

$$\left. + \frac{l+1/2}{2(Q+1)(2Q+3)}W^{(100)}_{100}W^{(100)}_{100}W^{(1 k \kappa)}_{-1\rho\pi}\right\}$$

$$\times |l(Q+1)L'S'(M_Q = -Q-1)M_{L'}M_{S'}\rangle. \qquad (16.5)$$

In (16.4) and (16.5) ranks K, k, κ are arbitrary save for the fact that they must obey the appropriate triangle conditions. Relation (16.5) enables us to construct wave functions characterized by seniority quantum number v, if the quantities with $v' = v - 2$ are known. To establish a similar formula relating wave functions to v and $v' = v - 1$, we must in (16.4) substitute operator $a^{(qls)}$ for tensor $T^{(K k \kappa)}$. Reasoning along the same lines as in deriving (16.5), we shall get

$$|lQLSM_QM_LM_S) = \mathcal{N}' \left[\frac{2Q+2}{2Q+1}\right]^{1/2} \sum_{mM_{L'}\mu M_{S'}} \begin{bmatrix} l & L' & L \\ m & M_{L'} & M_L \end{bmatrix}$$

$$\times \begin{bmatrix} s & S' & S \\ \mu & M_{S'} & M_S \end{bmatrix} \left\{ a_{m\mu}^{(ls)} - \frac{(l+1/2)^{1/2}}{Q+1} W_{10}^{(10)} \tilde{a}_{m\mu}^{(ls)} \right\}$$

$$\times |l(Q+\tfrac{1}{2})L'S'(M_Q = -Q-\tfrac{1}{2})M_{L'}M_{S'}). \qquad (16.6)$$

Expressions (16.5) and (16.6) include only wave functions where the number of electrons equals the seniority number.

Using relationships (16.5) and (16.6) and taking into account the tensorial structure of the vacuum state we find that the second-quantized operators $\hat{\varphi}(l^v)^{v(LS)}$, that produce from the vacuum the function with $N = v = 2$ and $N = v = 3$, have the form

$$\hat{\varphi}(l^2)^{2(LS)} = \sqrt{2}\,[1 - \delta(LS, 00)] \left[a^{(ls)} \times a^{(ls)}\right]^{(LS)}; \qquad (16.7)$$

$$\hat{\varphi}(l^3)^{3(LS)} = N_3 \left\{ \left[a^{(ls)} \times \left[a^{(ls)} \times a^{(ls)}\right]^{(L_2'S_2')}\right]^{(L_3S_3)} \right.$$

$$\left. + A_3 \left[a^{(ls)} \times \left[a^{(ls)} \times a^{(ls)}\right]^{(00)}\right]^{(L_3S_3)} \right\}. \qquad (16.8)$$

Here N_3 is a normalization factor, and

$$A_3 = \delta(L_3S_3, ls)[L_2', S_2']^{1/2}/(2l). \qquad (16.9)$$

In general, the operator $\hat{\varphi}(l^N)^{\alpha(LS)}$ can be expressed in terms of a linear combination of creation operators that provides a classification of the wave function produced in the additional quantum numbers α. For example, in relationship (16.8) there first appear the linear combinations of tensorial products of creation operators that provide classification of the wave functions produced from the vacuum in the seniority quantum number $v = 3$ (quasispin $Q = l - 1$).

The wave function with $N = v = 4$ can be constructed in two ways. So, applying (16.5) twice gives

$$\hat{\varphi}(l^4)^{4(L_4S_4)} = \mathcal{N}_4 \left\{ \left[\left[a^{(ls)} \times a^{(ls)}\right]^{(k_2'\kappa_2')} \times \left[a^{(ls)} \times a^{(ls)}\right]^{(k_2\kappa_2)}\right]^{(L_4S_4)} \right.$$

$$\left. + A_4 \left[a^{(ls)} \times a^{(ls)}\right]^{(00)} \left[a^{(ls)} \times a^{(ls)}\right]^{(L_4S_4)} \right\}, \qquad (16.10)$$

where $\mathcal{N}_4$ is a normalization factor, and

$$A_4 = \frac{1+(-1)^{L_4+S_4}}{3-2l}\left([l,s,k_2,\kappa_2,k_2',\kappa_2']^{1/2}\begin{Bmatrix} l & k_2' & l \\ k_2 & l & L_4 \end{Bmatrix}\right.$$

$$\left.\times\begin{Bmatrix} s & \kappa_2' & s \\ \kappa_2 & s & S_4 \end{Bmatrix} - \frac{2l+1}{2l}[k_2,\kappa_2]^{1/2}\delta(k_2\kappa_2,k_2'\kappa_2')\delta(L_4S_4,00)\right). \quad (16.11)$$

If, as an alternative, we use (16.6), then we obtain another representation of the operator that gives rise to the wave function with $v = 4$

$$\hat{\varphi}(l^4)^{4(L_4S_4)} = N_4'\left\{\left[a^{(ls)}\times\left[a^{(ls)}\times\left[a^{(ls)}\times a^{(ls)}\right]^{(k_2\kappa_2)}\right]^{(k_3\kappa_3)}\right]^{(L_4S_4)}\right.$$

$$\left.+ A_4'\left[a^{(ls)}\times a^{(ls)}\right]^{(00)}\times\left[a^{(ls)}\times a^{(ls)}\right]^{(L_4S_4)}\right\}, \quad (16.12)$$

where N_4' is the normalization factor, and

$$A_4' = \frac{1}{4l-2}\left[\frac{\delta(k_3\kappa_3,ls)}{l}(2l-(2l+1)\delta(L_4S_4,00))\right.$$

$$+(-1)^{l+s+k_3+\kappa_3}[l,s,k_2,\kappa_2,k_3,\kappa_3]^{1/2}\left(2\begin{Bmatrix} l & l & L_4 \\ l & k_3 & k_2 \end{Bmatrix}\right.$$

$$\left.\left.\times\begin{Bmatrix} s & s & S_4 \\ s & \kappa_3 & K_2 \end{Bmatrix} + \frac{\delta(k_2\kappa_2,L_4S_4)}{[k_2,\kappa_2]}\right)\right]. \quad (16.13)$$

16.2 Reduced coefficients (subcoefficients) of fractional parentage

As has been shown, second-quantized operators can be expanded in terms of triple tensors in the spaces of orbital, spin and quasispin angular momenta. The wave functions of a shell of equivalent electrons (15.46) are also classified using the quantum numbers L, S, Q, M_L, M_S, M_Q of the three commuting angular momenta. Therefore, we can apply the Wigner–Eckart theorem (5.15) in all three spaces to the matrix elements of any irreducible triple tensorial operator $T^{(K k \kappa)}$ defined relative to wave functions (15.46)

$$\left(l\alpha QLSM_QM_LM_S\,|\,T^{(K k\kappa)}_{\Pi\rho\pi}\,|\,l\alpha'Q'L'S'M_Q'M_L'M_S'\right)$$

$$= (-1)^{2(K+k+\kappa)}[Q,L,S]^{-1/2}\begin{bmatrix} Q' & K & Q \\ M_Q' & \Pi & M_Q \end{bmatrix}\begin{bmatrix} L' & k & L \\ M_L' & \rho & M_L \end{bmatrix}$$

$$\times\begin{bmatrix} S' & \kappa & S \\ M_S' & \pi & M_S \end{bmatrix}\left(l\alpha QLS\,|||\,T^{(K k\kappa)}\,|||\,l\alpha'Q'L'S'\right), \quad (16.14)$$

where the last factor is the completely reduced matrix (reduced submatrix) element. The special designation introduced for it – three vertical lines instead of the conventional two – is used here because in the literature the Wigner–Eckart theorem is normally employed only to single out the dependence of the matrix elements on projections of the orbital and spin angular momenta. The designation introduced here thus stresses that the Wigner–Eckart theorem has been applied once more, and the conventional reduced matrix (submatrix) elements are related to the quantities that enter into (16.14) by

$$\left(l\alpha QLSM_Q\| T_{\Pi}^{(Kk\kappa)}\|l\alpha' Q'L'S'M'_Q\right) = (-1)^{2K}$$

$$\times (2Q+1)^{-1/2} \begin{bmatrix} Q' & K & Q \\ M'_Q & \Pi & M_Q \end{bmatrix} \left(l\alpha QLS\|\|T^{(Kk\kappa)}\|\|l\alpha' Q'L'S'\right). \quad (16.15)$$

Specifically, this relationship holds for the electron creation operator $a_{m\mu}^{(ls)}$ which, by (15.49), is a component $v = 1/2$ of the triple tensor $a_{vm\mu}^{(qls)}$. If we take into consideration that the submatrix element of the creation operator is expressed in terms of the CFP from (15.21), we have

$$\left(l^N\alpha QLS\|l^{N-1}(\alpha_1 Q_1 L_1 S_1)l\right) = (-1)^{N-1}(N[Q,L,S])^{-1/2}$$

$$\times \begin{bmatrix} Q_1 & 1/2 & Q \\ M_{Q_1} & 1/2 & M_Q \end{bmatrix} \left(l\alpha QLS\|\|a^{(qls)}\|\|l\alpha_1 Q_1 L_1 S_1\right). \quad (16.16)$$

We refer to the last factor in this expression as the reduced coefficient (or subcoefficient) of fractional parentage (SCFP) [91]. It follows from (16.16) that these coefficients can be expressed fairly simply in terms of CFP for the term with $N = v$ that occur for the first time:

$$\left(l\alpha QLS\|\|a^{(qls)}\|\|l\alpha' Q'L'S'\right) = (l\alpha QLS\|\|l(\alpha' Q'L'S')l)$$

$$= (-1)^v \{2v(Q+1)[L,S]\}^{1/2} \left(l^v\alpha vLS\|l^{v-1}(\alpha'(v-1)L'S')l\right), \quad (16.17)$$

and their number equals the number of CFP for the terms that occur for the first time for a given l^N.

The transposition condition for SCFP is

$$\left(l\alpha QLS\|\|a^{(qls)}\|\|l\alpha' Q'L'S'\right) = (-1)^{l+Q+L+S-Q'-L'-S'}$$

$$\times \left(l\alpha' Q'L'S'\|\|a^{(qls)}\|\|l\alpha QLS\right). \quad (16.18)$$

In Chapters 9 and 15 we derived the set of equations (9.9), (9.10), (15.22)–(15.24) for the sums of products of CFP. Similar sets can be derived for the sums of products of SCFP. With this in view we consider the submatrix elements of the triple tensors $W^{(Kk\kappa)}$ (15.52). On the one hand, they can be expressed, using (5.16), in terms of the submatrix elements of the operators $a^{(qls)}$

$$\left(l\alpha QLS\||| W^{(Kk\kappa)}\|||l\alpha' Q'L'S'\right) = (-1)^{Q+L+S+Q'+L'+S'+K+k+\kappa}$$

$$\times [K,k,\kappa]^{1/2} \sum_{\alpha'' Q'' L'' S''} \left(l\alpha QLS\||a^{(qls)}\||l\alpha'' Q'' L'' S''\right)$$

$$\times \left(l\alpha'' Q'' L'' S''\||a^{(qls)}\||l\alpha' Q'L'S'\right)$$

$$\times \begin{Bmatrix} q & q & K \\ Q' & Q & Q'' \end{Bmatrix} \begin{Bmatrix} l & l & k \\ L' & L & L'' \end{Bmatrix} \begin{Bmatrix} s & s & \kappa \\ S' & S & S'' \end{Bmatrix}. \quad (16.19)$$

On the other hand, the submatrix elements of tensors $W^{(Kk\kappa)}$ at certain values of their ranks have simple algebraic expressions determined by the quantum numbers of l^N configuration. So, when $K = k = \kappa = 0$, by (16.19), we obtain, using (15.54),

$$\sum_{\alpha'' Q'' L'' S''} \left(l\alpha QLS\||a^{(qls)}\||l\alpha'' Q'' L'' S''\right) \left(l\alpha' QLS\||a^{(qls)}\||l\alpha'' Q'' L'' S''\right)$$

$$= \delta(\alpha, \alpha')[l, s, Q, L, S], \quad (16.20)$$

and at $K = 1$, $k = \kappa = 0$, using (15.56), we have

$$\sum_{\alpha'' Q'' L'' S''} (-1)^{Q+Q''-s} \left(l\alpha QLS\||a^{(qls)}\||l\alpha'' Q'' L'' S''\right)$$

$$\times \left(l\alpha' QLS\||a^{(qls)}\||l\alpha'' Q'' L'' S''\right) \begin{Bmatrix} q & q & 1 \\ Q & Q' & Q'' \end{Bmatrix}$$

$$= \delta(\alpha, \alpha') 2\sqrt{\frac{2}{3}}[L, S] \{Q(Q+1)(2Q+1)\}^{1/2}. \quad (16.21)$$

Further, at $K = \kappa = 0$, $k = 1$, from (15.57) we have

$$\sum_{\alpha'' Q'' L'' S''} (-1)^{L''+L+l+1} \left(l\alpha QLS\||a^{(qls)}\||l\alpha'' Q'' L'' S''\right)$$

$$\times \left(l\alpha' QLS\||a^{(qls)}\||l\alpha'' Q'' L'' S''\right) \begin{Bmatrix} l & l & 1 \\ L & L' & L'' \end{Bmatrix}$$

$$= \delta(\alpha, \alpha') 2[Q, S] \left\{ \frac{L(L+1)(2L+1)}{l(l+1)(2l+1)} \right\}^{1/2}, \quad (16.22)$$

and at $K = k = 0$, $\kappa = 1$, by (15.58),

$$\sum_{\alpha'' Q'' L'' S''} (-1)^{S+S''-q} \left(l\alpha QLS\||a^{(qls)}\||l\alpha'' Q'' L'' S''\right)$$

$$\times \left(l\alpha' QLS\||a^{(qls)}\||l\alpha'' Q'' L'' S''\right) \begin{Bmatrix} s & s & 1 \\ S & S' & S'' \end{Bmatrix}$$

$$= \delta(\alpha, \alpha') 2\sqrt{\frac{2}{3}}[L, Q] \{S(S+1)(2S+1)\}^{1/2}. \quad (16.23)$$

Table 16.1.   Numerical values of the subcoefficients of fractional parentage $\left(p\ ^{2S+1}_{v}L^{2Q+1} \middle|\middle|\middle| p\ \left(^{2S'+1}_{v'}L'^{2Q'+1} \right) p \right)$

$Q'L'S'$ ＼ QLS	$^4_3S^1$	$^2_1P^3$	$^2_3D^1$
$^1_0S^4$		$-2\sqrt{2\cdot 3}$	
$^3_2P^2$	$-2\sqrt{2\cdot 3}$	$-3\sqrt{2\cdot 3}$	$-\sqrt{2\cdot 3\cdot 5}$
$^1_2D^2$		$-\sqrt{2\cdot 3\cdot 5}$	$\sqrt{2\cdot 3\cdot 5}$

In addition, we can use the fact (see (15.54)) that tensors $W^{(K k \kappa)}$ for an odd sum of all ranks are zero. Then in the case of tensors $W^{(110)}$, $W^{(011)}$ and $W^{(101)}$, (16.19) becomes

$$\sum_{\alpha''Q''L''S''} (-1)^{Q''+L''} \left(l\alpha QLS \middle|\middle|\middle| a^{(qls)} \middle|\middle|\middle| l\alpha''Q''L''S'' \right)^2$$

$$\times \begin{Bmatrix} q & q & 1 \\ Q & Q & Q'' \end{Bmatrix} \begin{Bmatrix} l & l & 1 \\ L & L & L'' \end{Bmatrix} = 0; \qquad (16.24)$$

$$\sum_{\alpha''Q''L''S''} (-1)^{L''+S''} \left(l\alpha QLS \middle|\middle|\middle| a^{(qls)} \middle|\middle|\middle| l\alpha''Q''L''S'' \right)^2$$

$$\times \begin{Bmatrix} l & l & 1 \\ L & L & L'' \end{Bmatrix} \begin{Bmatrix} s & s & 1 \\ S & S & S'' \end{Bmatrix} = 0; \qquad (16.25)$$

and

$$\sum_{\alpha''Q''L''S''} (-1)^{Q''+S''} \left(l\alpha QLS \middle|\middle|\middle| a^{(qls)} \middle|\middle|\middle| l\alpha''Q''L''S'' \right)^2$$

$$\times \begin{Bmatrix} q & q & 1 \\ Q & Q & Q'' \end{Bmatrix} \begin{Bmatrix} s & s & 1 \\ S & S & S'' \end{Bmatrix} = 0. \qquad (16.26)$$

The last relation, together with (16.20), corresponds to the set of equations (9.9) and (9.10), which is generally used to find (by the recurrence method on the number of electrons N) numerical values of CFP. Using formulas (16.24)–(16.26), we can construct a system of equations that enable us, using the recurrence (in seniority number v) method, to find numerical values of SCFP. However, if the CFP are known, then the numerical values of the SCFP are determinable in the simplest way from (16.17). Tables 16.1 and 16.2 summarize the numerical values of the latter for the p- and d-electrons, taken from [91]. The coefficients that are related to the transposition condition (16.18) are omitted.

Knowing the values of SCFP, we can, from (16.19), find the values of

Table 16.2. Numerical values of the subcoefficients of fractional parentage $\left(d\ {}^{2S+1}_{v}L^{2Q+1}\|\|d\ \left({}^{2S'+1}_{v'}L'^{2Q'+1}\right)d\right)$

$v'Q'L'S'$ \ $vQLS$	$^6_5S^1$	$^2_5S^1$	$^4_3P^3$	$^2_3P^3$	$^2_1D^5$	$^2_3D^3$	$^4_5D^1$	$^2_5D^1$
$^1_0S^6$					$-2\sqrt{3\cdot5}$			
$^1_4S^2$						$-2\sqrt{3}$		$2\sqrt{2}$
$^3_2P^4$			$\dfrac{8\sqrt{2\cdot3}}{\sqrt{5}}$	$-\dfrac{2\sqrt{3\cdot7}}{\sqrt{5}}$	$-3\sqrt{2\cdot5}$	$\sqrt{2\cdot3\cdot7}$		
$^3_4P^2$			$\dfrac{2\sqrt{3\cdot7}}{\sqrt{5}}$	$\sqrt{2\cdot3\cdot5}$		$-2\sqrt{3}$	$-2\cdot3$	$-3\sqrt{2}$
$^5_4D^2$	$-2\sqrt{3\cdot5}$	$-3\sqrt{2\cdot5}$					$5\sqrt{2}$	
$^3_4D^2$		$-2\sqrt{3}$	$\sqrt{2\cdot3\cdot7}$	$2\sqrt{3}$		$\dfrac{2\cdot3\sqrt{2\cdot5}}{\sqrt{7}}$	$\dfrac{3\sqrt{2\cdot3\cdot5}}{\sqrt{7}}$	$\dfrac{2\sqrt{3\cdot5}}{\sqrt{7}}$
$^1_2D^4$				$-2\cdot3$	$-5\sqrt{2}$	$-\dfrac{3\sqrt{2\cdot3\cdot5}}{\sqrt{7}}$		
$^1_4D^2$		$2\sqrt{2}$		$3\sqrt{2}$		$-\dfrac{2\sqrt{3\cdot5}}{\sqrt{7}}$		$\dfrac{3\sqrt{2\cdot5}}{\sqrt{7}}$
$^3_2F^4$			$\dfrac{4\sqrt{3\cdot7}}{\sqrt{5}}$	$\dfrac{4\sqrt{2\cdot3}}{\sqrt{5}}$	$-\sqrt{2\cdot3\cdot5\cdot7}$	$-3\sqrt{2}$		
$^3_4F^2$			$\dfrac{4\sqrt{3\cdot7}}{\sqrt{5}}$	$-\sqrt{2\cdot3\cdot5}$		$-3\sqrt{2}$	$2\sqrt{2\cdot3}$	$-3\sqrt{3}$
$^1_4F^2$				$-3\sqrt{2}$		$\sqrt{2\cdot3\cdot5}$		$-\sqrt{5}$
$^3_4G^2$						$\dfrac{3\sqrt{2\cdot3\cdot5}}{\sqrt{7}}$	$-\dfrac{2\cdot3\sqrt{2\cdot5}}{\sqrt{7}}$	$-\dfrac{3\sqrt{5}}{\sqrt{7}}$
$^1_2G^4$					$-3\sqrt{2\cdot5}$	$\dfrac{5\sqrt{2\cdot3}}{\sqrt{7}}$		
$^1_4G^2$						$-\dfrac{\sqrt{2\cdot3\cdot11}}{\sqrt{7}}$		$\dfrac{3\sqrt{11}}{\sqrt{7}}$
$^3_4H^2$								
$^1_4I^2$								

the submatrix elements of operators $W^{(Kk\kappa)}$. Table 16.3 lists the numerical values of these reduced submatrix elements for p^N configurations of electrons. It does not provide the reduced submatrix elements of operators $W^{(Kkk)}$ that may be worked out from those given in the tables using the transposition conditions for them

$$\left(l\alpha QLS\|\|W^{(Kk\kappa)}\|\|l\alpha'Q'L'S'\right) = (-1)^{Q+L+S-Q'-L'-S'}$$
$$\times \left(l\alpha'Q'L'S'\|\|W^{(Kk\kappa)}\|\|l\alpha QLS\right). \quad (16.27)$$

Similar data for the d^N configuration may be found in [91]. Applying such tables we can directly find the matrix elements of one-particle operators corresponding to physical quantities. The matrix elements of two-particle

Table 16.2. (continued)

$v'Q'L'S'$ \\ $vQLS$	$^4_3F^3$	$^2_3F^3$	$^2_5F^1$	$^2_3G^3$	$^4_5G^1$	$^2_5G^1$	$^2_3H^3$	$^2_4I^1$
$^1_0S^6$								
$^1_4S^2$								
$^3_2P^4$	$\frac{4\sqrt{3\cdot7}}{\sqrt5}$	$-\frac{4\sqrt{3\cdot7}}{\sqrt5}$						
$^3_4P^2$	$-\frac{4\sqrt{2\cdot3}}{\sqrt5}$	$-\sqrt{2\cdot3\cdot5}$	$-3\sqrt2$					
$^5_4D^2$	$-\sqrt{2\cdot3\cdot5\cdot7}$				$3\sqrt{2\cdot5}$			
$^3_4D^2$	$-3\sqrt2$	$3\sqrt2$	$-\sqrt{2\cdot3\cdot5}$	$\frac{3\sqrt{2\cdot3\cdot5}}{\sqrt7}$	$-\frac{5\sqrt{2\cdot3}}{\sqrt7}$	$-\frac{\sqrt{2\cdot3\cdot11}}{\sqrt7}$		
$^1_2D^4$		$2\sqrt{2\cdot3}$		$\frac{2\cdot3\sqrt{2\cdot5}}{\sqrt7}$				
$^1_4D^2$		$3\sqrt3$	$\sqrt5$	$-\frac{3\sqrt5}{\sqrt7}$		$-\frac{3\sqrt{11}}{\sqrt7}$		
$^3_2F^4$	$-\frac{8\sqrt{3\cdot7}}{\sqrt5}$	$-\frac{2\sqrt{3\cdot7}}{\sqrt5}$		$-2\cdot3\sqrt3$			$2\sqrt{3\cdot11}$	
$^3_4F^2$	$\frac{2\sqrt{3\cdot7}}{\sqrt5}$	$-\frac{\sqrt{3\cdot5\cdot7}}{2}$	$\frac{3\sqrt7}{2}$	$\frac{9\sqrt3}{2}$	$2\sqrt{3\cdot5}$	$-\frac{\sqrt{3\cdot5\cdot11}}{2}$	$\sqrt{3\cdot11}$	
$^1_4F^2$		$\frac{3\sqrt7}{2}$	$\frac{\sqrt{3\cdot5\cdot7}}{2}$	$-\frac{3}{2\sqrt5}$		$\frac{3\sqrt{11}}{2}$	$\frac{3\sqrt{11}}{\sqrt5}$	
$^3_4G^2$	$-2\cdot3\sqrt3$	$-\frac{9\sqrt3}{2}$	$\frac{3}{2\sqrt5}$	$\frac{3\cdot9\sqrt{11}}{2\sqrt{5\cdot7}}$	$\frac{2\cdot3\sqrt{11}}{\sqrt7}$	$\frac{3\cdot13}{2\sqrt7}$	$-\frac{3\sqrt{3\cdot11}}{\sqrt5}$	$-\frac{3\sqrt{2\cdot13}}{\sqrt5}$
$^1_2G^4$		$2\sqrt{3\cdot5}$		$-\frac{2\cdot3\sqrt{11}}{\sqrt7}$			$-2\sqrt{3\cdot11}$	
$^1_4G^2$		$\frac{\sqrt{3\cdot5\cdot11}}{2}$	$-\frac{3\sqrt{11}}{2}$	$\frac{3\cdot13}{2\sqrt7}$		$\frac{3\cdot9\sqrt5}{2\sqrt{7\cdot11}}$	$-\sqrt3$	$\frac{3\sqrt{2\cdot13}}{\sqrt{11}}$
$^3_4H^2$	$-2\sqrt{3\cdot11}$	$\sqrt{3\cdot11}$	$\frac{3\sqrt{11}}{\sqrt5}$	$\frac{3\sqrt{3\cdot11}}{\sqrt5}$	$-2\sqrt{3\cdot11}$	$\sqrt3$	$\frac{\sqrt{2\cdot3\cdot11\cdot13}}{\sqrt5}$	$-\frac{\sqrt{2\cdot3\cdot7\cdot13}}{\sqrt5}$
$^1_4I^2$				$-\frac{3\sqrt{2\cdot13}}{\sqrt5}$		$\frac{3\sqrt{2\cdot13}}{\sqrt{11}}$	$\frac{\sqrt{2\cdot3\cdot7\cdot13}}{\sqrt5}$	$\frac{\sqrt{2\cdot5\cdot7\cdot13}}{\sqrt{11}}$

operators, e.g. (15.91), can be found from (5.16) in terms of the sums of the products of the reduced matrix elements of tensors $W^{(Kkk)}$.

Relationship (16.16) can easily be generalized to the case of CFP with σ detached electrons. To do this, we take into account the tensorial structure (16.2) of the operator that enters into (15.28) and apply relationship (16.15) to the relevant submatrix element in (15.28). This yields

$$
\left(l^{N-\sigma}(\alpha_1Q_1L_1S_1)l^\sigma(\alpha_2Q_2L_2S_2)\|l^N\alpha QLS\right) = (-1)^{(N-\sigma)\sigma}
$$

$$
\times \left(\frac{\sigma!(N-\sigma)!}{N![L,S,Q]}\right)^{1/2}\begin{bmatrix} Q_1 & \sigma/2 & Q \\ M_{Q_1} & \sigma/2 & M_Q \end{bmatrix}
$$

$$
\times \left(l\alpha_1Q_1L_1S_1\|\|\hat{\varphi}(l^\sigma)^{\alpha_2v_2(\frac{\sigma}{2}L_2S_2)}\|\|l\alpha QLS\right). \tag{16.28}
$$

The last factor in this expression will be referred to as the SCFP with σ

Table 16.3. Numerical values of the reduced submatrix elements

$$\left(p\ ^{2S+1}_{v}L^{2Q+1}|||W^{(Kk\kappa)}|||p\ ^{2S'+1}_{v'}L'^{2Q'+1}\right)$$

QLS	$Q'L'S'$	$W^{(120)}$	$W^{(111)}$	$W^{(021)}$
${}^{1}_{0}S^{4}$	${}^{3}_{2}P^{2}$		$-2\cdot3\sqrt{2}$	
	${}^{1}_{2}D^{2}$	$-2\sqrt{2\cdot5}$		
${}^{3}_{2}P^{2}$	${}^{3}_{2}P^{2}$	$-3\sqrt{2\cdot5}$	$2\cdot3\sqrt{3}$	$2\sqrt{3\cdot5}$
	${}^{1}_{2}D^{2}$		$-3\sqrt{2\cdot5}$	$3\sqrt{2\cdot5}$
${}^{1}_{2}D^{2}$	${}^{1}_{2}D^{2}$	$\sqrt{2\cdot5\cdot7}$		
${}^{4}_{3}S^{1}$	${}^{2}_{1}P^{3}$		$2\cdot3\sqrt{2}$	
	${}^{2}_{3}D^{1}$			$2\sqrt{2\cdot5}$
${}^{2}_{1}P^{3}$	${}^{2}_{1}P^{3}$	$2\sqrt{3\cdot5}$	$2\cdot3\sqrt{3}$	$-3\sqrt{2\cdot5}$
	${}^{2}_{3}D^{1}$	$3\sqrt{2\cdot5}$	$-3\sqrt{2\cdot5}$	
${}^{2}_{3}D^{1}$	${}^{2}_{3}D^{1}$			$\sqrt{2\cdot5\cdot7}$

detached electrons. Its transposition property is written as

$$\left(l\alpha_1Q_1L_1S_1|||\hat{\varphi}(l^\sigma)^{\alpha_2v_2(\frac{\sigma}{2}L_2S_2)}||||l\alpha QLS\right) = (-1)^{L+S+Q+\sigma^2/2}$$
$$\times(-1)^{L_2-S_2-L_1-S_1-Q_1}\left(l\alpha QLS|||\hat{\varphi}(l^\sigma)^{\alpha_2v_2(\frac{\sigma}{2}L_2S_2)}||||l\alpha_1Q_1L_1S_1\right). \quad (16.29)$$

For CFP with two detached electrons (see (15.32)), instead of (16.28) we obtain:

$$\left(l^{N-2}(\alpha_1Q_1L_1S_1)l^2(L_2S_2)\|l^N\alpha QLS\right) = N(N-1)[L,S,Q]^{-1/2}$$
$$\times\begin{bmatrix}Q_1 & 1 & Q \\ M_{Q_1} & 1 & M_Q\end{bmatrix}\left(l\alpha QLS|||W^{(1L_2S_2)}||||l\alpha_1Q_1L_1S_1\right), \quad (16.30)$$

i.e. this CFP is proportional to the reduced submatrix element of triple tensor $W^{(1L_2S_2)}$. On the other hand, in terms of these triple tensors, we can express operators U^k and V^{k1} in accordance with (15.60). Applying relation (16.15) to the relevant reduced matrix elements gives

$$\left(l^N\alpha QLS\|U^k\|l^N\alpha'Q'L'S'\right) = -[k,Q]^{-1/2}\begin{bmatrix}Q' & 1 & Q \\ M_Q & 0 & M_Q\end{bmatrix}$$
$$\times\left(l\alpha QLS|||W^{(1k0)}||||l\alpha'Q'L'S'\right) \quad (k\text{ is even}), \quad (16.31)$$

$$\left(l^N\alpha QLS\|U^k\|l^N\alpha'Q'L'S'\right) = -[k,Q]^{-1/2}\left(l\alpha QLS|||W^{(0k0)}||||l\alpha'Q'L'S'\right)$$
$$(k\text{ is odd}), \quad (16.32)$$

$$\left(l^N\alpha QLS\|V^{k1}\|l^N\alpha'Q'L'S'\right) = -\frac{1}{2}[k,Q]^{-1/2}\left(l\alpha QLS\||W^{(0k1)}|||l\alpha'Q'L'S'\right)$$

$$(k \text{ is even}), \qquad (16.33)$$

$$\left(l^N\alpha QLS\|V^{k1}\|l^N\alpha'Q'L'S'\right) = -\frac{1}{2}[k,Q]^{-1/2}\begin{bmatrix} Q' & 1 & Q \\ M_Q & 0 & M_Q \end{bmatrix}$$

$$\times \left(l\alpha QLS\||W^{(1k1)}|||l\alpha'Q'L'S'\right) \qquad (k \text{ is odd}). \qquad (16.34)$$

Comparison of (16.30) with (16.31) and (16.34) suggests that (cf. [105]):

$$\left(l^N\alpha QLS\|U^{L_2}\|l^N\alpha'Q'L'S'\right)\begin{bmatrix} Q' & 1 & Q \\ M_{Q'} & 1 & M_Q \end{bmatrix}$$

$$= -\left(\frac{N(N-1)[L,S]}{2L_2+1}\right)^{1/2}\left(l^N\alpha QLS\|l^{N-2}(\alpha'Q'L'S')l^2(L_20)\right)$$

$$\times \begin{bmatrix} Q' & 1 & Q \\ M_Q & 0 & M_Q \end{bmatrix} \qquad (L_2 \text{ is even}), \qquad (16.35)$$

$$2\left(l^N\alpha QLS\|V^{L_21}\|l^N\alpha'Q'L'S'\right)\begin{bmatrix} Q' & 1 & Q \\ M_{Q'} & 1 & M_Q \end{bmatrix}$$

$$= -\left(\frac{N(N-1)[L,S]}{2L_2+1}\right)^{1/2}\left(l^N\alpha QLS\|l^{N-2}(\alpha'Q'L'S')l^2(L_21)\right)$$

$$\times \begin{bmatrix} Q' & 1 & Q \\ M_{Q'} & 0 & M_Q \end{bmatrix} \qquad (L_2 \text{ is odd}). \qquad (16.36)$$

It follows from the explicit form of the Clebsch–Gordan coefficients that

$$2\left[2(2L_2+1)Q(N,v_1)\right]^{1/2}\left(l^N\alpha vLS\|V^{L_21}\|l^N\alpha_1v_1L_1S_1\right)$$

$$= N(N-1)[L,S]^{1/2}\left[Q(N,v_1) - Q(N+2,v)\right]$$

$$\times \left(l^N\alpha vLS\|l^{N-2}(\alpha_1v_1L_1S_1)l^2(^3L_2)\right). \qquad (16.37)$$

Here $Q(N,v)$ is defined by formula (9.17). A similar expression for U^{L_2} differs only by factor 2 on the right side.

To summarize, the CFP with two detached electrons are expressed in terms of submatrix elements of the operators U^k and V^{k1}, so their tabulation is unnecessary.

16.3 Relationship between states of partially and almost filled shells

Since the number of electrons in the l^N configuration is uniquely determined by the value of the z-projection of quasispin momentum, then the quasispin method provides us with a common approach to the spin-angular parts of the wave functions of partially and almost filled shells that differ only by the sign of that z-projection. The phase relations between various quantities will then uniquely follow from the symmetry properties of the conventional Clebsch–Gordan coefficients. Specifically, if in the case of partially or almost filled shells we go over to subcoefficients of fractional parentage with σ detached electrons [see (16.28)] and compare the resultant expressions, using transposition properties (16.29), we shall arrive at

$$\left(l^{4l+2-\sigma-N}(\alpha_1 Q_1 L_1 S_1)l^\sigma(\alpha_2 Q_2 L_2 S_2)\|l^{4l+2-N}\alpha QLS\right)$$

$$= (-1)^{L_1+S_1+L_2+S_2+Q_1-Q-L-S-\sigma^2/2}$$

$$\times \left\{\frac{(4l+2-N-\sigma)!(N+\sigma)![L_1,S_1]}{(4l+2-N)!N![L,S]}\right\}^{1/2}$$

$$\times \left(l^N(\alpha QLS)l^\sigma(\alpha_2 Q_2 L_2 S_2)\|l^{N+\sigma}\alpha_1 Q_1 L_1 S_1\right), \qquad (16.38)$$

hence, at $N = 0$,

$$\left(l^{4l+2-\sigma}(\alpha QLS)l^\sigma(\alpha QLS)\|l^{4l+2}\,{}^1_0 S\right) = (-1)^{(\sigma^2-v)/2}$$

$$\times \{(4l+2-\sigma)!\sigma![L,S]/(4l+2)!\}^{1/2}. \qquad (16.39)$$

Usually, the phase factor here is chosen in some arbitrary manner, fixing the relationships between the partially and almost filled shells [14, 23]. As is seen from (16.38), in the quasispin method these relationships are given rigorously and uniquely.

In exactly the same way, for operators U^k and V^{k1} we obtain the relationship between the submatrix elements defined relative to the wave functions of partially and almost filled shells

$$\left(l^{4l+2-N}\alpha v LS\|U^k\|l^{4l+2-N}\alpha'v'L'S'\right) = (-1)^{k+(v-v')/2}$$

$$\times \left(l^N\alpha v LS\|U^k\|l^N\alpha'v'L'S'\right);$$

$$\left(l^{4l+2-N}\alpha v LS\|V^{k1}\|l^{4l+2-N}\alpha'v'L'S'\right) = (-1)^{k+(v-v')/2}$$

$$\times \left(l^N\alpha v LS\|V^{k1}\|l^N\alpha'v'L'S'\right). \qquad (16.40)$$

The connection between appropriate quantities for partially and almost filled shells can be established by another method – by going over from the particle representation to the hole one. According to the results of Chapters 2, 13 and 14, the wave function of the completely filled shell is at the same time the vacuum state for holes

$$|_h0) = |l^{4l+2}\,{}^1S). \qquad (16.41)$$

The wave functions for almost filled shells will then be determinable in terms of the relevant quantities for hole states

$$|l^{4l+2-N}\alpha LS M_L M_S) = (-1)^{\eta(\alpha,L,S,N)}|_h l^N \alpha LS M_L M_S). \qquad (16.42)$$

Postulation of one or other phase factor η leads to certain phase relations between the CFP of hole and particle states, thus accounting for the emergence of various ambiguities and contradictions [14, 23, 106].

But this phase factor can be selected so that the signs of the CFP for almost filled shells are the same as in the quasispin method. It is worth recalling here that finite transformations generated by quasispin operators define the passage to quasiparticles. In much the same way, in the quasispin space of a shell of equivalent electrons the unitary transformations

$$|l_{qpl}^{N'}\alpha QLSM_LM_S) = \sum_{M_Q} D_{M_{Q'},M_Q}^{(Q)}|l^N\alpha QLSM_LM_S) \tag{16.43}$$

produce the wave functions for N' quasiparticles. Specifically, rotation through 180° clockwise about the Oy axis can be used as a definition for the transformation from particle to hole representation. Then

$$|_h l^N\alpha QLSM_LM_S) = (-1)^{Q-M_Q}|l^{4l+2-N}\alpha QLSM_LM_S), \tag{16.44}$$

i.e. the phase factor which in (16.42) defines the relationship between particle and hole states will, in this approach, have the following form (discarding the insignificant general phase factor):

$$\eta = (N+v)/2. \tag{16.45}$$

Rotation of any irreducible tensor in quasispin space transforms it according to the D-matrix of relevant dimensionality. For example, unitary transformations of $a^{(qls)}$ yield the creation operators

$$\alpha_{m\mu}^{(ls)} = D_{1/2,1/2}^{(1/2)}a_{m\mu}^{(ls)} + (-1)^{l+s-m-\mu}D_{1/2,-1/2}^{(1/2)}a_{-m-\mu}^{(ls)\dagger} \tag{16.46}$$

and the annihilation operators

$$\alpha_{m\mu}^{(ls)\dagger} = (-1)^{l+s+m+\mu}D_{-1/2,1/2}^{(1/2)}a_{-m-\mu}^{(ls)} + D_{-1/2,-1/2}^{(1/2)}a_{m\mu}^{(ls)\dagger} \tag{16.47}$$

for quasiparticles. Substituting (16.46) and (16.47) into (15.51) and taking into account the property of D-functions, we see that for new operators $\alpha^{(qls)}$ anticommutation relations (15.51) remain valid.

In the special case of the rotation through 180° clockwise about the Oy axis we have

$$\alpha_{\frac{1}{2}m\mu}^{(qls)} = a_{-\frac{1}{2}m\mu}^{(qls)}; \qquad \alpha_{-\frac{1}{2}m\mu}^{(qls)} = -a_{\frac{1}{2}m\mu}^{(qls)}, \tag{16.48}$$

i.e. in quasispin space the two components of tensor $\alpha^{(qls)}$ are interchangeable. Then,

$$W_{\Pi'\rho\pi}^{(Kk\kappa)} = (-1)^{K-\Pi'}W_{-\Pi'\rho\pi}^{(Kk\kappa)} \tag{16.49}$$

and

$$\hat{\varphi}(_h l^N)^{\alpha(LS)} = \hat{\bar{\varphi}}(l^N)^{\alpha(LS)}. \tag{16.50}$$

The expressions provided above enable us to work out relationships between matrix elements of any operators defined for partially and almost filled shells. Specifically, bearing in mind the unitarity of the particle-to-hole transformation, we can directly see that formulas (16.38) and (16.39) derived earlier by the quasispin method, hold for CFP.

16.4 Transposition of spin and quasispin quantum numbers

The analogy between spin and quasispin ($q = s = 1/2$) enabled Judd to introduce another unitary transformation – the operation $\hat{R}$ [12]

$$\hat{R}a_{vm\mu}^{(qls)}\hat{R}^{-1} = a_{\mu mv}^{(qls)}, \tag{16.51}$$

which means that the projection of the quasispin rank of operator $a^{(qls)}$ assumes the value of the projection of the spin rank of that operator, and vice versa. Hence,

$$\hat{R}W_{\Pi\rho\pi}^{(Kk\kappa)}\hat{R}^{-1} = W_{\pi\rho\Pi}^{(\kappa kK)}, \tag{16.52}$$

i.e. both the values of ranks in quasispin and spin spaces, and their projections are interchanged. Using the tensorial properties of wave functions in these spaces, we get

$$\hat{R}|l\alpha QLSM_QM_LM_S) = (-1)^\xi|l\alpha SLQM_SM_LM_Q),$$

or, from (15.42) and (9.22),

$$\hat{R}|l^N\alpha vLSM_LM_S) = (-1)^\xi|l^{N_S}\alpha v_S LQM_LM_Q), \tag{16.53}$$

where $v_S = -2S + 2l + 1$; $N_S = 2M_S + 2l + 1$.

A relation of the type (16.53) between states has been noted by Racah [24] who established that two states $\alpha v_1 LS_1$ and $\alpha v_2 LS_2$ appear to be coupled, if

$$v_1 + 2S_2 = v_2 + 2S_1 = 2l + 1.$$

For the matrix elements of $a^{(qls)}$, considering that $\hat{R}$ is a unitary operator, we obtain

$$\left(l^N\alpha QLSM_QM_LM_S|a_{vm\mu}^{(qls)}|l^{N'}\alpha'Q'L'S'M_{Q'}M_{L'}M_{S'}\right)$$

$$= \left(l^N\alpha QLSM_QM_LM_S|\hat{R}^{-1}\hat{R}a_{vm\mu}^{(qls)}\hat{R}^{-1}\hat{R}|l^{N'}\alpha'Q'L'S'M_{Q'}M_{L'}M_{S'}\right)$$

$$= (-1)^{\xi+\xi'}\left(l^{N_S}\alpha SLQM_SM_LM_Q|a_{vm\mu}^{(qls)}|l^{N'_S}\alpha'S'L'Q'M_{S'}M_{L'}M_{Q'}\right).$$

Applying the Wigner–Eckart theorem in all three spaces, we establish the property of SCFP in relation to interchanges of the spin and quasispin quantum numbers

$$\left(l\alpha QLS|||a^{(qls)}||||l\alpha'Q'L'S'\right) = (-1)^\varphi\left(l\alpha SLQ|||a^{(qls)}||||l\alpha'S'L'Q'\right). \tag{16.54}$$

Relationships for CFP corresponding to (16.54) can be derived from (16.16).

The problem of finding the phase factors in the above formulas is worthy of special attention. Simple arguments [12] show that in (16.53) this factor depends only on Q and S, i.e. $\xi = \xi(Q, S)$, although there is no way of finding a more detailed expression. The point is that the phase relations between wave functions are dependent on the phases of CFP. The latter are usually derived by solving the set of equations (9.10) and (9.9) ignoring their symmetry properties under transposition of Q and S. Consequently, the phase factor in (16.53) in general cannot be determined without changing signs in the available tables of fractional parentage coefficients.

For practical purposes, it is only sufficient to find the phase φ in equation (16.54) relating the SCFP.

In [91] ξ was given by

$$\xi = l + \delta(2Q + 1, 2n)(Q' - S) - \delta(2S + 1, 2n)(S' - Q) \qquad (16.55)$$

$$(n \text{ is a natural number}).$$

This expression for f-electrons can be derived using the phase relations established for isoscalar parts of factorized CFP with different parities of the seniority number [24]. It turned out [91] that phase (16.55) provides sign relations between the CFP in the tables for d- and f-electrons, but it is unsuitable for the p-electrons. In this connection, in what follows all the relationships derived using the symmetry properties under transposition of the quantum numbers of spin and quasispin are provided up to the sign.

Reasoning along the same lines as in the derivation of (16.54), we find for the submatrix elements of operator $W^{(K k \kappa)}$

$$\left(l\alpha QLS \left\lVert\!\right\lVert W^{(Kk\kappa)} \left\lVert\!\right\lVert l\alpha' Q'L'S' \right) = \pm \left(l\alpha SLQ \left\lVert\!\right\lVert W^{(Kk\kappa)} \left\lVert\!\right\lVert l\alpha' S'L'Q' \right). \qquad (16.56)$$

Using (16.31)–(16.34), we can establish from (16.56) the following relations between the submatrix elements of irreducible tensorial operators U^k and V^{k1} [92]:

$$\left(l^N \alpha v LS \lVert U^k \rVert l^N \alpha' v L'S \right) = \pm \left[(2S + 1)/(2Q + 1) \right]^{1/2}$$

$$\times \left(l^{N_S} \alpha v_S LQ \lVert U^k \rVert l^{N_S} \alpha' v_S L'Q \right) \qquad (k \text{ is odd}); \qquad (16.57)$$

$$\left(l^N \alpha v LS \lVert V^{k1} \rVert l^N \alpha' v' L'S' \right) = \pm \left[(2S + 1)/(2Q + 1) \right]^{1/2}$$

$$\times \begin{bmatrix} Q' & 1 & Q \\ M_Q & 0 & M_Q \end{bmatrix} \begin{bmatrix} S' & 1 & S \\ M_S & 0 & M_S \end{bmatrix}^{-1}$$

$$\times \left(l^{N_S} \alpha v_S LQ \lVert V^{k1} \rVert l^{N_S} \alpha' v'_S L'Q' \right) \qquad (k \text{ is odd}). \qquad (16.58)$$

$$\left(l^N \alpha v L S \| U^k \| l^N \alpha' v' L' S\right) = \pm \left[(2S+1)/(2Q+1)\right]^{1/2}$$

$$\times \left(l^{Ns} \alpha v_S L Q \| V^{k1} \| l^{Ns} \alpha' v_S L' Q'\right)$$

$$\times \begin{bmatrix} Q' & 1 & Q \\ M_Q & 0 & M_Q \end{bmatrix} \qquad (k \text{ is even}). \qquad (16.59)$$

Equation (16.59) is interesting in that it relates the submatrix elements of different operators. So, at $Q = Q'$, we obtain the simple relationship

$$\left(l^N \alpha v L S \| U^k \| l^N \alpha' v L' S\right) = \pm \left[\frac{(N-2l-1)^2(2S+1)}{Q(Q+1)(2Q+1)}\right]^{1/2}$$

$$\times \left(l^{Ns} \alpha v_S L Q \| V^{k1} \| l^{Ns} \alpha' v_S L' Q\right) \qquad (k \text{ is even}), \qquad (16.60)$$

that clearly shows, for example, that the submatrix elements of U^k, diagonal with respect to v (k is even), must be zero for the half-filled shell.

A similar treatment is possible, if we take into account (16.30), for the coefficients of fractional parentage with two detached electrons. Specifically, for an odd L_2 we have

$$\left(l^N \alpha v L S \| l^{N-2}(\alpha' v' L' S') l^2(^3 L_2)\right) = \pm \begin{bmatrix} Q' & 1 & Q \\ M_Q & 1 & M_Q \end{bmatrix}$$

$$\times \begin{bmatrix} S' & 1 & S \\ M_S' & 1 & M_S \end{bmatrix}^{-1} \left(l^{Ns} \alpha v_S L Q \| l^{Ns-2}(\alpha' v_S' L' Q') l^2(^3 L_2)\right). \qquad (16.61)$$

Some other expressions of this type and their consequences are given detailed consideration in [92].

16.5 Algebraic expressions for some specific CFP

The method of CFP is an elegant tool for the construction of wave functions of many-electron systems and the establishment of expressions for matrix elements of operators corresponding to physical quantities. Its major drawback is the need for numerical tables of CFP, normally computed by the recurrence method, and the presence in the matrix elements of multiple sums with respect to quantum numbers of states that are not involved directly in the physical problem under consideration. An essential breakthrough in this respect may be finding algebraic expressions for the CFP and for the matrix elements of the operators of physical quantities. For the latter, in a number of special cases, this can be done using the eigenvalues of the Casimir operators [90], however, it would be better to have sufficiently simple but universal formulas for the CFP themselves.

For some special cases such formulas have long been known. Racah, for example, derived relationship (16.39) for the CFP detaching σ electrons from the filled shell, and also simple formula (9.16) for the special case of the two-electron CFP in which two detached electrons form state $_0^1S$. In the quasispin method this relation can be obtained directly from equation (16.30), which in the case under consideration is written as

$$\left(l^N QLS \| l^{N-2}(Q'LS)l^2(_0^1S)\right) = i2N(N-1)[L,S,Q,l]^{-1/2}$$

$$\times \begin{bmatrix} Q' & 1 & Q \\ M_{Q'} & 1 & M_Q \end{bmatrix} \left(lQLS \| \|Q^{(1)}\| \|lQ'LS\right). \quad (16.62)$$

CFP (9.11) also have a simple algebraic form. In the previous paragraph we discussed the behaviour of coefficients of fractional parentage in quasispin space and their symmetry under transposition of spin and quasispin quantum numbers. The use of these properties allows one, from a single CFP, to find pertinent quantities in the interval of occupation numbers for a given shell for which a given state exists [92].

Equations of this kind can also be derived for the special cases of reduced matrix elements of operators composed of irreducible tensors. Then, using the relation between CFP and the submatrix element of irreducible tensorial operators established in [105], we can obtain several algebraic expressions for two-electron CFP [92]. Unfortunately such algebraic expressions for CFP do not embrace all the required values even for the p^N shell, which imposes constraints on their practical uses by preventing analytic summation of the matrix elements of operators of physical quantities. It has turned out, however, that there exist more general and effective methods to establish algebraic expressions for CFP, which do not feature the above-mentioned disadvantages.

16.6   Algebraic expressions for CFP of $l^N vLS$ shells

Results more general than those provided above have been obtained in [107] which relied on antisymmetrized wave functions to construct the states with a given orbital and spin angular momenta by vectorial coupling of momenta. Then, after the coordinates in the wave functions were ordered, the expressions corresponding to the CFP were separated. This technique enabled algebraic formulas for CFP to be found only for those simple cases where the shell contained no repeating terms. But even with three equivalent electrons (l^3), repeating terms occur beginning with the d-shell, i.e. in [107] the problem of finding algebraic expressions for CFP appeared to be solved only for the p-electrons. Now, following [108], we shall try to give a more general treatment of the problem – for those repeating terms whose classification requires only the seniority quantum number, i.e. where the terms of the l^N shell are uniquely described by the quantum numbers vLS. This can be done using the second-quantization and quasispin method, by consistently relying on the tensorial properties of wave functions and in such a way arriving at their explicit form.

In Chapter 15, for the CFP with σ detached electrons, we obtained a relationship (15.27) whose right side has the form of a vacuum average of a certain product of second-quantized operators $\hat{\varphi}$. To obtain algebraic formulas for CFP, it is necessary to compute this vacuum average by transposing all the annihilation operators to the right side of the creation operators. So, for $N = 3$, we take into account (for non-repeating terms) the explicit form of operators (15.2) and (15.5), which produce pertinent wave functions out of vacuum, and find (cf. [107])

$$\left(l^2(L_2 S_2)l \| l^3 LS\right) = (-1)^{l+s-L_3-S_3}\left(1 + (-1)^{L_2+S_2}\right)$$

$$\times \left(1 + (-1)^{L_2'+S_2'}\right)\frac{1}{4\sqrt{3}}$$

$$\times \frac{\delta(L_2 S_2, L_2' S_2') + 2[L_2, S_2, L_2', S_2']\begin{Bmatrix} l & l & L_2 \\ l & L_3 & L_2' \end{Bmatrix}\begin{Bmatrix} s & s & S_2 \\ s & S_3 & S_2' \end{Bmatrix}}{\left(1 + 2[L_2, S_2]\begin{Bmatrix} l & l & L_2' \\ l & L_3 & L_2' \end{Bmatrix}\begin{Bmatrix} s & s & S_2' \\ s & S_3 & S_2' \end{Bmatrix}\right)^{1/2}}. \quad (16.63)$$

Similar expressions can also be obtained at $N = 4$ [108]. To establish similar algebraic expressions for the repeating terms vLS only requires that the appropriate second-quantized operators (16.7), (16.8), (16.10) and (16.11) be used. For example, instead of (16.63), we have

$$\left(l^2(v_2 L_2 S_2)l \| l^3 v_3 L_3 S_3\right) = (-1)^{L_2+S_2+l+s-L_3-S_3}$$

$$\times \mathcal{N}_3(2/3)^{1/2}\left[\delta(L_2 S_2, L_2' S_2') + 2[L_2, S_2, L_2', S_2']^{1/2}\right.$$

$$\times \left.\left(\begin{Bmatrix} l & l & L_2' \\ l & L_3 & L_2 \end{Bmatrix}\begin{Bmatrix} s & s & S_2' \\ s & S_3 & S_2 \end{Bmatrix} - \frac{\delta(L_3 S_3, ls)}{(2l+1)(2l-1)}\right)\right]. \quad (16.64)$$

Using (16.16) and the above relations, we can work out algebraic expressions for SCFP, and hence for CFP, in the entire interval of the number of electrons in the shell existing for given v. Taking account of the symmetry of CFP under transposition of spin and quasispin quantum numbers further expands the number of such expressions. Formulas of this kind can be established also for larger v, but with $v = 5$ and above they become so unwieldy and difficult to handle that this limits their practical uses. They may be found in [108].

The technique used here to find algebraic expressions for CFP can also be applied to derive the recurrence relation that generalizes the well-known Redmond formula [109]. To this end, it is only required to write

(15.27) as the integral not depending on projections

$$\left(l^{N-\sigma}(\alpha_1 L_1 S_1)l^{\sigma}(\alpha_2 L_2 S_2)\|l^N \alpha L S\right)$$

$$= (-1)^{\sigma(N-\sigma)+L_1+S_1+L_2+S_2-L-S}\sqrt{\frac{\sigma!(N-\sigma)!}{N!}}$$

$$\times \left(l^N \alpha L S M_L M_S\bigl| \left[\hat{\varphi}(l^{\sigma})^{\alpha_2(L_2 S_2)} \times \hat{\varphi}(l^{N-\sigma})^{\alpha_1(L_1 S_1)}\right)\right]_{M_L M_S}^{(LS)}, \quad (16.65)$$

and then to represent the wave functions in the recurrent form (16.6). If then we transpose the annihilation operators to the right of the creation operators and recouple the momenta, we are able to express the resultant quantity in terms of CFP. We thus arrive at the recurrence relation

$$\left(l^{v-1}((v-1)L_1 S_1)l\|l^v v L S\right) = (-1)^{L_1+S_1-l-s-L-S+v} N v^{-1/2}$$

$$\times \left\{\delta(L_1 S_1, L_1' S_1') - \sum_{L_2 S_2}(v-1)[L_1, S_1, L_1', S_1']^{1/2}\right.$$

$$\times \left(l^{v-1}(v-1)L_1' S_1'\|l^{v-2}((v-2)L_2 S_2)l\right)$$

$$\times \left(l^{v-1}(v-1)L_1 S_1\|l^{v-2}((v-2)L_2 S_2)l\right)$$

$$\times \left.\left[(-1)^{2S}\begin{Bmatrix} l & L_1' & L \\ l & L_1 & L_2 \end{Bmatrix}\begin{Bmatrix} s & S_1' & S \\ s & S_1 & S_2 \end{Bmatrix} + \frac{\delta(LS, L_2 S_2)}{2(2l+3-v)[L,S]}\right]\right\}.$$

$$(16.66)$$

The term $L_1' S_1'$ can be chosen in an arbitrary manner, and the normalization factor is found from the normalization condition for coefficients of fractional parentage at fixed momenta L_1', S_1' and L, S. Equation (16.66) holds for repeating terms that are uniquely classified by the seniority quantum number v, but for non-repeating terms (when $\delta(L_2 S_2, L S) = 0$) that equation becomes the conventional Redmond formula [109].

The above recurrence relations allow the numerical values of CFP to be readily found and the availability of the above algebraic formulas for them makes it possible to establish similar algebraic expressions for other quantities in atomic theory, including the CFP with two detached electrons, the submatrix elements of irreducible tensors, and also the matrix elements of the operators of physical quantities [110].

16.7 Algebraic expressions for CFP of the $f^N(u_1 u_2)v L S$ shell

It is well known that even for three f-electrons the classification of terms by the seniority number turns out to be insufficient. In this case, the exclusive group G_2 has been applied with success [24], although it is still impossible to achieve a one-to-one classification of all the terms of the f^N configuration. The above method of deriving algebraic expressions for the CFP in the case of three and more f-electrons is no longer suitable. To extend it to cover the states of the f-shell we try to fuse the quasispin method with the methods of the G_2 group in such a way that, on the one hand, we can exploit the algebraic potentialities of the former method, and, on the other hand, preserve the conventional classification of the terms of the f-shell. To achieve this, we have to discuss the methodology of constructing wave functions that transform themselves by irreducible representations of the G_2 group.

We need the following scalar of the G_2 group, i.e. an operator that commutes with all the generators of this group:

$$X^{(3/2\ 0\ 3/2)} = \left[a^{(qls)} \times \left[a^{(qls)} \times a^{(qls)} \right]^{(1l1)} \right]^{(3/2\ 0\ 3/2)}. \tag{16.67}$$

The method used to establish the algebraic formulas for CFP is, in broad outline, as follows: if at least one wave function is available that is characterized by the quantum numbers (u_1u_2) and L, then all the remaining functions with the same quantum numbers can be constructed using irreducible tensorial products of the known wave function and scalar $X^{(3/2\ 0\ 3/2)}$:

$$\varphi_{M_QM_LM_S}^{(Q\ L\ S)}(u_1u_2) = \mathcal{N}'' \left[X^{(3/2\ 0\ 3/2)} \times \varphi^{(Q_1LS_1)}(u_1u_2) \right]_{M_QM_LM_S}^{(Q\ L\ S)}. \tag{16.68}$$

Accordingly, for each (u_1u_2) the problem comes down to the construction of one wave function with a given L. At $(u_1u_2) = (00), (10), (20)$ and (11) these quantities are known – these are the wave functions of one and two equivalent electrons with $v = 0, 1, 2$. In order to construct wave functions with other (u_1u_2) we make use of the fact that at a given v the wave function with maximal L has a uniquely defined characteristic of the G_2 group.

Wave functions with the same Q, S, (u_1u_2), but different L, can be obtained using one of the generators of the G_2 group, namely $W^{(050)}$, as follows:

$$\varphi_{M_QM_LM_S}^{(Q\ L\ S)}(u_1u_2) = \mathcal{N}''' \left[W^{(050)} \times \varphi^{(QL_1S)}(u_1u_2) \right]_{M_QM_LM_S}^{(Q\ L\ S)}. \tag{16.69}$$

If the use of $W^{(050)}$, due to the triangle condition, does not yield the desired term, formula (16.69) must be applied once more. Note, that the normalization factors $\mathcal{N}''$ and $\mathcal{N}'''$ are dependent on both the initial and final term. The method just described makes it possible to construct wave functions for those terms of the f-shell that can be classified uniquely using the characteristics of irreducible representations of the G_2 group.

The recurrence relations for CFP of the f-shell that generalize the Redmond formula (a counterpart of (16.66)) have been established in [111].

Their use makes it possible to determine the CFP for terms $(u_1u_2)vLS$ while having the pertinent quantities for terms with a smaller v. A separate relation occurs in the case where some representation of the G_2 group emerges for the first time at a given v. These equations are not provided here because they are relatively cumbersome.

The procedure for finding algebraic expressions for CFP differs from that utilized earlier only in its uses of tensors $X^{(3/2\ 0\ 3/2)}$ and $W^{(050)}$ instead of tensor $a^{(qls)}$. For terms described by a wave function of the type $|l^3(u_1u_2)3L_3S_3)$, the following algebraic expression holds for CFP with one detached electron:

$$\left(l^N(u_1u_2)3L_3S_3\|l^{N-1}((u_1u_2)2L_2S_2)l\right) = (-1)^{l+s+L_3+S_3-L_2-S_2}$$

$$\times \frac{\mathcal{N}_3}{\sqrt{5N}}\begin{bmatrix} 5/2 & 1/2 & 3 \\ M_Q & 1/2 & M_{Q'} \end{bmatrix}\left(1+(-1)^{L_2+S_2}\right)$$

$$\times (1-\delta(L_3S_3,ls))3^{1/2}\left[\delta(L_2S_2,l1)+2(3[l,L_2,S_2])^{1/2}\right.$$

$$\times \left\{\begin{matrix} l & l & L_2 \\ l & L_3 & l \end{matrix}\right\}\left\{\begin{matrix} s & s & S_2 \\ s & S_3 & 1 \end{matrix}\right\}\right]$$

$$\times \left[1+2\cdot3(2l+1)\left\{\begin{matrix} l & l & l \\ l & L_2 & l \end{matrix}\right\}\right.$$

$$\times \left.\left\{\begin{matrix} s & s & 1 \\ s & S_3 & 1 \end{matrix}\right\}\right]^{-1/2}, \qquad ((u_1u_2)\neq(21)). \qquad (16.70)$$

A similar expression for CFP with $(u_1u_2)=(21)$ reads

$$\left(l^N(21)3L_3S_3\|l^{N-1}((u_1''u_2'')2L_2S_2)l\right) = (-1)^{l+s+L_3+S_3-L_2-S_2}$$

$$\times \frac{\mathcal{N}_3\sqrt{3}}{\sqrt{5N}}\begin{bmatrix} 5/2 & 1/2 & 3 \\ M_Q & 1/2 & M_{Q'} \end{bmatrix}\left(1+(-1)^{L_2+S_2}\right)$$

$$\times \left[(-1)^{L_2'}\delta(L_2S_2,L_2'S_2')\left\{\begin{matrix} 5 & l & l \\ L_2' & L_3 & L_3' \end{matrix}\right\}+(-1)^{L_3+L_3'+L_2+L_2'+1}\right.$$

$$\times 2[L_2',L_2]^{1/2}\delta(S_2,S_2')\left\{\begin{matrix} 5 & L_2' & L_2 \\ l & L_3 & L_3' \end{matrix}\right\}\left\{\begin{matrix} 5 & L_2' & L_2 \\ l & l & l \end{matrix}\right\}$$

$$+2[L_2,S_2,L_2',S_2']^{1/2}\left\{\begin{matrix} s & s & S_2' \\ s & S_3 & S_2 \end{matrix}\right\}\left\{(-1)^{L_2'}\left\{\begin{matrix} 5 & l & l \\ L_2' & L_3 & L_3' \end{matrix}\right\}\right.$$

$$\times \left\{\begin{matrix} l & l & L_2 \\ l & L_3 & L_2' \end{matrix}\right\}+(-1)^{L_2}\left\{\begin{matrix} L_3' & l & L_2 \\ l & l & L_2' \end{matrix}\right\}\left\{\begin{matrix} L_3' & l & L_2 \\ l & L_3 & 5 \end{matrix}\right\}$$

$$\left.\left.+(-1)^{S_2+S_2'}\left\{\begin{matrix} l & l & L_2 \\ l & L_2' & l \\ 5 & L_3' & L \end{matrix}\right\}\right\}\right]. \qquad (16.71)$$

Thus we see that even at $v = 3$ the expressions for CFP become rather involved. In a similar way, we can construct CFP for higher v, but the form of the resultant formulas will be more complex. For example, at $v = 4$ we obtain eight expressions for the CFP. They contain $3nj$-coefficients of high order. Expressions for the CFP with $S = 2$ and any $Q(v)$ can, however, be derived using the symmetry properties under transposition of Q and S.

The above technique for constructing the CFP of the f-shell in principle solves the problem for the terms which, to be additionally classified, only require that the G_2 group be applied, i.e. essentially for the overwhelming majority of terms of the f^N configuration.

17

Tensorial properties and quasispin
of complex configurations

17.1 Irreducible tensors in the space of complex configurations

So far we have considered only one shell of equivalent electrons, but the mathematical techniques discussed can be translated fairly simply to the case of complex atomic configurations. With LS coupling the wave functions of $n_1 l_1^{N_1} n l_2^{N_2} \dots n_u l^{N_u}$ configuration are normally constructed by the vectorial coupling of orbital and spin momenta of all the shells:

$$L = \mathbf{L}(l_1) + \mathbf{L}(l_2) + \dots + \mathbf{L}(l_u),$$
$$\mathbf{S} = \mathbf{S}(l_1) + \mathbf{S}(l_2) + \dots + \mathbf{S}(l_u). \tag{17.1}$$

Individual operator terms in these sums are given by (14.15) and (14.16), and here we have introduced additional subscripts of the operators to indicate the space of shells in which they are defined.

According to the data of Part 3, the resultant wave function, in addition to orbital $\mathbf{L}$ and spin $\mathbf{S}$ angular momenta of the configuration, is characterized by a set of intermediate momenta and their coupling scheme. Only for $n_1 l^{N_1} n_2 l^{N_2}$ configuration do we not need these additional characteristics, therefore to simplify our discussion below we shall confine ourselves to these configurations. A generalization for any number of shells will present no difficulties.

The irreducible components of the operators of orbital and spin angular momenta of the $n_1 l^{N_1} n_2 l^{N_2}$ configuration have the form

$$L_\rho^{(1)} = L_\rho^{(1)}(l_1) + L_\rho^{(1)}(l_2), \tag{17.2}$$
$$S_\rho^{(1)} = S_\rho^{(1)}(l_1) + S_\rho^{(1)}(l_2). \tag{17.3}$$

When dealing with complex configurations, it is necessary, apart from one-shell tensors, to consider the irreducible tensorial products of creation and annihilation operators for electrons from different shells

$$\left[a^{(l_1s)} \times \tilde{a}^{(l_2s)}\right]^{(k\kappa)}_{\rho\pi} = \sum_{m_1m_2\mu_1\mu_2} \begin{bmatrix} l_1 & l_2 & k \\ m_1 & m_2 & \rho \end{bmatrix} \begin{bmatrix} s & s & \kappa \\ \mu_1 & \mu_2 & \pi \end{bmatrix} a^{(l_1s)}_{m_1\mu_1} \tilde{a}^{(l_2s)}_{m_2\mu_2}.$$
(17.4)

In the group-theoretical treatment of mixed configurations, we may, in analogy with the case of one shell, introduce the basis tensors for two shells of equivalent electrons [101]

$$W^{(k\kappa)}_{\rho\pi}(l_1, l_2) = \sum_{i=1}^{N} w^{(k\kappa)}_{i\rho\pi}(l_1, l_2),$$
(17.5)

where unit one-electron operators are given by their submatrix elements

$$\left(l_a \| w^{(k\kappa)}(l_1, l_2) \| l_b\right) = \delta(l_1, l_a)\delta(l_2, l_b)[k, \kappa]^{1/2}.$$
(17.6)

Next we shall represent (17.5) in the second-quantized form

$$W^{(k\kappa)}_{\rho\pi}(l_1, l_2) = \sum_{\substack{m_1m_2 \\ \mu_1\mu_2}} a^{(l_1s)}_{m_1\mu_1} a^{(l_2s)\dagger}_{m_2\mu_2} \left(l_am_1\mu_1 | w^{(k\kappa)}_{\rho\pi}(l_1, l_2) | l_bm_2\mu_2\right)$$
(17.7)

and apply Wigner–Eckart theorem (5.15) to the one-electron matrix element. Summation over the projections will then give the expressions

$$W^{(k\kappa)}_{\rho\pi}(l_1, l_2) = -\left[a^{(l_1s)} \times \tilde{a}^{(l_2s)}\right]^{(k\kappa)}_{\rho\pi},$$
(17.8)

i.e. tensorial product (17.4) is (up to the sign) the second-quantized form of operator $W^{(k\kappa)}(l_1, l_2)$. At $l_1 = l_2$, tensor (17.8) becomes the double tensorial operator (14.30) defined in the space of one shell. Using (6.2)–(6.6), we find the Hermitian-conjugated operator to be

$$W^{(k\kappa)\dagger}_{\rho\pi}(l_1, l_2) = (-1)^{l_1+l_2+\rho+\pi} W^{(k\kappa)}_{-\rho-\pi}(l_2, l_1),$$

i.e. operators $W^{(k\kappa)}$ are not Hermitian. In this connection, the use is generally made [101] of linear combinations of these tensors that meet the Hermitian condition

$$\pm W^{(k\kappa)}(l_1, l_2) = W^{(k\kappa)}(l_1, l_2) \pm (-1)^{k+\kappa+l_1} W^{(k\kappa)}(l_2, l_1).$$
(17.9)

Further we shall show that such linear combinations have a certain rank in the quasispin space of $n_1 l^{N_2} n_2 l^{N_2}$ configuration. The commutation relations between tensors (17.8) or (17.9) are completely defined by the anticommutation relations for creation and annihilation operators. These relations can be written as

$$\left\{a^{(l_1s)}_{m_1\mu_1}, \tilde{a}^{(l_2s)}_{m_2\mu_2}\right\} = (-1)^{l_1+s-m_2-\mu_2}\delta(n_1l_1, n_2l_2)\delta(m_1\mu_1, -m_2-\mu_2),$$
(17.10)

$$\left\{a^{(l_1s)}_{m_1\mu_1}, a^{(l_2s)}_{m_2\mu_2}\right\} = \left\{\tilde{a}^{(l_1s)}_{m_1\mu_1}, \tilde{a}^{(l_2s)}_{m_2\mu_2}\right\} = 0$$
(17.11)

(recall that the principal quantum number n is included in implicit form in the symbols for second-quantization operators).

Alongside operators (17.8) we can also define the irreducible tensorial products of creation operators for electrons in different shells

$$\left[a^{(l_1 s)} \times a^{(l_2 s)}\right]_{\rho\pi}^{(k\kappa)} = (-1)^{l_1 + l_2 + k + \kappa + 1} \left[a^{(l_2 s)} \times a^{(l_1 s)}\right]_{\rho\pi}^{(k\kappa)}, \tag{17.12}$$

and of appropriate annihilation operators

$$\left[\tilde{a}^{(l_1 s)} \times \tilde{a}^{(l_2 s)}\right]^{(k\kappa)} = (-1)^{l_1 + l_2 + k + \kappa + 1} \left[\tilde{a}^{(l_2 s)} \times \tilde{a}^{(l_1 s)}\right]^{(k\kappa)}. \tag{17.13}$$

17.2 Tensorial properties of operators of physical quantities

In Chapter 14 we have shown how an expansion in terms of irreducible tensors in the spaces of orbital and spin angular momenta for one shell can be obtained for the operators corresponding to physical quantities. The tensors introduced above enable the terms of a similar expansion to be also defined in the space of a two-shell configuration. So, for the one-particle operator of the most general tensorial structure (14.51) we find, instead of (14.52),

$$F^{(k\kappa\gamma)}{}_{\Gamma}(l_1, l_2) = \sum_{(a,b=1,2)} \left(n_a l_a \| f^{(k\kappa)} \| n_b l_b\right) W^{(k\kappa\gamma)}{}_{\Gamma}(l_a, l_b), \tag{17.14}$$

where

$$W^{(k\kappa\gamma)}{}_{\Gamma}(l_a, l_b) = \sum_{\rho\pi} W^{(k\kappa)}_{\rho\pi}(l_a, l_b) \begin{bmatrix} k & \kappa & \gamma \\ \rho & \pi & \Gamma \end{bmatrix}, \tag{17.15}$$

and summation subscripts a and b independently take on values 1 and 2. And so, in general, the right side of (17.14) is the sum of four terms, two of which are defined in the space of each of shells $n_1 l_1^{N_1}$ and $n_2 l_2^{N_2}$, and two others that relate the configurations of the type $n_1 l_1^{N_1} n_2 l_2^{N_2} - n_1 l_1^{N_1-1} n_2 l_2^{N_2+1}$. In special cases some of these terms may be identically equal to zero, for example, with the electric dipole transition operator (see (4.12) at $k = 1$) the intrashell terms are zero, and with the kinetic and potential energy operators the intershell terms are zero (at $l_1 \neq l_2$) – either case follows directly from the explicit form of relevant one-electron reduced matrix elements.

Among all the possible two-particle operators for physical quantities for l^N configuration we have only considered in detail the electrostatic interaction operator for electrons; here too we shall confine ourselves to the examination of this operator. The explicit form of the two-electron matrix elements of the electrostatic interaction operator for electrons (the last term in (1.15)) for non-antisymmetric wave functions will be

$$(n_1 l_1 m_1 \mu_1 n_2 l_2 m_2 \mu_2 | V | n_1' l_1' m_1' \mu_1' n_2' l_2' m_2' \mu_2') = \sum_{k\rho} (-1)^{k-\rho} [l_1, l_2]^{-1/2}$$

$$\times \begin{bmatrix} l_1' & k & l_1 \\ m_1' & \rho & m_1 \end{bmatrix} \begin{bmatrix} l_2' & k & l_2 \\ m_2' & -\rho & m_2 \end{bmatrix} \left(n_1 l_1 n_2 l_2 \| V^{(kk)} \| n_1' l_1' n_2' l_2' \right), \quad (17.16)$$

where

$$\left(n_1 l_1 n_2 l_2 \| V^{(kk)} \| n_1' l_1' n_2' l_2' \right) = \left(l_1 \| C^{(k)} \| l_1' \right) \left(l_2 \| C^{(k)} \| l_2' \right)$$

$$\times R_k(n_1 l_1 n_1' l_1', n_2 l_2 n_2' l_2'); \quad (17.17)$$

$$R_k(n_1 l_1 n_1' l_1', n_2 l_2 n_2' l_2') = \iint_0^\infty \frac{r_<^k}{r_>^{k+1}} P(n_1 l_1 | r_1) P(n_1' l_1' | r_1)$$

$$\times P(n_2 l_2 | r_2) P(n_2' l_2' | r_2) dr_1 dr_2; \quad (17.18)$$

$$R_k(nlnl, n'l'n'l') = F_k(nl, n'l'); \quad (17.19)$$

$$R_k(nln'l', nln'l') = G_k(nl, n'l'). \quad (17.20)$$

Formula (17.16) is the most general form of the two-electron matrix element in which all four one-electron wave functions have different quantum numbers. We shall put it into general formula (13.23), whereupon the creation and annihilation operators will be rearranged to place side by side those second-quantization operators whose rank projections enter into the same Clebsch–Gordan coefficient. Summing over the projections then gives

$$V = -2^{-1/2} \sum_{\substack{k l_a l_b l_d \\ n_a n_b n_d}} (2l+1)^{-1/2} R_k(n_a l_a n_b l_b, n_b l_b n_d l_d) \left(l_a \| C^{(k)} \| l_b \right)^2$$

$$\times W^{(00)}(n_a l_a, n_d l_d) + \sum_{\substack{k l_a l_b l_c l_d \\ n_a n_b n_c n_d}} (2k+1)^{-1/2} \left(l_a \| C^{(k)} \| l_c \right) \left(l_b \| C^{(k)} \| l_d \right)$$

$$\times R_k(n_a l_a n_c l_c, n_b l_b n_d l_d) \left[W^{(k0)}(l_a, l_c) \times W^{(k0)}(l_b, l_d) \right]^{(00)}. \quad (17.21)$$

For one shell of equivalent electrons representation (14.65) has been found in which the irreducible tensorial products are constructed separately for creation operators and annihilation operators. Likewise, such a representation can also be found for two-shell configurations. In that case,

$$V = -\frac{1}{2} \sum_{\substack{(l_a, l_b, l_c, l_d = 1,2) \\ k L_{12} S_{12}}} \left(\left[a^{(l_a s)} \times a^{(l_b s)} \right]^{(L_{12} S_{12})} \right.$$

$$\times \left. \left[\tilde{a}^{(l_c s)} \times \tilde{a}^{(l_d s)} \right]^{(L_{12} S_{12})} \right) \left(n_a l_a n_b l_b L_{12} S_{12} \| V^{(kk)} \| n_c l_c n_d l_d L_{12} S_{12} \right), \quad (17.22)$$

where

$$\left(n_a l_a n_b l_b L_{12} S_{12} \| V^{(kk)} \| n_c l_c n_d l_d L_{12} S_{12}\right) = (-1)^{l_a + l_b + L_{12}}$$

$$\times \begin{Bmatrix} l_a & l_b & L_{12} \\ l_d & l_c & k \end{Bmatrix} \left(n_a l_a n_b l_b \| V^{(kk)} \| n_c l_c n_d l_d\right), \qquad (17.23)$$

where the last factor is given by (17.17). It is worth noting that if we do not limit the values of summation parameters a, b, c, d, then equations (17.21) and (17.22) will give the most general form of the electrostatic interaction operator for electrons, which is suitable for any number of electron shells in the configuration.

Just as in the case of one-particle operator (17.14), expressions (17.21) and (17.22) embrace already in operator form, the interaction terms, both the diagonal ones relative to the configurations, and the non-diagonal ones. Coming under the heading of diagonal terms are, first, the one-shell operators of electrostatic interaction of electrons, discussed in detail in Chapter 14. Second, the operators of direct

$$V(n_1 l_1 n_2 l_2, n_1 l_1 n_2 l_2) = - \sum_{k L_{12} S_{12}} \left(\left[a^{(l_1 s)} \times a^{(l_2 s)}\right]^{(L_{12} S_{12})}\right.$$

$$\times \left[\tilde{a}^{(l_1 s)} \times \tilde{a}^{(l_2 s)}\right]^{(L_{12} S_{12})}\right) \left(n_1 l_1 n_2 l_2 L_{12} S_{12} \| V^{(kk)} \| n_1 l_1 n_2 l_2 L_{12} S_{12}\right), \qquad (17.24)$$

and exchange

$$V(n_1 l_1 n_2 l_2, n_2 l_2 n_1 l_1) = - \sum_{k L_{12} S_{12}} \left(\left[a^{(l_1 s)} \times a^{(l_2 s)}\right]^{(L_{12} S_{12})}\right.$$

$$\times \left[\tilde{a}^{(l_2 s)} \times \tilde{a}^{(l_1 s)}\right]^{(L_{12} S_{12})}\right) \left(n_1 l_1 n_2 l_2 L_{12} S_{12} \| V^{(kk)} \| n_2 l_2 n_1 l_1 L_{12} S_{12}\right) \qquad (17.25)$$

electrostatic interaction of electrons between the shells.

These operators can be averaged in the same manner as in Chapter 14 where we have introduced the average operator of electrostatic interaction of electrons in a shell. The main departure of the case at hand is that the Pauli exclusion principle, owing to the fact that electrons from different shells are not equivalent, imposes constraints neither on the pertinent two-particle matrix elements nor on the number of possible pairing states, which equals $(4l_1 + 2)(4l_2 + 2)$. The averaged submatrix element of direct interaction between the shells will then be

$$\left(n_1 l_1 n_2 l_2 \| V_{av}^{(kk)} \| n_1 l_1 n_2 l_2\right) = \sum_{L_{12} S_{12}} [L_{12}, S_{12}]$$

$$\times \left(n_1 l_1 n_2 l_2 L_{12} S_{12} \| V^{(kk)} \| n_1 l_1 n_2 l_2 L_{12} S_{12}\right) / [(4l_1 + 2)(4l_2 + 2)]$$

$$= \delta(k, 0)[l_1, l_2]^{-1/2} \left(n_1 l_1 n_2 l_2 \| V^{(kk)} \| n_1 l_1 n_2 l_2\right). \qquad (17.26)$$

Note that the summation is carried out using (6.25) and (6.18). Replacing in (17.24) the two-particle submatrix element of the direct interaction by the averaged one and summing over the values of the momenta, we arrive at

$$V_{av}(n_1l_1n_2l_2, n_1l_1n_2l_2) = \hat{N}(l_1)\hat{N}(l_2)\delta(l_1, l_2)F_0(n_1l_1, n_2l_2). \quad (17.27)$$

Reasoning along the same lines, we also find the averaged operator of exchange electrostatic interaction between the shells

$$V_{av}(n_1l_1n_2l_2, n_2l_2n_1l_1) = \sum_k \hat{\tilde{g}}_k(l_1, l_2)G_k(n_1l_1, n_2l_2),$$

where the operator

$$\hat{\tilde{g}}_k(l_1, l_2) = -\hat{N}(l_1)\hat{N}(l_2)\left(l_1\|C^{(k)}\|l_2\right)^2/(2[l_1, l_2]). \quad (17.28)$$

Specifically, at $k = 0$

$$\hat{\tilde{g}}_0(l_1, l_2) = -\hat{N}(l_1)\hat{N}(l_2)\delta(l_1, l_2)/(2(2l + 1)). \quad (17.29)$$

If necessary, we can introduce other effective operators (e.g. operators averaged over some groups of pairing states).

The non-diagonal (with respect to configurations) operators that enter into (17.21) and (17.22) define the interaction between all the possible electron distributions over various two-shell configurations and give a non-zero contribution only for pertinent interconfiguration matrix elements. So, in the case of a superposition of configurations of the kind $n_1l_1^{N_1}n_2l_2^{N_2}$ $- n_1l_1^{N_1-1}n_2l_2^{N_2+1}$ it is necessary to consider only the following terms of expansion (17.22):

$$V(n_1l_1n_2l_2, n_2l_2n_2l_2) = -\sum_{kL_{12}S_{12}}\left(\left[a^{(l_1s)} \times a^{(l_2s)}\right]^{(L_{12}S_{12})}\right.$$

$$\times \left[\tilde{a}^{(l_2s)} \times \tilde{a}^{(l_2s)}\right]^{(L_{12}S_{12})}\right) \left(n_1l_1n_2l_2L_{12}S_{12}\| V^{(kk)}\|n_2l_2n_2l_2L_{12}S_{12}\right), \quad (17.30)$$

$$V(n_1l_1n_1l_1, n_1l_1n_2l_2) = -\sum_{kL_{12}S_{12}}\left(\left[a^{(l_1s)} \times a^{(l_1s)}\right]^{(L_{12}S_{12})}\right.$$

$$\times \left[\tilde{a}^{(l_1s)} \times \tilde{a}^{(l_2s)}\right]^{(L_{12}S_{12})}\right) \left(n_1l_1n_1l_1L_{12}S_{12}\| V^{(kk)}\|n_1l_1n_2l_2L_{12}S_{12}\right). \quad (17.31)$$

These formulas include irreducible tensorial products of second-quantization operators belonging both to one and the same shell and to different shells. The expressions for the operators can, if necessary, be transformed so that the ranks of second-quantization operators for one shell are coupled first. For example, the operator that enters into (17.30), can be represented as follows:

$$\left(\left[a^{(l_1 s)} \times a^{(l_2 s)}\right]^{(L_{12}S_{12})} \cdot \left[\tilde{a}^{(l_2 s)} \times a^{(l_2 s)}\right]^{(L_{12}S_{12})}\right)$$

$$= -\sum_{L'_{12}S'_{12}} (-1)^{l_1+l_2}[L_{12}, S_{12}][L'_{12}, S'_{12}]^{1/2}[l_1, s]^{-1/2} \begin{Bmatrix} l_2 & l_2 & L'_{12} \\ l_1 & l_2 & L_{12} \end{Bmatrix}$$

$$\times \begin{Bmatrix} s & s & S'_{12} \\ s & s & S_{12} \end{Bmatrix} \left(a^{(l_1 s)} \cdot \left[\left[a^{(l_2 s)} \times \tilde{a}^{(l_2 s)}\right]^{(L'_{12}S'_{12})} \times a^{(l_2 s)}\right]^{(l_1 s)}\right). \quad (17.32)$$

As we shall see later, the merits of representing operators in one form or another are mostly determined by the technique used to find their matrix elements.

17.3   Wave functions and matrix elements

It has already been said at the beginning of this chapter that the wave functions of complex configurations are built by vectorial coupling of orbital and spin momenta of individual shells. Then,

$$|l_1^{N_1} l_2^{N_2} \dots l_u^{N_u} \alpha_1 L_1 S_1 \alpha_2 L_2 S_2 \dots \alpha_u L_u S_u LS)$$

$$= \left[\hat{\varphi}(l_1^{N_1})^{\alpha_1(L_1 S_1)} \times \hat{\varphi}(l_2^{N_2})^{\alpha_2(L_2 S_2)} \times \dots \times \hat{\varphi}(l_u^{N_u})^{\alpha_u(L_u S_u)}\right]^{(LS)} |0). \quad (17.33)$$

So, for two shells we have

$$|n_1 l_1^{N_1} n_2 l_2^{N_2} \alpha_1 L_1 S_1 \alpha_2 L_2 S_2 LSM_L M_S)$$

$$= \left[\hat{\varphi}(n_1 l_1^{N_1})^{\alpha_1(L_1 S_1)} \times \hat{\varphi}(n_2 l_2^{N_2})^{\alpha_2(L_2 S_2)}\right]^{(LS)}_{M_L M_S} |0). \quad (17.34)$$

Bearing in mind that second-quantization operators from different shells anticommute, we can represent the conjugate wave function as follows:

$$(n_1 l_1^{N_1} n_2 l_2^{N_2} \alpha_1 L_1 S_1 \alpha_2 L_2 S_2 LSM_L M_S|$$

$$= (0| \left\{ \left[\hat{\varphi}(n_1 l_1^{N_1})^{\alpha_1(L_1 S_1)} \times \hat{\varphi}(n_2 l_2^{N_2})^{\alpha_2(L_2 S_2)}\right]^{(LS)}_{M_L M_S} \right\}^{\dagger}$$

$$= (-1)^{N_1 N_2 + \frac{N_1(N_1-1)}{2} + \frac{N_2(N_2+1)}{2} + L + S + M_L + M_S}$$

$$\times (0| \left[\hat{\tilde{\varphi}}(n_1 l_1^{N_1})^{\alpha_1(L_1 S_1)} \times \hat{\tilde{\varphi}}(n_2 l_2^{N_2})^{\alpha_2(L_2 S_2)}\right]^{(LS)}_{-M_L -M_S}, \quad (17.35)$$

where the pertinent one-shell operators are given by (15.10). Similarly, we can represent the wave function conjugate of (17.33). Examination of the structure of (17.34) and (17.35) shows that the submatrix elements of any irreducible tensorial product of creation and annihilation operators can be expressed in terms of quantities that are defined in the space of each of the shells separately. Formulas (5.20)–(5.22) for the submatrix elements of operators in the vectorial coupling scheme for the momenta of two wave functions can, almost without modifications, be applied to the vectorial coupling scheme for momenta of two shells. The only departure is that we must additionally introduce a phase factor depending on the number of permutations necessary to 'split' the pertinent matrix elements. These permutations are indicated by arrows in the following expression:

$$(0|\hat{\varphi}(n_1 l_1^{N_1})\hat{\varphi}(n_2 l_2^{N_2})\hat{A}(l_1)\hat{B}(l_2)\hat{\varphi}(n_1 l_1^{N_1'})\hat{\varphi}(n_2 l_2^{N_2'})|0),$$

where $\hat{A}(l_1)$ and $\hat{B}(l_2)$ are any second-quantization operators defined in the space of respective shells, and the irreducible tensorial products are not represented in an explicit form because the phase factor is dependent only on the number of permutations performed.

The submatrix elements of creation operators will thus be

$$\left(l_1^{N_1} l_2^{N_2}\alpha_1 L_1 S_1 \alpha_2 L_2 S_2 LS \|a^{(l_1 s)}\| l_1^{N_1-1} l_2^{N_2}\alpha_1' L_1' S_1' \alpha_2' L_2' S_2' L' S'\right)$$

$$= \delta(\alpha_2 L_2 S_2, \alpha_2' L_2' S_2')(-1)^{L_1+S_1+L_2+S_2+L'+S'+l_1+s}[L,S,L',S']^{1/2}$$

$$\times \begin{Bmatrix} L_1 & L & L_2 \\ L' & L_1' & l_1 \end{Bmatrix} \begin{Bmatrix} S_1 & S & S_2 \\ S' & S_1' & s \end{Bmatrix}$$

$$\times \left(l_1^{N_1}\alpha_1 L_1 S_1 \|a^{(l_1 s)}\| l_1^{N_1-1}\alpha_1' L_1' S_1'\right), \tag{17.36}$$

$$\left(l_1^{N_1} l_2^{N_2}\alpha_1 L_1 S_1 \alpha_2 L_2 S_2 LS \|a^{(l_2 s)}\| l_1^{N_1} l_2^{N_2-1}\alpha_1' L_1' S_1' \alpha_2' L_2' S_2' L' S'\right)$$

$$= \delta(\alpha_1 L_1 S_1, \alpha_1' L_1' S_1')(-1)^{N_1+L_1+S_1+L_2'+S_2'+L+S+l+s}[L,S,L',S']^{1/2}$$

$$\times \begin{Bmatrix} L_2 & L & L_1 \\ L' & L_2' & l_2 \end{Bmatrix} \begin{Bmatrix} S_2 & S & S_1 \\ S' & S_2' & s \end{Bmatrix}$$

$$\times \left(l_2^{N_2}\alpha_2 L_2 S_1 \|a^{(l_2 s)}\| l^{N_2-1}\alpha_2' L_2' S_2'\right). \tag{17.37}$$

The submatrix elements that enter into the left sides of these equations can be expressed in terms of complex CFP [112]. Using these coefficients the many-shell wave function for N electrons ($N = N_1 + N_2 + \ldots + N_u$) is composed of the antisymmetric wave functions of $(N-1)$ electrons

$$|l_1^{N_1}\ldots l_u^{N_u} LS) = \sum_{i=1}^{u} \sum_{L'S'} |l_1^{N_1}\ldots l_i^{N_i-1}\ldots l_u^{N_u}(L'S')l_i LS)$$

$$\times \left(l_1^{N_1}\ldots l_i^{N_i-1}\ldots l_u^{N_u}(L'S')l_i\| l_1^{N_1}\ldots l_i^{N_i}\ldots l_u^{N_u} LS\right). \tag{17.38}$$

In Chapter 15 the one-shell CFP was related to the pertinent submatrix element of the creation operator (formula (15.21)) using an expansion of antisymmetric wave functions in terms of one-determinant functions. This reasoning can be reproduced here to yield

$$\left(l_1^{N_1}\ldots l_i^{N_i}\ldots l_u^{N_u}LS\,\|a^{(l_i s)}\|\,l_1^{N_1}\ldots l_i^{N_i-1}\ldots l_u^{N_u}L'S'\right)=(-1)^N$$
$$\times (N[L,S])^{1/2}\left(l_1^{N_1}\ldots l_i^{N_i-1}\ldots l_u^{N_u}(L'S')l_i\|l_1^{N_1}\ldots l_i^{N_i}\ldots l_u^{N_u}LS\right).\,(17.39)$$

It is to be noted that the complex CFP constitutes no qualitatively new quantity. In particular, formulas (17.36) and (17.37) show that for two shells this quantity is given by the product of the conventional CFP and 6j-coefficients. This can be accounted for by the fact that the procedures of antisymmetrization of the wave function and the vectorial coupling of the momenta of individual shells are independent.

In an analogous way, we can consider the submatrix elements of irreducible tensorial products of second-quantization operators. So operator (14.40) will be

$$\left(l_1^{N_1}l_2^{N_2}\alpha_1 L_1 S_1 \alpha_2 L_2 S_2 LS\,\|\left[a^{(l_1 s)}\times a^{(l_1 s)}\right]^{(L_{12}S_{12})}\|\right.$$
$$\times\left.l_1^{N_1-2}l_2^{N_2}\alpha_1'L_1'S_1'\alpha_2'L_2'S_2'L'S'\right)=\delta(\alpha_2 L_2 S_2,\alpha_2'L_2'S_2')$$
$$\times(-1)^{L_1+S_1+L_2+S_2+L'+S'+L_{12}+S_{12}}[L,S,L',S']^{1/2}$$
$$\times\begin{Bmatrix}L_1 & L & L_2\\ L' & L_1' & L_{12}\end{Bmatrix}\begin{Bmatrix}S_1 & S & S_2\\ S' & S_1' & S_{12}\end{Bmatrix}$$
$$\times\left(l_1^{N_1}\alpha_1 L_1 S_1\|\left[a^{(l_1 s)}\times a^{(l_1 s)}\right]^{(L_{12}S_{12})}\|l_1^{N_1-2}\alpha_1'L_1'S_1'\right),\quad(17.40)$$

and tensor (17.12)

$$\left(l_1^{N_1}l_2^{N_2}\alpha_1 L_1 S_1 \alpha_2 L_2 S_2 LS\,\|\left[a^{(l_1 s)}\times a^{(l_2 s)}\right]^{(L_{12}S_{12})}\|\right.$$
$$\times\left.l_1^{N_1-1}l_2^{N_2-1}\alpha_1'L_1'S_1'\alpha_2'L_2'S_2'L'S'\right)=(-1)^{N_1-1}[L,S,L',S',L_{12},S_{12}]^{1/2}$$
$$\times\left(l^{N_1}\alpha_1 L_1 S_1\|a^{(l_1 s)}\|l^{N_1-1}\alpha_1'L_1'S_1'\right)\left(l^{N_2}\alpha_2 L_2 S_2\|a^{(l_2 s)}\|l^{N_2-1}\alpha_2'L_2'S_2'\right)$$
$$\times\begin{Bmatrix}L_1' & L_2' & L'\\ l_1 & l_2 & L_{12}\\ L_1 & L_2 & L\end{Bmatrix}\begin{Bmatrix}S_1' & S_2' & S'\\ s & s & S_{12}\\ S_1 & S_2 & S\end{Bmatrix}.\qquad(17.41)$$

The left sides of these equations differ from complex CFP only by simple factors, whereas the right sides include quantities proportional to one-shell CFP. Accordingly, in this case too, we can do without introducing complex CFP.

The method presented enables expressions to be found for submatrix elements of irreducible tensorial products of second-quantization operators for configurations of any complexity. This method provides a unified approach both to diagonal and non-diagonal (relative to the configuration) matrix elements of operators of physical quantities.

17.4   Operators in quasispin space of separate shells

Since the wave function of a complex configuration is constructed by vectorial coupling of the orbital and spin momenta of individual shells, one-shell quantum numbers $\alpha_i v_i L_i S_i$ (just like the intermediate momenta) are additional characteristics for the wave function of entire configuration (17.33). Specifically, the two-shell wave function

$$|l_1^{N_1} l_2^{N_2} \alpha_1 v_1 L_1 S_1 \alpha_2 v_2 L_2 S_2 L S M_L M_S)$$
$$= |l_1 l_2 \alpha_1 Q_1 L_1 S_1 \alpha_2 Q_2 L_2 S_2 L S M_L M_S M_{Q_1} M_{Q_2}) \tag{17.42}$$

has the quantum numbers $\alpha_i v_i L_i S_i$ $(i = 1, 2)$ of terms of both shells as additional characteristics. Creation and annihilation operators are the components of an irreducible tensor of rank $q = 1/2$ in the space of quasispin of each shell individually

$$a_{\frac{1}{2}m\mu}^{(ql_is)} = a_{m\mu}^{(l_is)}, \qquad a_{-\frac{1}{2}m\mu}^{(ql_is)} = \tilde{a}_{m\mu}^{(l_is)}. \tag{17.43}$$

Operators corresponding to physical quantities can also be expanded in terms of irreducible tensors in the quasispin space of each individual shell. To this end, it is sufficient to go over to tensors (17.43) and next to provide their direct product in the quasispin space of individual shells. This procedure can conveniently be carried out for a representation of operators such that the orbital and spin ranks of all the one-shell tensors are coupled directly. Here we shall provide the final result for the two-particle operator of general form (14.57)

$$G = [1 + P(1 \leftrightarrow 2)] \sum_{i=1}^{7} G_i, \tag{17.44}$$

$$G_1 = b_{1122}^{k_2 \kappa_2 k_2' \kappa_2'} \sum_{pL_1S_1L_2S_2} (-1)^{k-\rho} [L_1, S_1, L_2, S_2]^{1/2}$$

$$\times \begin{Bmatrix} l_1 & L_1 & l_1 \\ l_2 & L_2 & l_2 \\ k_2' & k & k_2 \end{Bmatrix} \begin{Bmatrix} s & S_1 & s \\ s & S_2 & s \\ \kappa_2' & k & \kappa_2 \end{Bmatrix}$$

$$\times \left[W_1^{(1L_1S_1)}(l_1) \times W_{-1}^{(1L_2S_2)}(l_2) \right]_{\rho-\rho}^{(kk)}, \tag{17.45}$$

$$G_2 = \frac{1}{2} b_{1221}^{k_2 \kappa_2 k_2' \kappa_2'} \sum_{K_1L_1S_1K_2L_2S_2} (-1)^{k_2'+\kappa_2'+l_1+l_2+L_2+S_2+k-\rho} [L_1, L_2, S_1, S_2]^{1/2}$$

$$\times \begin{Bmatrix} l_1 & L_1 & l_1 \\ l_2 & L_2 & l_2 \\ k_2' & k & k_2 \end{Bmatrix} \begin{Bmatrix} s & S_1 & s \\ s & S_2 & s \\ \kappa_2' & k & \kappa_2 \end{Bmatrix}$$

$$\times \left[W_0^{(K_1L_1S_1)}(l_1) \times W_0^{(K_2L_2S_2)}(l_2) \right]_{\rho-\rho}^{(kk)}, \tag{17.46}$$

$$G_3 = -\frac{b_{1212}^{k_2\kappa_2 k_2'\kappa_2'}}{[k_2, k_2', \kappa_2, \kappa_2']^{1/2}} \sum_{K_1 K_2} \left[W_0^{(K_1 k_2 \kappa_2)}(l_1) \times W_0^{(K_2 k_2'\kappa_2')}(l_2) \right]_{\rho-\rho}^{(kk)}, \quad (17.47)$$

$$G_4 = \frac{1}{2} b_{2111}^{k_2\kappa_2 k_2'\kappa_2'} \sum_{KK_1 L_1 S_1} (-1)^{l_2+s-L_1-S_1} \left[\frac{L_1, S_1, K_1}{k_2', \kappa_2', K} \right]^{1/2}$$
$$\times \left\{ \begin{matrix} k_2 & k_2' & k \\ L_1 & l_2 & l_1 \end{matrix} \right\} \left\{ \begin{matrix} \kappa_2 & \kappa_2' & k \\ S_1 & s & s \end{matrix} \right\}$$
$$\times \left[a_{\frac{1}{2}}^{(ql_2 s)} \times W^{(K k_2'\kappa_2')}(l_1) \times a^{(ql_1 s)} \right]_{-\frac{1}{2}}^{(K_1 L_1 S_1)} \right]_{\rho-\rho}^{(kk)}, \quad (17.48)$$

$$G_5 = \frac{1}{2} b_{1211}^{k_2\kappa_2 k_2'\kappa_2'} \sum_{KK_1 L_1 S_1} (-1)^{l_2+s-L_1-S_1+k_2+\kappa_2+k_2'+\kappa_2'} \left[\frac{L_1, S_1, K_1}{k_2, \kappa_2, K} \right]^{1/2}$$
$$\times \left\{ \begin{matrix} k_2' & k_2 & k \\ L_1 & l_2 & l_1 \end{matrix} \right\} \left\{ \begin{matrix} \kappa_2' & \kappa_2 & k \\ S_1 & s & s \end{matrix} \right\}$$
$$\times \left[a_{\frac{1}{2}}^{(ql_2 s)} \times \left[W^{(K k_2 \kappa_2)}(l_1) \times a^{(ql_1 s)} \right]_{-\frac{1}{2}}^{(K_1 L_1 S_1)} \right]_{\rho-\rho}^{(kk)}, \quad (17.49)$$

$$G_6 = \frac{1}{2} b_{2212}^{k_2\kappa_2 k_2'\kappa_2'} \sum_{KK_1 L_1 S_1} (-1)^{k_2+k_2'+\kappa_2+\kappa_2'} \left[\frac{K_2, L_2, S_2}{K, k_2', \kappa_2'} \right]^{1/2}$$
$$\times \left\{ \begin{matrix} k_2' & k_2 & k \\ l_1 & L_2 & l_2 \end{matrix} \right\} \left\{ \begin{matrix} \kappa_2' & \kappa_2 & k \\ s & S_2 & s \end{matrix} \right\}$$
$$\times \left[a_{-\frac{1}{2}}^{(ql_1 s)} \times \left[a^{(ql_2 s)} \times W^{(K k_2'\kappa_2')}(l_2) \right]_{-\frac{1}{2}}^{(K_2 L_2 S_2)} \right]_{\rho-\rho}^{(kk)}, \quad (17.50)$$

$$G_7 = \frac{1}{2} b_{1112}^{k_2\kappa_2 k_2'\kappa_2'} \sum_{KK_1 L_1 S_1} (-1)^{k_2+k_2'+\kappa_2+\kappa_2'} \left[\frac{K_1, L_1, S_1}{K, k_2, \kappa_2} \right]^{1/2}$$
$$\times \left\{ \begin{matrix} k_2 & k_2' & k \\ l_2 & L_1 & l_1 \end{matrix} \right\} \left\{ \begin{matrix} \kappa_2 & \kappa_2' & k \\ s & S_1 & s \end{matrix} \right\}$$
$$\times \left[a_{-\frac{1}{2}}^{(ql_2 s)} \times \left[a^{(ql_1 s)} \times W^{(K k_2'\kappa_2')}(l_1) \right]_{\frac{1}{2}}^{(K_1 L_1 S_1)} \right]_{\rho-\rho}^{(kk)}. \quad (17.51)$$

In the above equations we have introduced

$$b_{ijtr}^{k_2\kappa_2 k_2'\kappa_2'} = -\frac{1}{2} \left(n_i l_i n_j l_j \| f^{(k_2\kappa_2 k_2'\kappa_2')} \| n_t l_t n_r l_r \right), \quad (17.52)$$

and $P(1 \leftrightarrow 2)$ is the operator of transposition of subscripts 1 and 2.

It has been shown in the previous section how the submatrix elements of irreducible tensorial products of creation and annihilation operators can be expressed in terms of pertinent one-shell submatrix elements. The submatrix elements of the operators G_1–G_7 are also defined in terms of the same quantities. Since the quasispin ranks from different shells that enter into these operators are not interrelated, the methods described in Chapter 15 can be used for each of the one-shell submatrix elements independently.

17.5 Superposition of configurations in quasispin space

The operators of orbital, spin and quasispin angular momenta of the two-shell configuration are expressed in terms of sums of one-shell triple tensors (15.52):

$$L_\rho^{(1)}(l_1, l_2) = -i3^{-1/2}\left\{[l_1(l_1 + 1)(2l_1 + 1)]^{1/2}W_{0\rho 0}^{(010)}(l_1)\right.$$
$$\left. + [l_2(l_2 + 1)(2l_2 + 1)]^{1/2}W_{0\rho 0}^{(010)}(l_2)\right\}, \qquad (17.53)$$

$$S_\rho^{(1)}(l_1, l_2) = -\frac{i}{2}\left\{(2l_1 + 1)^{1/2}W_{00\rho}^{(001)}(l_1) + (2l_2 + 1)^{1/2}W_{00\rho}^{(001)}(l_2)\right\}, \quad (17.54)$$

$$Q_\rho^{(1)}(l_1, l_2) = -\frac{i}{2}\left\{(2l_1 + 1)^{1/2}W_{\rho 00}^{(100)}(l_1) + (2l_2 + 1)^{1/2}W_{\rho 00}^{(100)}(l_2)\right\}. \quad (17.55)$$

The wave function of the two-shell configuration (17.42) corresponds to the representation of uncoupled quasispin momenta of individual shells. The eigenfunction of the square of the operator of total quasispin and its z-projection can be written as follows in the scheme of the vectorial coupling of momenta in quasispin space:

$$|(l_1 + l_2)^N\alpha_1Q_1L_1S_1\alpha_2Q_2L_2S_2QLSM_LM_S)$$
$$= \sum_{M_{Q_1}M_{Q_2}}\begin{bmatrix}Q_1 & Q_2 & Q \\ M_{Q_1} & M_{Q_2} & M_Q\end{bmatrix}\left|l_1^{N_1}l_2^{N_2}\alpha_1Q_1L_1S_1\alpha_2Q_2L_2S_2LSM_LM_S\right).(17.56)$$

It will represent a two-shell multi-configuration wave function in which the coefficients of expansion in configurations are the conventional Clebsch–Gordan coefficients.

The group-theoretical methods of constructing the multi-configuration wave functions (the so-called mixed configurations) invariably attract a great deal of interest. Mixed configurations of the type $(l_1 + l_2)^N$ and $(d+s)^N$ were studied in detail earlier [113–116]. These configurations were covered by the group-theoretical techniques used in the case of one shell of equivalent electrons. In [113, 114] mixed states were classified by the chain of subgroups $U_{12} \supset SU_6 \times SU_2$; $SU_6 \supset R_6 \supset R_5 \times R_1 \supset R_3$. The multi-configuration wave functions provided in [114] coincide, with some exceptions, with the functions derived in the vectorial coupling scheme for quasispin momenta (17.56). These exceptions are explained by the fact that the irreducible representations of R_5 group, used in [114], are symmetrical under transposition of the quantum numbers of spin S and quasispin Q, whereby the wave functions of the d-shell with these characteristics appear simultaneously in the expansion of mixed configuration. But when configurations are mixed in the vectorial coupling scheme for quasispin momenta, the quantum numbers of terms of individual shells will be the same for all configurations that enter into one multi-configuration wave function. Morrison [116] has shown that using a slightly different subgroup chain

$$R_{8l_1+8l_2+8} \supset SU_2^Q \times Sp_{4l_1+2} \times Sp_{4l_2+2}, \tag{17.57}$$

where Sp_{4l_1+2} and Sp_{4l_2+2} are the symplectic groups, irreducible representations have characteristics that correspond to the seniority quantum numbers of individual shells. Using further reduction scheme (14.38) for each of the shells the SU_2^Q group in this scheme will be a group of total quasispin of both shells, and the wave functions will exactly correspond to expansion (17.56).

The concept of the vectorial coupling of quasispin momenta was first applied to the nucleus to study the short-range pairing nucleonic interaction [117]. For interactions of that type the quasispin of the system is a sufficiently good quantum number. In atoms there is no such interaction – the electrons are acted upon by electrostatic repulsion forces, for which the quasispin quantum number is not conserved. Therefore, in general, the Hamiltonian matrix defined in the basis of wave functions (17.56) is essentially non-diagonal.

In certain special cases the approximate symmetries in atoms are sufficiently well explained using the quasispin formalism. In particular, the quasispin technique can be utilized to describe fairly accurately configuration mixing for doubly excited states of the two-electron atom. In the quasispin basis the energy matrix of the electrostatic interaction operator of such configurations is nearly diagonal, and hence the quantum number of total quasispin Q is approximately 'good'.

Note that for doubly excited states of helium similar multi-configuration wave functions can be obtained within the framework of a totally different group-theoretical method suggested in [118–120] and based on the properties of the symmetry group of the hydrogen atom, R_4. Although these methods are totally different, the weight coefficients of wave functions $|2(s+p)^2 Q\,^1S)$ coincide up to general sign.

17.6   Tensors in the space of total quasispin and their submatrix elements

The representation of the wave function in the scheme of coupled quasispin momenta calls for the introduction of operators (17.43) whose quasispin ranks from different shells would be coupled as well.

Remembering the quasispin properties of creation and annihilation operators, we introduce the triple tensor

$$W_{\Pi\rho\pi}^{(Kk\kappa)}(l_1, l_2) = \left[a^{(ql_1s)} \times a^{(ql_2s)} \right]_{\Pi\rho\pi}^{(Kk\kappa)}. \qquad (17.58)$$

At $K = \Pi = 0$

$$W_{0\rho\pi}^{(0k\kappa)}(l_1, l_2) = -^{(-)} W_{\rho\pi}^{(k\kappa)}(l_1, l_2), \qquad (17.59)$$

and at $K = 1, \Pi = 0$

$$W_{0\rho\pi}^{(1k\kappa)}(l_1, l_2) = {}^{(+)} W_{\rho\pi}^{(k\kappa)}(l_1, l_2). \qquad (17.60)$$

The tensors on the right sides of (17.59) and (17.60) are given by (17.9) and are generally introduced for mixed configurations by linear combinations of the tensors $W_{\rho\pi}^{(k\kappa)}(l_1, l_2)$ and $W_{\rho\pi}^{(k\kappa)}(l_2, l_1)$. Accordingly, these tensors are special cases of the triple tensor $W_{\Pi\rho\pi}^{(Kk\kappa)}$ at a fixed projection $\Pi = 0$. These tensors, in addition to the triple tensors inside the shell introduced in Chapter 15, will be the basic standard quantities in our further treatment.

Now, following Chapter 15, we shall examine tensorial products of two-shell operators. Consider the product $\left[W^{(Kk\kappa)}(l_1) \times W^{(K'\kappa'\kappa')}(l_2) \right]^{(K_1 k_1 \kappa_1)}$. Exchanging the operators $a^{(ql_1s)}$ and $a^{(ql_2s)}$ and recoupling the momenta yields [121]

$$\left[W^{(Kk\kappa)}(l_1) \times W^{(K'k'\kappa')}(l_2) \right]^{(K_1 k_1 \kappa_1)}$$

$$= - \sum_{K_2 K_2' k_2 k_2' \kappa_2 \kappa_2'} [K, K', K_2, K_2', k, k', k_2, k_2', \kappa, \kappa', \kappa_2, \kappa_2']^{1/2}$$

$$\times \begin{Bmatrix} q & q & K \\ q & q & K' \\ K_2 & K_2' & K_1 \end{Bmatrix} \begin{Bmatrix} l_1 & l_1 & k \\ l_2 & l_2 & k' \\ k_2 & k_2' & k_1 \end{Bmatrix} \begin{Bmatrix} s & s & \kappa \\ s & s & \kappa' \\ \kappa_2 & \kappa_2' & \kappa_1 \end{Bmatrix}$$

$$\times \left[W^{(K_2 k_2 \kappa_2)}(l_1, l_2) \times W^{(K_2' k_2' \kappa_2')}(l_1, l_2) \right]^{(K_1 k_1 \kappa_1)}. \qquad (17.61)$$

We next put the orbital ranks on the left side of (17.61) equal to zero ($k = k' = k_1 = 0$), and then construct a set of equations by selecting the values of spin and quasispin ranks so that on the left side of this equation we have operators with known eigenvalues. In the case of complete scalars $K_1 = k_1 = \kappa_1$, selecting the values of ranks the same as they were in Chapter 15 and solving the resultant set of equations for the sum of a given structure of spin and quasispin ranks, we find

$$\sum_k (2k + 1)^{1/2} \left[W^{(0k0)}(l_1, l_2) \times W^{(0k0)}(l_1, l_2) \right]^{(000)}$$

$$= (\mathbf{Q}_1 \cdot \mathbf{Q}_2) + (\mathbf{S}_1 \cdot \mathbf{S}_2) - [l_1, l_2]/4, \qquad (17.62)$$

$$\sum_k [3(2k+1)]^{1/2} \left[W^{(0k1)}(l_1, l_2) \times W^{(0k1)}(l_1, l_2) \right]^{(000)}$$

$$= 3(\mathbf{Q}_1 \cdot \mathbf{Q}_2) - (\mathbf{S}_1 \cdot \mathbf{S}_2) - 3[l_1, l_2]/4, \qquad (17.63)$$

$$\sum_k [3(2k+1)]^{1/2} \left[W^{(1k0)}(l_1, l_2) \times W^{(1k0)}(l_1, l_2) \right]^{(000)}$$

$$= -(\mathbf{Q}_1 \cdot \mathbf{Q}_2) + 3(\mathbf{S}_1 \cdot \mathbf{S}_2) - 3[l_1, l_2]/4, \qquad (17.64)$$

$$\sum_k (2k+1)^{1/2} \left[W^{(1k1)}(l_1, l_2) \times W^{(1k1)}(l_1, l_2) \right]^{(000)}$$

$$= -(\mathbf{Q}_1 \cdot \mathbf{Q}_2) - (\mathbf{S}_1 \cdot \mathbf{S}_2) - 3[l_1, l_2]/4. \qquad (17.65)$$

Similar expressions can be established for non-zero ranks K_1, k_1, κ_1 as well. So, for example, at $k_1 = \kappa_1 = 0$, by the same procedure as in the derivation of (15.82) and (15.83), we arrive at

$$\sum_k (2k+1)^{1/2} \left[W^{(1k0)}(l_1, l_2) \times W^{(1k0)}(l_1, l_2) \right]^{(200)}_{\Pi 00} = 2 \left[Q^{(1)}(l_1) \times Q^{(1)}(l_2) \right]^{(2)}_{\Pi},$$

$$(17.66)$$

$$\sum_k (2k+1)^{1/2} \left[W^{(1k1)}(l_1, l_2) \times W^{(1k1)}(l_1, l_2) \right]^{(200)}_{\Pi 00}$$

$$= 2\sqrt{3} \left[Q^{(1)}(l_1) \times Q^{(1)}(l_2) \right]^{(2)}_{\Pi}. \qquad (17.67)$$

Expressions (17.62)–(17.67) are suitable for shells with any values of l_1 and l_2. Similar formulas for tensorial products with fixed k are obtainable for mixed configuration $(l+s)^N$ only when the left sides of (17.62)–(17.67) have the term with $k = 0$.

Using (17.61) we can additionally obtain the following relationships for the tensorial products that are complete scalars in orbital, spin and quasispin spaces:

$$\sum_k k(k+1)[3(2k+1)]^{1/2} \left\{ \left[W^{(1k0)}(l_1, l_2) \times W^{(1k0)}(l_1, l_2) \right]^{(000)} \right.$$

$$+ \sqrt{3} \left[W^{(1k1)}(l_1, l_2) \times W^{(1k1)}(l_1, l_2) \right]^{(000)} \right\} = -6(\mathbf{L}_1 \cdot \mathbf{L}_2)$$

$$- 3[l_1(l_1+1) + l_2(l_2+1)][l_1, l_2] - 4[l_1(l_1+1) + l_2(l_2+1)]$$

$$\times (\mathbf{Q}_1 \cdot \mathbf{Q}_2), \qquad (17.68)$$

$$\sum_k k(k+1)(2k+1)^{1/2} \left\{ \left[W^{(0k0)}(l_1, l_2) \times W^{(0k0)}(l_1, l_2) \right]^{(000)} \right.$$

$$+ \sqrt{3} \left[W^{(1k0)}(l_1, l_2) \times W^{(1k0)}(l_1, l_2) \right]^{(000)} \right\} = -2(\mathbf{L}_1 \cdot \mathbf{L}_2)$$

$$- 4[l_1(l_1+1) + l_2(l_2+1)](\mathbf{S}_1 \cdot \mathbf{S}_2) - 3[l_1(l_1+1) + l_2(l_2+1)]$$

$$\times [l_1, l_2], \qquad (17.69)$$

$$\sum_k k(k+1)[3(2k+1)]^{1/2}\left\{\left[W^{(0k1)}(l_1,l_2)\times W^{(0k1)}(l_1,l_2)\right]^{(000)}\right.$$

$$+\sqrt{3}\left.\left[W^{(1k1)}(l_1,l_2)\times W^{(1k1)}(l_1,l_2)\right]^{(000)}\right\}=-6(\mathbf{L}_1\cdot\mathbf{L}_2)$$

$$-4[l_1(l_1+1)+l_2(l_2+1)](\mathbf{S}_1\cdot\mathbf{S}_2)-3[l_1(l_1+1)+l_2(l_2+1)]$$

$$\times[l_1,l_2]. \tag{17.70}$$

These equations can be used to establish additional analytical relationships when dealing with the matrix elements of operators of physical quantities in the case of complex electron configurations.

Now let us represent the two-particle operator of general form (14.57) in terms of the irreducible tensors in the space of total quasispin for the two-shell configuration

$$G=[1+P(1\leftrightarrow 2)]\sum_{i=1}^{4}G'_i, \tag{17.71}$$

$$G'_1=c_{1,2;1,2}b_{1122}^{k_2\kappa_2k'_2\kappa'_2}+c_{1,1;2,2}b_{1212}^{k_2\kappa_2k'_2\kappa'_2}, \tag{17.72}$$

$$G'_2=b_{1221}^{k_2\kappa_2k'_2\kappa'_2}\left\{c_{1,2;2,1}-\sum_{K'}2^{-1/2}\begin{Bmatrix}l_1&l_1&k\\k_2&k'_2&l_2\end{Bmatrix}\begin{Bmatrix}s&s&k\\\kappa_2&\kappa'_2&s\end{Bmatrix}\right.$$

$$\left.\times W_{0\rho-\rho}^{(K'kk)}(l_1)\right\}, \tag{17.73}$$

$$G'_3=c_{1,1;2,1}\left\{b_{1121}^{k_2\kappa_2k'_2\kappa'_2}+(-1)^{k_2+\kappa_2+k'_2+\kappa'_2}b_{2111}^{k_2\kappa_2k'_2\kappa'_2}\right\}, \tag{17.74}$$

$$G'_4=c_{1,2;1,1}\left\{b_{1121}^{k_2\kappa_2k'_2\kappa'_2}+(-1)^{k_2+\kappa_2+k'_2+\kappa'_2}b_{1222}^{k_2\kappa_2k'_2\kappa'_2}\right\}, \tag{17.75}$$

where

$$c_{i,j;t,r}=\frac{1}{2}\sum_{KK'K''}[k_2,k'_2,\kappa_2,\kappa'_2]^{1/2}\begin{bmatrix}K&K'&K''\\0&0&0\end{bmatrix}$$

$$\times\left[W^{(Kk_2\kappa_2)}(l_i,l_j)\times W^{(K'k'_2\kappa'_2)}(l_t,l_r)\right]_{0\rho-\rho}^{(K''kk)}. \tag{17.76}$$

These operators must be supplemented by intrashell operators whose expansion has been derived in Chapter 15. Comparison of the expansion of the operator in the quasispin space of two shells (17.71) with that of the same operator in the quasispin space of each shell individually (17.44) shows that many terms of expansion (17.71) have a much simpler tensorial structure and a smaller number of operators being summed up than those of (17.44).

This is accounted for by the fact that for a mixed configuration we need not separate the creation and annihilation operators of different shells, it would be sufficient to couple them to form irreducible tensorial products in the way that is most suitable for each expansion term.

These tensorial products have certain ranks with respect to the total orbital, spin and quasispin angular momenta of both shells. We shall now proceed to compute the matrix elements of such tensors relative to multi-configuration wave functions defined according to (17.56). Applying the Wigner–Eckart theorem in quasispin space to the submatrix element of tensor $T^{(K k \kappa)}$ gives

$$
\begin{aligned}
\Big(&(l_1+l_2)^N \alpha_1 Q_1 L_1 S_1 \alpha_2 Q_2 L_2 S_2 Q L S \| T_\Pi^{(Kk\kappa)} \| (l_1+l_2)^{N'} \\
&\times \alpha_1' Q_1' L_1' S_1' \alpha_2' Q_2' L_2' S_2' Q' L' S' \Big) = (-1)^{2K}(2Q+1)^{-1/2} \\
&\times \begin{bmatrix} Q' & K & Q \\ M_Q' & \Pi & M_Q \end{bmatrix} \\
&\times \Big((l_1+l_2)\alpha_1 Q_1 L_1 S_1 \alpha_2 Q_2 L_2 S_2 Q L S \||| T^{(Kk\kappa)} \||| (l_1+l_2) \\
&\times \alpha_1' Q_1' L_1' S_1' \alpha_2' Q_2' L_2' S_2' Q' L' S' \Big),
\end{aligned}
\tag{17.77}
$$

where the last factor is a matrix element reduced relative to all three spaces. And so the dependence of the matrix element on the total number of electrons in the mixed configuration is again separated to yield one Clebsch–Gordan coefficient.

According to the foregoing discussion two-shell operators are expressed in terms of irreducible products of tensors $W^{(Kk\kappa)}(l_i, l_j)$ and $W^{(Kk\kappa)}(l_i)$. Using a standard technique, we can represent the submatrix elements of these products in terms of relevant quantities defined for these tensors. We shall, therefore, confine ourselves to consideration of the submatrix element of tensor $W^{(Kk\kappa)}(l_1, l_2)$. By its definition (17.58),

$$
\begin{aligned}
\Big(&(l_1+l_2)\alpha_1 Q_1 L_1 S_1 \alpha_2 Q_2 L_2 S_2 Q L S \||| W^{(Kk\kappa)}(l_1,l_2) \||| (l_1+l_2) \\
&\times \alpha_1' Q_1' L_1' S_1' \alpha_2' Q_2' L_2' S_2' Q' L' S' \Big) = [Q, L, S, Q', L', S', K, k, \kappa]^{1/2} \\
&\times \begin{Bmatrix} Q_1' & Q_2' & Q' \\ q & q & K \\ Q_1 & Q_2 & Q \end{Bmatrix} \begin{Bmatrix} L_1' & L_2' & L' \\ l_1 & l_2 & k \\ L_1 & L_2 & L \end{Bmatrix} \begin{Bmatrix} S_1' & S_2' & S' \\ s & s & \kappa \\ S_1 & S_2 & S \end{Bmatrix} \\
&\times \Big(l_1\alpha_1 Q_1 L_1 S_1 \||| a^{(q l_1 s)} \||| l_1 \alpha_1' Q_1' L_1' S_1' \Big) \\
&\times \Big(l_2\alpha_2 Q_2 L_2 S_2 \||| a^{(q l_2 s)} \||| l_2 \alpha_2' Q_2' L_2' S_2' \Big),
\end{aligned}
\tag{17.78}
$$

where the last two factors are the reduced coefficients (subcoefficients) of fractional parentage of respective shells introduced in Chapter 16.

The submatrix element of $W^{(Kk\kappa)}(l_2, l_1)$ can be derived, according to

(17.78), from

$$W_{\Pi\rho\pi}^{(Kk\kappa)}(l_2, l_1) = (-1)^{l_1+l_2+k+\kappa+1} W_{\Pi\rho\pi}^{(Kk\kappa)}(l_1, l_2).$$ (17.79)

It has been shown earlier (see Chapters 15 and 16) that the technique relying on the tensorial properties of operators and wave functions in quasispin, orbital and spin spaces is an alternative but more convenient one than the method of higher-rank groups. It is more convenient not only for classification of states, but also for theoretical studies of interactions in equivalent electron configurations. The results of this chapter show that the above is true of more complex configurations as well.

18

Isospin in the theory of an atom

18.1 Isospin

For an additional degree of freedom of a particle that defines one of the two possible states in which it exists, we shall apply now the concept of isospin (in analogy with the theory of the nucleus where the isospin doublet includes protons and neutrons, which are treated as two states of the same particle – the nucleon [122]). For a pair of states (α, β) we introduce the isospin operators

$$t^{(1)}_{+1}(\alpha, \beta) = -i2^{-1/2}a^\dagger_\alpha a_\beta; \qquad (18.1)$$

$$t^{(1)}_{-1}(\alpha, \beta) = i2^{-1/2}a^\dagger_\beta a_\alpha; \qquad (18.2)$$

$$t^{(1)}_0(\alpha, \beta) = \frac{i}{2}(\hat{n}_\alpha - \hat{n}_\beta) = \frac{i}{2}(a_\alpha a^\dagger_\alpha - a_\beta a^\dagger_\beta). \qquad (18.3)$$

Using the anticommutation relations (13.15) we can readily verify that these operators obey the conventional commutation relations (14.2) for the irreducible components of the angular momentum operator. Further, from the definition

$$t_z(\alpha, \beta) = -it^{(1)}_0(\alpha, \beta) = (\hat{n}_\alpha - \hat{n}_\beta)/2 \qquad (18.4)$$

we see that the operator t_z has a zero eigenvalue when the pair of states (α, β) is vacant or filled with two particles. The operator has an eigenvalue of $\pm 1/2$ when only one particle is in the pairing state (α, β). Thus, if a pair of states is occupied by 0 or 2 particles, then the isospin of the state is $t = 0$, and if by one particle, then $t = 1/2$. Examining the commutation relations for the electron creation operators a_α and a_β with the isospin operators, we can establish that the latter behave as the components of a tensor of rank $\tau = 1/2$ in isospin space (see (14.13))

$$\left[t^{(1)}_\rho(\alpha, \beta), a^{(\tau)}_m \right] = i\sqrt{\tau(\tau+1)} \begin{bmatrix} \tau & 1 & \tau \\ m & \rho & m+\rho \end{bmatrix} a^{(\tau)}_{m+\rho}, \qquad (18.5)$$

where

$$a_m^{(\tau)} = \begin{cases} a_\alpha & \text{at} \quad m = 1/2, \\ a_\beta & \text{at} \quad m = -1/2. \end{cases} \tag{18.6}$$

The annihilation operators $a_\alpha^\dagger$ and $a_\beta^\dagger$ do not form an irreducible tensor, but the operators

$$\tilde{a}_m^{(\tau)} = (-1)^{\tau-m} a_{-m}^{(\tau)\dagger} = \begin{cases} a_\beta^\dagger & \text{at} \quad m = 1/2, \\ -a_\alpha^\dagger & \text{at} \quad m = -1/2, \end{cases} \tag{18.7}$$

do, by (13.40), and from the anticommutation relations (13.15) it follows that

$$\left\{ a_m^{(\tau)}, \tilde{a}_{m'}^{(\tau)} \right\} = (-1)^{\tau-m} \delta(m, m') = 2^{-1/2} \begin{bmatrix} \tau & \tau & 0 \\ m & m' & 0 \end{bmatrix}; \tag{18.8}$$

$$\left\{ a_m^{(\tau)}, a_{m'}^{(\tau)} \right\} = \left\{ \tilde{a}_m^{(\tau)}, \tilde{a}_{m'}^{(\tau)} \right\} = 0. \tag{18.9}$$

The isospin operators can be expressed in terms of the irreducible tensors (18.6) and (18.7)

$$t_{+1}^{(1)} = -i2^{-1/2} \left[a^{(\tau)} \times \tilde{a}^{(\tau)} \right]_1^{(1)};$$

$$t_{-1}^{(1)} = -i2^{-1/2} \left[a^{(\tau)} \times \tilde{a}^{(\tau)} \right]_{-1}^{(1)}; \tag{18.10}$$

$$t_0^{(1)} = -i2^{-1/2} \left[a^{(\tau)} \times \tilde{a}^{(\tau)} \right]_0^{(1)}.$$

All these operators commute with the particle number operator in the pairing state

$$\hat{N}(\alpha, \beta) = \hat{n}_\alpha + \hat{n}_\beta = -2^{1/2} \left[a^{(\tau)} \times \tilde{a}^{(\tau)} \right]_0^{(0)}, \tag{18.11}$$

which is a scalar in isospin space. Finite rotations in isospin space will then leave the number of particles in the pairing state (α, β) unchanged. Under such rotations, according to the general relation (5.9), the creation and annihilation operators are transformed as follows:

$$\alpha_{m'}^{(\tau)} = \sum_m D_{m',m}^{(1/2)} a_m^{(\tau)}; \qquad \tilde{\alpha}_{m'}^{(\tau)} = \sum_m D_{m',m}^{(1/2)} \tilde{a}_m^{(\tau)}. \tag{18.12}$$

The anticommutation relations (18.8) and (18.9) for the new second-quantization operators $\alpha_m^{(\tau)}$ and $\tilde{\alpha}_m^{(\tau)}$ also hold, which can be readily verified by computing the anticommutators involved.

All four creation and annihilation operators for electrons in the pairing state (α, β) can be expressed via the tensor $a_{vm}^{(q\tau)}$ in (15.38) at various values of the projections v and m. The anticommutation relations (18.8) and

(18.9) can then be written as

$$\left\{a^{(q\tau)}_{\nu m}, a^{(q\tau)}_{\nu'm'}\right\} = (-1)^{q+\tau-\nu-m}\delta(m\nu, -m'-\nu')$$

$$= 2\begin{bmatrix} q & q & 0 \\ \nu & \nu' & 0 \end{bmatrix}\begin{bmatrix} \tau & \tau & 0 \\ m & m' & 0 \end{bmatrix}. \qquad (18.13)$$

Any products of creation and annihilation operators for electrons in a pairing state can be expanded in terms of irreducible tensors in the space of quasispin and isospin. So, for the operators (18.10) and (15.35) we have, respectively,

$$t^{(1)}_\rho(\alpha, \beta) = -\frac{i}{2}\left[a^{(q\tau)} \times a^{(q\tau)}\right]^{(01)}_{0\rho}; \qquad (18.14)$$

$$Q^{(1)}_\rho(\alpha, \beta) = -\frac{i}{2}\left[a^{(q\tau)} \times a^{(q\tau)}\right]^{(10)}_{\rho 0}, \qquad (18.15)$$

i.e. the quasispin operator is a scalar in isospin space, and the isospin operator is a scalar in quasispin space, and these operators commute with each other. The scalar product of the operators

$$\left[a^{(q\tau)} \times a^{(q\tau)}\right]^{(00)} = (\hat{N} + {}_h\hat{N})/2 = -1 \qquad (18.16)$$

does not change under rotations in quasispin or isospin spaces. Since the number of particles in a pairing state is determined by the z-projection of the quasispin angular momentum, then under rotations in quasispin space this number is not preserved. The tensors $a^{(q\tau)}_{\nu m}$ under finite rotations in that space are transformed as follows:

$$\alpha^{(q\tau)}_{\nu'm} = \sum_\nu D^{(1/2)}_{\nu',\nu}a^{(q\tau)}_{\nu m}, \qquad (18.17)$$

and for the new tensors $\alpha^{(q\tau)}_{\nu'm}$ the anticommutation relations (18.13) remain valid. Operators $\alpha^{(q\tau)}_{\nu'm}$ at a fixed value of the projection ν' are creation and annihilation operators for quasiparticles.

Both isospin operators and quasispin operators of various pairing states commute with each other, therefore we can use the vectorial coupling of momenta of individual pairing states and introduce the isospin and quasispin of the N-particle state.

18.2   Isospin basis and its properties

Let us take a closer look at the concept of isospin referring to two shells of equivalent electrons with the same orbital quantum numbers. Our reasoning will partially follow the work [123].

The concept of isospin was introduced by Heisenberg who suggested that the proton and neutron can be treated as two states of the nucleon in isospin space [122]. The mathematical technique of isospin was then applied to classify the states in nuclear spectroscopy. Its effectiveness is there determined by physical considerations such as the independence of the strong interaction of whether the particles are protons or neutrons. This made it possible in the shell model of the nucleus to view the two shells (proton and neutron shells) with the same momenta as one nucleon shell with an additional degree of freedom in isospin space.

Mathematically, it is a rather straightforward exercise to use this approach in theoretical treatments of the electron configuration $n_1 l^{N_1} n_2 l^{N_2}$ in atomic spectroscopy. But here the physical meaning of the isospin formalism is in need of some additional explanation. The general procedure for constructing the wave functions of complex electron configurations is to use the vectorial coupling of momenta of individual shells and then to diagonalize the energy matrix in this basis. Utilization of the concept of isospin yields a new alternative basis in which it is possible to define all the standard quantities of the theory of an atom (unit tensors, coefficients of fractional parentage, etc.) and to compute the required characteristics. In some cases the isospin basis may appear to be preferable to the conventional one. It is these questions that we are going to address in this chapter.

Let the quantities $a_{m\mu}^{(ls)}$ and $a_{m\mu}^{(ls)\dagger} = (-1)^{l+s+m+\mu} \tilde{a}_{-m-\mu}^{(ls)}$ be creation and annihilation operators for the electrons in the $n_1 l^{N_1}$ shell; and $b_{m\mu}^{(ls)}$ and $b_{m\mu}^{(ls)\dagger} = (-1)^{l+s+m+\mu} \tilde{b}_{-m-\mu}^{(ls)}$ be creation and annihilation operators for the electrons in the $n_2 l^{N_2}$ shell. We shall now introduce the isospin angular momentum operator $T_\rho^{(1)}$ given by

$$T_1^{(1)} = i(2l+1)^{1/2} \left[b^{(ls)} \times \tilde{a}^{(ls)} \right]^{(00)}, \tag{18.18}$$

$$T_{-1}^{(1)} = -i(2l+1)^{1/2} \left[a^{(ls)} \times \tilde{b}^{(ls)} \right]^{(00)}, \tag{18.19}$$

and

$$T_0^{(1)} = iT_z = \left[(2l+1)/2 \right]^{1/2} \left(\left[a^{(ls)} \times \tilde{a}^{(ls)} \right]^{(00)} - \left[b^{(ls)} \times \tilde{b}^{(ls)} \right]^{(00)} \right) = \frac{i}{2}(\hat{N}_2 - \hat{N}_1), \tag{18.20}$$

where $\hat{N}_1$ and $\hat{N}_2$ are the particle number operators for the $n_1 l^{N_1}$ and $n_2 l^{N_2}$ shells, respectively. It follows from (18.18)–(18.20) that the operator $T_1^{(1)}$ annihilates an electron in the first shell and produces an electron in the second one, and the operator $T_{-1}^{(1)}$ does the reverse. This leaves the total number of electrons $N = N_1 + N_2$ of the $n_1 l^{N_1} n_2 l^{N_2}$ configuration unchanged.

The operators (18.18)–(18.20) obey the conventional commutation rela-

tions for the irreducible components of the angular momentum operator
(14.2), and their commutation relations with creation and annihilation operators show that they both are the components of an irreducible tensor
of rank τ in some additional (isospin) space, and

$$a^{(\tau ls)}_{\frac{1}{2}m\mu} = b^{(ls)}_{m\mu}, \qquad \tilde{a}^{(\tau ls)}_{\frac{1}{2}m\mu} = \tilde{a}^{(ls)}_{m\mu},$$
$$a^{(\tau ls)}_{-\frac{1}{2}m\mu} = a^{(ls)}_{m\mu}, \qquad \tilde{a}^{(\tau ls)}_{-\frac{1}{2}m\mu} = -\tilde{b}^{(ls)}_{m\mu}. \tag{18.21}$$

Consequently, in the $n_1 l^{N_1} n_2 l^{N_2}$ configuration the creation and annihilation operators for electrons

$$a^{(\tau ls)}_{\eta m\mu}, \quad \tilde{a}^{(\tau ls)}_{\eta m\mu} = (-1)^{\tau-\eta+l-m+s-\mu} a^{(\tau ls)\dagger}_{-\eta-m-\mu} \tag{18.22}$$

are irreducible tensors of ranks $\tau = 1/2$, l and $s = 1/2$ in isospin, orbital
and spin spaces, respectively. They obey the anticommutation relations

$$\left\{ a^{(\tau ls)}_{\eta m\mu}, a^{(\tau ls)}_{\eta' m' \mu'} \right\} = \left\{ \tilde{a}^{(\tau ls)}_{\eta m\mu}, \tilde{a}^{(\tau ls)}_{\eta' m' \mu'} \right\} = 0,$$
$$\left\{ a^{(\tau ls)}_{\eta m\mu}, \tilde{a}^{(\tau ls)}_{\eta' m' \mu'} \right\} = (-1)^{\tau-\eta+l-m+s-\mu} \delta(\eta m\mu, -\eta' - m' - \mu'). \tag{18.23}$$

Out of the operators (18.22) we can compose the following triple irreducible tensors:

$$U^{(\gamma\kappa\kappa)}_{\Gamma\rho\pi} = \left[a^{(\tau ls)} \times \tilde{a}^{(\tau ls)} \right]^{(\gamma\kappa\kappa)}_{\Gamma\rho\pi}, \tag{18.24}$$

$$U^{(\gamma\kappa\kappa)\dagger}_{\Gamma\rho\pi} = (-1)^{\Gamma+\rho+\pi} U^{(\gamma\kappa\kappa)}_{-\Gamma-\rho-\pi}, \tag{18.25}$$

in terms of which we can express, for example, the operators of isospin,
orbital and spin momenta of the configuration, as well as the particle
number operator $\hat{N}$

$$T^{(1)}_\rho = i(2l+1)^{1/2} U^{(100)}_{\rho 00}, \qquad L^{(1)}_\rho = i[4l(l+1)(2l+1)/3]^{1/2} U^{(010)}_{0\rho 0},$$
$$S^{(1)}_\rho = i(2l+1)^{1/2} U^{(001)}_{00\rho}, \qquad \hat{N} = 2(2l+1)^{1/2} U^{(000)}_{000}. \tag{18.26}$$

Thus, with the $n_1 l^{N_1} n_2 l^{N_2}$ configuration we can do without the wave
functions derived using the vectorial coupling of momenta of individual
shells and use the functions characterized by eigenvalues of the commuting
operators $\hat{N}$, $\mathbf{T}^2$, T_z, $\mathbf{L}^2$, L_z, $\mathbf{S}^2$, S_z

$$|n_1 n_2 (ll)^N \alpha T L S M_T M_L M_S) = |n_1 n_2 l^{N_1} l^{N_2} \alpha T L S M_L M_S), \tag{18.27}$$

where α is an additional index used to classify the repeating terms;
$M_T = (N_2 - N_1)/2$. The operators (18.18) and (18.19) here only change
their projection M_T, i.e. the eigenvalues of the operator (18.20).

The result of acting with $\mathbf{T}^2$ on the wave function (18.27) is the multi-
plication of the latter by the eigenvalue of this operator

$$\mathbf{T}^2 |n_1 n_2 (ll)^N \alpha T L S M_T M_L M_S) = T(T+1) |n_1 n_2 (ll)^N \alpha T L S M_T M_L M_S), \tag{18.28}$$

where T defines the numerical value of the isospin angular momentum and takes on the values $N/2$, $N/2 - 1, \ldots, 1/2$ or 0 depending on the parity of $N/2$.

For the $n_1 l^{N_1} n_2 l^{N_2}$ configuration the following basis was suggested [124]:

$$|n_1 l^{N_1} n_2 l^{N_2} \beta i LS) = |n_1 l^{N_1+N_2-\sigma} n_2 l^\sigma \to n_1 l^{N_1} n_2 l^{N_2} \beta i LS), \qquad (18.29)$$

where i is the number of electrons in the second shell of the configuration series $n_1 l^x n_2 l^y$ ($x + y = N_1 + N_2$, $x \geq N_1$, $n_1 < n_2$) for which this term first appears, and β labels the repeating terms. The wave function $|n_1 l^{N_1} n_2 l^{N_2} \beta i LS)$ follows from the function $|n_1 l^{N_1+N_2-\sigma} n_2 l^\sigma \beta i LS)$ by replacing $R(n_1 l|r)$ by $R(n_2 l|r)$ in the latter $N_2 - \sigma$ radial orbitals, which actually corresponds to the $(N_2 - \sigma)$-multiple action of the operator $T_1^{(1)}$ on the wave function $|n_1 l^{N_1+N_2-\sigma} n_2 l^\sigma \beta i LS)$. Note that

$$T = (N - 2i)/2. \qquad (18.30)$$

It is seen from (18.30) that corresponding to terms with $i = 0$, i.e. the ones that first appear in the $n_1 l^{N_1}$ configuration, is the isospin $T = N/2$ that is a maximally possible value at a given N. The values of the isospin projection M_T show the interval of the $n_1 l^x n_2 l^y$ ($x+y = N$) configurations in which this term exists. And corresponding to the terms that first appear in the $n_1 l^{N-1} n_2 l$ ($i = 1$) configuration is $T = (N - 2)/2$; these enter into the configurations $n_1 l^{N-1} nl$, $n_1 l^{N-2} n_2 l^2, \ldots, n_1 l n_2 l^{N-1}$. In a similar way, the bases (18.27) and (18.29) are related at other values of i. Put another way, because of the constraint $N_1 \geq N_2$ imposed in [124] the wave functions (18.29) constitute a part ($M_T \leq 0$) of the functions (18.27). The basis (18.27), within the framework of the concept of the isospin, has a more rigorous interpretation than the basis (18.29).

For the basis (18.27) to be used effectively in practical computations an adequate mathematical tool is required that would permit full account to be taken of the tensorial properties of wave functions and operators in their spaces. In particular, matrix elements can now be defined using the Wigner–Eckart theorem (5.15) in all three spaces, so that the submatrix element will be given by

$$\left(n_1 n_2 (ll)^N \alpha_1 T_1 L_1 S_1 \| F^{(\gamma k \kappa)} \| n_1 n_2 (ll)^{N'} \alpha_2 T_2 L_2 S_2 \right)$$
$$= (-1)^{T_1 + L_1 + S_1 + \gamma + k + \kappa - T_2 - L_2 - S_2} [T_1, L_1, S_1]^{1/2}$$
$$\times \left(n_1 n_2 (ll)^N \alpha_1 T_1 L_1 S_1 M_{T_1} M_{L_1} M_{S_1} \Big| \Big[F^{(\gamma k \kappa)} \right.$$
$$\left. \times |n_1 n_2 (ll)^{N'} \alpha_2 T_2 L_2 S_2 \right) \Big]_{M_{T_1} M_{L_1} M_{S_1}}^{(T_1 L_1 S_1)}. \qquad (18.31)$$

Antisymmetric wave functions for N electrons ($0 \leq N \leq 8l + 4$) in an isospin basis can be constructed out of antisymmetric functions of $N - \sigma$

and σ electrons using fractional parentage coefficients with σ detached electrons, which in second-quantization form are given as

$$\left((ll)^{N-\sigma}(\alpha_1 T_1 L_1 S_1)(ll)^{\sigma}(\alpha_2 T_2 L_2 S_2)\|(ll)^N \alpha T L S\right)$$

$$= (-1)^{\sigma(N-\sigma)+2(L_2+S_2+T_2)} \left(\sigma!(N-\sigma)!/(N![L,S,T])\right)^{1/2}$$

$$\times \left((ll)^N \alpha T L S \| \hat{\varphi}(l^{\sigma})^{\alpha_2(T_2 L_2 S_2)} \|(ll)^{N-\sigma} \alpha_1 T_1 L_1 S_1\right). \qquad (18.32)$$

Here $\hat{\varphi}(l^{\sigma})^{\alpha_2(T_2 L_2 S_2)}$ is the second-quantized operator producing the relevant normalized wave function out of vacuum, i.e. it simply generalizes the relationship (15.4) to the case of tensors with an additional (isospin) rank. Since the vacuum state is a scalar in isospin space (unlike quasispin space), the expressions for wave functions and matrix elements of standard quantities in the spaces of orbital and spin momenta can readily be generalized by the addition of a third (isospin) rank to two ranks in appropriate formulas of Chapter 15.

Just as for one shell of equivalent electrons, for an isospin basis there exists a symmetry between the states of the partially filled ($N \leq 4l+2$) and almost filled ($N > 4l+2$) configurations. We shall define the wave function for the almost filled configuration in the form

$$|n_1 n_2 (ll)^{8l+4-N} \alpha T L S) = (-1)^{\eta(\alpha,T,L,S)} \hat{\tilde{\varphi}}(l^N)^{\alpha(TLS)} |_h 0), \qquad (18.33)$$

$$|_h 0) = |n_1 n_2 (ll)^{8l+4} T = 0, L = 0, S = 0), \qquad (18.34)$$

where $\eta(\alpha, T, L, S)$ defines the phase relations between the partially and almost filled configurations. The pertinent CFP will then be

$$\left((ll)^{8l+4-N-\sigma}(\alpha_1 T_1 L_1 S_1)(ll)^{\sigma}(\alpha_2 T_2 L_2 S_2)\|(ll)^{8l+4-N} \alpha T L S\right)$$

$$= (-1)^{T_1+L_1+S_1+T_2+L_2+S_2+T+L+S+\eta(\alpha_1,T_1,L_1,S_1)+\eta(\alpha,T,L,S)+\sigma(\sigma+1)/2}$$

$$\times \left\{ \frac{(8l+4-N-\sigma)!(N+\sigma)![T_1,L_1,S_1]}{(8l+4-N)!N![T,L,S]} \right\}^{1/2}$$

$$\times \left((ll)^N(\alpha T L S)(ll)^{\sigma}(\alpha_2 T_2 L_2 S_2)\|(ll)^{N+\sigma} \alpha_1 T_1 L_1 S_1\right). \qquad (18.35)$$

Wave functions in an isospin basis can be represented as a linear combination of appropriate quantities obtained by vectorial coupling of momenta of individual shells

$$|n_1 n_2 (ll)^N \alpha T L S M_T M_L M_S)$$

$$= \sum_{\alpha_1 L_1 S_1 \alpha_2 L_2 S_2} |n_1 n_2 l^{N_1} l^{N_2} \alpha_1 L_1 S_1 \alpha_2 L_2 S_2 L S M_L M_S)$$

$$\times \left(l^{N_1} l^{N_2} \alpha_1 L_1 S_1 \alpha_2 L_2 S_2 L S |(ll)^N \alpha T L S M_T\right). \qquad (18.36)$$

To establish the expansion coefficients we multiply (18.36) by the quantity that is the Hermitian conjugate of the function

$$|n_1 n_2 l^{N_1} l^{N_2} \alpha_1 L_1 S_1 \alpha_2 L_2 S_2 L S) $$
$$= \left[\hat{\varphi}(l^{N_1})^{\alpha_1 (L_1 S_1)} \times \hat{\varphi}(l^{N_2})^{\alpha_2 (L_2 S_2)} \right]^{(LS)} |0), \qquad (18.37)$$

then integrate over all variables and apply the Wigner–Eckart theorem in all three spaces. According to (18.32) we then have

$$\left(l^{N_1} l^{N_2} \alpha_1 L_1 S_1 \alpha_2 L_2 S_2 L S | (ll)^N \alpha T L S M_T \right)$$
$$= (-1)^{N/2-T} \left\{ N! (2T+1)/(N/2-T)!(N/2+T+1)! \right\}^{1/2}$$
$$\times \left((ll)^{N_1} (\alpha_1 T_1 L_1 S_1)(ll)^{N_2} (\alpha_2 T_2 L_2 S_2) \| (ll)^N \alpha T L S \right) \qquad (18.38)$$

where

$$T_1 = N_1/2, \quad T_2 = N_2/2,$$

i.e. the coefficients of the linear combination (18.36) (the transformation matrix from conventional to isospin basis) are proportional to the fractional parentage coefficients in isospin basis with N_2 detached electrons. At $T = N/2$ the latter become conventional one-shell coefficients of fractional parentage

$$\left(l^{N_1} l^{N_2} \alpha_1 L_1 S_1 \alpha_2 L_2 S_2 L S | (ll)^N \alpha (T = N/2) L S M_T \right)$$
$$= \left(l^{N_1} (\alpha_1 L_1 S_1) l^{N_2} (\alpha_2 L_2 S_2) \| l^N \alpha L S \right). \qquad (18.39)$$

According to the general relationship (5.9), rotations in isospin space transform the electron creation operators by the D-matrix of rank 1/2. If we go over from these operators to the one-electron wave functions they produce, then we shall have the unitary transformation of radial orbitals

$$\begin{pmatrix} R'(n_1 l | r) \\ R'(n_2 l | r) \end{pmatrix} = \begin{pmatrix} u & v \\ -v^* & u^* \end{pmatrix} \begin{pmatrix} R(n_1 l | r) \\ R(n_2 l | r) \end{pmatrix}, \qquad (18.40)$$

where $uu^* + vv^* = 1$. We can go over to the orthogonal transformations of these orbitals by considering only rotations in one plane of isospin space.

Under rotations in isospin space the wave functions (18.27) are transformed as follows:

$$|(ll)^N \alpha T L S M_T') = \sum_{M_T} D^{(T)}_{M_T', M_T} |(ll)^N \alpha T L S M_T), \qquad (18.41)$$

and as the wave functions

$$|n_1 n_2 (ll)^N \alpha T = 0 L S M_T = 0) \equiv |n_1 n_2 l^{N/2} l^{N/2} \alpha T = 0 L S) \qquad (18.42)$$

are scalars in isospin space, they are invariant under transformations of the radial orbitals (18.40). When the wave functions (18.42) can be constructed from the functions (18.37) at one fixed value of momenta of individual shells, the functions (18.42) show this invariance as well. These functions include, for example, wave functions $|n_1 n_2 pp\,^3S)$, $|n_1 n_2 pp\,^1P)$, $|n_1 n_2 pp\,^3D)$, or functions $|n_1 n_2 p^2 p^2\,^3P\,^3P\,^5G)$, $|n_1 n_2 p^2 p^2\,^3P\,^3P\,^5D)$, $|n_1 n_2 p^2 p^2\,^1D\,^1D\,^1G)$. The scalar nature of wave functions of this kind in isospin space graphically illustrates their invariance under transformations of radial orbitals [45].

An isospin basis possesses another important physical property – it expands the applicability of the Brillouin theorem to the excited $n_1 l^{N_1} n_1 l^{N_2}$ configurations [45, 124].

18.3   Additional classification of terms in isospin basis

Along with the above advantages of an isospin basis as compared with a conventional one, obtained by vectorial coupling of momenta of individual shells, is the disadvantage of the separation of repeating states. The isospin quantum number T in the general case is insufficient for unambiguous classification of allowed terms ^{2S+1}L. This problem is absent for s-electrons, but then the isospin characteristic is superfluous in general, since in this configuration there are no repeating terms. For the s-electrons

$$\mathbf{T}^2 + \mathbf{S}^2 \stackrel{s}{=} 2\left[U^{(101)} \times U^{(101)}\right]^{(000)} \stackrel{s}{=} \hat{N}(4-\hat{N})/2. \tag{18.43}$$

But with p-electrons already in the $n_1 p^2 n_2 p$ configuration of the isospin basis, two ^2P terms appear with the same $T = 1/2$. As the number of electrons (or the value of l) increases, the number of repeating terms grows. They can be labelled by applying groups of higher ranks: by requiring that pertinent wave functions be transformed by irreducible representations of these groups, we can use the characteristics of these representations as additional quantum numbers.

In Chapter 14 we have already discussed the group-theoretical method of classification of the states of a shell of equivalent electrons. Remembering that second-quantization operators in isospin space have an additional degree of freedom, we can approach the classification of states in isospin basis in exactly the same way.

The set of $(8l+4)^2$ operators $U_{\Gamma\rho\pi}^{(\gamma k\kappa)}$ is complete relative to the commutation operation

$$\left[U_{\Gamma\rho\pi}^{(\gamma k\kappa)}, U_{\Gamma'\rho'\pi'}^{(\gamma'k'\kappa')}\right] = (-1)^{\gamma+k+\kappa+\Gamma+\rho+\pi}$$

$$\times \sum_{\gamma_1 k_1 \kappa_1 \Gamma_1 \rho_1 \pi_1} \left\{(-1)^{\gamma+k+\kappa+\gamma'+k'+\kappa'} - (-1)^{\gamma_1+k_1+\kappa_1}\right\}$$

$$\times [\gamma, k, \kappa, \gamma_1, k_1, \kappa_1]^{1/2} \begin{Bmatrix} \tau & \tau & \gamma_1 \\ \gamma & \gamma' & \tau \end{Bmatrix} \begin{Bmatrix} l & l & k_1 \\ k & k' & l \end{Bmatrix} \begin{Bmatrix} s & s & \kappa_1 \\ \kappa & \kappa' & s \end{Bmatrix}$$

$$\times \begin{bmatrix} \gamma_1 & \gamma & \gamma' \\ \Gamma_1 & -\Gamma & \Gamma' \end{bmatrix} \begin{bmatrix} k_1 & k & k' \\ \rho_1 & -\rho & \rho' \end{bmatrix} \begin{bmatrix} \kappa_1 & \kappa & \kappa' \\ \pi_1 & -\pi & \pi' \end{bmatrix} U_{\Gamma_1 \rho_1 \pi_1}^{(\gamma_1 k_1 \kappa_1)}. \qquad (18.44)$$

These operators are generators of the U_{8l+4} group, and the characteristics of irreducible representations of the latter correspond only to the number of electrons N in the $(ll)^N$ configuration. The possible subgroup chains that preserve the classification according to the quantum numbers L, S and T are given by the following reduction schemes:

$$U_{8l+4} \begin{array}{c} \nearrow SU_2^T \times Sp_{4l+2} \searrow \\ \\ \searrow SU_4 \times SU_{2l+1} \nearrow \end{array} SU_2^T \times SU_2^S \times (R_{2l+1} \to R_3^L), \qquad (18.45)$$

which, besides the groups contained in the scheme (14.38), include the SU_2^T group generated by isospin angular momentum operators and the SU_4 group of unitary unimodular matrices of the fourth order. The generators of the SU_4 group are

$$U_{\rho 00}^{(100)}, \qquad U_{00\rho}^{(001)}, \qquad U_{\gamma 0\pi}^{(101)}, \qquad (18.46)$$

and the irreducible representations are characterized by a set of three parameters (P, P', P''). Table 18.1 classifies the terms of the partially filled $n_1 p^{N_1} n_2 p^{N_2}$ configurations by irreducible representations of the SU_4 group, from which it is seen that with this classification scheme nearly all the terms of these configurations are described unambiguously (there only remain two double terms at $N = 6$).

The other subgroup chain in (18.45) includes the Sp_{4l+2} group, whose generators are the tensors $U_{0\rho\pi}^{(0k\kappa)}$ with odd sum of ranks $k + \kappa$. Since these generators are scalars in isospin space, they commute with isospin angular momentum operators, forming the direct product of the two groups $SU_2^T \times Sp_{4l+2}$. For the quasispin group to be introduced instead of the Sp_{4l+2} in the case of a shell of equivalent electrons, in Chapter 15 we have considered the operators that change the number of particles in a wave function. In the case of isospin basis, these operators are the tensors

$$V_{\Gamma\rho\pi}^{(\gamma k\kappa)} = \left[a^{(\tau l s)} \times a^{(\tau l s)} \right]_{\Gamma\rho\pi}^{(\gamma k\kappa)}, \qquad (18.47)$$

$$\tilde{V}_{\Gamma\rho\pi}^{(\gamma k\kappa)} = (-1)^{\Gamma+\rho+\pi} V_{-\Gamma-\rho-\pi}^{(\gamma k\kappa)} = \left[\tilde{a}^{(\tau l s)} \times \tilde{a}^{(\tau l s)} \right]_{\Gamma\rho\pi}^{(\gamma k\kappa)}. \qquad (18.48)$$

and by the anticommutation relations the sum of ranks in the formulas (18.47) and (18.48) is an odd number, since otherwise these operators are identically equal to zero. The commutation relations for them are given in [123].

Separating the operators that are scalars in orbital and spin spaces

Table 18.1. Classification of the terms of the $n_1 p^{N_1} n_2 p^{N_2}$ configuration by irreducible representations of the SU_4 group

N	L	(P, P', P'')	$(2T+1, 2S+1)$
0	S	(000)	(11)
1	P	$\left(\frac{1}{2}\frac{1}{2}\frac{1}{2}\right)$	(22)
2	SD	(100)	$(13)(31)$
	P	(111)	$(11)(33)$
3	PF	$\left(\frac{1}{2}\frac{1}{2}-\frac{1}{2}\right)$	(22)
	PD	$\left(\frac{3}{2}\frac{1}{2}\frac{1}{2}\right)$	$(22)(24)(42)$
	S	$\left(\frac{3}{2}\frac{3}{2}\frac{3}{2}\right)$	$(22)(44)$
4	SDG	(000)	(11)
	PDF	(110)	$(13)(31)(33)$
	SD	(200)	$(11)(15)(51)(33)$
	P	(211)	$(13)(31)(33)(35)(53)$
5	$PDFG$	$\left(\frac{1}{2}\frac{1}{2}\frac{1}{2}\right)$	(22)
	PDF	$\left(\frac{3}{2}\frac{1}{2}-\frac{1}{2}\right)$	$(22)(24)(42)$
	SD	$\left(\frac{3}{2}\frac{3}{2}\frac{1}{2}\right)$	$(22)(24)(42)(44)$
	P	$\left(\frac{5}{2}\frac{1}{2}\frac{1}{2}\right)$	$(22)(24)(42)(26)(62)(44)$
6	SD^2FG	(100)	$(13)(31)$
	PF	(111)	$(11)(33)$
	PF	$(11-1)$	$(11)(33)$
	PD	(210)	$(13)(31)(33)^2(15)(51)(35)(53)$
	S	(300)	$(13)(31)(35)(53)(17)(71)$

(a similar procedure for one shell of equivalent electrons produces the quasispin classification), we obtain the ten operators

$$U^{(100)}_{\rho 00}, \quad V^{(100)}_{\Gamma' 00}, \quad V^{(100)}_{\Gamma 00}, \quad U^{(000)}_{000}. \tag{18.49}$$

These operators play the role of generators of the five-dimensional quasispin group Sp_4, which can be readily verified by comparing their commu-

Table 18.2. Classification of the terms of the $n_1 p^{N_1} n_2 p^{N_2}$ configuration by irreducible representations of the Sp_4 group

N	$(v, 2t+1)$	$^{2S+1}L^{2T+1}$
1	(1,2)	$^2P^2$
2	(0,1)	$^1S^3$
	(2,1)	$^3S^1\ {}^1P^1\ {}^3D^1$
	(2,3)	$^3P^3\ {}^1D^3$
3	(1,2)	$^2P^2\ {}^2P^4$
	(3,2)	$^2S^2\ {}^2P^2\ {}^4P^2\ {}^2D^2\ {}^4D^2\ {}^2F^2$
	(3,4)	$^4S^4\ {}^2D^4$
4	(0,1)	$^1S^1\ {}^1S^5$
	(2,1)	$^3S^3\ {}^1P^3\ {}^3D^3$
	(2,3)	$^3P^1\ {}^3P^3\ {}^3P^5\ {}^1D^1\ {}^1D^3\ {}^1D^5$
	(4,1)	$^1S^1\ {}^5S^1\ {}^3P^1\ {}^1D^1\ {}^3D^1\ {}^5D^1\ {}^3F^1\ {}^1G^1$
	(4,3)	$^1P^3\ {}^3P^3\ {}^5P^3\ {}^3D^3\ {}^1F^3\ {}^3F^3$
5	(1,2)	$^2P^2\ {}^2P^4\ {}^2P^6$
	(3,2)	$^2S^2\ {}^2P^2\ {}^4P^2\ {}^2D^2\ {}^4D^2\ {}^2F^2$
		$^2S^4\ {}^2P^4\ {}^4P^4\ {}^2D^4\ {}^4D^4\ {}^2F^4$
	(3,4)	$^4S^2\ {}^4S^4\ {}^2D^2\ {}^2D^4$
	(5,2)	$^2P^2\ {}^4P^2\ {}^6P^2\ {}^2D^2\ {}^4D^2\ {}^2F^2\ {}^4F^2\ {}^2G^2$
6	(0,1)	$^1S^3\ {}^1S^7$
	(2,1)	$^3S^1\ {}^3S^5\ {}^1P^1\ {}^1P^5\ {}^3D^1\ {}^3D^5$
	(2,3)	$2(^3P^3)\ {}^3P^5\ 2(^1D^3)\ {}^1D^5$
	(4,1)	$^1S^3\ {}^5S^3\ {}^3P^3\ {}^1D^3\ {}^3D^3\ {}^5D^3\ {}^3F^3\ {}^1G^3$
	(4,3)	$^1P^1\ {}^1P^3\ {}^3P^1\ {}^3P^3\ {}^5P^1\ {}^5P^3\ {}^3D^1\ {}^3D^3\ {}^1F^1\ {}^1F^3\ {}^3F^1\ {}^3F^3$
	(6,1)	$^3S^1\ {}^7S^1\ {}^3D^1\ {}^5D^1\ {}^1F^1\ {}^3G^1$

tation relations with the standard commutators for the generators of this group.

The algebra of the Sp_4 group coincides with the algebra of the rotation group of five-dimensional Euclidean space R_5, i.e. these groups are locally isomorphic. The irreducible representations of the Sp_4 group can be characterized by a set of two parameters (v, t) where the seniority quantum number v for the five-dimensional quasispin group indicates the number of electrons in a configuration, at which this term appears for the first time, and t is the quantum number of reduced isospin (t is given by the isospin quantum number T at $N = v$). The generators of the Sp_4 group commute with the generators of the Sp_{4l+2} group, i.e. the direct product of these groups $Sp_4 \times Sp_{4l+2}$ occurs. Moreover, these groups are complementary [125] to each other in the sense that in the case under consideration the characteristics of their irreducible representations have a one-to-one correspondence. Table 18.2 classifies the terms of the configuration $n_1 p^{N_1} n_2 p^{N_2}$ ($N = N_1 + N_2 \leq 4l+2$) in isospin basis according to irreducible representations of the Sp_4 group: it is seen that, as with the SU_4 group (Table 18.1), there remain only two repeating terms at $N = 6$. In both tables only the wave functions of a partially filled configuration are classified, since due to the symmetry of particle and hole states the wave functions for an almost filled configuration are classified in the same way.

If a specific group is chosen to classify the states additionally, then in the pertinent basis we can define CFP and also, by applying a relation of the type (18.38), the transformation matrices between isospin and conventional bases. This enables us to compute the physical characteristics (e.g. interaction energy) in a new basis and to establish the area of its applicability.

18.4   Electron interaction energy in isospin basis

For the isospin basis to be used effectively it is necessary to establish the irreducible tensorial form of operators of physical quantities in isospin angular momentum space. In the general case, one-electron operators can be expressed directly in terms of linear combinations of the operators (18.26), but a similar procedure for two-electron operators is much more complex. By way of example, we shall find the expansion in isospin space of the configuration of one of the most important two-electron operators – the operator of electrostatic interaction energy for electrons. Proceeding from the general expression for a two-electron operator in the second-quantization representation (13.23) and by going over, using (18.21), to triple tensors, after some transformations and relating the orbital, spin and isospin momenta we obtain

$$\hat{V} = \sum_k \hat{F}^k, \tag{18.50}$$

where

$$\hat{F}^k = \frac{(l\|C^{(k)}\|l)^2}{(2k+1)^{1/2}} \left\{ \frac{1}{2} \sum_{T_1 T_2 T} \begin{bmatrix} T_1 & T_2 & T \\ 0 & 0 & 0 \end{bmatrix} \left[U^{(T_1 k 0)} \times U^{(T_2 k 0)} \right]_{000}^{(T00)} \right.$$

$$\times \left\{ R_k(n_2 l n_2 l, n_2 l n_2 l) + (-1)^T R_k(n_1 l n_1 l, n_1 l n_1 l) \right.$$

$$+ (-1)^{T_1} 2R_k(n_1 l n_1 l, n_2 l n_2 l)\} - 2 \sum_T \begin{bmatrix} 1 & 1 & T \\ -1 & 1 & 0 \end{bmatrix}$$

$$\times \left[U^{(1k0)} \times U^{(1k0)} \right]_{000}^{(T00)} R_k(n_1 l n_2 l, n_1 l n_2 l)$$

$$- \frac{(-1)^k}{2} \left(\frac{2k+1}{2l+1} \right)^{1/2} \sum_T U_{000}^{(T00)} \{ R_k(n_2 l n_2 l, n_2 l n_2 l)$$

$$+ (-1)^T R_k(n_1 l n_1 l, n_1 l n_1 l) + (-1)^T 2R_k(n_1 l n_2 l, n_1 l n_2 l) \}$$

$$+ \sqrt{2} \sum_{T_2 T (M_T = \pm 1)} (-1)^{(T_2 + T)(1 + M_T)/2} \begin{bmatrix} 1 & T_2 & T \\ -1 & 0 & -1 \end{bmatrix}$$

$$\times \left[U^{(1k0)} \times U^{(T_2 k0)} \right]_{M_T 00}^{(T00)} \{ R_k(n_1 l n_2 l, n_2 l n_2 l)$$

$$+ (-1)^{T_2} R_k(n_1 l n_1 l, n_1 l n_2 l) \} + (-1)^k \left[\frac{2(2k+1)}{2l+1} \right]^{1/2}$$

$$\times \left\{ U_{100}^{(100)} R_k(n_1 l n_1 l, n_1 l n_2 l) - U_{-100}^{(100)} R_k(n_1 l n_2 l, n_2 l n_2 l) \right\}$$

$$+ \sum_{M_T = \pm 2} \left[U^{(1k0)} \times U^{(1k0)} \right]_{M_T 00}^{(200)} R_k(n_1 l n_2 l, n_1 l n_2 l) \Bigg\}. \tag{18.51}$$

The radial integrals in (18.51) are defined according to (17.19) and (17.20). It is remarkable that the formula (18.51) encompasses, already in operator form (unlike the conventional approach [14]), interactions not only within the configuration but also between all possible electron distributions in various configurations determined by the projection of isospin angular momentum. So, the matrix elements of the tensorial operators with isospin projection $M_T = 0$ in (18.51) are non-zero only when the configurations of the initial and final states coincide, i.e. these operators correspond to the electrostatic interaction of electrons within a configuration. The tensorial operators with $M_T = \pm 1$ and $M_T = \pm 2$ correspond to a superposition of configurations of the type $n_1 l^{N_1} n_2 l^{N_2} - n_1 l^{N_1 \pm 1} n_2 l^{N_2 \mp 1}$ and $n_1 l^{N_1} n_2 l^{N_2} - n_1 l^{N_1 \pm 2} n_2 l^{N_2 \mp 2}$, respectively. Consequently, if we take into account the tensorial properties of operators and wave functions in quasi-spin space, we shall have a unified approach to part of the correlation effects.

At $k = 0$, it follows from (18.51) that

$$\hat{F}^0 = \frac{1}{2} \left\{ \hat{N}_1(\hat{N}_1 - 1)R_0(n_1 l n_1 l, n_1 l n_1 l) + \hat{N}_2(\hat{N}_2 - 1)R_0(n_2 l n_2 l, n_2 l n_2 l) \right\}$$

$$+ \left\{ (T^{(1)} \cdot T^{(1)}) + (T_0^{(1)})^2 - \frac{1}{2} \hat{N} - (T_{-1}^{(1)})^2 - (T_1^{(1)})^2 \right\} R_0(n_1 l n_2 l, n_1 l n_2 l)$$

$$+ \hat{N}_1 \hat{N}_2 R_0(n_1 l n_1 l, n_2 l n_2 l) - i\sqrt{2} \left\{ (\hat{N}_1 - 1)T_{-1}^{(1)} - \hat{N}_1 T_1^{(1)} \right\}$$

$$\times R_0(n_1 l n_2 l, n_2 l n_2 l) + i\sqrt{2} \left\{ (\hat{N}_2 - 1)T_1^{(1)} - \hat{N}_2 T_{-1}^{(1)} \right\} R_0(n_1 l n_2 l, n_1 l n_1 l),$$

$$(18.52)$$

so that the matrix elements of the operators at the integrals $R_0(n_i l n_i l, n_j l n_j l)$ yield conventional expressions [14] for the coefficients $f_0(l^{N_i})$, $f_0(l^{N_i}, l^{N_j})$. The coefficients of the exchange integrals of electrostatic interaction are equal to the matrix elements of the operator

$$\hat{g}_0(l^{N_1}, l^{N_2}) = (T^{(1)} \cdot T^{(1)}) + (T_0^{(1)})^2 - \hat{N}/2, \qquad (18.53)$$

i.e. are expressed in terms of the operators whose eigenvalues are known and have a simple algebraic form, but the main thing about them is that they are diagonal relative to all the quantum numbers, which is not the case in conventional basis.

Thus, in the limiting case, where the expansion of the electrostatic interaction operator in terms of the multipoles (see (19.6)) includes only the central-symmetric part (i.e. only the terms with $k = 0$), dependent on the term in (18.52) is only the summand with the operator T^2. The eigenvalues of the operator T^2, according to (18.28), are equal to $T(T + 1)$, i.e. in this approximation we obtain the spectrum of energy levels 'rotational' with respect to isospin.

When introducing a new basis it may always be asked whether new quantum numbers are 'good', at least approximately, i.e. to what degree is the energy matrix diagonal in this basis? The fact that the matrix element of the operator (18.53) of the main exchange integral of electrostatic interaction is diagonal is a first argument in favour of the new basis. A second argument is the expansion of the applicability of the Brillouin theorem. But the most effective criterion of suitability of a basis is the computation of the energy matrix and its subsequent diagonalization. Analysis of the weights of wave functions of intermediate coupling, as described above and their comparison with calculations carried out in other possible bases enable us to select the optimal basis and coupling scheme for momenta.

It follows from the general behaviour of isospin basis that the isospin quantum number will be 'good' if an electron belonging to one shell interacts with another electron belonging to the same shell in exactly the same way as with an electron in another shell. In so doing, we should include only the part of electron–electron interaction that depends on the term. This is described by ($k > 0$)

$$F_k(n_1 l, n_1 l) = F_k(n_2 l, n_2 l) = F_k(n_1 l, n_2 l) + G_k(n_1 l, n_2 l). \qquad (18.54)$$

In the general case, this equation does not hold, of course. But for

Table 18.3. Squares of the largest weights in conventional (α^2_{max}) and isospin (β^2_{max}) bases for the $n_1 p^{N_1} n_2 p^{N_2}$ configurations

N_1, N_2	^{2s+1}L	$^{2s_1+1}L_1,\ ^{2s_2+1}L_2$	$^{2s'_1+1}L_1,\ ^{2s'_2+1}L_2$	T, T'	n_1, n_2	α^2_{max}	β^2_{max}
3, 1	^1D	$^2P,\quad ^2P$	$^2D,\quad ^2P$	2, 1	2, 3	0.548	0.907
					3, 4	0.500	0.933
					4, 5	0.525	0.946
2, 1	^2D	$^1D,\quad ^2P$	$^3P,\quad ^2P$	$\frac{3}{2}, \frac{1}{2}$	2, 3	0.705	0.956
					3, 4	0.637	0.981
					4, 5	0.595	0.991
2, 2	^3F	$^1D,\quad ^3P$	$^3P,\quad ^1D$	1, 0	2, 3	0.784	0.911
					3, 4	0.717	0.950
					4, 5	0.679	0.967
1, 2	^2D	$^2P,\quad ^3P$	$^2P,\quad ^1D$	$\frac{3}{2}, \frac{1}{2}$	2, 3	0.505	1.000
					3, 4	0.518	1.000
					4, 5	0.530	0.999
1, 3	^1D	$^2P,\quad ^2D$	$^2P,\quad ^2P$	2, 1	2, 3	0.755	1.000
					3, 4	0.773	0.999
					4, 5	0.798	0.997

specific configurations less rigorous constraints are often sufficient. So, it follows from the consideration of non-diagonal matrix elements of the electrostatic energy operator in isospin basis that for configurations of the type $n_1 l^{4l+1} n_2 l^{N_2}$ and $n_1 l n_2 l^{N_2}$ ($N_2 \geq 2$) a sufficient criterion of the accuracy of the isospin quantum number is the respective conditions

$$F_k(n_1 l, n_1 l) = F_k(n_1 l, n_2 l) + G_k(n_1 l, n_2 l) \qquad (k > 0), \qquad (18.55)$$
$$F_k(n_2 l, n_2 l) = F_k(n_1 l, n_2 l) + G_k(n_1 l, n_2 l) \qquad (k > 0). \qquad (18.56)$$

The calculations of these integrals show that there are notable cases for $n_1 l n_2 l^{N_2}$ configurations at $n_2 = n_1 + 1$ where the isospin quantum number is fairly 'good'. A similar inference can be made from comparison of the matrix of the operator of electrostatic interaction energy for electrons in isospin and conventional bases. For the above type of multiply excited configuration in certain cases the electrostatic energy matrix in isospin basis is nearly diagonal. This is also testified by the weights of the wave functions of isospin basis derived after the energy matrix has been diagonalized: as a rule, the weight of the function of one of the sets of quantum numbers for the pure coupling scheme is close to unity.

By way of example, Table 18.3 provides the squares of the largest weight coefficients α^2_{max}, β^2_{max} of the wave functions that are obtained using radial hydrogen orbitals after the matrix of the electrostatic interaction operator has been diagonalized in a conventional basis

$$\Psi = \alpha |n_1 n_2 l^{N_1} l^{N_2} L_1 S_1 L_2 S_2 L S) + \alpha' |n_1 n_2 l^{N_1} l^{N_2} L'_1 S'_1 L'_2 S'_2 L S) \qquad (18.57)$$

and isospin basis

$$\Psi = \beta |n_1 n_2 l^{N_1} l^{N_2} TLS) + \beta' |n_1 n_2 l^{N_1} l^{N_2} T'LS). \qquad (18.58)$$

When the isospin quantum number is 'good', the states characterized by a larger value of the isospin T have a higher energy, which is mainly conditioned by the presence of the 'rotational' term $T(T+1)G_0(n_1 l, n_2 l)$ in diagonal (in isospin basis) matrix elements of the electrostatic energy operator.

Calculations carried out with hydrogen radial orbitals indicate that the supermultiplet basis SU_4 in many cases is more diagonal than the basis in which an additional classification of states is achieved using the quantum numbers (v, t) of the five-dimensional quasispin group.

To sum up: the potentialities of the isospin method are not exhausted by the results stated above. There is a deep connection between orthogonal transformations of radial orbitals and rotations in isospin space (see (18.40) and (18.41)). This shows that the tensorial properties of wave functions and operators in isospin space must be dominant in the Hartree–Fock method. This issue is in need of further consideration.

Part 5

Matrix Elements of the Energy Operator

19

The energy of a shell
of equivalent electrons

19.1   Expression of the energy operator in terms of irreducible tensors

While calculating the energy spectra of atoms or ions one ought to be able to find numerical expressions for all terms of the energy operator (e.g. (1.16) or (2.1)) with regard to the wave functions of the system under consideration. The total matrix element of each term of the energy operator in the case of complex electronic configurations will consist of matrix elements, describing the interaction inside each shell (in relativistic case – subshell) of equivalent electrons as well as between these shells. However, as a rule, it is impossible to find directly formulas for the matrix elements of the corresponding operators. For this purpose each operator must be expressed in terms of the so-called irreducible tensors. This will enable us, further on, to find their matrix elements, to exploit very efficiently the mathematical apparatus of the angular momentum theory, second-quantization as well as the methods of the coefficients of fractional parentage (CFP) and the irreducible tensorial operators, composed of the unit tensors.

In practice, the transformation of any operator to irreducible form means in atomic spectroscopy that we employ the spherical coordinate system (Fig. 5.1), present all quantities in the form of tensors of corresponding ranks (scalar is a zero rank tensor, vector is a tensor of the first rank, etc.) and further on express them, depending on the particular form of the operator, in terms of various functions of radial variable, the angular momentum operator $L^{(1)}$, spherical functions $C^{(k)}$ (2.13), as well as the Clebsch–Gordan and $3nj$-coefficients. Below we shall illustrate this procedure by the examples of operators (1.16) and (2.1). Formulas (1.15), (1.18)–(1.22) present concrete expressions for each term of Eq. (1.16). It is convenient to divide all operators (1.15), (1.18)–(1.22) into two groups. The first group is composed of one-electron operators (1.18), the first two terms T and P of (1.15) and the first terms of (1.20), (1.21), acting on the coordinates of each electron separately. The rest of the terms compose the second group. These are two-electron operators, they simultaneously act on the coordinates of two electrons. The expression of one-electron operators in terms of irreducible tensors is much simpler than that of two-electron ones. Let us consider them first.

Further on, while considering this Hamiltonian, we shall exploit atomic units. In order to transform the first term of (1.15) into an irreducible form we use the equality

$$p_q^{(1)} = -i\nabla_q^{(1)}, \tag{19.1}$$

where

$$\nabla^{(1)} = C^{(1)}\frac{\partial}{\partial r} + \frac{i\sqrt{2}}{r}\left[C^{(1)} \times L^{(1)}\right]^{(1)} = C^{(1)}\frac{\partial}{\partial r} + \frac{1}{r}\nabla^{*(1)}. \tag{19.2}$$

Then

$$(p_i^{(1)})^2 = -\Delta_i, \tag{19.3}$$

where

$$\Delta = \left(\nabla^{(1)} \cdot \nabla^{(1)}\right) = \frac{1}{r^2}\left[\frac{\partial}{\partial r}\left(r^2\frac{\partial}{\partial r}\right) - \frac{2}{3}\left(L^{(1)} \cdot L^{(1)}\right)\right.$$

$$\left. + \sqrt{\frac{2}{3}}\left(C^{(2)} \cdot L^{(2)}\right)\right] = \frac{1}{r^2}\left[\frac{\partial}{\partial r}\left(r^2\frac{\partial}{\partial r}\right) - \left(L^{(1)} \cdot L^{(1)}\right)\right]. \tag{19.4}$$

Thus, by using the most convenient expression for the Laplacian, the kinetic energy operator T of the atomic electrons in the irreducible form is as follows:

$$T = -\frac{1}{2}\sum_i \frac{1}{r_i^2}\left[\frac{\partial}{\partial r_i}\left(r_i^2\frac{\partial}{\partial r_i}\right) - \left(L_i^{(1)} \cdot L_i^{(1)}\right)\right]. \tag{19.5}$$

In obtaining Eq. (19.4) we have made use of the so-called theorem of the addition of spherical functions (5.5).

The operator P (second term in (1.15)) is a complete scalar and does not require any changes. The Coulomb interaction operator (the last item in Eq. (1.15)), after expanding $1/r_{12}$ in terms of spherical functions, in an irreducible form is equal to

$$Q = \sum_{i>j}\sum_k \frac{r_<^k}{r_>^{k+1}}\left(C_i^{(k)} \cdot C_j^{(k)}\right). \tag{19.6}$$

Here $r_<$ $(r_>)$ is the smaller (larger) of r_1 and r_2. The irreducible form of relativistic correction (1.18) is found from the condition

$$\mathbf{p}^4 = \Delta \cdot \Delta \tag{19.7}$$

and formula (19.4).

Transformation of the orbit–orbit (1.19), spin–orbit and spin–spin interaction operators to the irreducible form is connected with complicated mathematical manipulations, therefore, we shall present here only final results. The details of this procedure may be found in [14] and in the original papers quoted there. The irreducible form of the orbit–orbit interaction operator is as follows:

$$
\begin{aligned}
H_2 = \frac{\alpha^2}{2} \sum_{i>j} \sum_{k} & \left\{ \left[(k+1)(k+2) \left(C_i^{(k+1)} \cdot C_j^{(k+1)} \right) \right. \right. \\
& - k(k-1) \left(C_i^{(k-1)} \cdot C_j^{(k-1)} \right) \left] \frac{a_k}{2k+1} \frac{\partial}{\partial r_i} \frac{\partial}{\partial r_j} + 2i(1+P_{ij}) \right. \\
& \times \left[\sqrt{(k+1)(k+2)} \left(\left[C_i^{(k+1)} \times L_i^{(1)} \right]^{(k+1)} \cdot C_j^{(k+1)} \right) \right. \\
& - \sqrt{k(k-1)} \left(\left[C_i^{(k-1)} \times L_i^{(1)} \right]^{(k-1)} \cdot C_j^{(k-1)} \right) \left] \frac{1}{2k+1} \left(\frac{1}{2} \frac{\partial a_k}{\partial r_i} + \frac{a_k}{r_i} \right) \frac{\partial}{\partial r_j} \right. \\
& + \left[\left(\left[C_i^{(k+1)} \times L_i^{(1)} \right]^{(k+1)} \cdot \left[C_j^{(k+1)} \times L_j^{(1)} \right]^{(k+1)} \right) \right. \\
& - \left(\left[C_i^{(k-1)} \times L_i^{(1)} \right]^{(k-1)} \cdot \left[C_j^{(k-1)} \times L_j^{(1)} \right]^{(k-1)} \right) \left] \frac{(k-1)(k+2)}{2k+1} \frac{a_k}{r_i r_j} \right. \\
& \left. - 2 \left(\left[C_i^{(k-1)} \times L_i^{(1)} \right]^{(k)} \cdot \left[C_j^{(k-1)} \times L_j^{(1)} \right]^{(k)} \right) \frac{2k-1}{k+1} \frac{a_k}{r_i r_j} \right\}.
\end{aligned}
\tag{19.8}
$$

Here

$$
a_k = \frac{r_<^k}{r_>^{k+1}},
\tag{19.9}
$$

and P_{ij} is the operator of the permutation of the indices i and j.

Replacing the Dirac δ-function of vectorial argument by the corresponding value of scalar variable, for H_3' in (1.20) we find

$$
H_3' = Z \frac{\alpha^2}{8} \sum_i \frac{\delta(r_i)}{r_i^2}.
\tag{19.10}
$$

Having in mind that spherical functions compose a full set of functions, we get the following irreducible form of the correction H_3'' in (1.20):

$$
H_3'' = -\frac{\alpha^2}{4} \sum_{i>j} \frac{\delta(r_{ij})}{r^2} \sum_k (2k+1) \left(C_i^{(k)} \cdot C_j^{(k)} \right).
\tag{19.11}
$$

In a similar fashion we find the irreducible form of the operator of spin–contact interaction H_5' in (1.22)

$$
H_5' = -\frac{2}{3} \alpha^2 \sum_{i>j} \frac{\delta(r_{ij})}{r^2} \left(s_i^{(1)} \cdot s_j^{(1)} \right) \sum_k (2k+1) \left(C_i^{(k)} \cdot C_j^{(k)} \right).
\tag{19.12}
$$

It is interesting to notice that in formulas (19.11), (19.12) the radial part of the operators does not depend on the summation parameter k.

The operator of spin–orbit interaction H_4 (1.21) may be written in the following irreducible form:

$$H_4 = \sum_i H^{so}(i) + \sum_{i>j} H^{soo}(i,j) = \frac{Z\alpha^2}{2} \sum_i \frac{1}{r_i^3}(\mathbf{L}_i \cdot \mathbf{s}_i)$$

$$-\frac{\alpha^2}{2} \sum_{i>j} \frac{1}{r_{ij}^3} \{([\mathbf{L}_{ij} + 2\mathbf{L}_{ji}] \cdot \mathbf{s}_i) + ([\mathbf{L}_{ji} + 2\mathbf{L}_{ij}] \cdot \mathbf{s}_j)\}, \quad (19.13)$$

where $\mathbf{L}_{ij}$ is the angular momentum of the ith electron with respect to the jth one. The first sum in (9.13) is a one-electron operator, already written in irreducible form. It is frequently called the spin–own-orbit interaction operator, whereas the second sum represents a complex two-electron operator, called the spin–other-orbit interaction. For the second sum we have

$$H^{soo} = \frac{\alpha^2}{\sqrt{2}} \sum_{kKK'} \frac{i}{\sqrt{(2k+1)(K'+k+1)}} \sum_{i>j}(1 + P_{ij})$$

$$\times \left\{ \left(\left[P_i^{(K1K')} \times C_j^{(k)}\right]^{(1)} \cdot \left[s_i^{(1)} + 2s_j^{(1)}\right] \right) \frac{1}{r_i} \left(D_i^K \frac{r_<^k}{r_>^{k+1}} \right) \right.$$

$$\left. + \left(\left[C_i^{(K1K')} \times C_j^{(k)}\right]^{(1)} \cdot \left[s_i^{(1)} + 2s_j^{(1)}\right] \right) \left(D_i^K \frac{r_<^k}{r_>^{k+1}} \right) \frac{\partial}{\partial r_i} \right\}. \quad (19.14)$$

Here

$$P^{(K1K')} = i\sqrt{2}a_{KK'} \left[C^{(K)} \times \left[C^{(1)} \times L^{(1)}\right]^{(1)}\right]^{(K')}, \quad (19.15)$$

$$C^{(K1K')} = a_{KK'} \left[C^{(K)} \times C^{(1)}\right]^{(K')}, \quad (19.16)$$

where

$$a_{KK'} = i^{k+1-K} \sqrt{(2k+1)(K'+k+1)(2K+1)(2K'+1)}$$

$$\times \begin{bmatrix} k & 1 & K \\ 0 & 0 & 0 \end{bmatrix} \begin{Bmatrix} k & 1 & K \\ 1 & K' & 1 \end{Bmatrix}, \quad (19.17)$$

and

$$D^K = \frac{\partial}{\partial r} + \begin{cases} (k+1)/r & (K = k-1) \\ -k/r & (K = k+1) \end{cases}. \quad (19.18)$$

The summation limits over parameters K and K' in (19.14) are defined by the conditions of non-zero values of the Clebsch–Gordan and $6j$-coefficients. The spin–spin operator H_5'' (the second sum in (1.22)) may be written as

$$H_5'' = \alpha^2 \sum_{i>j} \left[\left(s_i^{(1)} \cdot \nabla_i^{(1)} \right) \left(s_j^{(1)} \cdot \nabla_j^{(1)} \right) \frac{1}{r_{ij}} \right] \qquad (19.19)$$

and in irreducible form

$$H_5'' = \frac{\alpha^2}{\sqrt{5}} \sum_{i>j} \sum_{kKK'} \frac{\sqrt{k(k+1)(2k+3)(2k-1)}}{(2k+1)\sqrt{2k+1}}$$

$$\times \left(\left[C_i^{(K)} \times C_j^{(K')} \right]^{(2)} \cdot \left[s_i^{(1)} \times s_j^{(1)} \right]^{(2)} \right) \left(D_i^k D_j^{K'} \frac{r_<^k}{r_>^{k+1}} \right). \quad (19.20)$$

Here

$$D_i^k D_j^{K'} \frac{r_<^k}{r_>^{k+1}} = -(2k+1)^2 \begin{cases} r_i^{k-1}/r_j^{k+2} & (r_i \leq r_j, \ K = k-1, \ K' = k+1), \\ r_j^{k-1}/r_i^{k+2} & (r_j \leq r_i, \ K = k+1, \ K' = k-1). \end{cases}$$
$$(19.21)$$

An interesting peculiarity of the irreducible expressions for the operators of the spin–other-orbit and spin–spin interactions (formulas (19.14) and (19.20), respectively) is that the resulting ranks of the tensors in the orbital and spin spaces are fixed and equal to one for (19.14) and to two for (19.20).

Thus, we have expressed the non-relativistic Hamiltonian of a many-electron atom with relativistic corrections of order α^2 in the framework of the Breit operator (formulas (1.15), (1.18)–(1.22)) in terms of the irreducible tensorial operators (second term in (1.15), formulas (19.5)–(19.8), (19.10)–(19.14), (19.20), respectively).

19.2   Matrix elements of the energy operator in the case of a shell l^N

Let us proceed by obtaining expressions for matrix elements of the operators, considered in the previous section, in the case of shell l^N. Let us assume also that LS coupling takes place in it and start with the one-electron operators, utilizing their irreducible forms.

The energy operators are scalar, therefore, their matrix elements, according to the Wigner–Eckart theorem (5.15), are diagonal with respect to the total momenta and do not depend on their projections. For these reasons we shall skip projections further on. The expression for the matrix elements of the sum of operators (19.5) and P in (1.15) is simply equal to its one-electron matrix element, multiplied by N, i.e.

$$\left(nl^N \alpha LS \left| (T + P) \right| nl^N \alpha' L'S' \right) = \delta(\alpha LS, \alpha' L'S') N I(nl), \qquad (19.22)$$

where

$$I(nl) = -\frac{1}{2} \int\limits_0^\infty P(nl|r) \left[\frac{d^2}{dr^2} + \frac{2Z}{r} - \frac{l(l+1)}{r^2} \right] P(nl|r)dr. \qquad (19.23)$$

For the configuration consisting of several shells, the corresponding matrix element will be equal to the sum of the quantities of the kind (19.22). For H_1 and H_3' we have

$$(nl^N \alpha LS|H_1|nl^N \alpha' L'S') = -\delta(\alpha LS, \alpha' L'S')\frac{\alpha^2}{8} N\{I''(nl|0)$$

$$+ 2l(l+1)I'(nl|-2) + l(l+1)[l(l+1) - 6]I(nl|-4)\}, \quad (19.24)$$

$$(nl^N \alpha LS|H_3'|nl^N \alpha' L'S') = \delta(\alpha LS, \alpha' L'S')\frac{\alpha^2}{8} ZN \left| \frac{P(nl|r)}{r} \right|^2_{r=0}. \quad (19.25)$$

Here

$$I(nl|-\delta) = \int\limits_0^\infty \frac{1}{r^\delta} P^2(nl|r)dr, \qquad (19.26)$$

$$I'(nl|-\beta) = \int\limits_0^\infty \frac{1}{r^\beta} P'^2(nl|r)dr, \qquad (19.27)$$

$$I''(nl|0) = \int\limits_0^\infty P''^2(nl|r)dr. \qquad (19.28)$$

Formula (19.24) has been obtained by replacing the integral of $\psi^* \Delta\Delta\psi$ by the integral of $(\Delta\psi)^2$. The generalization of Eqs. (19.24) and (19.25) to cover the cases of the more complex configurations is similar to the case of formula (19.22). The expressions presented show that matrix elements of the one-electron operators T, P, H_1 and H_3' do not depend on the quantum numbers of the configuration but rather N, n and l. Therefore, they do not change mutual positions of the terms, they only shift all levels by the same value. These operators are important in the theory of electronic transitions, when one has to know as accurately as possible the energy differences between various configurations. Accounting for the relativistic corrections allows one to improve this quantity.

Let us proceed to the two-electron operators. The corresponding procedure is essentially based on the exploitation of the tensors, composed of unit tensorial operators (see Chapter 5). Let us consider first the electrostatic interaction (19.6). Using (5.42) we find (notice that further on in this chapter $N \leq 2l + 1$, i.e. shell l^N is considered to be partially or half-filled)

$$(nl^N \alpha LS|Q|nl^N \alpha' LS) = \sum_{k=0}^{2l} f_k(l^N \alpha LS, l^N \alpha' LS) F^k(nl, nl), \qquad (19.29)$$

where

$$f_k(l^N \alpha LS, l^N \alpha' LS) = \frac{1}{2}(l\|C^{(k)}\|l)^2$$

$$\times \left\{ \frac{1}{(l\|u^k\|l)^2} \left(l^N \alpha LS|(U^k \cdot U^k)|l^N \alpha' LS \right) - \frac{\delta(\alpha, \alpha')N}{2l+1} \right\}. \qquad (19.30)$$

Here the radial integral

$$F^k(nl, n'l') = \int\limits_0^\infty\!\!\!\int P^2(nl|r_1) \frac{r_<^k}{r_>^{k+1}} P^2(n'l'|r_2) dr_1 dr_2. \qquad (19.31)$$

The symbol of integration of the function $r_<^k/r_>^{k+1}$ means

$$\int\limits_0^\infty\!\!\!\int \dots \frac{r_<^k}{r_>^{k+1}} \dots dr_1 dr_2 = \int\limits_0^\infty \dots dr_1 \left\{ \frac{1}{r_1^{k+1}} \int\limits_0^{r_1} \dots r_2^k dr_2 \right.$$

$$\left. + r_1^k \int\limits_{r_1}^\infty \dots \frac{1}{r_2^{k+1}} dr_2 \right\}, \qquad (19.32)$$

i.e. the whole integration region is divided into two parts, corresponding to the values $r_2 < r_1$ and $r_2 > r_1$. A scalar product in (19.30) is defined according to the equality

$$(\alpha jm|(T^{(k)} \cdot U^{(k)})|\alpha' j'm') = \frac{\delta(jm, j'm')}{\sqrt{2j+1}} (\alpha j\|(T^{(k)} \cdot U^{(k)})\|\alpha' j)$$

$$= \frac{\delta(jm, j'm')}{2j+1} \sum_{\alpha'' j''} (-1)^{j-j''} (\alpha j\| T^{(k)} \|\alpha'' j'')(\alpha'' j''\| U^{(k)} \|\alpha' j). \qquad (19.33)$$

A submatrix element of the tensor U^k has the definition

$$(l^N \alpha LS \| U^k \| l^N \alpha' L'S') = \delta(S, S')N \sqrt{(2L+1)(2S+1)(2L'+1)}$$

$$\times \sum_{\alpha_1 L_1 S_1} (-1)^{L_1+l+L+k} (l^N \alpha LS \| l^{N-1}(\alpha_1 L_1 S_1)l)$$

$$\times (l^{N-1}(\alpha_1 L_1 S_1)l \| l^N \alpha' L'S') \begin{Bmatrix} L_1 & l & L \\ k & L' & l \end{Bmatrix}. \qquad (19.34)$$

Let us also present, for completeness, the corresponding expression for the submatrix element of the tensor V^{k1} which will be needed further:

$$(l^N \alpha LS \| V^{k1} \| l^N \alpha' L'S') = N\sqrt{\frac{3}{2}(2L+1)(2S+1)(2L'+1)(2S'+1)}$$

$$\times \sum_{\alpha_1 L_1 S_1} (-1)^{L_1+S_1+L+S-l-s+k}(l^N \alpha LS \| l^{N-1}(\alpha_1 L_1 S_1)l)$$

$$\times (l^{N-1}(\alpha_1 L_1 S_1)l \| l^N \alpha' L'S') \begin{Bmatrix} L_1 & l & L \\ k & L' & l \end{Bmatrix} \begin{Bmatrix} S_1 & s & S \\ 1 & S' & s \end{Bmatrix}. \quad (19.35)$$

If $k = 0$, then (19.30), for any $0 \le N \le 4l+2$, becomes equal to

$$f_0 = \delta(\alpha, \alpha')N(N-1)/2. \quad (19.36)$$

For the almost filled shell, making use of the relationship $(k_1 + k_2 > 0)$

$$\left(l^N \alpha v LS \| F^{k_1 k_2} \| l^N \alpha' v' L'S'\right) = (-1)^{k_1+k_2+1+(v-v')/2}$$

$$\times \left(l^{4l+2-N} \alpha v LS \| F^{k_1 k_2} \| l^{4l+2-N} \alpha' v' L'S'\right), \quad (19.37)$$

where the tensor $F^{k_1 k_2}$ in special cases may be equal to U^k or V^{k1}, we obtain the following relationship between the coefficients f_k, connecting the supplementary shells $(k > 0)$:

$$f_k(l^{4l+2-N} \alpha v LS, l^{4l+2-N} \alpha' v' LS) = (-1)^{(v-v')/2} f_k(l^N \alpha v LS, l^N \alpha' v' LS)$$

$$- \delta(\alpha v, \alpha' v')(2l+1-N)(2l+1)^{-1}(l \| C^{(k)} \| l)^2. \quad (19.38)$$

For the closed shell we find from Eq. (19.38)

$$f_k(l^{4l+2} \, {}_0^1 S, l^{4l+2} \, {}_0^1 S) = -(l \| C^{(k)} \| l)^2. \quad (19.39)$$

The complete tables of the numerical values of coefficients (19.30) for p-, d- and f-shells may be found in [14, 97].

Making use of the properties of the eigenvalues of Casimir operators, mentioned in Chapter 5, we are in a position to find a number of interesting features of the matrix elements of the Coulomb interaction operator. Thus, it has turned out that for the p^N shell there exists an extremely simple algebraic expression for this matrix element

$$(np^N LS | Q | np^N LS) = \frac{N(N-1)}{2} F^0(np, np)$$

$$- \frac{1}{50}[3L(L+1) + 12S(S+1) + 5N(N-4)]F^2(np, np). \quad (19.40)$$

There is the following relationship between the coefficients f_k (for arbitrary N):

$$2 \sum_{k=0}^{2l} \frac{2k+1}{(l\|C^{(k)}\|l)^2} f_k(l^N \alpha v LS, l^N \alpha' v' LS)$$

$$= \delta(\alpha v, \alpha' v')[Nl - v(4l + 4 - v)/4 - S(S+1) - N(N-4)/2]. \quad (19.41)$$

An interesting pecularity of this formula is its diagonality with respect to all quantum numbers. This feature allows us to establish a very simple relationship between non-diagonal f^n coefficients, namely,

$$\sum_{k>0}^{2l} \frac{2k+1}{(l\|C^{(k)}\|l)^2} f_k^n(l^N \alpha v LS, l^N \alpha' v' LS) = 0, \quad (19.42)$$

which, for the special case of the d-shell, leads to the equality

$$5 f_2^n(d^N v LS, d^N v' LS) = -9 f_4^n(d^N v LS, d^N v' LS). \quad (19.43)$$

The use of properties (19.41) in (19.30) allows one to extract from the coefficient (matrix element) its diagonal part. For the d-shell we have

$$(nd^N v LS|Q|nd^N v' LS) = f_2(d^N v LS, d^N v' LS)$$

$$\times \left[F^2(nd, nd) - \frac{5}{9} F^4(nd, nd) \right] + \delta(v, v') \frac{N(N-1)}{2} F^0(nd, nd)$$

$$- \delta(v, v') \frac{5}{63} [S(S+1) - v(12-v)/4 + 7N(N-6)/10]$$

$$\times F^4(nd, nd). \quad (19.44)$$

It contains only one coefficient f_2 having a non-diagonal part, whose calculation requires numerical values of the submatrix elements of the operator U^2.

We discussed in detail the properties of the matrix elements of the electrostatic energy operator for shell l^N. The corresponding expressions for the remaining two-electron operators may be found in a similar way, therefore, here we shall present only final results. For the case of relativistic corrections H_2, H_3'' and H_5' to the Coulomb energy (formulas (19.8), (19.11) and (19.12), respectively) we have

$$(nl^N \alpha v LS|H_2|nl^N \alpha' v' LS)$$

$$= \delta(v, v') \sum_{k=0}^{2l-2} m_k(l^N \alpha v LS, l^N \alpha' v' LS) M^k(nl, n'l'), \quad (19.45)$$

where

$$m_k = -\frac{4(2k+1)(2k+3)}{k+2} l(l+1)(2l+1)(l\|C^{(k)}\|l)^2 \begin{Bmatrix} k & 1 & k+1 \\ l & l & l \end{Bmatrix}^2$$

$$\times \left\{ (l^N \alpha v LS|(U^{k+1} \cdot U^{k+1})|l^N \alpha' v' LS) - \frac{\delta(\alpha, \alpha')N}{2l+1} \right\}, \quad (19.46)$$

and

$$M^k(nl, n'l') = \frac{\alpha^2}{4} \int_0^\infty r^{-k-3} P^2(nl|r) dr \int_0^r r_1^k P^2(n'l'|r_1) dr_1; \quad (19.47)$$

$$(nl^N \alpha v LS | H'_5 | nl^N \alpha' v LS) = -2\delta(v, v')(nl^N \alpha v LS | H''_3 | nl^N \alpha' v' LS)$$

$$= \delta(v, v') \left\{ \sum_{k=0}^{2l} (2k+1)[(l\|C^{(k)}\|l)/(l\|u^k\|l)]^2 \right.$$

$$\left. \times \left(l^N \alpha v LS |(U^k \cdot U^k)| l^N \alpha' v' LS \right) - \delta(\alpha, \alpha')N(2l+1) \right\} R_2(nl). \quad (19.48)$$

Here the integral $R_2(nl)$ is the special case $(n_1 l_1 = n_2 l_2)$ of the integral

$$R_2(n_1 l_1, n_2 l_2) = \frac{\alpha^2}{4} \int_0^\infty r^{-2} P^2(n_1 l_1 | r) P^2(n_2 l_2 | r) dr. \quad (19.49)$$

It follows from Eq. (19.48) that the angular part of the corrections H''_3 and H'_5 is proportional to the corresponding expression for the electrostatic energy. Therefore, these terms may be accounted for by some corrections to the integrals F^k, namely,

$$(nl^N \alpha v LS |(Q + H''_3 + H'_5)| nl^N \alpha' v' LS)$$

$$= \sum_k f_k(l^N \alpha v LS, l^N \alpha' v' LS)[F^k(nl, nl) + \delta(v, v')(2k+1)R_2(nl)]. \quad (19.50)$$

Again, using the properties of the Casimir operators, we can establish the following simple algebraic expressions for these corrections ($0 \le N \le 4l+2$, for s-electrons the orbit–orbit interaction vanishes):

$$(p^N LS | H_2 | p^N LS) = -2[L(L+1) - 2N]M^0, \quad (19.51)$$

$$(d^N v LS | H_2 | d^N v LS) = -2L(L+1)\left(M^0 - \frac{6}{49}M^2\right)$$

$$- \frac{120}{49}g(R_5)M^2 + 12N\left(M^0 + \frac{2}{7}M^2\right), \quad (19.52)$$

$$(f^N \alpha LS | H_2 | f^N \alpha LS) = -2L(L+1)\left(M^0 - \frac{5}{121}M^4\right)$$

$$+ 8g(G_2)\left(M^2 - \frac{35}{121}M^4\right)$$

$$- 8g(R_7)M^2 + 8N\left(3M^0 + M^2 + \frac{5}{11}M^4\right), \quad (19.53)$$

$$(p^N LS |(H''_3 + H'_5)| p^N LS)$$

$$= \frac{3}{10}[5N - L(L+1) - 4S(S+1)]R_2(p), \quad (19.54)$$

$$(d^N vLS|(H_3'' + H_5')|d^N vLS)$$
$$= \frac{5}{7} \left[\frac{7}{2} N - \frac{v}{4}(12 - v) - S(S + 1) \right] R_2(d). \tag{19.55}$$

The expression for quantity $g(R_5)$ in (19.52) follows directly from (5.40). Thus, for p- and d-shells we have simple algebraic formulas for the coefficients of radial integrals (actually, for matrix elements) of all relativistic corrections of the order α^2 to the Coulomb energy. For the f-shell such a formula exists only in the case of the orbit–orbit interaction. For the almost filled shell we find

$$(nl^{4l+2-N} \alpha vLS|(H_3'' + H_5')|nl^{4l+2-N} \alpha' vLS)$$
$$= (nl^N \alpha vLS|(H_3'' + H_5')|nl^N \alpha' vLS)$$
$$+ \delta(\alpha, \alpha')(2l + 1)(2l + 1 - N)R_2(nl). \tag{19.56}$$

The particular case of the closed shell follows from (19.56), if $N = 0$.

Operators H_4 and H_5'', corresponding to the spin–orbit and spin–spin interactions, are in charge of the fine structure of the terms. As a rule, operator H_4 plays the main role. The one-electron part of (19.13) is often called the spin–own-orbit interaction operator. In the case of many-electron atoms it is also called the simplified operator of the spin–orbit interaction.

The dependence of the matrix elements of the above-mentioned interactions on the total momentum J is defined by the formula

$$(LSJ|H^a|L'S'J) = (-1)^{L+S+J+a}(LS\|H^a\|L'S') \begin{Bmatrix} L & S & J \\ S' & L' & a \end{Bmatrix}. \tag{19.57}$$

Here $a = 1$ and 2 for the spin–orbit and spin–spin interactions, respectively. The submatrix element of the spin–own-orbit interaction operator equals

$$(nl^N \alpha vLS \| H^{so} \| nl^N \alpha' v' L' S')$$
$$= \sqrt{l(l + 1)(2l + 1)}(l^N \alpha vLS \| V^{11} \| l^N \alpha' v' L' S')\eta(nl), \tag{19.58}$$

where

$$\eta(nl) = \int_0^\infty \xi(r)P^2(nl|r)dr. \tag{19.59}$$

It is interesting to emphasize that the diagonal matrix elements of the spin–own-orbit interaction vanish for half-filled shells.

The submatrix element of the energy of the spin–other-orbit interaction operator is

$$(nl^N \alpha v LS \| H^{soo} \| nl^N \alpha' v' L' S')$$
$$= \sum_k m'_{k-1}(l^N \alpha v LS, l^N \alpha' v' L' S') M^{k-1}(nl, nl), \qquad (19.60)$$

where

$$m'_{k-1}(l^N \alpha v LS, l^N \alpha' v' L' S') = -2\sqrt{6} \sum_K a(Kll') \sqrt{\frac{2K+1}{K+k+1}}$$

$$\times (l\|C^{(K)}\|l)^2 \left\{ \frac{1}{(l\|u^K\|l)} \sum_{\alpha'' v'' L''} (-1)^{L+L'} \begin{Bmatrix} k & K & 1 \\ L' & L & L'' \end{Bmatrix} \right.$$

$$\times \left[\frac{1}{\sqrt{2S'+1}}(l^N \alpha v LS \| V^{k1} \| l^N \alpha'' v'' L'' S') \right.$$

$$\times (l^N \alpha'' v'' L'' S' \| U^K \| l^N \alpha' v' L' S')$$

$$+ \frac{2}{\sqrt{2S+1}}(l^N \alpha v LS \| U^k \| l^N \alpha'' v'' L'' S)$$

$$\times (l^N \alpha'' v'' L'' S \| V^{K1} \| l^N \alpha' v' L' S') \Bigg]$$

$$\left. -3 \begin{Bmatrix} k & K & 1 \\ l & l & l \end{Bmatrix} (l^N \alpha v LS \| V^{11} \| l^N \alpha' v' L' S') \right\}. \qquad (19.61)$$

Here

$$a(Kll') = -\frac{\sqrt{2k+1}}{2\sqrt{3}}$$

$$\times \begin{cases} [(l+l'+k+2)(l+l'-k) \\ \times(-l+l'+k+1)(l-l'+k+1)]^{1/2} \quad (K=k+1) \\ [(l+l'+k+1)(l+l'-k+1) \\ \times(-l+l'+k)(l-l'+k)]^{1/2} \quad (K=k-1). \end{cases} \qquad (19.62)$$

The submatrix element of the energy of spin–spin interaction operator $H'' = H^{ss}$ (19.20) may be written in the form

$$(nl^N \alpha v LS | H^{ss} | nl^N \alpha' v' L' S') = \sum_k m''_{k-1}(l^N \alpha v LS, l^N \alpha' v' L' S') M^{k-1}(nl, nl),$$

$$(19.63)$$

where

$$m''_{k-1}(l^N \alpha v LS, l^N \alpha' v' L' S') = \delta(v, v') 4\sqrt{5}(-1)^{L+S+L'+S'}$$

$$\times \sqrt{k(k+1)(2k-1)(2k+1)(2k+3)}(l\|C^{(k-1)}\|l)(l\|C^{(k+1)}\|l)$$

$$\times (l\|u^k\|l)^{-1} \sum_{\alpha'' L'' S''} \begin{Bmatrix} 1 & 1 & 2 \\ S & S' & S'' \end{Bmatrix} \begin{Bmatrix} k-1 & k+1 & 2 \\ L' & L & L'' \end{Bmatrix}$$

$$\times (l^N \alpha v LS \| V^{k-11} \| l^N \alpha'' v L'' S'')(l^N \alpha'' v L'' S'' \| V^{k+11} \| l^N \alpha' v L' S'). \quad (19.64)$$

The diagonality of formula (19.64) with respect to v is conditioned by the corresponding properties of the submatrix elements of the tensors V^{k1} with an odd sum of ranks ($k + 1$ is odd).

19.3 Relativistic Breit operator and its matrix elements

Let us consider in a similar way the relativistic atomic Hamiltonian in Breit approximation (2.1). Terms H^1, H^2 and H^3 are one-electron operators. It is very easy to find their irreducible forms. Operators H^2 and H^3 are complete scalars and do not require any transformations, whereas for H_i^1 we have to substitute (19.1) and (19.2) in (2.2). This gives

$$H_i^1 = -ic \left(\alpha_i^{(1)} \cdot C_i^{(1)} \right) \frac{\partial}{\partial r_i} + \sqrt{2} \frac{c}{r_i} \left(\alpha_i^{(1)} \cdot \left[C_i^{(1)} \times L_i^{(1)} \right]^{(1)} \right). \qquad (19.65)$$

The energy operator of electrostatic interaction has the same expression as in the non-relativistic case ($H_6^4 = Q$, where Q is defined by (1.15)). Its irreducible form is given by (19.6). In order to find the irreducible form of the operator of magnetic interactions H_{ij}^m, we make use of expansion (19.6) and then transform the coupling scheme of tensors to one in which the operators acting on one and the same coordinates, would be directly coupled into a tensorial product. This gives finally

$$H_{ij}^m = -\sum_{kK} \frac{r_<^k}{r_>^{k+1}} \left(\left[C_i^{(k)} \times \alpha_i^{(1)} \right]^{(K)} \cdot \left[C_j^{(k)} \times \alpha_j^{(1)} \right]^{(K)} \right). \qquad (19.66)$$

It is evident that the corresponding expression for operator H^5 will differ from (19.66) by the multiplier $1/2$.

The most complicated issue is the transformation of the operator of retarding interactions H^r to the irreducible form. The final result is as follows:

$$H_{ij}^r = \sum_k \left\{ \left[\frac{k}{2k-1} \left(\left[C_i^{(k)} \times \alpha_i^{(1)} \right]^{(k-1)} \cdot \left[C_j^{(k)} \times \alpha_j^{(1)} \right]^{(k-1)} \right) \right. \right.$$
$$+ \frac{k+1}{2k+3} \left(\left[C_i^{(k)} \times \alpha_i^{(1)} \right]^{(k+1)} \cdot \left[C_j^{(k)} \times \alpha_j^{(1)} \right]^{(k+1)} \right) \right] \frac{r_<^k}{r_>^{k+1}}$$
$$+ \frac{\sqrt{(k+1)(k+2)(2k+1)(2k+5)}}{2(2k+3)}$$
$$\times \left[\left(\left[C_i^{(k)} \times \alpha_i^{(1)} \right]^{(k+1)} \cdot \left[C_j^{(k+2)} \times \alpha_j^{(1)} \right]^{(k+1)} \right) \left[\frac{r_i^{k+2}}{r_j^{k+3}} - \frac{r_i^k}{r_j^{k+1}} \right] \right.$$
$$\left. \left. + \left(\left[C_i^{(k+2)} \times \alpha_i^{(1)} \right]^{(k+1)} \cdot \left[C_j^{(k)} \times \alpha_j^{(1)} \right]^{(k+1)} \right) \left[\frac{r_j^{k+2}}{r_i^{k+3}} - \frac{r_j^k}{r_i^{k+1}} \right] \right] \right\}. \qquad (19.67)$$

Let us find expressions for the matrix elements of these operators for a subshell of equivalent electrons with respect to the relativistic wave functions, in a one-electron case defined by (2.15), and for a subshell, by (9.8).

The matrix element of one-electron energy operators $(H^1+H^2+H^3)$ for a subshell of equivalent electrons will be equal (remember that the matrix elements of the operators with respect to the relativistic wave functions are denoted as $\langle \ | \ | \ \rangle$ to distinguish them clearly from the case of non-relativistic functions, when we use simple brackets $(\ |\ |\ )$):

$$\left\langle nlj^N\alpha J|(H^1+H^2+H^3)|nlj^N\alpha' J'\right\rangle = \delta(\alpha J,\alpha' J')NE(nlj), \qquad (19.68)$$

where

$$E(nlj) = \left\langle nlj|(H^1+H^2+H^3)|nlj\right\rangle$$

$$= \int_0^\infty \left\{ cg(\lambda') \left[\frac{df(\lambda)}{dr} + \frac{1+\kappa}{r}f(\lambda)\right] - (c^2-V)g^2(\lambda') \right.$$

$$\left. - cf(\lambda) \left[\frac{dg(\lambda')}{dr} + \frac{1-\kappa}{r}g(\lambda')\right] + (c^2+V)f^2(\lambda) \right\}r^2dr. \quad (19.69)$$

Here

$$\lambda = nlj, \qquad \lambda' = nl'j, \qquad \kappa = (-1)^{l+s+j}(j+1/2). \qquad (19.70)$$

Exploiting wave function (9.8) and equality (7.3), we find the following expression for the relativistic matrix element of the Coulomb interaction:

$$\left\langle nlj^N\alpha J|H^e|nlj^N\alpha' J\right\rangle = \sum_{k_{even}} f_k(j^N\alpha\alpha' J)$$

$$\times [1 + P(\lambda,\lambda')][R_k(\lambda\lambda,\lambda\lambda) + R_k(\lambda\lambda',\lambda\lambda')], \qquad (19.71)$$

in which operator $P(\lambda,\lambda')$ performs transposition $\lambda \leftrightarrow \lambda'$ in all multipliers and terms in its right side, whereas the radial integral is defined below, i.e.

$$R_k(\lambda_1\lambda_2,\lambda_3\lambda_4) = (l_1l_3k)(l_2l_4k)$$

$$\times \iint_0^\infty f(\lambda_1|r_1)f(\lambda_2|r_2)\frac{r_<^k}{r_>^{k+1}}f(\lambda_3|r_1)f(\lambda_4|r_2)r_1^2r_2^2dr_1dr_2. \quad (19.72)$$

Here and further on the symbol (abc) means that parameters a, b and c obey the triangular condition with even perimeter. If in the radial integral there is $\lambda_i' = n_il_i'j_i$ instead of $\lambda_i = n_il_ij_i$, then this indicates that in the corresponding parts of the integral function f and quantum number l must be replaced by g and l', respectively. Coefficient f_k has the form

$$f_k(j^N\alpha\alpha' J) = \frac{2j+1}{\sqrt{2J+1}} \begin{bmatrix} k & j & j \\ 0 & 1/2 & 1/2 \end{bmatrix}^2 (j^N\alpha J\|T^{kk0}\|j^N\alpha' J), \qquad (19.73)$$

where $(j^N \alpha J \| T^{kk0} \| j^N \alpha' J)$ is defined by (7.11). For $k = 0$ Eq. (19.73) gives

$$f_0(j^N \alpha \alpha' J) = \delta(\alpha, \alpha') N(N-1)/2. \qquad (19.74)$$

It follows from Eq. (19.73) that the coefficient f_k for a subshell does not depend on orbital quantum number l, the summation runs over $0, 2, \ldots, 2j-1$. Coefficients f_k for complementary subshells j^N and j^{2j+1-N}, according to equality (7.6), are connected with each other by the relation

$$f_k(j^{2j+1-N} \alpha v \alpha' v' J) = (-1)^{(v-v')/2} f_k(j^N \alpha v \alpha' v' J)$$
$$- \delta(\alpha v, \alpha' v') \frac{2j+1-2N}{2} \begin{bmatrix} k & j & j \\ 0 & 1/2 & 1/2 \end{bmatrix}^2. \qquad (19.75)$$

Formula (19.73) for the closed subshell acquires the form

$$f_k(j^{2j+1} 00) = -\frac{2j+1}{2} \begin{bmatrix} k & j & j \\ 0 & 1/2 & 1/2 \end{bmatrix}^2. \qquad (19.76)$$

As in a non-relativistic case, making use of the eigenvalues of the corresponding Casimir operators, we find an interesting relationship

$$\sum_{k_{even}=0}^{2j-1} (2k+1) f_k(j^N \alpha v \alpha' v' J) \Big/ \begin{bmatrix} k & j & j \\ 0 & 1/2 & 1/2 \end{bmatrix}^2$$
$$= \delta(\alpha v, \alpha' v') \frac{2j+1}{4} [2N(2-N) - 2v(j+1) + v(v-1) + 2jN], \qquad (19.77)$$

which is very convenient for the simplification of formulas and the checking of numerical values of coefficients f_k.

The formula for a matrix element of the energy operator of magnetic interactions (its irreducible form is presented by Eq. (19.66)) can be found in a similar way to that for the Coulomb interaction. It is

$$\left\langle nl j^N \alpha J |H^m| nl j^N \alpha' J \right\rangle = \sum_{k_{odd}} d_k(j^N \alpha \alpha' J) R_k(\lambda \lambda, \lambda' \lambda'), \qquad (19.78)$$

where

$$d_k(j^N \alpha \alpha' J) = -\frac{4(2j+1)^3}{k(k+1)\sqrt{2J+1}} \begin{bmatrix} k & j & j \\ 0 & 1/2 & 1/2 \end{bmatrix}^2 (j^N \alpha J \| T^{kk0} \| j^N \alpha' J). \qquad (19.79)$$

Index k runs here through odd values $1, 3, \ldots, 2j$. Expression (19.79) is diagonal with respect to seniority quantum number v, i.e.

$$d_k(j^N \alpha v \alpha' v' J) = \delta(v, v') d_k(j^N \alpha v \alpha' v J). \qquad (19.80)$$

This is because of the diagonality of the submatrix elements of operator T^k (k is odd) with respect to v.

For the majority of interesting cases of electronic configurations (s-, p-, d- and f-electrons), quantum numbers v and J are sufficient to classify unambiguously the energy levels. Therefore, the magnetic interactions inside subshell nlj^N will be described by completely diagonal matrices. The relationship for these matrix elements for complementary shells may be presented in the following general form:

$$\left\langle nlj^{N_1}\alpha J|H^m|nlj^{N_1}\alpha'J\right\rangle = \left\langle nlj^{N_2}\alpha J|H^m|nlj^{N_2}\alpha'J\right\rangle$$

$$+\sum_{k_{odd}}\delta(\alpha,\alpha')\frac{2(N_1-N_2)(2j+1)^2}{k(k+1)}\begin{bmatrix}k & j & j\\0 & 1/2 & 1/2\end{bmatrix}^2 R_k(\lambda\lambda,\lambda'\lambda'). \quad (19.81)$$

For the filled subshell Eq. (19.79) turns into

$$d_k(j^{2j+1}00) = \delta(\alpha,\alpha')\frac{2(2j+1)^3}{k(k+1)}\begin{bmatrix}k & j & j\\0 & 1/2 & 1/2\end{bmatrix}^2. \quad (19.82)$$

The analogue of equality (19.77) here will be

$$\sum_{k_{odd}}k(k+1)(2k+1)d_k(j^N\alpha v\alpha'v'J)\bigg/\begin{bmatrix}k & j & j\\0 & 1/2 & 1/2\end{bmatrix}^2$$

$$= \delta(\alpha v,\alpha'v')(2j+1)^3[v(v-1)+2(N-v)(j+1)]. \quad (19.83)$$

Finally, a similar consideration of the energy operator of retarding interactions (19.67) leads to the equality

$$\left\langle nlj^N\alpha J|H^r|nlj^N\alpha'J\right\rangle = 0, \quad (19.84)$$

i.e. the contribution of the retarding interactions to the relativistic energy is zero for the case when the interaction is inside a subshell of equivalent electrons.

20

Interaction energy of
two shells in LS coupling

While studying the energy spectra or other spectral quantities of atoms and ions having complex electronic configurations, one ought to consider the expressions for the matrix elements of the operators both within each shell of equivalent electrons and between each pair of these shells. For example, in order to find the energy spectrum of the ground configuration $1s^2 2s^2$ of the beryllium atom, we have to calculate the interaction energy in each shell $1s^2$ and $2s^2$ as well as between them. The last case will be discussed in this chapter. If there are more shells, then according to the two-particle character of interelectronic interactions, we have to account for this interaction between all possible pairs of shells.

Let us notice that momenta of each shell may be coupled into total momenta by various coupling schemes. Therefore, here, as in the case of two non-equivalent electrons, coupling schemes (11.2)–(11.5) are possible, only instead of one-electronic momenta there will be the total momenta of separate shells. To indicate this we shall use the notation LS, LK, JK and JJ. Some peculiarities of their usage were discussed in Chapters 11 and 12 and will be additionally considered in Chapter 30. Therefore, here we shall restrict ourselves to the case of LS coupling for non-relativistic and JJ (or jj) coupling for relativistic wave functions. We shall not indicate explicitly the parity of the configuration, consisting of several shells, because it is simply equal to the sum of parities of all shells.

As was mentioned in previous chapters, the wave function of an atom must be antisymmetric with respect to the transposition of the coordinates (sets of quantum numbers) of each pair of electrons. The antisymmetric wave function of a shell of equivalent electrons may be constructed with the help of the CFP (formulas (9.7) and (9.8) for LS and jj couplings, respectively). Antisymmetrization of the atomic wave function with respect to electrons belonging to different shells may be ensured with the help of generalized CFP [14] or by imposing certain phase conditions on the

235

wave function while transposing the above-mentioned electrons. Let us illustrate this for the case of the configuration consisting of two non-equivalent electrons. The corresponding antisymmetric wave function (the writing of two principal quantum numbers beside $n_1 n_2$ indicates this) will be as follows (compare with (10.6)):

$$\psi(n_1 n_2 l_1 s_1 l_2 s_2 LSJM|x_1 x_2) = N\{\psi(n_1 l_1 s_1 n_2 l_2 s_2 LSJM|x_1 x_2)$$
$$+ (1)^{l_1+l_2-L-S}\psi(n_2 l_2 s_2 n_1 l_1 s_1 LSJM|x_1 x_2)\}, \quad (20.1)$$

where

$$N = \begin{cases} 1/\sqrt{2} & (n_1 l_1 \neq n_2 l_2) \\ 1/2 & (n_1 l_1 = n_2 l_2) \end{cases}. \quad (20.2)$$

For the special case of two equivalent electrons formula (20.1) turns into (9.1).

While evaluating the matrix element of a two-particle operator with respect to antisymmetric wave functions of the kind (20.1) two types of items occur: the first ones contain the wave functions of coupled momenta characterized by an identical arrangement of quantum numbers $n_1 l_1 s_1$ and $n_2 l_2 s_2$, whereas the second ones are characterized by different arrangement. Therefore, the corresponding matrix element is usually divided in two parts – direct and exchange. The occurrence of an exchange part is a completely quantum-mechanical effect, having no analogues in classical mechanics.

Matrix elements of the operators of the interaction energy between two shells of equivalent electrons may be expressed, with the aid of the CFP, in terms of the corresponding two-electron quantities. Substituting in such formula the explicit expression for the two-electron matrix element, after a number of mathematical manipulations and using the definition of submatrix elements of operators composed of unit tensors, we get convenient expressions for the matrix elements in the case of two shells of equivalent electrons. The corresponding details may be found in [14], here we present only final results.

Let us start with the one-electron operators. Their intershell matrix elements are small or vanish. Therefore, the total matrix element of such an operator is simply equal to the sum of corresponding quantities describing that interaction within each shell. Thus, for operators T and P (1.15) we have:

$$\left(n_1 n_2 l_1^{N_1} l_2^{N_2} \alpha_1 L_1 S_1 \alpha_2 L_2 S_2 LS|(T+P)|n_1 n_2 l_1^{N_1} l_2^{N_2} \alpha_1' L_1' S_1' \alpha_2' L_2' S_2' L'S'\right)$$
$$= \delta(\alpha_1 L_1 S_1 \alpha_2 L_2 S_2 LS, \alpha_1' L_1' S_1' \alpha_2' L_2' S_2' L'S')$$
$$\times [N_1 I_1(n_1 l_1) + N_2 I_2(n_2 l_2)], \quad (20.3)$$

where $I(n_i l_i)$ is defined according to (19.23). This formula can, in a trivial way, be generalized for any number of shells. In the same way the matrix elements of the one-electron operators of relativistic corrections H_1 (1.18) as well as of the one-electron part H_3' of operator H_3 (1.20) may be expressed in terms of the corresponding one-shell quantities.

In the case of the spin–own-orbit interaction (the first item in (1.21) or the first sum in irreducible form (19.13)) there is the following equality, illustrating that this interaction also breaks up into the sum of intrashell terms:

$$
\left(n_1 n_2 l_1^{N_1} l_2^{N_2} \alpha_1 L_1 S_1 \alpha_2 L_2 S_2 L S \middle| H^{so} \middle| n_1 n_2 l_1^{N_1} l_2^{N_2} \alpha_1' L_1' S_1' \alpha_2' L_2' S_2' L' S' \right)
$$

$$
= (-1)^{L'+S+J} \begin{Bmatrix} L & S & J \\ S' & L' & 1 \end{Bmatrix} \sqrt{(2L+1)(2L'+1)(2S+1)(2S'+1)}
$$

$$
\times \Bigg\{ (-1)^{L_1+L_2+L'+S_1+S_2+S'} \delta(\alpha_2 L_2 S_2, \alpha_2' L_2' S_2') \sqrt{l_1(l_1+1)(2l_1+1)}
$$

$$
\times \begin{Bmatrix} L_1 & L & L_2 \\ L' & L_1' & 1 \end{Bmatrix} \begin{Bmatrix} S_1 & S & S_2 \\ S' & S_1' & 1 \end{Bmatrix} \left(l_1^{N_1} \alpha_1 L_1 S_1 \| V^{11} \| l_1^{N_1} \alpha_1' L_1' S_1' \right)
$$

$$
\times \eta(n_1 l_1) + (-1)^{L_1+L_2'+L+S_1+S_2'+S} \delta(\alpha_1 L_1 S_1, \alpha_1' L_1' S_1')
$$

$$
\times \sqrt{l_2(l_2+1)(2l_2+1)} \begin{Bmatrix} L_2 & L & L_1 \\ L' & L_2' & 1 \end{Bmatrix} \begin{Bmatrix} S_2 & S & S_1 \\ S' & S_2' & 1 \end{Bmatrix}
$$

$$
\times \left(l_2^{N_2} \alpha_2 L_2 S_2 \| V^{11} \| l_2^{N_2} \alpha_2' L_2' S_2' \right) \eta(n_2 l_2) \Bigg\}. \tag{20.4}
$$

It equals zero for completely filled (closed) shells.

20.1 Electrostatic interaction

Electrostatic interaction energy operator Q in (1.15) or in irreducible form (19.6) may formally be written for the case of two shells of equivalent electrons as

$$
Q = Q_{11} + Q_{22} + Q_{12}^d + Q_{12}^e, \tag{20.5}
$$

where the first two items represent the Coulomb interaction within each shell and the remaining two stand for the direct (Q_{12}^d) and exchange (Q_{12}^e) parts of this interaction between shells. Operator Q_{11} acts only on the coordinates of the wave function of the first shell. Therefore, its matrix element with respect to the wave functions of both shells will be diagonal relative to the quantum numbers of the second shell, and vice versa. Thus, for the electrostatic interaction inside both shells we have (here and further on in this chapter, if it is not indicated otherwise, $N_1 \leq 2l_1 + 1$,

$N_2 \leq 2l_2 + 1$):

$$\left(n_1 n_2 l_1^{N_1} l_2^{N_2} \alpha_1 L_1 S_1 \alpha_2 L_2 S_2 LS | (Q_{11} + Q_{22}) | n_1 n_2 l_1^{N_1} l_2^{N_2} \alpha_1' L_1' S_1' \alpha_2' L_2' S_2' L'S'\right)$$

$$= \delta(LS, L'S')\delta(L_1 S_1, L_1' S_1')\delta(\alpha_2 L_2 S_2, \alpha_2' L_2' S_2')$$

$$\times \left(n_1 l_1^{N_1} \alpha_1 L_1 S_1 \| Q_{11} \| n_1 l_1^{N_1} \alpha_1' L_1 S_1\right) + \delta(LS, L'S')\delta(L_2 S_2, L_2' S_2')$$

$$\times \delta(\alpha_1 L_1 S_1, \alpha_1' L_1' S_1') \left(n_2 l_2^{N_2} \alpha_2 L_2 S_2 \| Q_{22} \| n_2 l_2^{N_2} \alpha_2' L_2 S_2\right). \tag{20.6}$$

Here the matrix elements inside each shell may be calculated using formula (19.29). For the interaction between electrons belonging to different shells $l_1^{N_1}$ and $l_2^{N_2}$, operator (19.6) may be written in the form

$$\sum_k \sum_{j=N_1+1}^{N} \sum_{i=1}^{N_1} \frac{r_<^k}{r_>^{k+1}} \left(C_i^{(k)} \cdot C_j^{(k)}\right), \tag{20.7}$$

and the corresponding matrix elements for Q_{12}^d and Q_{12}^e will be

$$\left(n_1 n_2 l_1^{N_1} l_2^{N_2} \alpha_1 L_1 S_1 \alpha_2 L_2 S_2 LS | Q_{12}^d | n_1 n_2 l_1^{N_1} l_2^{N_2} \alpha_1' L_1' S_1' \alpha_2' L_2' S_2' L'S'\right)$$

$$= \delta(LS, L'S')\delta(S_1 S_2, S_1' S_2')$$

$$\times \sum_k f_k(l_1^{N_1} l_2^{N_2} \alpha_1 L_1 \alpha_1' L_1' S_1, \alpha_2 L_2 \alpha_2' L_2' S_2, LS) F^k(n_1 l_1, n_2 l_2), \tag{20.8}$$

$$\left(n_1 n_2 l_1^{N_1} l_2^{N_2} \alpha_1 L_1 S_1 \alpha_2 L_2 S_2 LS | Q_{12}^e | n_1 n_2 l_1^{N_1} l_2^{N_2} \alpha_1' L_1' S_1' \alpha_2' L_2' S_2' L'S'\right)$$

$$= \delta(LS, L'S') \sum_k g_k(l_1^{N_1} l_2^{N_2} \alpha_1 L_1 S_1 \alpha_1' L_1' S_1', \alpha_2 L_2 S_2 \alpha_2' L_2' S_2', LS)$$

$$\times G^k(n_1 l_1, n_2 l_2). \tag{20.9}$$

The radial integral of the direct part of interaction F^k is defined by (19.31), whereas that of the exchange part is as follows:

$$G^k(nl, n'l') = \int_0^\infty\!\!\!\int P(nl|r_1)P(n'l'|r_1)\frac{r_<^k}{r_>^{k+1}}P(nl|r_2)P(n'l'|r_2)dr_1 dr_2. \tag{20.10}$$

Coefficients f_k and g_k at radial integrals F^k and G^k in formulas (20.8) and (20.9) are the matrix elements of the operators, respectively,

$$(l_1\|C^{(k)}\|l_1)(l_2\|C^{(k)}\|l_2)(U_{N_1}^k \cdot U_{N_2}^k), \tag{20.11}$$

$$-(l_1\|C^{(k)}\|l_2)^2 \sum_x (-1)^x (2x+1) \begin{Bmatrix} l_1 & l_2 & k \\ l_2 & l_1 & x \end{Bmatrix} (l_1\|u^x\|l_1)^{-1}$$

$$\times (l_2\|u^x\|l_2)^{-1} \left[\frac{1}{2}(U_{N_1}^x \cdot U_{N_2}^x) + 2(V_{N_1}^{x1} \cdot V_{N_2}^{x1})\right]. \tag{20.12}$$

Here x acquires values from 0 to the smaller of $2l_1$, $2l_2$.

The coefficients at the radial integrals in (20.8) have the expressions (for brevity we shall skip further the quantum numbers in the left side of the equalities):

$$f_0 = \delta(\alpha_1 L_1 S_1 \alpha_2 L_2 S_2, \alpha'_1 L'_1 S'_1 \alpha'_2 L'_2 S'_2) N_1 N_2, \qquad (20.13)$$

and for $k > 0$

$$f_k(l_1^{N_1} l_2^{N_2}) = (-1)^{L'_1 + L_2 + L} \frac{(l_1 || C^{(k)} || l_1)(l_2 || C^{(k)} || l_2)}{\sqrt{(2S_1 + 1)(2S_2 + 1)}}$$

$$\times \begin{Bmatrix} L_1 & L_2 & L \\ L'_2 & L'_1 & k \end{Bmatrix} \left(l_1^{N_1} \alpha_1 L_1 S_1 || U^k || l_1^{N_1} \alpha'_1 L'_1 S_1 \right)$$

$$\times \left(l_2^{N_2} \alpha_2 L_2 S_2 || U^k || l_2^{N_2} \alpha'_2 L'_2 S_2 \right). \qquad (20.14)$$

$$g_k(l_1^{N_1} l_2^{N_2}) = -(l_1 || C^{(k)} || l_2)^2 \sum_{x>0} (-1)^x (2x + 1) \begin{Bmatrix} l_1 & l_2 & k \\ l_2 & l_1 & x \end{Bmatrix}$$

$$\times \left\{ (-1)^{L'_1 + L_2 + L} \frac{\delta(S_1 S_2, S'_1 S'_2)}{2\sqrt{(2S_1 + 1)(2S_2 + 1)}} \begin{Bmatrix} L_1 & L_2 & L \\ L'_2 & L'_1 & x \end{Bmatrix} \right.$$

$$\times \left(l_1^{N_1} \alpha_1 L_1 S_1 || U^x || l_1^{N_1} \alpha'_1 L'_1 S_1 \right) \left(l_2^{N_2} \alpha_2 L_2 S_2 || U^x || l_2^{N_2} \alpha'_2 L'_2 S_2 \right)$$

$$+ 2(-1)^{L'_1 + L_2 + L + S'_1 + S_2 + S} \begin{Bmatrix} L_1 & L_2 & L \\ L'_2 & L'_1 & x \end{Bmatrix} \begin{Bmatrix} S_1 & S_2 & S \\ S'_2 & S'_1 & 1 \end{Bmatrix}$$

$$\left. \times \left(l_1^{N_1} \alpha_1 L_1 S_1 || V^{x1} || l_1^{N_1} \alpha'_1 L'_1 S'_1 \right) \left(l_2^{N_2} \alpha_2 L_2 S_2 || V^{x1} || l_2^{N_2} \alpha'_2 L'_2 S'_2 \right) \right\}$$

$$+ \delta(\alpha_1 L_1 S_1 \alpha_2 L_2 S_2, \alpha'_1 L'_1 S'_1 \alpha'_2 L'_2 S'_2)(l_1 || C^{(k)} || l_2)^2 \times$$

$$\times \frac{2 \left(S_1(S_1 + 1) + S_2(S_2 + 1) - S(S + 1) \right) - N_1 N_2}{2(2l_1 + 1)(2l_2 + 1)}. \qquad (20.15)$$

Making use of relationship (19.37), these formulas may be easily generalized to cover the case of almost filled shells. Expression (20.13) is valid for any degree of occupation of the shells. If one shell, for example, the first one, is almost filled, then, according to (19.37)

$$f_k(l_1^{4l_1 + 2 - N_1} l_2^{N_2}) = -(-1)^{(v_1 - v'_1)/2} f_k(l_1^{N_1} l_2^{N_2}) \qquad (k > 0), \qquad (20.16)$$

$$g_k(l_1^{4l_1 + 2 - N_1} l_2^{N_2}) = (-1)^{(v_1 - v'_1)/2} g_k(l_1^{N_1} l_2^{N_2})$$

$$+ (-1)^{(v_1 - v'_1)/2 + L'_1 + L_2 + L}(l_1 || C^{(k)} || l_2)^2 \sum_{x_e > 0} (2x + 1) \begin{Bmatrix} l_1 & l_2 & k \\ l_2 & l_1 & x \end{Bmatrix}$$

$$\times \frac{\delta(S_1 S_2, S'_1 S'_2)}{\sqrt{(2S_1 + 1)(2S_2 + 1)}} \begin{Bmatrix} L_1 & L_2 & L \\ L'_2 & L'_1 & x \end{Bmatrix} \left(l_1^{N_1} \alpha_1 L_1 S_1 || U^x || l_1^{N_1} \alpha'_1 L'_1 S_1 \right)$$

$$\times \left(l_2^{N_2} \alpha_2 L_2 S_2 || U^x || l_2^{N_2} \alpha'_2 L'_2 S_2 \right) - 4(-1)^{(v_1 - v'_1)/2 + L'_1 + L_2 + L + S'_1 + S_2 + S}$$

$$\times (l_1||C^{(k)}||l_2)^2 \sum_{x_o}(2x+1)\begin{Bmatrix} l_1 & l_2 & k \\ l_2 & l_1 & x \end{Bmatrix}\begin{Bmatrix} L_1 & L_2 & L \\ L_2' & L_1' & x \end{Bmatrix}$$

$$\times \begin{Bmatrix} S_1 & S_2 & S \\ S_2' & S_1' & 1 \end{Bmatrix}\left(l_1^{N_1}\alpha_1 L_1 S_1||V^{x1}||l_1^{N_1}\alpha_1' L_1' S_1'\right)$$

$$\times \left(l_2^{N_2}\alpha_2 L_2 S_2||V^{x1}||l_2^{N_2}\alpha_2' L_2' S_2'\right) + \delta(\alpha_1 L_1 S_1 \alpha_2 L_2 S_2, \alpha_1' L_1' S_1' \alpha_2' L_2' S_2')$$

$$\times (l_1||C^{(k)}||l_2)^2 \frac{N_1 N_2 - (2l_1+1)N_2}{(2l_1+1)(2l_2+1)}. \tag{20.17}$$

The first summation in (20.17) is over the even $x > 0$ values (x_e), whereas the second one is over odd x (x_o). If both shells are almost filled, then

$$f_k(l_1^{4l_1+2-N_1} l_2^{4l_2+2-N_2}) = (-1)^{(v_1-v_1'+v_2-v_2')/2} f_k(l_1^{N_1} l_2^{N_2}) \qquad (k>0), \tag{20.18}$$

$$g_k(l_1^{4l_1+2-N_1} l_2^{4l_2+2-N_2}) = (-1)^{(v_1-v_1'+v_2-v_2')/2} g_k(l_1^{N_1} l_2^{N_2})$$
$$+ \delta(\alpha_1 L_1 S_1 \alpha_2 L_2 S_2, \alpha_1' L_1' S_1' \alpha_2' L_2' S_2')$$
$$\times (l_1||C^{(k)}||l_2)^2\left(-2 + \frac{N_1}{2l_1+1} + \frac{N_2}{2l_2+1}\right). \tag{20.19}$$

If the first shell is completely filled, then the corresponding expression for f_k follows directly from (20.13), whereas for g_k we have

$$g_k(l_1^{4l_1+2} l_2^{N_2}) = -\delta(\alpha_2 L_2 S_2, \alpha_2' L_2' S_2')(l_1||C^{(k)}||l_2)^2 N_2/(2l_2+1). \tag{20.20}$$

If both shells are almost filled, then again for f_k formula (20.13) is valid, whereas g_k is given by

$$g_k(l_1^{4l_1+2} l_2^{4l_2+2}) = -2(l_1||C^{(k)}||l_2)^2. \tag{20.21}$$

Making use of the formulas presented in this section we are in a position to write the expression for the energy of the electrostatic interaction between any number of shells. In this case only a number of terms' items increase in the corresponding formulas. However, we have to notice that these terms must be diagonal with respect to the quantum numbers of those shells on which they do not act. Let us also recall that in all formulas containing at least one almost filled shell, it is meant that the seniority quantum number v is excluded from α. This is necessary for the reconciliation of phase conditions between the corresponding quantities.

If some shell $l_i^{N_i}$ consists of only one electron ($N_i = 1$, $L_i = l_i$, $S_i = s = 1/2$), then the electrostatic interaction in it equals zero, whereas in the formulas describing this interaction with other shells, the corresponding submatrix elements of operators U^k and V^{k1} turn into

$$(l^1 ls||u^k||l^1 ls) = \sqrt{2}, \qquad (l^1 ls||v^{k1}||l^1 ls) = \sqrt{3/2}. \tag{20.22}$$

In a similar manner the particular case of two-electron matrix elements may be deduced from the two-shell expressions presented above.

Corresponding expressions for matrix elements of relativistic corrections to the electrostatic energy and to two-electron parts of magnetic interactions are rather cumbersome and are not presented here. They may be found in [14]. In Chapter 27 we describe the simplified method of taking them into account.

Accounting for the properties of the seniority (quasispin) quantum numbers, we are able to express the matrix elements of the energy operator in terms of the corresponding quantities for the electronic configuration, for which this term has occurred for the first time $(l_i^{N_i}\alpha_i v_i L_i S_i \rightarrow l_i^{N_{v_i}}\alpha_i v_i L_i S_i)$. This question is considered in detail in [102]. The utilization of the relationship between the CFP and the submatrix elements of the operators, composed of unit tensors, which was established in [105], allows one to find a large number of new expressions for the above-mentioned matrix elements.

20.2 Relativistic energy operator

The relativistic analogue of formula (20.1) will be

$$\psi(n_1 n_2 l_1 j_1 l_2 j_2 J M | x_1 x_2) = T\{\psi(n_1 l_1 j_1 n_2 l_2 j_2 J M | x_1 x_2)$$
$$-(-1)^{j_1+j_2-J}\psi(n_2 l_2 j_2 n_1 l_1 j_1 J M | x_1 x_2)\}, \quad (20.23)$$

where $T = 1/\sqrt{2}$ for $n_1 l_1 j_1 \neq n_2 l_2 j_2$ and $T = 1/2$ for $n_1 l_1 j_1 = n_2 l_2 j_2$. Matrix elements of the one-electron energy operators for configurations consisting of several subshells are simply equal to the sum of the corresponding intrasubshell quantities (19.68), i.e.

$$\left\langle n_1 \ldots n_k l_1 j_1^{N_1} \ldots l_k j_k^{N_k} A | (H^1 + H^2 + H^3) | n_1 \ldots n_k l_1 j_1^{N_1} \ldots l_k j_k^{N_k} A' \right\rangle$$

$$= \delta(A, A') \sum_{i=1}^{k} N_i E(n_i l_i j_i), \quad (20.24)$$

where A denotes the set of all quantum numbers describing the configuration considered. Two-electron terms of operator H in a relativistic case may be divided into the parts (20.5), corresponding to the interaction already considered in Chapter 19 inside each subshell as well as to the direct and exchange terms of the intersubshell interactions. The first ones are (compare with (20.6)):

$$\left\langle n_1 n_2 l_1 j_1^{N_1} l_2 j_2^{N_2} \alpha_1 J_1 \alpha_2 J_2 J | (H_{11} + H_{22}) | n_1 n_2 l_1 j_1^{N_1} l_2 j_2^{N_2} \alpha_1' J_1' \alpha_2' J_2' J' \right\rangle$$

$$= \delta(J, J') \Big[\delta(J_1 \alpha_2 J_2, J_1' \alpha_2' J_2') \left\langle n_1 l_1 j_1^{N_1} \alpha_1 J_1 | H_{11} | n_1 l_1 j_1^{N_1} \alpha_1' J_1 \right\rangle$$

$$+ \delta(\alpha_1 J_1 J_2, \alpha_1' J_1' J_2') \left\langle n_2 l_2 j_2^{N_2} \alpha_2 J_2 | H_{22} | n_2 l_2 j_2^{N_2} \alpha_2' J_2 \right\rangle \Big]. \quad (20.25)$$

As was mentioned at the beginning of this chapter, the matrix element of the interaction between subshells may be expressed in terms of the CFP with one detached electron and two-electron matrix elements of the operator considered. The corresponding formula for jj coupling is as follows:

$$\left\langle n_1 n_2 l_1 j_1^{N_1} l_2 j_2^{N_2} \alpha_1 J_1 \alpha_2 J_2 J | H_{12} | n_1 n_2 l_1 j_1^{N_1} l_2 j_2^{N_2} \alpha_1' J_1' \alpha_2' J_2' J' \right\rangle$$

$$= \delta(J, J') N_1 N_2 \sqrt{[J_1, J_2, J_1', J_2']} \sum_{\bar{\alpha}_1 \bar{J}_1 \bar{\alpha}_2 \bar{J}_2 J_{12} j_{12}} [\bar{J}_{12}, j_{12}]$$

$$\times \left(j_1^{N_1} \alpha_1 J_1 \| j_1^{N_1-1} (\bar{\alpha}_1 \bar{J}_1) j_1 \right) \left(j_1^{N_1-1} (\bar{\alpha}_1 \bar{J}_1) j_1 \| j_1^{N_1} \alpha_1' J_1' \right)$$

$$\times \left(j_2^{N_2} \alpha_2 J_2 \| j_2^{N_2-1} (\bar{\alpha}_2 \bar{J}_2) j_2 \right) \left(j_2^{N_2-1} (\bar{\alpha}_2 \bar{J}_2) j_2 \| j_2^{N_2} \alpha_2' J_2' \right)$$

$$\times \begin{Bmatrix} \bar{J}_1 & \bar{J}_2 & \bar{J}_{12} \\ j_1 & j_2 & j_{12} \\ J_1 & J_2 & J \end{Bmatrix} \begin{Bmatrix} \bar{J}_1 & \bar{J}_2 & \bar{J}_{12} \\ j_1 & j_2 & j_{12} \\ J_1' & J_2' & J \end{Bmatrix}$$

$$\times \langle n_1 n_2 l_1 j_1 l_2 j_2 j_{12} | H_{12} | n_1 n_2 l_1 j_1 l_2 j_2 j_{12} \rangle. \tag{20.26}$$

However, formulas of the kind (20.26) are rather inconvenient for calculations. Therefore, one has usually to insert explicit expressions for two-electron matrix elements, to perform, where it turns out to be possible, the summations necessary and to find finally the representation of the energy matrix element of the interaction between two subshells in the form of direct and exchange parts. Thus, for the electrostatic interaction we find

$$\left\langle n_1 n_2 l_1 j_1^{N_1} l_2 j_2^{N_2} \alpha_1 J_1 \alpha_2 J_2 J | Q_{12}^d | n_1 n_2 l_1 j_1^{N_1} l_2 j_2^{N_2} \alpha_1' J_1' \alpha_2' J_2' J \right\rangle$$

$$= \sum_{k_{even}} f_k(j_1^{N_1} j_2^{N_2}) \left[1 + P(\lambda_i, \lambda_i') \right] \left[R_k(\lambda_1 \lambda_2, \lambda_1 \lambda_2) + R_k(\lambda_1 \lambda_2', \lambda_1 \lambda_2') \right], \tag{20.27}$$

$$\left\langle n_1 n_2 l_1 j_1^{N_1} l_2 j_2^{N_2} \alpha_1 J_1 \alpha_2 J_2 J | Q_{12}^e | n_1 n_2 l_1 j_1^{N_1} l_2 j_2^{N_2} \alpha_1' J_1' \alpha_2' J_2' J \right\rangle$$

$$= \sum_k g_k(j_1^{N_1} j_2^{N_2}) \left[1 + P(\lambda_i, \lambda_i') \right] \left[R_k(\lambda_1 \lambda_2, \lambda_2 \lambda_1) + R_k(\lambda_1 \lambda_2', \lambda_2 \lambda_1') \right]. \tag{20.28}$$

Here the radial integrals are defined according to (19.72), whereas their coefficients are equal to

$$f_k(j_1^{N_1} j_2^{N_2}) = (-1)^{j_1+j_2} A_k(j_1^{N_1} j_2^{N_2}), \tag{20.29}$$

$$g_k(j_1^{N_1} j_2^{N_2}) = \frac{(-1)^{k+1}}{2j_2+1} \begin{bmatrix} k & j_1 & j_2 \\ 0 & 1/2 & 1/2 \end{bmatrix}^2 B_k(j_1^{N_1} j_2^{N_2}), \tag{20.30}$$

where

$$A_k(j_1^{N_1} j_2^{N_2}) = (-1)^{j_1+j_2+J_1'+J_2+J} \sqrt{[j_1, j_2]}$$

$$\times \begin{bmatrix} k & j_1 & j_1 \\ 0 & 1/2 & 1/2 \end{bmatrix} \begin{bmatrix} k & j_2 & j_2 \\ 0 & 1/2 & 1/2 \end{bmatrix} \begin{Bmatrix} J_2' & J_2 & k \\ J_1 & J_1' & J \end{Bmatrix}$$

$$\times \left(j_1^{N_1} \alpha_1 J_1 \| T^k \| j_1^{N_1} \alpha_1' J_1' \right) \left(j_2^{N_2} \alpha_2 J_2 \| T^k \| j_2^{N_2} \alpha_2' J_2' \right), \tag{20.31}$$

$$B_k(j_1^{N_1} j_2^{N_2}) = B_k^0(j_1^{N_1} j_2^{N_2}) + B_k^1(j_1^{N_1} j_2^{N_2})$$

$$= (-1)^k \delta(\alpha_1 J_1 \alpha_2 J_2 J, \alpha_1' J_1' \alpha_2' J_2' J') N_1 N_2$$

$$+ (-1)^{j_1+j_2+J_1'+J_2'+J} [j_1, j_2] \sum_{x>0} (-1)^x (2x+1) \begin{Bmatrix} j_1 & j_2 & k \\ j_2 & j_1 & x \end{Bmatrix}$$

$$\times \begin{Bmatrix} J_2' & J_2 & x \\ J_1 & J_1' & J \end{Bmatrix} \left(j_1^{N_1} \alpha_1 J_1 \| T^x \| j_1^{N_1} \alpha_1' J_1' \right)$$

$$\times \left(j_2^{N_2} \alpha_2 J_2 \| T^x \| j_2^{N_2} \alpha_2' J_2' \right). \tag{20.32}$$

It is worth emphasizing that coefficients (20.29) and (20.30) do not depend on orbital quantum numbers. These numbers define only the parity of summation index k, following from the conditions of the non-vanishing of radial integrals in (20.27) and (20.28). For the direct term, parameter k acquires even values, whereas for the exchange part the parity of k equals the parity of sum $l_1 + l_2$.

If one or two subshells are almost filled, then the following equalities are valid:

$$f_k(j_1^{2j_1+1-N_1} j_2^{N_2}) = (-1)^{1+(v_1-v_1')/2} f_k(j_1^{N_1} j_2^{N_2}), \tag{20.33}$$

$$f_k(j_1^{2j_1+1-N_1} j_2^{2j_2+1-N_2}) = (-1)^{(v_1-v_1'+v_2-v_2')/2} f_k(j_1^{N_1} j_2^{N_2}), \tag{20.34}$$

$$B_k^1(j_1^{2j_1+1-N_1} j_2^{N_2}) = (-1)^{1+(v_1-v_1')/2} B_k^1(j_1^{N_1} j_2^{N_2})$$

$$- (-1)^{J_1'+J_2+J+j_1+j_2+(v_1-v_1')/2} 2[j_1, j_2] \sum_{x_o} (2x+1) \begin{Bmatrix} j_1 & j_2 & k \\ j_2 & j_1 & x \end{Bmatrix}$$

$$\times \begin{Bmatrix} J_2' & J_1 & x \\ J_1 & J_1' & J \end{Bmatrix} \left(j_1^{N_1} \alpha_1 J_1 \| T^x \| j_1^{N_1} \alpha_1' J_1' \right)$$

$$\times \left(j_2^{N_2} \alpha_2 J_2 \| T^x \| j_2^{N_2} \alpha_2' J_2' \right), \tag{20.35}$$

$$g_k(j_1^{2j_1+1-N_1} j_2^{2j_2+1-N_2}) = (-1)^{(v_1-v_1'+v_2-v_2')/2} g_k(j_1^{N_1} j_2^{N_2})$$

$$- \delta(\alpha_1 J_1 \alpha_2 J_2 J, \alpha_1' J_1' \alpha_2' J_2' J')(2j_2+1)^{-1} \begin{bmatrix} k & j_1 & j_2 \\ 0 & 1/2 & 1/2 \end{bmatrix}^2$$

$$\times [(2j_1+1)(2j_2+1) - N_2(2j_1+1) - N_1(2j_2+1)]. \tag{20.36}$$

From these formulas and the definitions of f_k (20.29) and g_k (20.30) there follow directly simple expressions for the coefficients considered for one closed subshell, namely,

$$f_k(j_1^{2j_1+1} j_2^{N_2}) = \delta(k, 0)\delta(\alpha_2 J_2 J, \alpha_2' J_2' J')(2j_1+1)N_2, \tag{20.37}$$

$$g_k(j_1^{2j_1+1} j_2^{N_2}) = -\delta(\alpha_2 J_2 J, \alpha_2' J_2' J')N_2 \frac{2j_1+1}{2j_2+1} \begin{bmatrix} k & j_1 & j_2 \\ 0 & 1/2 & 1/2 \end{bmatrix}^2. \tag{20.38}$$

The case of both closed shells directly follows from (20.37) and (20.38) if we assume $N_2 = 2j_2 + 1$.

Let us consider in a similar manner the matrix elements of the operators of magnetic H^m and retarding H^r interactions. For them also, a formula of the sort (20.6), where the necessary matrix elements in the case of one subshell of equivalent electrons are defined by equations (19.78) and (19.84), is valid. Retarding interactions exist only between subshells, while inside them they, according to (19.84), vanish.

The matrix element of the operator of the energy of magnetic interactions H_{12}^m between two subshells may be found in a similar way to the case of electrostatic interaction. The final result is as follows for direct $^d H_{12}^m$ and exchange $^e H_{12}^m$ parts, respectively:

$$
\left\langle n_1 n_2 l_1 j_1^{N_1} l_2 j_2^{N_2} \alpha_1 J_1 \alpha_2 J_2 J \right|^d H_{12}^m \left| n_1 n_2 l_1 j_1^{N_1} l_2 j_2^{N_2} \alpha_1' J_1' \alpha_2' J_2' J \right\rangle
$$
$$
= \sum_{k_{odd}} (-1)^{l_1+l_2} d_k(j_1^{N_1} j_2^{N_2}) R_k(\lambda_1 \lambda_2, \lambda_1' \lambda_2'), \tag{20.39}
$$

$$
\left\langle n_1 n_2 l_1 j_1^{N_1} l_2 j_2^{N_2} \alpha_1 J_1 \alpha_2 J_2 J \right|^e H_{12}^m \left| n_1 n_2 l_1 j_1^{N_1} l_2 j_2^{N_2} \alpha_1' J_1' \alpha_2' J_2' J \right\rangle
$$
$$
= \sum_k g_k(j_1^{N_1} j_2^{N_2}) \left[1 + P(\lambda_i \lambda_i') \right] \left\{ \frac{k+1}{2k+3} R_{k+1}(\lambda_1 \lambda_2, \lambda_2' \lambda_1') \right.
$$
$$
- \frac{k+1}{2k+3} R_{k+1}(\lambda_1' \lambda_2, \lambda_2 \lambda_1') + \frac{k}{2k-1} R_{k-1}(\lambda_1 \lambda_2, \lambda_2' \lambda_1')
$$
$$
- \frac{k}{2k-1} R_{k-1}(\lambda_1' \lambda_2, \lambda_2 \lambda_1') + (-1)^{l_2+j_2-s} \left[(-1)^{j_1+j_2-k}(2j_1+1) \right.
$$
$$
\left. + 2j_2+1 \right] \left[\frac{1}{2k+3} R_{k+1}(\lambda_1' \lambda_2, \lambda_2 \lambda_1') - \frac{1}{2k-1} R_{k-1}(\lambda_1' \lambda_2, \lambda_2 \lambda_1') \right]
$$
$$
- \frac{1}{4k(k+1)} \left[(-1)^{j_1+j_2-k}(2j_1+1) + 2j_2+1 \right]^2
$$
$$
\times \left[\frac{k}{2k+3} R_{k+1}(\lambda_1 \lambda_2, \lambda_2' \lambda_1') + \frac{k}{2k+3} R_{k+1}(\lambda_1' \lambda_2, \lambda_2 \lambda_1') \right.
$$
$$
+ \frac{k+1}{2k-1} R_{k-1}(\lambda_1 \lambda_2, \lambda_2' \lambda_1') + \frac{k+1}{2k-1} R_{k-1}(\lambda_1' \lambda_2, \lambda_2 \lambda_1')
$$
$$
\left. \left. + R_k(\lambda_1 \lambda_2, \lambda_2' \lambda_1') + R_k(\lambda_1' \lambda_2, \lambda_2 \lambda_1') \right] \right\}. \tag{20.40}
$$

The coefficient d_k in Eq. (20.39) is equal to

$$
d_k(j_1^{N_1} j_2^{N_2}) = \frac{4(2j_1+1)(2j_2+1)}{k(k+1)} A_k(j_1^{N_1} j_2^{N_2}), \tag{20.41}
$$

where A_k is defined by (20.31) and $g_k(j_1^{N_1} j_2^{N_2})$ by (20.30). Thus, both the radial integrals and their coefficients in the matrix elements of the energy operators of electrostatic and magnetic interactions qualitatively have one and the same form. This considerably simplifies their calculation. Notice that for $k = 1$, coefficient (20.41) equals

$$d_1(j_1^{N_1} j_2^{N_2}) = (-1)^{j_1+j_2+1} \delta(\alpha_1 J_1 \alpha_2 J_2 J, \alpha_1' J_1' \alpha_2' J_2' J')(2j_1 + 1)(2j_2 + 1)$$
$$\times [J_1(J_1 + 1) + J_2(J_2 + 1) - J(J + 1)]/[4j_1(j_1 + 1)j_2(j_2 + 1)]. \quad (20.42)$$

Relationships between coefficients g_k for different degrees of occupation of the subshells (the cases of almost and completely filled subshells) are described by equalities (20.35), (20.36) and (20.38). Therefore, for magnetic interactions one has additionally to consider such conditions only in the case of coefficient d_k. Bearing in mind that k acquires only odd values and that the submatrix elements of operator T^k are diagonal with respect to seniority quantum number v, we find

$$d_k(j_1^{N_1} j_2^{N_2}) = d_k(j_1^{N_1+2n} j_2^{N_2+2m}), \quad (20.43)$$

where n and m are integer numbers; the quantum numbers in the left and right sides of equality (20.43), of course, must be the same. The direct part of the matrix element of the energy operator of magnetic interactions is diagonal with respect to seniority quantum numbers v_1 and v_2 of the separate subshells

$$d_k(j_1^{N_1} j_2^{N_2} \alpha_1 v_1 J_1 \alpha_2 v_2 J_2, \alpha_1' v_1' J_1' \alpha_2' v_2' J_2', J) = \delta(v_1 v_2, v_1' v_2')$$
$$\times d_k(j_1^{N_1} j_2^{N_2} \alpha_1 v_1 J_1 \alpha_2 v_2 J_2, \alpha_1' v_1 J_1' \alpha_2' v_2 J_2', J). \quad (20.44)$$

Coefficient d_k is equal to zero, when one or both subshells are closed, due to the condition that k is odd, i.e.

$$d_k(j_1^{2j_1+1} j_2^{N_2}) = d_k(j_1^{2j_1+1} j_2^{2j_2+1}) = 0. \quad (20.45)$$

The relationships for complementary shells follow from (20.43) if $N_1 + 2n = 2j_1 + 1 - N_1$ and $N_2 + 2m = 2j_2 + 1 - N_2$.

The direct part of the matrix element of the energy operator of the retarding interactions vanishes

$$\left\langle n_1 n_2 l_1 j_1^{N_1} l_2 j_2^{N_2} \alpha_1 J_1 \alpha_2 J_2 J \right|^d H_{12}^r | n_1 n_2 l_1 j_1^{N_1} l_2 j_2^{N_2} \alpha_1' J_1' \alpha_2' J_2' J \right\rangle = 0. \quad (20.46)$$

We do not present here separately the corresponding expression for the exchange part, because while calculating the energy spectrum we have to sum the contributions of all interactions; matrix elements of magnetic interactions have many common terms, and their sum may be considerably simplified. Thus, the direct part of the matrix element of the magnetic and retarding interactions between two subshells, due to (20.46), simply equals (20.39), whereas the exchange part is given by

$$\left\langle n_1 n_2 l_1 j_1^{N_1} l_2 j_2^{N_2} \alpha_1 J_1 \alpha_2 J_2 J \right| (^e H_{12}^m + {}^e H_{12}^r) | n_1 n_2 l_1 j_1^{N_1} l_2 j_2^{N_2} \alpha_1' J_1' \alpha_2' J_2' J \right\rangle$$
$$= -\sum_k g_k(j_1^{N_1} j_2^{N_2}) [1 + P(\lambda_i \lambda_i')] \left\{ \frac{k(k+1)}{2} \left[\frac{1}{2k+3} R_{k+1}(\lambda_1 \lambda_2, \lambda_2' \lambda_1') \right. \right.$$

$$-\frac{1}{2k+3}R_{k+1}(\lambda'_1\lambda_2,\lambda_2\lambda'_1)-\frac{1}{2k-1}R_{k-1}(\lambda_1\lambda_2,\lambda'_2\lambda'_1)$$

$$+\frac{1}{2k-1}R_{k-1}(\lambda'_1\lambda_2,\lambda_2\lambda'_1)\Big]+(-1)^{l_2+j_2+s}\frac{1}{4}\Big[(-1)^{j_1+j_2-k}(2j_1+1)$$

$$+2j_2+1\Big]\Big[\frac{3}{2k+3}R_{k+1}(\lambda'_1\lambda_2,\lambda_2\lambda'_1)-\frac{3}{2k-1}R_{k-1}(\lambda'_1\lambda_2,\lambda_2\lambda'_1)$$

$$+R^1_{k+1}(\lambda_1\lambda_2,\lambda'_2\lambda'_1)-R^2_{k+1}(\lambda_1\lambda_2,\lambda'_2\lambda'_1)-R^1_{k-1}(\lambda_1\lambda_2,\lambda'_2\lambda'_1)$$

$$+R^2_{k-1}(\lambda_1\lambda_2,\lambda'_2\lambda'_1)\Big]+\frac{1}{8k(k+1)}\Big[(-1)^{j_1+j_2-k}(2j_1+1)+2j_2+1\Big]^2$$

$$\times\Big[\frac{k(k+3)}{2k+3}\left(R_{k+1}(\lambda_1\lambda_2,\lambda'_2\lambda'_1)+R_{k+1}(\lambda'_1\lambda_2,\lambda_2\lambda'_1)\right)$$

$$+2\left(R_k(\lambda_1\lambda_2,\lambda'_2\lambda'_1)+R_k(\lambda'_1\lambda_2,\lambda_2\lambda'_1)\right)-\frac{(k-2)(k+1)}{2k-1}$$

$$\times\left(R_{k-1}(\lambda_1\lambda_2,\lambda'_2\lambda'_1)+R_{k-1}(\lambda'_1\lambda_2,\lambda_2\lambda'_1)\right)\Big]\Big]\Big\}. \tag{20.47}$$

It follows from (20.47) that the coefficients of the radial integrals of the exchange part of the retarding interactions may also be expressed in terms of quantities g_k, i.e. the exchange parts of the matrix elements of all two-electron operators $H^4 = Q$, H^m and H^r are defined by the same coefficient $g_k(j_1^{N_1}j_2^{N_2})$. In the case of the given l_1 and l_2, the parity of k for H^4 and H^r is the same and is equal to the parity of sum $l_1 + l_2$, whereas for H^m, k may be of any parity. The values of k for radial integrals in the exchange parts of the matrix elements of operators H^4 and H^r will be of the opposite parity. Notice that the following new kinds of integrals are used in (20.47):

$$R^1_k(\lambda_1\lambda_2,\lambda_3\lambda_4)=(l_1l_3k)(l_2l_4k)$$

$$\times\int\limits_0^\infty r_2^2dr_2\int\limits_0^{r_2}f(\lambda_1|r_1)f(\lambda_2|r_2)\frac{r_1^k}{r_2^{k+1}}f(\lambda_3|r_1)f(\lambda_4|r_2)r_1^2dr_1, \tag{20.48}$$

$$R^2_k(\lambda_1\lambda_2,\lambda_3\lambda_4)=(l_1l_3k)(l_2l_4k)$$

$$\times\int\limits_0^\infty r_2^2dr_2\int\limits_{r_2}^\infty f(\lambda_1|r_1)f(\lambda_2|r_2)\frac{r_2^k}{r_1^{k+1}}f(\lambda_3|r_1)f(\lambda_4|r_2)r_1^2dr_1. \tag{20.49}$$

Also,

$$R_k(\lambda_1\lambda_2,\lambda_3\lambda_4)=R^1_k(\lambda_1\lambda_2,\lambda_3\lambda_4)+R^2_k(\lambda_1\lambda_2,\lambda_3\lambda_4). \tag{20.50}$$

It is necessary to underline that the dependence of the matrix elements of the relativistic energy operator on orbital quantum numbers is contained only in phase multipliers and radial integrals. There are only a few types of radial integrals and their coefficients, which considerably simplifies their calculation. Radial integrals have no derivatives of radial orbitals, whereas the spin–angular parts consist only of the Clebsch–Gordan coefficients (which play here the same role as the quantities $(l||C^{(k)}||l')$ in the non-relativistic approach) and the $6j$-coefficients and submatrix elements of irreducible tensorial operators T^k.

While calculating the energy spectrum in relativistic approximation, the non-diagonal with respect to the configuration's matrix elements must also be taken into consideration [62]. The necessary expressions for four open subshells, corresponding to two non-relativistic shells of equivalent electrons, are presented in [126].

As was mentioned in Chapter 2, there exists another method of constructing the theory of many-electron systems in jj coupling, alternative to the one discussed above. It is based on the exploitation of non-relativistic or relativistic wave functions, expressed in terms of generalized spherical functions [28] (see Eqs. (2.15) and (2.18)). Spin–angular parts of all operators may also be expressed in terms of these functions (2.19). The dependence of the spin–angular part of the wave function (2.18) on orbital quantum number is contained only in the form of a phase multiplier, therefore this method allows us to obtain directly optimal expressions for the matrix elements of any operator. The coefficients of their radial integrals will not depend, except phase multipliers, on these quantum numbers. This is the case for both relativistic and non-relativistic approaches in jj coupling.

Let us discuss in conclusion the use of approach (2.14) for the calculation of energy spectra. Utilizing the following expansion of $\frac{1}{r_{ij}}\cos\omega_{AC}r_{ij}$ (compare with the real part of the exponent in (2.14))

$$\frac{\cos\omega_{AC}r_{ij}}{r_{ij}} = -\sum_k(2k+1)\omega_{AC}j_k(\omega_{AC}r_<)n_k(\omega_{AC}r_>)\left(C_i^{(k)}\cdot C_j^{(k)}\right), \quad (20.51)$$

where j_k and n_k are the spherical Bessel functions of the first and the second kinds, respectively, we can learn that the direct part of the matrix element of the quantity (2.14), due to the vanishing of the diagonal, with respect to configurations, matrix elements of the operator of the retarding interactions, coincides with the corresponding expressions for operator $H^4 + H^m$ (2.5), (2.10). The exchange terms follow from the corresponding expressions for $H^4 + H^m$, if we substitute in them instead of $r_<^k/r_>^{k+1}$, the coefficient of $(C_i^{(k)}\cdot C_j^{(k)})$ from (20.51). Thus, the utilization of approach (2.14) leads only to changes in the coefficients of the radial integrals occurring in the exchange terms of the matrix elements describing the electrostatic and magnetic interactions. All the remaining expressions do not require any alterations.

The possibility of performing the calculations in LS coupling preserving the main part of the relativistic effects will be discussed in Chapter 21.

21

Semi-empirical methods of calculation of the energy spectra

21.1 Least squares fitting

The methods of theoretical study of the energy spectra of atoms and ions, described in the previous chapters, do not always ensure the required accuracy. Therefore, in a number of cases, when we are not interested in the accuracy of the theory and when part of the data is available (e.g. experimental measurements of the energy levels, which are usually known to high precision), then we can utilize the latter for the so-called semi-empirical evaluation of wave functions and other spectroscopic characteristics. The most widespread is the semi-empirical method of least squares fitting. As a rule, it is used together with diagonalization of the total energy matrix, built in a certain coupling scheme, to find the whole energy spectrum and corresponding eigenfunctions of an atom or ion. The eigenfunctions are used further on to calculate the oscillator strengths, transition probabilities and other spectral characteristics. Let us sketch its main ideas.

As we have seen while considering the matrix elements of various operators, any expression obtained consists of radial integrals and coefficients. These coefficients can be calculated using the techniques of irreducible tensorial sets and CFP, whereas the values of radial integrals are found starting with analytical or numerical radial wave functions. Uncertainties of these quantities calculated in a given approximation (it is well known, for example, that the Hartree–Fock values of radial integrals of electrostatic interactions in some cases exceed the exact ones by 1.5 times) are the main reason for discrepancies between calculations and experimental measurements of spectral characteristics. In a semi-empirical approach we consider these integrals as unknown parameters which can be determined from experimental data by extrapolation or interpolation. The condition that they reproduce, with the accuracy required, the experimental data will allow us to find unknown levels of the configuration studied. The

248

theoretical values of the radial integrals may serve as the initial quantities of the parameters, found further on by an iterative procedure.

The Slater integrals of the energy of electrostatic interaction and the spin–orbit interaction constant are the main parameters searched. It is obvious that the number of known levels must be equal to or exceed the number of unknown parameters.

The simplest way to account for correlation and relativistic effects is by use of the so-called 'linear' theory of superposition-of-configurations [127] in which, instead of diagonalization of the energy matrix, additional parameters are introduced [127]. 'Non-linear' theory [128], which includes the diagonalization of the energy matrix as well as the correlation effects and spin–orbit interactions between the states of a given set of configurations, is more exact. In this approach, by introducing new parameters, one is in a position to account for the superposition of other important configurations, including the contribution of the continuous spectrum as well as the remaining relativistic effects [129].

However, the most efficient and general is the least squares fitting. Let us briefly describe it, following [130]. Let us assume that we have an atom or ion in the configuration having a discrete energy spectrum. A part of these levels $(E_1, E_2, \ldots, E_n)$ is known from experimental measurements, although with certain errors; and we are looking for the rest of the levels. Let us present energy values $\mathscr{E}_i$ of each level in the spectra as functions of a set of m parameters R_k:

$$\mathscr{E}_i = \mathscr{E}_i(R_1 R_2, \ldots, R_m), \qquad m < n. \tag{21.1}$$

The number m of these parameters must be smaller than the number of known levels n, i.e. $m < n$, otherwise the problem is unsolvable. If a set of parameters R_k exactly describes levels $\mathscr{E}_i$, and levels E_i are measured accurately and identified correctly, then for finding the true values of these parameters, it would be enough to solve a system of m equations with m unknowns. In practice, due to non-observance of the above-mentioned conditions, the obtained roots of the system of m equations will not satisfy the remaining $n - m$ equations. However, our purpose is to find values of parameters R_k, for which all equations are satisfied with the smallest error. Then the root-mean-square deviation of energy levels $\mathscr{E}_i$, found from those E_i measured, would be minimal.

First we have to choose the initial values of semi-empirical parameters R_k^0. Theoretical values of radial integrals of the interactions under consideration may serve for this purpose, or R_k^0 may be found from the above-mentioned system of equations. After that, let us change the parameters by a small quantity ΔR_k, i.e. let us define

$$R_k = R_k^0 + \Delta R_k. \tag{21.2}$$

Let us also assume that function $\mathscr{E}_i$ (21.1) satisfies the necessary requirements to expand it in a Taylor series in the vicinity of the point $(R_1^0, R_2^0, \ldots, R_m^0)$ retaining only terms of first order in powers of ΔR_k:

$$\mathscr{E}_i = \mathscr{E}_i(R_1^0, R_2^0, \ldots, R_m^0) + \sum_{k=1}^{m} \Delta R_k \frac{\partial \mathscr{E}_i(R_1^0, R_2^0, \ldots, R_m^0)}{\partial R_k}. \tag{21.3}$$

Then, in order to find the corrections we are looking for, we obtain the system of n linear equations with $m < n$ unknowns:

$$\left.\begin{array}{ccccccc}
\mathscr{E}_{1,1}^0 \Delta R_1 & + & \mathscr{E}_{1,2}^0 \Delta R_2 & + & \cdots & + & \mathscr{E}_{1,m}^0 \Delta R_m & = & \delta_1^0, \\
\mathscr{E}_{2,1}^0 \Delta R_1 & + & \mathscr{E}_{2,2}^0 \Delta R_2 & + & \cdots & + & \mathscr{E}_{2,m}^0 \Delta R_m & = & \delta_2^0, \\
\cdots\cdots & \cdots & \cdots\cdots & \cdots & \cdots & \cdots & \cdots\cdots & \cdots & \cdots \\
\mathscr{E}_{n,1}^0 \Delta R_1 & + & \mathscr{E}_{n,2}^0 \Delta R_2 & + & \cdots & + & \mathscr{E}_{n,m}^0 \Delta R_m & = & \delta_n^0,
\end{array}\right\} \tag{21.4}$$

in which the following notation is utilized:

$$\mathscr{E}_{i,k}^0 = \frac{\partial \mathscr{E}_i(R_1^0, R_2^0, \ldots, R_m^0)}{\partial R_k}, \qquad i = 1, 2, \ldots, n; \ k = 1, 2, \ldots, m, \tag{21.5}$$

$$\delta_i^0 = E_i - \mathscr{E}_i(R_1^0, R_2^0, \ldots, R_m^0). \tag{21.6}$$

The system of equations (21.4) may be solved by the least squares fitting method, i.e. by looking for such values of unknowns, for which sum S of the differences squared between the left and right parts of each equation (21.4) are minimal:

$$S = \sum_{i=1}^{n} \left(\delta_i^0 - \sum_{k=1}^{m} \mathscr{E}_{i,k}^0 \Delta R_k \right)^2. \tag{21.7}$$

Root-mean-square deviation σ, defined according to the formula

$$\sigma = \sqrt{\sum_i (E_i - \mathscr{E}_i)^2 / (n - m)^2}, \tag{21.8}$$

is usually adopted as the criterion of the correspondence of the calculated values of the energy levels to the known experimental data.

If there exists a minimum of function S, then corrections ΔR_k satisfy the system of m equations with the same number of unknowns:

$$\frac{\partial S}{\partial \Delta R_1} = 0, \quad \frac{\partial S}{\partial \Delta R_2} = 0, \quad \ldots, \quad \frac{\partial S}{\partial \Delta R_m} = 0. \tag{21.9}$$

Solving this system, we find new more accurate values of the parameters R_k^1. Using them further on as the initial ones, we repeat the procedure until the following interactions practically do not alter S and, thus, σ. In this way all possibilities of the refinement of the energy spectrum will be exhausted in the framework of a given number of parameters R_k and the form of function (21.1).

Thus, the least squares method includes the choice of the energy levels measured and initial values of semi-empirical parameters, forming the energy matrix, its diagonalization and finding the eigenfunctions and eigenvalues. Then evaluation of the root-mean-square deviation of calculated energy values from experimental ones follows and, if this deviation is large, the determination of the improved values of these semi-empirical parameters. This procedure is repeated until the given accuracy is achieved or further iterations will not improve the result.

If the orders of the energy matrix are large, and the values of non-diagonal matrix elements are comparable to the differences between the corresponding diagonal quantities, then after diagonalization of such matrices it is very difficult to attribute unambiguously sets of certain quantum numbers to the energy levels obtained, i.e. to identify them correctly. As is known, the order of arrangement of the eigenvalues of the energy operator do not necessarily coincide with the given position of the diagonal matrix elements in the matrix before diagonalization. The least squares fitting implies that each eigenvalue in all cases corresponds to one and the same experimental energy level, otherwise the root-mean-square deviation σ increases drastically. Therefore, one has to control the correctness of the classification of the quantities mentioned, particularly while observing an unexpected increase of σ.

Energy levels measured with low accuracy or identified incorrectly lead to slow convergence or even divergence of the iterative procedure. Such levels may be detected by the successive exclusion of separate experimentally measured levels from calculations and by analysis of the σ values obtained.

21.2 Accounting for relativistic and correlation effects as corrections

In a single-configurational non-relativistic approach, the integrals of electrostatic interactions and the constant of spin–orbit interactions compose the minimal set of semi-empirical parameters. Then for p^N and d^N shells we have two and three parameters, respectively. However, calculations show that such numbers of parameters are insufficient to achieve high accuracy of the theoretical energy levels. Therefore, we have to look for extra parameters, which would be in charge of the relativistic and correlation effects not yet described.

One of the easiest ways to improve the results is through the replacement of the matrix element of the energy operator of electrostatic interaction by some effective interaction, in which, together with the usual expression of the type (19.29), there are also terms containing odd k values. This means that we adopt some effective Hamiltonian, whose matrix elements of the operator of electrostatic interaction there has the integrals $F^k(nl, nl)$ (k is even) and parameters ϕ^k (k is odd). Further on we shall learn that the inclusion of such terms in the case of configurations l^N accounts for the main part of relativistic and correlation effects.

For p^N shells the effective Hamiltonian H_{eff} contains two parameters F_2 and ϕ_1, as well as the constant of spin–orbit interaction. The term with $k = 0$ causes a general shift of all levels, which is usually taken from experimental data in semi-empirical calculations, and can therefore be neglected. The coefficient at ϕ^1 is proportional to $L(L + 1)$. Therefore, to find the matrix elements of the effective Hamiltonian it is enough to add the term $\alpha L(L + 1)$ to the matrix elements of the energy of electrostatic and spin–orbit interactions. Here α stands for the extra semi-empirical parameter.

For the d^N shell we have two additional parameters (F^4 and ϕ^3) compared to the p^N shell. The quantity with ϕ^3 is expressed in terms of corrections $\alpha L(L + 1)$ and $\beta Q(N, v)$, where $Q(N, v)$ is defined by (9.17) and β is one more additional parameter we are searching for.

For the f^N shell we have to take into consideration terms containing expressions of the kind (5.36) and (5.37). As was shown in [127, 129], parameters α and β account for the superposition of all configurations which differ from ground l^N by two electrons. If the admixed configurations differ from the principal one by the excitation of one electron, then we have to introduce one extra parameter T, the coefficient of which will be described by the matrix element of the tensorial operator of the type $[U^k \times U^{k'} \times U^{k''}]^0$, for which, unfortunately, there is no known simple algebraic expression. This correction is important only for $N \geq 3$.

Thus, introducing parameters α, β and T we can account for the essential part of the correlation effects. However, it turned out that in the framework of the semi-empirical approach, all relativistic corrections of the second order of the Breit operator improving the relative positions of the terms, are also taken into consideration (operators H_2, H_3'' and H_5', described by formulas (1.19), (1.20) and (1.22), respectively). Indeed, as we have seen in Chapter 19, the effect of accounting for corrections H_3'' and H_5' in a general case may be taken into consideration by modifications of the integrals of electrostatic interaction, i.e. by representing them in form (19.50). It follows from the algebraic expressions for the matrix elements (19.51), (19.52) and (19.53) of the operator of orbit–orbit interaction that this interaction is also accounted for by these parameters. Sometimes a semi-empirical correction proportional to $S(S + 1)$ is introduced. However, as it follows from the known algebraic expressions for the matrix elements of the energy operator, it is already absorbed by them, therefore, its use is inefficient.

Thus, the least squares fitting procedure can be successfully applied to a large variety of ionized atoms, including fairly highly ionized ones.

For configurations of the type $l^N(\alpha_1 L_1 S_1)lLS$, along with semi-empirical calculations of the energy of the intrashell interactions considered above, we have to include parameters corresponding to integrals $F^k(l, l')$ and $G^k(l, l')$, as well as the term $AL(L + 1)$, where L represents the total orbital momentum of the whole configuration $l^N l'$. In a similar fashion more complex configurations may be treated. Along with the increase of the complexity of the configurations the number of necessary parameters increases rapidly. The expressions for the spin–angular coefficients become more and more complicated, and the non-diagonal matrix elements, with respect to the configurational quantum numbers, must also be accounted for.

Usually, for both theoretical and semi-empirical determination of energy spectra, radial integrals that do not depend on term energy of the configuration are used. More exact values of the energy levels are obtained while utilizing the radial wave functions, which depend on term. Therefore, there have been attempts to account for this dependence in semi-empirical calculations. Usually the Slater parameters are multiplied by the energy dependent coefficient

$$k = 1 - \eta(E - E_{\min})/(E_{\max} - E_{\min}), \qquad (21.10)$$

where η is the parameter varied, and E, $E_{\min}$, $E_{\max}$ are the energies of the searched, lowest and highest terms, respectively. However, such a method can be successfully applied only to comparatively simple configurations not containing the non-diagonal matrix elements of the operator of electrostatic energy (e.g. $3d^2 4s$).

The data of Table 21.1 illustrate the utilization of the least squares fitting method on the example of doubly ionized vanadium in configuration $3d^3$. The letters a, b, c and d denote four variants of the calculations, differing by the number of semi-empirical parameters used. In the first column E_0 is that part of the energy not depending on term (common shift of all levels), $F^2(dd)$, $F^4(dd)$, $\eta(d)$, α, β, T stand for the parameters discussed above, n denotes the number of the levels measured experimentally (here it coincides with the total number of levels in the spectrum of configuration $3d^3$), m indicates the number of free parameters, σ is root-mean-square deviation (21.8), and $\mu = \max|E_i - \mathscr{E}_i|$ describes the maximal deviation of the calculated energy value from the experimental one for the term indicated. Beside the first four parameters are their ratios (in brackets) with the Hartree–Fock integrals. They show that the Hartree–Fock radial integrals overestimate their real values.

It follows from Table 21.1 that successive accounting for parameters α, β and T quickly diminishes the root-mean-square deviation σ and

Table 21.1. Numerical values of various sets of semi-empirical parameters (in cm^{-1}) for VIII $3d^3$

	a	b	c	d
E_0	23763 (0.15)	23031 (0.15)	23380 (0.15)	24487 (0.16)
$F^2(dd)$	57321 (0.79)	57917 (0.80)	58744 (0.81)	59856 (0.81)
$F^4(dd)$	35528 (0.79)	34282 (0.76)	35235 (0.78)	35991 (0.80)
$\eta(d)$	151 (0.83)	125 (0.69)	124 (0.68)	166 (0.92)
α		58.6	51.8	27.1
β			-263	-459
T				-4.24
n	19	19	19	19
m	4	5	6	7
σ	558	239	173	28
μ	1125(^2H)	503(^2P)	295(^2P)	48(^2F)

maximal deviation μ (approximately 20 times for case d compared to a). Initially the largest deviations are observed for terms ^2H and ^2P, because these terms have the same energy if we account only for electrostatic interaction. Taking into consideration the spin–orbit interaction as well as the relativistic and correlation effects through parameters α, β and, especially, T, we are able to find their energy values fairly accurately.

However, the iterative process and finding the 'physical' minimum of expression (21.7) has some pecularities, the most clearly pronounced being the increase of the number of the parameters to be optimized. The problem becomes a multiextremal one. For finding the true minimum one has to adopt more and more accurate initial values of the parameters, otherwise the optimization procedure leads to incorrect values of the parameters describing the main physical interactions.

It is very important for semi-empirical calculations to have experimental data measured accurately and identified correctly. Otherwise, it is impossible to achieve good correspondence between calculated and measured quantities. To overcome these dangers one has to analyse the structure of the eigenfunctions, energy levels and oscillator strengths of electronic transitions along the isoelectronic sequences. As a rule, all these quantities vary fairly smoothly and, therefore, the occurrence of any inconsistency in their behaviour usually indicates that there are uncertainities or errors in the data utilized.

Having accurate semi-empirical values of the energy levels and eigenfunctions one is in a position to calculate wavelengths and oscillator strengths of allowed and forbidden electronic transitions in the intermediate coupling scheme. These data are used further to identify the experimentally measured wavelengths and to find numerical values of all energy levels of the atom or ion studied. Good correspondence of an experimentally measured level to its calculated value, as well as the obeying of the Ritz combination principle, serve as the main criteria for the correct identification of the experimentally measured levels. If no one coupling scheme is fairly accurate, then one has to use the wave functions of the intermediate coupling, starting with the closest to reality coupling scheme.

There are cases when correlation effects are rather large and cannot be accounted for as corrections. Excited configurations $1s^2 2s^2 2p^{N-1} nl$ ($n > 3$) and $1s^2 2s 2p^N n'l'$ may serve as such examples. Their averaged energies for certain Z values in an isoelectronic sequence become very close to each other or even intersect. In the vicinity of the intersection region the single-configurational approach is completely unfit. Then the least squares fitting must be combined with the superposition-of-configurations method, i.e. we have to form the energy matrix, accounting for the matrix elements of both configurations, including the non-diagonal ones with respect to them, and further on, minimize all parameters of both configurations (generalized method of least squares fitting [131]).

If relativistic effects are large and cannot be taken into consideration as corrections with respect to non-relativistic wave functions, then one has to exploit the relativistic version of the least squares fitting, starting with relativistic Hamiltonian (2.1) and relativistic wave functions. Although there is a great demand for such studies due to the rapid growth in investigations of the spectra of highly ionized atoms in astrophysical and laboratory high-temperature plasma [132, 133], the relativistic version is not sufficiently well developed so far.

The so-called method of orthogonal operators, mentioned in the Introduction, looks fairly promising in semi-empirical calculations [20,34,134]. Its main advantage is that the addition of new parameters practically does not change the former ones. As a rule this approach allows one to reduce the root-mean-deviation of calculated values from the measured ones by an order of magnitude or so in comparison with the conventional semi-empirical method [135]. Unfortunately it requires the calculation of complex matrix elements of many-electron operators.

Another method worthy of attention is the utilization of algebraic diagonalization of the low order energy matrices in semi-empirical calculations [136].

The semi-empirical methods described above cannot be utilized in cases when only a small number of energy levels are measured or part of them are identified incorrectly. Then the standard way to find the interaction parameters by the least squares fitting breaks down, because the parameters varied become unstable or acquire completely incorrect values, and the interaction procedure does not converge. In such cases the methods of interpolation or extrapolation can be used to determine acceptable values of the semi-empirical parameters, starting with known values for the neighbouring atoms or ions.

The simplest means of interpolation consists in the determination of the arithmetic average value of the quantities studied if these are known for two neighbouring ions from the left and right of the ion considered in the isoelectronic sequence. If the spectra of a number of ions in the isoelectronic sequence are known, and we intend to advance further, then we have to adopt the extrapolation principle: to find graphically the law of the dependence of the parameter considered on nuclear charge Z and to extend it according to the same law to the region of unknown Z values. This supplies us with a fairly accurate value of the parameter we are seeking. However, for this purpose we have to know the character of the dependence of the parameters describing the energy spectra on nuclear charge Z. These dependences may be found both theoretically and from experimental data.

21.3   Dependence of the energy operator matrix elements on nuclear charge Z

Dependence of the main terms of the Hamiltonian of a many-electron atom on nuclear charge Z may be easily revealed in the following fashion. Let us consider the Hamiltonian of the kind (1.15)

$$H = \sum_i \left(\frac{\mathbf{p}_i'^2}{2m} - \frac{Ze^2}{r_i} \right) + \sum_{i>j} \frac{e^2}{r_{ij}}. \tag{21.11}$$

Let us introduce new variable $\rho = Zr_i$ then $\mathbf{p}_i = Z^{-1}\mathbf{p}_i'$ and equation (21.11) becomes

$$H = Z^2 \sum_i \left(\frac{\mathbf{p}_i^2}{2m} - \frac{e^2}{\rho_i} \right) + Z \sum_{i>j} \frac{e^2}{\rho_{ij}} = Z^2(H_0 + Z^{-1}V), \tag{21.12}$$

whereas eigenvalue equation $H\psi = E\psi$ in this case will be

$$(H_0 + Z^{-1}V)\psi = (Z^{-2}E)\psi. \tag{21.13}$$

For large Z, term $Z^{-1}V$ will be small compared to H_0 and therefore we can use perturbation theory. If we expand the wave function and the energy in powers of $1/Z$, i.e.

$$\psi = \psi_0 + Z^{-1}\psi_1 + Z^{-2}\psi_2 + Z^{-3}\psi_3 + \ldots, \tag{21.14}$$

$$E = Z^2(E_0 + Z^{-1}E_1 + Z^{-2}E_2 + Z^{-3}E_3 + \ldots), \tag{21.15}$$

substitute these series into (21.13) and compare the coefficients at the same Z powers, then for the zero Z order we obtain the eigenvalue equation,

$$H_0\psi_0 = E_0\psi_0, \tag{21.16}$$

indicating that the zero-order wave functions are the hydrogenic ones. Adopting them we can easily obtain expansions of the other terms of the many-electron Hamiltonian in inverse powers of nuclear charge Z. Usually only three terms in the expansion (21.15) are retained. Following this procedure we obtain for the energies of the electrostatic, spin–orbit and spin–spin interactions, respectively:

$$E_{el} = Z(E_0 + Z^{-1}E_1 + Z^{-2}E_3 + \dots), \tag{21.17}$$

$$E_{so} = Z^4(E_0' + Z^{-1}E_1' + Z^{-2}E_3' + \dots), \tag{21.18}$$

$$E_{ss} = Z^3(E_0'' + Z^{-1}E_1'' + Z^{-2}E_3'' + \dots). \tag{21.19}$$

Such expansions are very useful while evaluating the relative contributions of separate terms of the Hamiltonian to the total energy. They are of particular importance for the estimation of the role of correlation and relativistic effects. Equation (21.17) shows that the correlation effects are proportional to the first, whereas the main relativistic effects (21.18) are proportional to the fourth power of the nuclear charge. Therefore, with the increase of Z the relativistic effects start to predominate very rapidly.

Calculations show that such a method is fairly efficient. Two orders of perturbation theory in powers of $1/Z$ give fairly accurate (as a rule, more exact than the single-configurational Hartree–Fock method) values of the energy levels, wavelengths and the characteristics of electronic transitions. Another advantage consists in the possibility of obtaining such quantities for the whole isoelectronic sequence from calculations. Its principal shortcoming is the fact that, due to computational difficulties and to the worsening of the convergence of the series with the increase of the number of electrons in the configuration, practically, it is applicable only to systems having up to 10 electrons (configurations $1s^{N_1}2s^{N_2}2p^{N_3}$). Moreover, this method usually fails for neutral and a-few-times-ionized atoms.

In this respect, the single-configurational Hartree–Fock method looks more promising and universal when combined with accounting for the relativistic effects in the framework of the Breit operator and for correlation effects by the superposition-of-configurations or by some other method (e.g. by solving the multi-configurational Hartree–Fock–Jucys equations (29.8), (29.9)).

The idea of expansion of physical quantities in powers of Z may also be applied to the semi-empirical parameters of the least squares fitting method as well as to the approximation of their theoretical, e.g. Hartree–

Fock, values. Each semi-empirical or theoretical parameter $R(Z)$ may be presented as [137]

$$R_p(Z) = a(Z - s)^q, \tag{21.20}$$

in which three parameters a, s, q must be found by the method of least squares fitting, minimizing the expression

$$\sum_Z [a(Z - s)^q / R(Z) - 1]^2. \tag{21.21}$$

Here the sum is over all Z values for which there is a known $R(Z)$ value. As expected, the values of the parameter q, obtained in this way, as a rule are very close to the corresponding quantities, following from the hydrogenic approximation: $q = 1$ for the integrals of the electrostatic interaction, 2 for average energies, 4 for spin–orbit parameter and $q = -k$ for mean values of r^k.

More accurate is the approximation

$$R_p(Z) = \sum_k a_k(Z - s)^k, \qquad k_{min} \leq k \leq k_{max}, \tag{21.22}$$

in which the values of s must be taken from (21.20) and a_k is found by least squares fitting, minimizing the expression

$$\sum_Z \left[[1/R(Z)] \sum_k a_k(Z - s)^k - 1 \right]. \tag{21.23}$$

The summation limits over k are found from the condition that the largest relative uncertainty of approximation does not exceed the given quantity, and the number of terms in expansion (21.23) are minimal.

Starting with the method described above, extensive tables of the numerical values of mean energies, integrals of electrostatic and constant of spin–orbit interactions are presented in [137] for the ground and a large number of excited configurations, for atoms of boron up to nobelium and their positive ions. They are obtained by approximation of the corresponding Hartree–Fock values by polynomials (21.20) and (21.22). Such data can be directly utilized for the calculation of spectral characteristics of the above-mentioned elements or they can serve as the initial parameters for semi-empirical calculations [138].

21.4   Semi-empirical method of model potential

In conclusion of this chapter let us discuss briefly the use of model (effective) potentials. They represent one more variety of simplified theoretical or semi-empirical calculations. Let us notice at once that these methods have a rather narrow domain of applicability. They are most frequently adopted to describe atoms having one electron above closed shells. Alkali atoms are the typical example. If we are interested only in the wave functions of the outer electron, then we can find them by solving a one-electron equation describing the motion of the electron in some effective field, created by nuclear charge and the charges of the electrons of closed shells. For this purpose we need the effective potential of this field. Usually a simple analytical expression for this potential is chosen, which is described by a finite number of free parameters and satisfies the condition of spherical symmetry. These parameters may be found from experimental data or from the more accurate theoretical calculations. This method may be exploited for both non-relativistic and relativistic wave functions. One of the simplest expressions for the potential,

$$V(r) = -\frac{Z}{r} + \frac{a(a+1)}{r^2} \qquad (21.24)$$

contains only one parameter a, found, for example, from the energy levels measured. Various versions of such semi-empirical calculations differ on the whole by the manner of use of experimental data and the character of the wave functions obtained.

One group is formed by the methods in which one and the same potential is adopted to calculate all electronic states whereas the energy parameter in the equations is made equal to the corresponding experimental quantity. Unfortunately, the wave functions obtained in this way are not eigenfunctions of the initial one-particle Hamiltonian. Therefore, they do not satisfy simultaneously the true boundary condition for $r = 0$ and $r = \infty$. Usually the solution having the true asymptote for $r = \infty$, is chosen. The methods, utilizing the Coulomb potential in the one-electron equation, as well as the quantum defect approach, belong to this group.

Another group of semi-empirical methods is formed by approaches in which the effective potential contains parameters modified for each state so that the experimental energy value used already becomes the eigenvalue of the corresponding equation. The modified potential itself is determined with the help of the wave functions of the core configuration, which in their turn are found adopting other methods (numerical Hartree–Fock radial orbitals, analytical functions, etc.). They can also be found by utilizing statistical distribution of the core charge (statistical Thomas–Fermi potential). Sometimes additional corrections accounting for the core polarization by the outer electron are exploited.

Certain semi-empirical methods are aimed at choosing the effective potential, allowing one to determine as accurately as possible values of oscillator strengths of electronic transitions; others, more complex, are also directed at achieving the best correspondence between calculated energy spectra and experimental ones.

Model potential methods and their utilization in atomic structure calculations are reviewed in [139], main attention being paid to analytic effective model potentials in the Coulomb and non-Coulomb approximations, to effective model potentials based on the Thomas–Fermi statistical model of the atom, as well as employing a self-consistent field core potential. Relativistic effects in model potential calculations are discussed there, too. Paper [140] has examples of numerous model potential calculations of various atomic spectroscopic properties.

The main advantage of the effective potential method consists in the relative simplicity of the calculations, conditioned by the comparatively small number of semi-empirical parameters, as well as the analytical form of the potential and wave functions; such methods usually ensure fairly high accuracy of the calculated values of the energy levels and oscillator strengths. However, these methods, as a rule, can be successfully applied only for one- and two-valent atoms and ions. Therefore, the semi-empirical approach of least squares fitting is much more universal and powerful than model potential methods; it combines naturally and easily the accounting for relativistic and correlation effects.

The energy levels and eigenfunctions, obtained in one or other semi-empirical approach, may be successfully used further on to find fairly accurate values of the oscillator strengths, electron transition probabilities, lifetimes of excited states, etc., of atoms and ions [18, 141–144].

22

Hyperfine structure of the energy spectra, isotopic and Lamb shift

22.1 Hyperfine interactions in non-relativistic approximation

The hyperfine structure (splitting) of energy levels is mainly caused by electric and magnetic multipole interactions between the atomic nucleus and electronic shells. From the known data on hyperfine structure we can determine the electric and magnetic multipole momenta of the nuclei, their spins and other parameters.

In a non-relativistic approximation the usual fine structure (splitting) of the energy terms is considered as a perturbation whereas the hyperfine splitting – as an even smaller perturbation, and they both are calculated as matrix elements of the corresponding operators with respect to the zero-order wave functions.

The representation of hyperfine interactions in the form of multipoles follows from expansion of the potentials of the electric and magnetic fields of a nucleus, conditioned by the distribution of nuclear charges and currents, in a series of the corresponding multipole momenta. It follows from the properties of the operators obtained with respect to the inversion operation that the nucleus can possess non-zero electric multipole momenta of the order $k = 0, 2, 4, \ldots$, and magnetic ones with $k = 1, 3, 5, \ldots$. The energy of the hyperfine interactions decreases rapidly with the increase of multipolarity, therefore, it is usually enough to take into consideration only the first non-zero term in the expansion over multipoles. However, in a number of cases, due to the existence of selection rules and other effects, the higher multipoles must be accounted for. Indeed, in [145] the hexadecapole nuclear momentum for ^{165}Ho was found.

As usual, the operators describing hyperfine interactions are to be expressed in terms of irreducible tensors. Then we are in a position to find formulas for their matrix elements. The corresponding operator, caused by the magnetic multipole interaction, may be presented in the following form [146] (in a.u.):

$$H^{(kk0)} = \mu_0 \mu_N \left(\left(\nabla^{(1)} \left\{ r^{-k-1} C_q^{(k)} \right\} \cdot \left(\frac{2}{k} L^{(1)} - 2S^{(1)} \right) \right)_J \cdot {}^n M^{(k)} \right) \quad (22.1)$$

where the nuclear magnetic multipole momentum is

$${}^n M^{(k)} = (-1)^k \left(\nabla^{(1)} \left\{ r^k C^k \right\} \cdot \left(\frac{2}{k+1} g_l L^{(1)} + g_s S^{(1)} \right) \right)_J . \quad (22.2)$$

Here μ_0 is the Bohr magneton, μ_N denotes the nuclear magneton, and g stands for the gyromagnetic ratios being equal to $g_l = 1$, $g_s = 5.58$ for the proton, and $g_l = 0$, $g_s = -3.82$ for the neutron.

Hyperfine interactions control the splitting of the usual atomic level with given J into a number of components, each of them corresponding to a certain value of vectorial sum $\mathbf{J} + \mathbf{I}$ ($\mathbf{I}$ is the angular momentum of the nucleus), describing total momentum of atom $\mathbf{F}$, i.e.

$$\mathbf{F} = \mathbf{J} + \mathbf{I}. \quad (22.3)$$

Due to the presence of the interaction between $\mathbf{J}$ and $\mathbf{I}$ they will not be exact quantum numbers, only their vectorial sum (total momentum $\mathbf{F}$) will give the exact quantum number. However, this interaction is fairly small, therefore, the hyperfine splitting may be considered separately for each level via calculation of the corresponding matrix elements. The irreducible form of operator (22.1) is as follows:

$$H^{(kk0)} = 2\mu_0 \mu_N r^{-k-2} \left(\left\{ \sqrt{\frac{2k-1}{k}} \left[C^{(k-1)} \times L^{(1)} \right]^{(k)} \right. \right.$$
$$\left. \left. + \sqrt{(2k+3)(k+1)} \left[C^{(k+1)} \times S^{(1)} \right]^{(k)} \right\} \cdot {}^n M^{(k)} \right). \quad (22.4)$$

Its matrix element is

$$\left(LSJIF \middle| H^{(kk0)} \middle| L'S'J'I'F \right) = (-1)^{F+I+J'+k} \mu_N \left(I \middle\| {}^n M^{(k)} \middle\| I' \right)$$
$$\times \begin{Bmatrix} J & I & F \\ I' & J' & k \end{Bmatrix} \left(LSJ \middle\| M^{(k)} \middle\| L'S'J' \right). \quad (22.5)$$

We need the expression for the final (atomic) submatrix element in (22.5). It may be found similarly to the case of the energy operator already considered. For a shell of equivalent electrons, it equals

$$\left(l^N \alpha LSJ \middle| M^{(k)} \middle| l^N \alpha' L'S'J' \right) = 2\mu_0 \sqrt{[J,J']}$$
$$\times \left((-1)^{L+S+J'+1} \frac{i\delta(S,S')}{2(2S+1)} \begin{Bmatrix} L & J & S \\ J' & L' & k \end{Bmatrix} \sqrt{(2l+k+1)(2l-k+1)} \right.$$

$$\times \left(l\|C^{(k-1)}\|l\right)\left(l^N\alpha LS\|U^k\|l^N\alpha'L'S\right)$$

$$+i\sqrt{(k+1)[k,k+1]}\left(l\|C^{(k+1)}\|l\right)\left(l^N\alpha LS\|V^{k+11}\|l^N\alpha'L'S'\right)$$

$$\times \left. \begin{Bmatrix} L & L' & k+1 \\ S & S' & 1 \\ J & J' & k \end{Bmatrix} \right) N(l,l|-k-2). \tag{22.6}$$

Here the radial integral has the following definition:

$$N(nl,n'l'|a) = \int_0^\infty r^a P(nl|r)P(n'l'|r)dr. \tag{22.7}$$

For two open shells we obtain

$$\left(l_1^{N_1}\alpha_1 L_1 S_1 l_2^{N_2}\alpha_2 L_2 S_2 LSJ\|M^{(k)}\|l_1^{N_1}\alpha_1' L_1' S_1' l_2^{N_2}\alpha_2' L_2' S_2' L'S'J'\right)$$

$$= 2\mu_0\delta(\alpha_1 L_1 S_1, \alpha_1' L_1' S_1')$$

$$\times \left[(-1)^{S+J'+L_1+L_2'+k+1} \frac{i\delta(S_2,S_2')}{2(2S_2+1)} \sqrt{(2l_2+k+1)(2l_2-k+1)} \right.$$

$$\times \sqrt{[L,L',J,J']}\left(l_2^{N_2}\alpha_2 L_2 S_2\|U^k\|l_2^{N_2}\alpha_2' L_2' \dot S_2\right)\left(l_2\|C^{(k-1)}\|l_2\right)$$

$$\times \begin{Bmatrix} L_2 & L & L_1 \\ L' & L_2' & k \end{Bmatrix} \begin{Bmatrix} L & J & S \\ J' & L' & k \end{Bmatrix}$$

$$+ (-1)^{L_1+L_2'+L+S+S_1+S_2'+1}i\sqrt{(k+1)[k,k+1,L,L',S,S',J,J']}$$

$$\times \left(l_2^{N_2}\alpha_2 L_2 S_2\|V^{k+11}\|l_2^{N_2}\alpha_2' L_2' S_2'\right)\left(l_2\|C^{(k+1)}\|l_2\right)$$

$$\times \left. \begin{Bmatrix} L_2 & L & L_1 \\ L' & L_2' & k+1 \end{Bmatrix} \begin{Bmatrix} S_2 & S & S_1 \\ S' & S_2' & 1 \end{Bmatrix} \begin{Bmatrix} L & L' & k+1 \\ S & S' & 1 \\ J & J' & k \end{Bmatrix} \right]$$

$$\times N(l_2,l_2|-k-2) + 2\mu_0\delta(\alpha_2 L_2 S_2, \alpha_2' L_2' S_2')$$

$$\times \left[(-1)^{S+J'+L+L'+L_1+L_2+k+1} \frac{i\delta(S_1,S_1')}{2(2S_1+1)} \sqrt{(2l_1+k+1)(2l_1-k+1)} \right.$$

$$\times \sqrt{[L,L',J,J']}\left(l_1^{N_1}\alpha_1 L_1 S_1\|U^k\|l_1^{N_1}\alpha_1' L_1' S_1\right)\left(l_1\|C^{(k-1)}\|l_1\right)$$

$$\times \begin{Bmatrix} L_1 & L & L_2 \\ L' & L_1' & k \end{Bmatrix} \begin{Bmatrix} L & J & S \\ J' & L' & 1 \end{Bmatrix}$$

$$+ (-1)^{L_1+L_2+L'+S_1+S_2+S'+1}i\sqrt{(k+1)[k,k+1,L,L',S,S',J,J']}$$

$$\times \left(l_1^{N_1}\alpha_1 L_1 S_1\|V^{k+11}\|l_1^{N_1}\alpha_1' L_1' S_1'\right)\left(l_1\|C^{(k+1)}\|l_1\right)$$

$$\times \left. \begin{Bmatrix} L_1 & L & L_2 \\ L' & L_1' & k+1 \end{Bmatrix} \begin{Bmatrix} S_1 & S & S_2 \\ S' & S_1' & 1 \end{Bmatrix} \begin{Bmatrix} L & L' & k+1 \\ S & S' & 1 \\ J & J' & k \end{Bmatrix} \right]$$

$$\times N(l_1,l_1|-k-2). \tag{22.8}$$

Numerical values of the submatrix elements of operators U^k and V^{k1} may be taken from [87, 97] or calculated via the quasispin technique.

The operator of the hyperfine structure, caused by electric multipole radiation, may be presented in the form

$$T^{(kk0)} = \left(O^{(k)} \cdot {}^n O^{(k)} \right), \tag{22.9}$$

where

$$O^{(k)} = \sum_j r_j^{-k-1} C_j^{(k)}, \tag{22.10}$$

$$^n O^{(k)} = r_n^k \, {}^n C^{(k)}. \tag{22.11}$$

Matrix elements of operator (22.9) may be expressed in terms of the corresponding submatrix element in accordance with formula (22.5). The latter for shell l^N is equal to

$$\left(l^N \alpha L S J \| O^{(k)} \| l^N \alpha' L' S' J' \right)$$
$$= (-1)^{L+S+J'+k} \delta(S,S') \left(l \| C^{(k)} \| l \right) \sqrt{[J,J']/[S]}$$
$$\times \left\{ \begin{matrix} L & S & J \\ J' & L' & k \end{matrix} \right\} \left(l^N \alpha L S \| U^k \| l^N \alpha' L' S \right) N(l,l|-k-1). \tag{22.12}$$

For two open shells we have

$$\left(l_1^{N_1} \alpha_1 L_1 S_1 l_2^{N_2} \alpha_2 L_2 S_2 L S J \| O^{(k)} \| l_1^{N_1} \alpha_1' L_1' S_1' l_2^{N_2} \alpha_2' L_2' S_2' L' S' J' \right)$$
$$= (-1)^{L+S+J'+k} \delta(S,S') \sqrt{[J,J']} \left\{ \begin{matrix} L & S & J \\ J' & L' & k \end{matrix} \right\}$$
$$\times \left\{ (-1)^{L_1+L_2'+L+k} \delta(\alpha_1 L_1 S_1 S_2, \alpha_1' L_1' S_1' S_2') \right.$$
$$\times \sqrt{[L,L']/[S_2]} \left\{ \begin{matrix} L_2 & L & L_1 \\ L' & L_2' & k \end{matrix} \right\} \left(l_2 \| C^{(k)} \| l_2 \right)$$
$$\times \left(l_2^{N_2} \alpha_2 L_2 S_2 \| U^k \| l_2^{N_2} \alpha_2' L_2' S_2 \right) N(l_2,l_2|-k-1)$$
$$+ (-1)^{L_1+L_2+L'+k} \delta(\alpha_2 L_2 S_2 S_1, \alpha_2' L_2' S_2' S_1') \sqrt{[L,L']/[S_1]}$$
$$\times \left\{ \begin{matrix} L_1 & L & L_2 \\ L' & L_1' & k \end{matrix} \right\} \left(l_1 \| C^{(k)} \| l_1 \right) \left(l_1^{N_1} \alpha_1 L_1 S_1 \| U^k \| l_1^{N_1} \alpha_1' L_1' S_1 \right)$$
$$\times N(l_1,l_1|-k-1) \Big\}. \tag{22.13}$$

In formulas (22.12) and (22.13) k acquires only even values: for $k = 2$ we have the usual electric quadrupole interaction, whereas for $k = 4$ we have the electric hexadecapole interaction, already observed in [145]. The expressions for the matrix elements of the hyperfine structure operators considered above for the closed shells follow straightforwardly from the formulas presented while using the relevant expressions for the submatrix elements of the operators U^k and V^{k1}.

22.2 Hyperfine structure in relativistic approximation

Practical studies of the hyperfine structure of the levels of atoms and ions reveal the importance of relativistic effects for this phenomenon. Therefore, we need the corresponding relativistic formulas as well. A relativistic hyperfine structure operator has the form

$$H_h = \sum_k \left({}_nT^{(k)} \cdot {}_eT^{(k)} \right) = \sum_{kq} (-1)^{k-q} \, {}_nT_q^{(k)} \, {}_eT_{-q}^{(k)}. \qquad (22.14)$$

Here ${}_nT_q^{(k)}$ is the component of the operator, acting on the coordinates of the nucleons,

$$ {}_nT_q^{(k)} = i^{-k} \sum_j g_{l_j} r_j^k C_{jq}^{(k)}, \qquad k \neq 0 \text{ is even}, \qquad (22.15)$$

$$ {}_nT_q^{(k)} = i^{-k} \mu_N \sum_j \left(\left(\nabla_j^{(1)} \left\{ r_j^k C_{jq}^{(k)} \right\} \right) \cdot \left(\frac{2}{k+2} g_{l_j} L^{(1)} + g_{s_j} S^{(1)} \right) \right),$$

$$k \text{ is odd.} \qquad (22.16)$$

The summation is over all nucleons. Operator $\nabla^{(1)}$ in (22.16) acts only on the quantity in curly brackets and makes a scalar product with the tensors $L^{(1)}$ and $S^{(1)}$. The second multiplier in (22.14) is the component of the operator, acting on the electronic coordinates,

$$ {}_eT_{-q}^{(k)} = -i^{-k} \sum_j r_j^{-k-1} C_{j,-q}^{(k)}, \qquad k \neq 0 \text{ is even}, \qquad (22.17)$$

$$ {}_eT_{-q}^{(k)} = i^{-k-1} \frac{1}{k} \sum_j r_j^{-k-1} \left(\alpha_j^{(1)} \cdot \left\{ L_j^{(1)} \right\} C_{j,-q}^{(k)} \right\}, \qquad k \text{ is odd.} \qquad (22.18)$$

In these formulas the summation is over the coordinates of all electrons. Operators $\alpha^{(1)}$ and $L^{(1)}$ in (22.18) compose a scalar product. However, $L^{(1)}$ acts only on $C^{(k)}$. This pecularity is indicated by the introduction of curly brackets. In order to find the matrix elements of operator (22.18), we have to transform it to the form

$$ {}_eT_{-q}^{(k)} = -i^{-k} \sqrt{\frac{k+1}{k}} \sum_j r_j^{-k-1} \left[C_j^{(k)} \times \alpha_j^{(1)} \right]_{-q}^{(k)}. \qquad (22.19)$$

As in the non-relativistic case, matrix elements of operator (22.19) are calculated using first-order perturbation theory

$$W_F = \langle \beta I J F M_F | H_h | \beta' I' J' F M_F \rangle$$

$$= \sum_k (-1)^{l'+J+F+k} \begin{Bmatrix} I & J & F \\ J' & I' & k \end{Bmatrix}$$

$$\times \left\langle \beta_1 I \| _n T^{(k)} \| \beta_1' I' \right\rangle \left\langle \alpha J \| _e T^{(k)} \| \alpha' J' \right\rangle. \tag{22.20}$$

Here β, β', β_1, β_1' and α, α' contain all additional quantum numbers of the nucleus and electronic configuration, respectively. The nuclear submatrix element in (22.20) is proportional to k-pole electric (k is even) or magnetic (k is odd) momenta of the nucleus. Similarly to the non-relativistic case, we shall not consider explicit expressions for nuclear submatrix elements because accurate wave functions of the nucleus are not known. That is why further on we present only the formulas for relativistic submatrix elements of the operators acting on the coordinates of the atomic electrons.

Submatrix elements of operators (22.17) and (22.18) for the case of a subshell of equivalent electrons, due to their one-electron character, are simply proportional to the corresponding one-electron quantities, where the submatrix elements of the operators, composed of unit tensors (7.2), serve as the coefficients of proportionality, i.e.

$$\left\langle nlj^N \alpha J \| _e T^{(k)} \| nlj^N \alpha' J' \right\rangle = (j^N \alpha J \| T^k \| j^N \alpha' J') \left\langle nlj \| _e T^{(k)} \| nlj \right\rangle. \tag{22.21}$$

One-electron submatrix elements in (22.21) for the even and odd k values are equal, correspondingly (let us recall that $\lambda = nlj$, $\lambda' = nl'j$):

$$\left\langle n_1 l_1 j_1 \| _e T^{(k)} \| n_2 l_2 j_2 \right\rangle$$

$$= (-1)^{j_1+j_2+(l_1-l_2)/2} \sqrt{2j_1+1} \begin{bmatrix} k & j_1 & j_2 \\ 0 & 1/2 & 1/2 \end{bmatrix} R_1^k(\lambda_1 \lambda_2), \tag{22.22}$$

$$\left\langle n_1 l_1 j_1 \| _e T^{(k)} \| n_2 l_2 j_2 \right\rangle = (-1)^{j_2+(l_1+l_2-1)/2}$$

$$\times \frac{\sqrt{2j_1+1}}{2k} \left\{ 2j_1 + 1 - (-1)^{j_1+j_2}(2j_2+1) \right\}$$

$$\times \begin{bmatrix} k & j_1 & j_2 \\ 0 & 1/2 & 1/2 \end{bmatrix} R_2^k(\lambda_1 \lambda_2). \tag{22.23}$$

Radial integrals are defined as follows:

$$R_1^k(\lambda_1 \lambda_2) = (l_1 l_2 k)(l_1' l_2' k)$$

$$\times \int_0^\infty \frac{1}{r^{k+1}} \left\{ f(\lambda_1 | r) f(\lambda_2 | r) + g(\lambda_1' | r) g(\lambda_2' | r) \right\} r^2 dr, \tag{22.24}$$

$$R_2^k(\lambda_1\lambda_2) = (l_1l_2'k)(l_1'l_2k)$$

$$\times \int\limits_0^\infty \frac{1}{r^{k+1}} \{f(\lambda_1|r)g(\lambda_2'|r) + g(\lambda_1'|r)f(\lambda_2|r)\}\, r^2dr. \quad (22.25)$$

For relativistic calculations of a hyperfine structure in the intermediate coupling, the non-diagonal with respect to configuration's submatrix elements are necessary too. The corresponding expressions may be found in [147].

The above-mentioned considerations are fairly general; the expressions are presented for the interaction of any multipolarity k. However, in practical calculations only the diagonal (with respect to all quantum numbers) matrix elements are usually taken into account, and the case $M_I = I$, $M_J = J$ and $q = 0$ is chosen. Then formula (22.20) may be rewritten as follows:

$$W_F \approx \sum_k A_k M_k(IJF), \quad (22.26)$$

where

$$A_k = \left\langle \beta_1 II|_n T_0^{(k)}|\beta_1 II \right\rangle \left\langle \alpha JJ|_e T_0^{(k)}|\alpha JJ \right\rangle, \quad (22.27)$$

$$M_k(IJF) = (-1)^{I+J+F+k} \begin{Bmatrix} I & J & F \\ J & I & K \end{Bmatrix}$$

$$\times \sqrt{(2I-k)!(2I+k+1)!(2J-k)!(2J+k+1)!}/[(2I)!(2J)!]. \quad (22.28)$$

Retaining in (22.26) only the first two terms in the sum over k, defining $A = -A_1/IJ$ and $B = 4A_2$, and utilizing the algebraic expressions for $6j$-coefficient, we obtain

$$W_F \approx \frac{A}{2}K - B\left[\frac{(I+1)(J+1)}{2(2I-1)(2J-1)} - \frac{3K(K+1)}{8I(2I-1)J(2J-1)}\right], \quad (22.29)$$

where

$$K = F(F+1) - I(I+1) - J(J+1). \quad (22.30)$$

The quantities A and B are usually called the constants of the magnetic dipole and electric quadrupole interactions. The nuclear matrix elements in (22.27) for $k = 1$ and 2 are proportional to the magnetic dipole μ and electric quadrupole Q nuclear momenta, respectively:

$$\mu = \left\langle \beta_1 II|_n T_0^{(1)}|\beta_1 II \right\rangle, \quad (22.31)$$

$$Q = 2\left\langle \beta_1 II|_n T_0^{(2)}|\beta_1 II \right\rangle. \quad (22.32)$$

Performing the calculations in the intermediate coupling scheme (*ic*), whose quantum numbers let us denote as γJ, we find for the constants A and B, correspondingly:

$$A = -\frac{\mu}{I\sqrt{J(J+1)(2J+1)}}\left\langle \gamma J\|_e T^{(1)}\|\gamma J\right\rangle_{ic},\qquad(22.33)$$

$$B = 7.1415 \cdot 10^{-8} Q\sqrt{\frac{J(2J-1)}{(J+1)(2J+1)(2J+3)}}\left\langle \gamma J\|_e T^{(2)}\|\gamma J\right\rangle_{ic}.\qquad(22.34)$$

Knowing the numerical values of μ and Q, for example, from experimental measurements, we can easily determine A and B, or from the known A and B we may find nuclear moments μ and Q, thus revealing information about the structure and properties of the nuclei [148].

In the conclusion of this section let us notice that a wealth of data on the applications of the relativistic self-consistent field method to the studies of the hyperfine structure of atomic levels is collected in [149]. Investigations of the hyperfine structure by the methods of perturbation theory are described in monograph [17].

Calculations of the hyperfine structure require account of both the relativistic and correlation effects. Therefore, only sophisticated theories can lead to good agreement of calculations with experimental data [150–153]. Modern experimental methods allow study of the hyperfine structure of practically any atom or ion, including extreme ionization degrees (e.g. hydrogen-like ions of very heavy elements). So, in [154] there were reported measurements of the 1s hyperfine splittings in $^{209}Bi^{82+}$, that showed results already in the optical region. The progress of experiments encourages theoreticians to model such systems mathematically and to use data obtained in this way as a probe of quantum-electrodynamical effects at high Z [50, 155].

22.3   Isotopic and Lamb shifts of the energy levels

Until now, while studying the energy of an atom we took into account only the fact that its nucleus has a charge, considering the nucleus itself as a pointlike and motionless object without inner structure and having infinitely large mass. However, real nuclei have finite mass, a certain volume and spatial form and they move with regard to the centre of masses of an atom. The electric field inside the nucleus is not Coulombic. For these reasons the energy levels of separate isotopes of a given atom will be shifted relative to each other, and isotopic shift of the levels occurs.

Let us consider first the isotopic shift of the energy levels, conditioned by the difference of masses of various isotopes (the mass effect, causing the so-called normal and specific shifts). However, the isotopic shift may also be caused by the differences of the radii and forms of separate isotopes (volume effect). Let us emphasize that for light atoms the isotopic shift is mainly conditioned by a mass effect, whereas for heavy ones $(Z > 60)$ it is by a volume effect. The isotopic shift due to the mass effect may be presented as a sum of two terms

$$\Delta E_m = \Delta E_n + \Delta E_s, \tag{22.35}$$

where normal shift

$$\Delta E_n = -Em/M. \tag{22.36}$$

Here m is the electron mass, M stands for the mass of a given isotope, and E denotes the total energy of an atom for motionless nucleus. The specific shift of energy is described by the matrix element

$$\Delta E_s = (\psi|\hat{S}|\psi), \tag{22.37}$$

where the corresponding two-electron operator equals

$$\hat{S} = \frac{1}{M} \sum_{i>j} \left(\nabla_i^{(1)} \cdot \nabla_j^{(1)} \right). \tag{22.38}$$

In (22.37) ψ is the wave function of an atom with motionless nucleus. The one-electronic submatrix element of the gradient operator $(nl\|\nabla^{(1)}\|n_1 l_1)$ is non-zero only for $l_1 = l \pm 1$. Therefore, the matrix element of operator (22.38) inside a shell of equivalent electrons vanishes and one has to account for this interaction only between shells. For the configuration, consisting of j closed and two open shells, it is defined by the following formula [156]:

$$\left(n_1 l_1^{4l_1+2} \dots n_j l_j^{4l_j+2} n_{j+1} l_{j+1}^{N_1}(\alpha_1 L_1 S_1) n_{j+2} l_{j+2}^{N_2}(\alpha_2 L_2 S_2) LS|\hat{S}| \right.$$

$$\left. \times n_1 l_1^{4l_1+2} \dots n_j l_j^{4l_j+2} n_{j+1} l_{j+1}^{N_1}(\alpha_1' L_1' S_1') n_{j+2} l_{j+2}^{N_2}(\alpha_2' L_2' S_2') LS \right)$$

$$= -\frac{1}{M} \left\{ \sum_{i=1}^{j} \left[\frac{N_1}{2l_{j+1}+1} \left(n_i l_i \|\nabla^{(1)}\| n_{j+1} l_{j+1} \right)^2 \right. \right.$$

$$\left. + \frac{N_2}{2l_{j+2}+1} \left(n_i l_i \|\nabla^{(1)}\| n_{j+2} l_{j+2} \right)^2 \right] + 2 \sum_{i,i'}^{j} \left(n_i l_i \|\nabla^{(1)}\| n_{i'} l_{i'} \right)^2$$

$$+ g \left(l_{j+1}^{N_1} l_{j+2}^{N_2}; \alpha_1 L_1 S_1, \alpha_1' L_1' S_1'; \alpha_2 L_2 S_2, \alpha_2' L_2' S_2'; LS \right)$$

$$\times \left(n_{j+1} l_{j+1} \|\nabla^{(1)}\| n_{j+2} l_{j+2} \right)^2 \right\}, \tag{22.39}$$

where

$$g\left(l_{j+1}^{N_1}l_{j+2}^{N_2};\alpha_1 L_1 S_1,\alpha_1' L_1' S_1';\alpha_2 L_2 S_2,\alpha_2' L_2' S_2';LS\right)$$

$$= (-1)^{L_1'+L_2+L+S_1'+S_2'+S} \sum_{x=0}^{\min(2l_1,2l_2)} \sum_{x'=0}^{1} (-1)^x (2x+1)(2x'+1)$$

$$\times \begin{Bmatrix} l_1 & l_2 & 1 \\ l_2 & l_1 & x \end{Bmatrix} \begin{Bmatrix} L_1 & L_2 & L \\ L_2' & L_1' & x \end{Bmatrix} \begin{Bmatrix} S_1 & S_2 & S \\ S_2' & S_1' & x' \end{Bmatrix}$$

$$\times \left(l_{j+1}^{N_1}\alpha_1 L_1 S_1 \| V^{xx'} \| l_{j+1}^{N_1}\alpha_1' L_1' S_1'\right)$$

$$\times \left(l_{j+2}^{N_2}\alpha_2 L_2 S_2 \| V^{xx'} \| l_{j+2}^{N_2}\alpha_2' L_2' S_2'\right), \tag{22.40}$$

and

$$\left(nl\|\nabla^{(1)}\|n_1 l-1\right) = \sqrt{l}\left[-l\int_0^\infty P(nl|r)\frac{1}{r}P(n_1 l-1|r)dr\right.$$

$$\left.+\int_0^\infty P(nl|r)\frac{d}{dr}P(n_1 l-1|r)dr\right]$$

$$= -\sqrt{l}\left[l\int_0^\infty P(nl|r)\frac{1}{r}P(n_1 l-1|r)dr\right.$$

$$\left.+\int_0^\infty P(n_1 l-1|r)\frac{d}{dr}P(nl|r)dr\right], \tag{22.41}$$

$$\left(nl\|\nabla^{(1)}\|n_1 l+1\right) = \sqrt{l+1}\left[(l+1)\int_0^\infty P(nl|r)\frac{1}{r}P(n_1 l+1|r)dr\right.$$

$$\left.+\int_0^\infty P(nl|r)\frac{d}{dr}P(n_1 l+1|r)dr\right]$$

$$= \sqrt{l+1}\left[(l+1)\int_0^\infty P(nl|r)\frac{1}{r}P(n_1 l+1|r)dr\right.$$

$$\left.-\int_0^\infty P(n_1 l+1|r)\frac{d}{dr}P(nl|r)dr\right]. \tag{22.42}$$

The following condition is valid:

$$\left(nl\|\nabla^{(1)}\|n_1 l\pm 1\right) = \left(n_1 l\pm 1\|\nabla^{(1)}\|nl\right). \tag{22.43}$$

The summation in (22.39) over i runs through all closed shells, whereas

$l_{i'} = l_i \pm 1$. The estimations show that normal and specific shifts have the same order of magnitude.

It is much more difficult to take into account the influence of finite dimensions and form of the nucleus (volume effect) on the atomic energy levels, because we do not know exactly the nuclear volume, or its form, or the character of the distribution of the charge in it. Therefore, in such cases one sometimes finds it by subtracting its part (22.35) from the experimentally measured total isotopic shift. Further on, having the value of the shift caused by the volume effect, we may extract information on the structure and properties of the nucleus itself. For the approximate determination of the isotope shift, connected with the differences δr_0 of the nuclear radii of two isotopes, the following formula may be used [15]:

$$\delta E = \frac{4\pi a_0^3}{Z} |\psi_s(0)|^2 \frac{\gamma + 1}{[\Gamma(2\gamma + 1)]^2} B(\gamma) \left(\frac{2Z r_0}{a_0} \right)^{2\gamma} \frac{\delta r_0}{r_0} Ry. \tag{22.44}$$

Here $\gamma = \sqrt{1 - \alpha^2 Z^2}$, $\Gamma(\beta)$ is the gamma-function, a_0 stands for the Bohr radius, and $\psi_s(0)$ denotes the value of the non-relativistic wave function of an atom at the origin of the coordinates. Multiplier $B(\gamma)$ describes the distribution of the proton charge in the nucleus. For a homogeneous distribution of charge on the surface and in all volume of the spherically-symmetric nucleus, this multiplier $B(\gamma)$ is, respectively,

$$B(\gamma) = \frac{2\gamma^2(2 - \gamma)}{(\gamma + 1)(2 + \gamma)}, \qquad B(\gamma) = \frac{3}{(2\gamma + 1)(2\gamma + 3)}. \tag{22.45}$$

In single-configurational approximation $\psi(0) \neq 0$ only for s electrons. That is why in (22.44) the symbol ψ has subscript s. The whole dependence of the shift under investigation on the pecularities of the electronic shells of an atom is contained only in the multiplier $\psi_s(0)$. Unfortunately, formula (22.44) does not account for the deviations of the shape of the nucleus from spherical symmetry. Therefore it is unfit for non-spherical nuclei. The accuracy of the determination of all these quantities may be improved by accounting for both the correlation and relativistic effects [157, 158]. A universal program to compute isotope shifts in atomic spectra is described in [159].

The atomic Hamiltonian considered in previous chapters may be made more accurate by adding to it a number of quantum-electrodynamical corrections. It has turned out that the most important among them are the radiative terms, leading to the so-called Lamb shift of an energy level. It is conditioned by the interaction of an electron with the radiation field. The contribution of the Lamb shift increases rapidly in the isoelectronic sequence. Therefore accounting for it is essential when calculating or identifying the spectra of highly ionized atoms. Accurate calculations of the Lamb shift are very complicated and time consuming even for the

simplest atoms or ions. That is why one usually utilizes simplified formulas
or makes tables of their numerical values. Extensive information on such
procedures is presented in [50, 160–163].

As in the case of a hyperfine structure, there is a wealth of new
experimental data on Lamb shift measurements of such exotic systems
as lithium-like uranium U^{89+}, again encouraging the relevant calculations
[164–166].

23

Quasispin and isospin for relativistic matrix elements

23.1 Relativistic approach and quasispin for one subshell

As we have already seen in Chapters 11 and 12, the realization of one or another coupling scheme in the many-electron atom is determined by the relation between the spin–orbit and non-spherical parts of electrostatic interactions. As the ionization degree of an atom increases, the coupling scheme changes gradually from LS to jj coupling. The latter, for highly ionized atoms, occurs even within the shell of equivalent electrons (see Chapter 31).

With jj coupling, the spin–angular part of the one-electron wave function (2.15) is obtained by vectorial coupling of the orbital and spin–angular momenta of the electron. Then the total angular momenta of individual electrons are added up. In this approach a shell of equivalent electrons is split into two subshells with $j = l \pm 1/2$. The shell structure of electronic configurations in jj coupling becomes more complex, but is compensated for by a reduction in the number of electrons in individual subshells.

The energy spectrum of atoms and ions with jj coupling can be found using the relativistic Hamiltonian of N-electron atoms (2.1)–(2.7). Its irreducible tensorial form is presented in Chapter 19. The relativistic one-electron wave functions are four-component spinors (2.15). They are the eigenfunctions of the total angular momentum operator for the electron and are used to determine one-electron and two-electron matrix elements of relativistic interaction operators. These matrix elements, in the representation of occupation numbers, are the parameters that enter into the expansions of the operators corresponding to physical quantities (see general expressions (13.22) and (13.23)).

Consequently, with second quantization, the approach using Hamiltonian (2.1)–(2.7) and relativistic wave functions (2.15) differs from the approach using Hamiltonian (1.16)–(1.22) and the non-relativistic wave functions only by the way of finding the one- and two-electron matrix elements in coordinate representation. The tensorial properties of second-quantization operators in both cases are the same, since they are determined only by the fact that the relevant mathematics is based on jj coupling.

The relationships describing the tensorial properties of wave functions, second-quantization operators and matrix elements in the space of total angular momentum J can readily be obtained by the use of the results of Chapters 14 and 15 with the more or less trivial replacement of the ranks of the tensors l and s by j and the corresponding replacement of various factors and $3nj$-coefficients. Therefore, we shall only give a sketch of the uses of the quasispin method for jj coupling, following mainly the works [30, 167, 168]. For a subshell of equivalent electrons, the creation and annihilation operators $a_m^{(j)}$ and $a_m^{(j)\dagger}$ are the components of the same tensor of rank $q = 1/2$

$$a_{\frac{1}{2}m}^{(qj)} = a_m^{(j)}; \qquad a_{-\frac{1}{2}m}^{(qj)} = \tilde{a}_m^{(j)} = (-1)^{j-m} a_{-m}^{(j)\dagger} \tag{23.1}$$

in the space of quasispin angular momentum

$$Q_\rho^{(1)} = -\frac{i\sqrt{2j+1}}{2\sqrt{2}} \left[a^{(qj)} \times a^{(qj)} \right]_{\rho 0}^{(10)}. \tag{23.2}$$

The property of Hermitian conjugation for these operators has the form

$$a_{\rho m}^{(qj)\dagger} = (-1)^{q+\rho+j-m} a_{-\rho-m}^{(qj)}. \tag{23.3}$$

From the anticommutation properties of these operators

$$a_{\rho m}^{(qj)} a_{\rho' m'}^{(qj)} + a_{\rho' m'}^{(qj)} a_{\rho m}^{(qj)} = (-1)^{q+j+\rho-m} \delta(\rho m, -\rho' - m') \tag{23.4}$$

we obtain the following relationship:

$$T_{\Pi\rho}^{(Kk)} = \left[a^{(qj)} \times a^{(qj)} \right]_{\Pi\rho}^{(Kk)} = \begin{cases} 0 & \text{if } K + k \text{ is even;} \\ -\sqrt{(2j+1)/2} & \text{if } K = k = 0. \end{cases} \tag{23.5}$$

In this method tensors of such a type are the basic operators, and the operators for the pertinent physical quantities will be given in terms of irreducible tensorial products of these tensors. Specifically, unit tensor T_ρ^k (7.4) will be

$$T_\rho^k = -\frac{1}{\sqrt{2(2k+1)}} T_{0\rho}^{(Kk)}, \tag{23.6}$$

where the required rank K (zero or unity) is derived from the condition that $K + k$ is an odd number. The operators of total and quasispin angular momenta will, respectively, be given by

$$J_\rho^{(1)} = -i\sqrt{j(j+1)(2j+1)/6}\,T_{0\rho}^{(01)};$$ (23.7)

$$Q_\rho^{(1)} = -i\frac{\sqrt{2j+1}}{2\sqrt{2}}\,T_{\rho 0}^{(10)}.$$ (23.8)

Utilizing the tensorial properties of operators in quasispin space, we can, in particular, find an expansion of the scalar products of the operators T^k in terms of irreducible tensors in quasispin space

$$\left(T^k \cdot T^k\right) = -\frac{1}{2\sqrt{2k+1}}\left[T^{(0k)} \times T^{(0k)}\right]^{(00)} \qquad (k \text{ is odd});$$ (23.9)

$$\left(T^k \cdot T^k\right) = -\frac{1}{2\sqrt{3(2k+1)}}\left\{\left[T^{(1k)} \times T^{(1k)}\right]^{(00)}\right.$$
$$\left. -\sqrt{2}\left[T^{(1k)} \times T^{(1k)}\right]^{(20)}\right\} \qquad (k \text{ is even}).$$ (23.10)

Following the technique put forward in Chapter 15, we shall now express the tensorial products of the operators $T^{(Kk)}$ in terms of operators whose eigenvalues are known. If now we make use of definition (23.5), commute once the operators $a^{(qj)}$ and recouple the momenta, we arrive at

$$\left[T^{(Kk)} \times T^{(K'k')}\right]_{\Pi\rho}^{(K''k'')} = -\sum_{K_1 K_2 k_1 k_2}\sqrt{[K,K',K_1,K_2,k,k',k_1,k_2]}$$

$$\times \begin{Bmatrix} q & q & K \\ q & q & K' \\ K_1 & K_2 & K'' \end{Bmatrix}\begin{Bmatrix} j & j & k \\ j & j & k' \\ k_1 & k_2 & k'' \end{Bmatrix}\left[T^{(K_1 k_1)} \times T^{(K_2 k_2)}\right]_{\Pi\rho}^{(K''k'')}$$

$$-(-1)^{K''+k''}\sqrt{[K,K',k,k']}\begin{Bmatrix} K & K' & K'' \\ q & q & q \end{Bmatrix}\begin{Bmatrix} k & k' & k'' \\ j & j & j \end{Bmatrix}T_{\Pi\rho}^{(K''k'')}.$$ (23.11)

From this equation, at various specific values of the ranks K, k, K', k', K'', k'', we can work out a set of equations for sums of a definite parity of ranks of the total angular momentum. The values of the ranks are selected to yield operators with known eigenvalues on the left side. For complete scalars $K'' = k'' = 0$, we get

$$\sum_k \sqrt{2k+1}\left[T^{(0k)} \times T^{(0k)}\right]^{(00)} = 4Q^2 - (2j+1)(2j+3)/4;$$ (23.12)

$$\sqrt{3}\sum_k \sqrt{2k+1}\left[T^{(1k)} \times T^{(1k)}\right]^{(00)} = -4Q^2 - 3(2j+1)(2j-1)/4;$$ (23.13)

$$\sum_k (-1)^k \sqrt{2k+1}\begin{Bmatrix} j & j & 1 \\ j & j & k \end{Bmatrix}\left[T^{(0k)} \times T^{(0k)}\right]^{(00)}$$

$$= \frac{1}{j(j+1)(2j+1)}J^2 + \frac{1}{2};$$ (23.14)

$$\sum_k (-1)^k \sqrt{3(2k+1)} \begin{Bmatrix} j & j & 1 \\ j & j & k \end{Bmatrix} \left[T^{(1k)} \times T^{(1k)} \right]^{(00)}$$

$$= \frac{3}{j(j+1)(2j+1)} \mathbf{J}^2 - \frac{3}{2}, \tag{23.15}$$

and at $K'' = 2, k = 0$

$$\sum_{k>0} \sqrt{2k+1} \left[T^{(1k)} \times T^{(1k)} \right]^{(20)}_{\Pi 0} = \frac{16(j+1)}{2j+1} Q^{(2)}_{\Pi}; \tag{23.16}$$

$$\sum_{k>0} \sqrt{2k+1} \begin{Bmatrix} j & j & 1 \\ j & j & k \end{Bmatrix} \left[T^{(1k)} \times T^{(1k)} \right]^{(20)}_{\Pi 0} = \frac{8}{(2j+1)^2} Q^{(2)}_{\Pi}, \tag{23.17}$$

where $Q^{(2)}_{\Pi}$ is the special case $k = 2$ of the generalized operator of the quasispin (e.g. [169]):

$$Q^{(k)}_m = \left[\cdots \left[\left[Q^{(1)} \times Q^{(1)} \right]^{(2)} \times Q^{(1)} \right]^{(3)} \cdots \right]^{(k)}_m . \tag{23.18}$$

Interesting theorems for tensorial products of such operators are discussed in [170].

Combining the above formulas, we can work out analytical expressions for the sums of the scalar products $(T^k \cdot T^k)$ which are generally obtained using the Casimir operators of the unitary group U_{2j+1} and the symplectic group Sp_{2j+1}

$$\sum_{k=1} (2k+1) \left(T^k \cdot T^k \right) = -\frac{8(j+1)}{2j+1} Q_z Q_z + (2j+1)/2; \tag{23.19}$$

$$\sum_{k=1} (2k+1) \left(T^k \cdot T^k \right) = -2\mathbf{Q}^2 + (2j+1)(2j+3)/8. \tag{23.20}$$

All the expressions derived so far are suitable for subshells with any value of the quantum number j. For j to be concretized, we can obtain additional relations from (23.11), which makes it possible to find, in certain cases, similar formulas for individual terms of the sums. With the scalar products $\left(T^k \cdot T^k \right)$ such expressions can be found [18] using the Casimir operators of appropriate groups (see Chapters 5, 15 and 18). Since these scalar products at odd k differ from the irreducible tensorial products $T^{(Kk)}$ only by a factor, we shall list the results that have been obtained for the first time (the number over the equality sign indicates for which j-shell a given equality occurs):

$$\sqrt{3 \cdot 5} \left[T^{(12)} \times T^{(12)} \right]^{(00)} \overset{3/2}{=} \frac{2}{3} \mathbf{J}^2 - 10; \tag{23.21}$$

$$\sqrt{5} \left[T^{(12)} \times T^{(12)} \right]^{(20)} \overset{3/2}{=} 10 \dot{Q}^{(2)}; \tag{23.22}$$

$$\sqrt{3 \cdot 5} \left[T^{(12)} \times T^{(12)} \right]^{(00)} \overset{5/2}{=} \frac{1}{7} (3\mathbf{J}^2 + \frac{5^2}{3 \cdot 7} \mathbf{Q}^2 - 5 \cdot 9^2/4); \quad (23.23)$$

$$\sqrt{3} \left[T^{(14)} \times T^{(14)} \right]^{(00)} \overset{5/2}{=} -\frac{1}{7} (\mathbf{J}^2 + 9\mathbf{Q}^2 + 3 \cdot 11/4); \quad (23.24)$$

$$\sqrt{5} \left[T^{(12)} \times T^{(12)} \right]^{(20)} \overset{5/2}{=} \frac{10}{3} Q^{(2)}; \quad (23.25)$$

$$\left[T^{(14)} \times T^{(14)} \right]^{(20)} \overset{5/2}{=} 2Q^{(2)}. \quad (23.26)$$

For operators corresponding to physical quantities, we can also obtain an expansion in terms of irreducible tensors in quasispin space. Specifically, for two-particle operators (13.23) that are scalars with respect to the total momentum J

$$G^{(kk)} = \sum_{i>j=1}^{N} g_{ij}^{(kk)} = \sum_{i>j=1}^{N} g(r_i, r_j) \left[g_i^{(k)} \times g_j^{(k)} \right]^{(0)} \quad (23.27)$$

we can find two different expansions of the same type as in LS coupling (see (15.91) and (15.92))

$$G^{(kk)} = -\frac{1}{2} \sum_{JK} (nl\,jnl\,jJ \| g_{12} \| nl\,jnl\,jJ)$$

$$\times \begin{bmatrix} 1 & 1 & K \\ 1 & -1 & 0 \end{bmatrix} \left[T^{(1J)} \times T^{(1J)} \right]_{00}^{(K0)}; \quad (23.28)$$

$$G^{(kk)} = \frac{1}{4} (nl\,jnl\,j \| g_{12} \| nl\,jnl\,j) \left\{ \sum_{K_1 K_2 K} \frac{1}{\sqrt{2k+1}} \begin{bmatrix} K_1 & K_2 & K \\ 0 & 0 & 0 \end{bmatrix} \right.$$

$$\times \left. \left[T^{(K_1 k)} \times T^{(K_2 k)} \right]_{00}^{(K0)} - (-1)^k \frac{\sqrt{2}}{\sqrt{[k,j]}} T_{00}^{(K0)} \right\}. \quad (23.29)$$

Concretizing j and using the above results, the part of (23.29) that is scalar in both spaces can be expressed in terms of the operators with known eigenvalues.

Unlike LS coupling, additional classification of states of the subshell of equivalent electrons in jj coupling in the cases of practical interest causes no problems. In fact, it is well known [18] that for the degenerate states of the j^N configuration, at $j \leq 7/2$, to be classified it is sufficient only to use additionally the seniority quantum number v.

The wave function for N electrons with a given quantum number is an eigenfunction of the operators $\mathbf{Q}^2$ and Q_z

$$\left| j^N \alpha v J M_J \right\rangle \equiv |jQJM_Q M_J\rangle, \quad (23.30)$$

where $Q = (2j + 1 - 2v)/4$, $M_Q = -(2j + 1 - 2N)/4$.

As in the case of *LS* coupling, the tensorial properties of wave functions and second-quantization operators in quasispin space enable us to separate, using the Wigner–Eckart theorem, the dependence of the submatrix elements on the number of electrons in the subshell into the Clebsch–Gordan coefficient. If then we use the relation of the submatrix element of the creation operator to the CFP

$$\left(j^N\alpha QJ\|a^{(j)}\|j^{N-1}\alpha'Q'J'\right) = (-1)^N\sqrt{N(2J+1)}\left(j^N\alpha vJ\|j^{N-1}(\alpha'v'J')j\right),$$
(23.31)

go over to tensor $a^{(qj)}$ and use the Wigner–Eckart theorem, we obtain the reduced coefficients (subcoefficients) of fractional parentage (CFP reduced in quasispin space)

$$\left(j\alpha QJ\|\|a^{(qj)}\|\|j\alpha'Q'J'\right)$$

$$= \frac{(-1)^{N+1}\sqrt{N[Q,J]}}{\begin{bmatrix} Q' & q & Q \\ M_{Q'} & \frac{1}{2} & M_Q \end{bmatrix}}\left(j^N\alpha vJ\|j^{N-1}(\alpha'v'J')j\right).$$
(23.32)

Table 23.1 summarizes numerical values of subcoefficients of fractional parentage (SCFP) for subshells with $j = 5/2$ and $7/2$, taken from [167]. The simplest way of expressing the SCFP is in terms of the CFP for the subshell j^v term that occurs for the first time

$$\left(j\alpha QJ\|\|a^{(qj)}\|\|j\alpha'Q'J'\right)$$

$$= (-1)^v\sqrt{2v(Q+1)(2J+1)}\left(j^v\alpha QJ\|j^{v-1}(\alpha'Q'J')j\right).$$
(23.33)

The transposition condition for these coefficients is

$$\left(j\alpha QJ\|\|a^{(qj)}\|\|j\alpha'Q'J'\right)$$

$$= (-1)^{Q-Q'+J-J'+j-1/2}\left(j\alpha'Q'J'\|\|a^{(qj)}\|\|j\alpha QJ\right).$$
(23.34)

For the SCFP we can derive expressions similar to the set of equations (16.20) and (16.24)–(16.26), namely

$$\sum_{\alpha''J''}\left(j\alpha QJ\|\|a^{(qj)}\|\|j\alpha''(Q+1/2)J''\right)$$

$$\times \left(j\alpha''(Q+1/2)J''\|\|a^{(qj)}\|\|j\alpha'(Q+1)J'\right)$$

$$\times \begin{Bmatrix} j & j & J_0 \\ J' & J & J'' \end{Bmatrix} = 0 \quad (J_0 \text{ is odd});$$
(23.35)

$$\sum_{\alpha''J''}(-1)^{2J''}\left(j\alpha QJ\|\|a^{(qj)}\|\|j\alpha''(Q+1/2)J''\right)$$

$$\times \left(j\alpha'(Q+1)J'\|\|a^{(qj)}\|\|j\alpha''(Q+1/2)J''\right) = 0;$$
(23.36)

Table 23.1. Numerical values of subcoefficients of fractional parentage for $j = 5/2$ and $7/2$

$$\left(5/2(v)QJ|||a^{(qj)}|||5/2(v')Q'J'\right)$$

(v)QJ \ (v')Q'J'	(0) 3/2 0	(2) 1/2 2	(2) 1/2 4
(1) 1 5/2	$2\sqrt{2\cdot3}$	$-\sqrt{2\cdot3\cdot5}$	$-3\sqrt{2\cdot3}$
(3) 0 3/2	0	$\dfrac{2\sqrt{2\cdot3\cdot5}}{\sqrt7}$	$-\dfrac{2\cdot2\cdot\sqrt3}{\sqrt7}$
(3) 0 9/2	0	$-\dfrac{3\sqrt{2\cdot5}}{\sqrt7}$	$\dfrac{\sqrt{2\cdot3\cdot5\cdot11}}{\sqrt7}$

$$\left(7/2(v)QJ|||a^{(qj)}|||7/2(v')Q'J'\right)$$

(v)QJ \ (v')Q'J'	(3) 1/2 3/2	(3) 1/2 5/2	(1) 3/2 7/2	(3) 1/2 9/2	(3) 1/2 11/2	(3) 1/2 15/2
(0) 2 0	0	0	$-2\sqrt{2\cdot5}$	0	0	0
(2) 1 2	$-\dfrac{3\sqrt{2\cdot3}}{\sqrt7}$	$\sqrt{3\cdot11}$	$2\sqrt{2\cdot5}$	$\dfrac{\sqrt{5\cdot13}}{\sqrt7}$	$\sqrt{2\cdot3\cdot5}$	0
(4) 0 2	$-\dfrac{2\sqrt{2\cdot11}}{\sqrt7}$	-1	0	$-\dfrac{3\sqrt{3\cdot5\cdot13}}{\sqrt{7\cdot11}}$	$\dfrac{2\sqrt{2\cdot5}}{\sqrt{11}}$	0
(2) 1 4	$\dfrac{3\sqrt{2\cdot11}}{\sqrt7}$	$\dfrac{2\cdot3}{\sqrt{11}}$	$2\cdot3\sqrt2$	$-\dfrac{2\cdot3\cdot5\sqrt5}{\sqrt{7\cdot11}}$	$-\dfrac{3\sqrt{2\cdot13}}{\sqrt{11}}$	$-\dfrac{2\cdot3\sqrt{2\cdot5}}{\sqrt{11}}$
(4) 0 4	$\dfrac{\sqrt{2\cdot3\cdot13}}{\sqrt{5\cdot7}}$	$\dfrac{2\sqrt{3\cdot13}}{\sqrt5}$	0	$\dfrac{2\cdot3\sqrt3}{\sqrt{7\cdot13}}$	$\sqrt{2\cdot3\cdot5}$	$-\dfrac{2\cdot2\sqrt{2\cdot3}}{\sqrt{13}}$
(4) 0 5	$\dfrac{\sqrt{2\cdot3\cdot11}}{\sqrt5}$	$-\dfrac{7}{\sqrt5}$	0	$-3\sqrt3$	$\dfrac{\sqrt{2\cdot5\cdot13}}{\sqrt7}$	$\dfrac{2\sqrt{2\cdot17}}{\sqrt7}$
(2) 1 6	0	$-\dfrac{\sqrt{3\cdot5\cdot13}}{\sqrt{11}}$	$2\sqrt{2\cdot13}$	$\dfrac{7\sqrt5}{\sqrt{11}}$	$-\dfrac{4\sqrt{3\cdot13}}{\sqrt{11}}$	$\dfrac{2\cdot3\sqrt{2\cdot17}}{\sqrt{11}}$
(4) 0 8	0	0	0	$\dfrac{4\sqrt{2\cdot5\cdot17}}{\sqrt{11\cdot13}}$	$\dfrac{4\cdot3\sqrt{17}}{\sqrt{7\cdot11}}$	$\dfrac{2\sqrt{2\cdot3\cdot17\cdot19}}{\sqrt{7\cdot13}}$

$$\sum_{\alpha''Q''J''}\left(j\alpha QJ|||a^{(qj)}|||j\alpha''Q''J''\right)\left(j\alpha'QJ|||a^{(qj)}|||j\alpha''Q''J''\right)$$
$$=\delta(\alpha,\alpha')[j,Q,J], \tag{23.37}$$

which enable one to work out the values of the SCFP by recurrence relations in seniority quantum number v, and to derive their algebraic expressions [168] by the methods described in Chapter 16.

Let us turn now to the totally reduced (reduced in both spaces) matrix elements of tensor $T^{(Kk)}_{\Pi\rho}$ for which we obtain:

$$
\left(j\alpha QJ\|\|T^{(Kk)}\|\|j\alpha'Q'J'\right) = (-1)^{Q+Q'+k+K+J+J'}\sqrt{[k,K]}
$$
$$
\times \sum_{\alpha''Q''J''} \left(j\alpha QJ\|\|a^{(qj)}\|\|j\alpha''Q''J''\right)\left(j\alpha''Q''J''\|\|a^{(qj)}\|\|j\alpha'Q'J'\right)
$$
$$
\times \begin{Bmatrix} q & q & K \\ Q' & Q & Q'' \end{Bmatrix}\begin{Bmatrix} j & j & k \\ J' & J & J'' \end{Bmatrix}. \tag{23.38}
$$

We have used here its inner structure (23.5). The values of the totally reduced matrix (reduced submatrix) elements can be found from pertinent quantities for the conventional submatrix elements of operators T^k using the relationships

$$
\left(j\alpha QJ\|\|T^{(1k)}\|\|j\alpha'Q'J'\right) = \frac{\sqrt{2[Q,k]}}{\begin{bmatrix} Q' & 1 & Q \\ M_{Q'} & 0 & M_Q \end{bmatrix}}
$$
$$
\times \left(j^N\alpha QJM_Q\| T^k \| j^N\alpha'Q'J'M_{Q'}\right);
$$
$$
(k \text{ is even}). \tag{23.39}
$$
$$
\left(j\alpha QJ\|\|T^{(0k)}\|\|j\alpha'QJ'\right) = -\sqrt{2[Q,k]}
$$
$$
\times \left(j^N\alpha QJM_Q\| T^k \| j^N\alpha'QJ'M_{Q'}\right);
$$
$$
(k \text{ is odd}). \tag{23.40}
$$

These reduced submatrix elements have the following transposition property:

$$
\left(j\alpha QJ\|\|T^{(Kk)}\|\|j\alpha'Q'J'\right) = (-1)^{J'-J+Q'-Q}\left(j\alpha'Q'J'\|\|T^{(Kk)}\|\|j\alpha QJ\right). \tag{23.41}
$$

The irreducible tensorial products of the double tensors can be found from

$$
\left(j\alpha QJ\|\|\left[T^{(Kk)}\times T^{(K'k')}\right]^{(K''k'')}\|\|j\alpha'Q'J'\right) = (-1)^{Q+Q'+K''+k''+J+J'}
$$
$$
\times\sqrt{[K'',k'']}\sum_{\alpha''Q''J''}\left(j\alpha QJ\|\|T^{(Kk)}\|\|j\alpha''Q''J''\right)
$$
$$
\times \left(j\alpha''Q''J''\|\|T^{(K'k')}\|\|j\alpha Q'J'\right)\begin{Bmatrix} K & K' & K'' \\ Q' & Q & Q'' \end{Bmatrix}\begin{Bmatrix} k & k' & k'' \\ J' & J & J'' \end{Bmatrix}. \tag{23.42}
$$

Likewise, we can rewrite other equations that yield the totally reduced matrix elements of complex tensorial products.

Note that the conventional submatrix elements $(j^N\alpha QJ\| T^{(Kk)}_{\Pi}\| j^{N'}\alpha'Q'J')$ at $\Pi = 0$ are proportional to the submatrix elements of the irreducible tensors T^k and at $\Pi = \pm 1$, to the CFP with two detached electrons

$$\left(j\alpha QJ ||| \left[a^{(qj)} \times a^{(qj)} \right]^{(1J_2)} ||| j\alpha_1 Q_1 J_1 \right) = \frac{\sqrt{N(N-1)[Q,J]}}{\begin{bmatrix} Q_1 & 1 & Q \\ M_{Q_1} & 1 & M_Q \end{bmatrix}}$$

$$\times \left(j^N \alpha v J \| j^{N-2}(\alpha_1 v_1 J_1) j^2 (J_2) \right), \quad (23.43)$$

i.e. both these quantities are just special cases of the same common quantity – the matrix element of the tensor $T^{(Kk)}$.

23.2 Quasispin for complex electronic configurations

With jj coupling the number of electrons in the subshell is smaller than with LS coupling, but the number of subshells increases, and the configuration becomes more complex. Furthermore, in relativistic treatments it becomes necessary at the very beginning to apply the technique of superposition of configurations. Accordingly in the theory of atoms and ions, even with a small number of electrons, for jj coupling it becomes necessary to deal with complex electronic configurations, which call for an adequate mathematical method.

We shall now turn to the mathematical techniques that rely on the tensorial properties of the quasispin space of two subshells of equivalent electrons. They can be readily generalized to more complex configurations.

The wave function of two subshells is generally constructed by vectorial coupling of the total angular momenta of each subshell

$$\left| j_1^{N_1} j_2^{N_2} \alpha_1 v_1 J_1 \alpha_2 v_2 J_2 J M \right\rangle = \left| j_1^{N_1} j_2^{N_2} \alpha_1 Q_1 J_1 \alpha_2 Q_2 J_2 J M \right\rangle. \quad (23.44)$$

Clearly, for operators corresponding to physical quantities we can derive expansions that enable us to make use of their tensorial properties in quasispin space. Specifically, operator (23.27) will be

$$G = [1 + P(1 \leftrightarrow 2)] \left\{ \sum_J (-1)^k \sqrt{\frac{2J+1}{2k+1}} \begin{Bmatrix} j_1 & j_2 & k \\ j_2 & j_1 & J \end{Bmatrix} \left((-1)^{j_1+j_2+J} \right. \right.$$

$$\times \left[T_1^{(1J)}(j_1) \times T_{-1}^{(1J)}(j_2) \right]^{(0)} b_{1122}^k$$

$$+ \frac{1}{2} \sum_{Q_1 Q_2} \left[T_0^{(Q_1 J)}(j_1) \times T_0^{(Q_2 J)}(j_2) \right]^{(0)} b_{1221}^k \right)$$

$$- \frac{1}{2} b_{1212}^k \sum_{J Q_1 Q_2} \left[T_0^{(Q_1 J)}(j_1) \times T_0^{(Q_2 J)}(j_2) \right]^{(0)} + \frac{1}{2} \sum_{Q_1 Q} \frac{(-1)^{k+j_1+j_2}}{2k+1}$$

$$\times \sqrt{\frac{2Q_1+1}{2Q+1}} \left(\left[a_{\frac{1}{2}}^{(qj_1)} \times \left[T^{(Qk)}(j_2) \times a^{(qj_2)} \right]^{(Q_1 j_1)} \right]_{-\frac{1}{2}}^{(0)} \right.$$

$$\times \left[b_{1222}^k + b_{2122}^k \right] + \left[\left[a^{(qj_2)} \times T^{(Qk)}(j_2) \right]_{\frac{1}{2}}^{(Q_1)} \times a_{-\frac{1}{2}}^{(qj_1)} \right]^{(0)}$$

$$\times \left[b_{2212}^k + b_{2221}^k \right] \Big) \Big\}, \tag{23.45}$$

where the one-shell double tensors $T^{(Kk)}(j_1)$ are given by (23.5) and

$$b_{rstp}^k = -\frac{1}{2} \left(j_r \| f^{(k)} \| j_t \right) \left(j_s \| f^{(k)} \| j_p \right) (n_r l_r j_r n_s l_s j_s | f_{12} | n_t l_t j_t n_p l_p j_p), \tag{23.46}$$

$P(1 \leftrightarrow 2)$ is the operator of permutation of subscripts of the first and second subshells.

There is another way of looking at the tensorial properties of operators and wave functions in the quasispin space of the entire two-shell configuration. If now we introduce the basis tensors for two subshells of equivalent electrons

$$T_\rho^k(j_1, j_2) = \sum_{i=1}^N {}_i t_\rho^k(j_1, j_2), \tag{23.47}$$

where the unit tensors are given by

$$\left(j_i \| t^k(j_1, j_2) \| j_m \right) = \delta(j_i, j_1) \delta(j_2, j_m)(2k + 1), \tag{23.48}$$

and then go over to the second-quantization representation, we arrive at

$$T_\rho^k(j_1, j_2) = - \left[a^{(j_1)} \times a^{(j_2)} \right]_\rho^{(k)}. \tag{23.49}$$

Generally we do not deal with tensors (23.47) themselves, but with their linear combinations

$$^+T_\rho^k(j_1, j_2) = - \Big\{ \left[a^{(j_1)} \times \tilde{a}^{(j_2)} \right]_\rho^{(k)} + (-1)^{j_1 + j_2 - k}$$

$$\times \left[a^{(j_2)} \times \tilde{a}^{(j_1)} \right]_\rho^{(k)} \Big\}, \tag{23.50}$$

$$^-T_\rho^k(j_1, j_2) = - \Big\{ \left[a^{(j_1)} \times \tilde{a}^{(j_2)} \right]_\rho^{(k)} - (-1)^{j_1 + j_2 - k}$$

$$\times \left[a^{(j_2)} \times \tilde{a}^{(j_1)} \right]_\rho^{(k)} \Big\}. \tag{23.51}$$

Considering the tensorial properties of the electron creation and annihilation operators in quasispin space, we shall introduce the double tensor

$$T_{\Pi\rho}^{(Kk)}(j_1, j_2) = \left[a^{(qj_1)} \times a^{(qj_2)} \right]_{\Pi\rho}^{(Kk)}. \tag{23.52}$$

For specific ranks ($K = 0$, $\Pi = 0$) we have

$$T_{0\rho}^{(0k)}(j_1, j_2) = -\frac{1}{\sqrt{2}} + T_\rho^k(j_1, j_2); \qquad (23.53)$$

and for $K = 1$, $\Pi = 0$,

$$T_{0\rho}^{(1k)}(j_1, j_2) = -\frac{1}{\sqrt{2}} - T_\rho^k(j_1, j_2). \qquad (23.54)$$

Thus, the double tensor, defined by (23.52), is a convenient standard quantity for studies of mixed configurations. Let us turn now to the properties of the tensorial products of two-shell operators (23.52). Proceeding in the same way as in derivation of (23.11), we arrive at

$$\left[T^{(Kk)}(j_1) \times T^{(K'k')}(j_2) \right]_{\Pi\rho}^{(K''k'')}$$

$$= -\sum_{K_1 K_2 k_1 k_2} \sqrt{[K, k, K', k', K_1, k_1, K_2, k_2]} \begin{Bmatrix} q & q & K \\ q & q & K' \\ K_1 & K_2 & K'' \end{Bmatrix}$$

$$\times \begin{Bmatrix} j_1 & j_1 & k \\ j_2 & j_2 & k' \\ k_1 & k_2 & k'' \end{Bmatrix} \left[T^{(K_1 k_1)}(j_1, j_2) \times T^{(K_2 k_2)}(j_1, j_2) \right]_{\Pi\rho}^{(K''k'')}. \qquad (23.55)$$

At some specific values of the ranks on the left side of (23.55) we can get a set of equations for sums with the same structure of the ranks of total angular momentum. For the operators that are scalars in the spaces of quasispin and total angular momenta ($K'' = k'' = 0$), we have

$$\sum_k \sqrt{2k+1} \left[T^{(0k)}(j_1, j_2) \times T^{(0k)}(j_1, j_2) \right]^{(00)}$$

$$= 4(\mathbf{Q}_1 \cdot \mathbf{Q}_2) - (2j_1 + 1)(2j_2 + 1)/4 \qquad (23.56)$$

and

$$\sum_k \sqrt{3(2k+1)} \left[T^{(1k)}(j_1, j_2) \times T^{(1k)}(j_1, j_2) \right]^{(00)}$$

$$= -4(\mathbf{Q}_1 \cdot \mathbf{Q}_2) - 3(2j_1 + 1)(2j_2 + 1)/4. \qquad (23.57)$$

Subscripts 1 and 2 of the operators indicate a subshell acted upon with the operator.

Using the properties of the tensors $T^{(Kk)}(j_i)$ and of the $3nj$-coefficients (see Chapter 6), we find the additional relationships for the sums of double tensors

$$\sum_k k(k+1)\sqrt{2k+1} \left[T^{(0k)}(j_1, j_2) \times T^{(0k)}(j_1, j_2) \right]^{(00)}$$

$$= -2(\mathbf{J}_1 \cdot \mathbf{J}_2) + [j_1(j_1 + 1) + j_2(j_2 + 1)]$$

$$\times [4(\mathbf{Q}_1 \cdot \mathbf{Q}_2) - (2j_1 + 1)(2j_2 + 1)/4]; \qquad (23.58)$$

$$\sum_k k(k+1)\sqrt{2k+1}\left[T^{(1k)}(j_1, j_2) \times T^{(1k)}(j_1, j_2)\right]^{(00)}$$

$$= -6(\mathbf{J}_1 \cdot \mathbf{J}_2) - [j_1(j_1 + 1) + j_2(j_2 + 1)]$$
$$\times [4(\mathbf{Q}_1 \cdot \mathbf{Q}_2) + 3(2j_1 + 1)(2j_2 + 1)/4]. \qquad (23.59)$$

For $K = 2$, $k = 0$ we have

$$\sum_k \sqrt{2k+1}\left[T^{(1k)}(j_1, j_2) \times T^{(1k)}(j_1, j_2)\right]^{(20)} = 8\left[Q_1^{(1)} \times Q_2^{(1)}\right]^{(2)} \qquad (23.60)$$

and

$$\sum_k k(k+1)\sqrt{2k+1}\left[T^{(1k)}(j_1, j_2) \times T^{(1k)}(j_1, j_2)\right]^{(20)}$$

$$= 8[j_1(j_1 + 1) + j_2(j_2 + 1)]\left[Q_1^{(1)} \times Q_2^{(1)}\right]^{(2)}. \qquad (23.61)$$

Equations (23.56)–(23.58) hold for any j_1 and j_2, but for some specific values of j_1 and j_2 these relationships are simplified, thereby enabling us to obtain analytical expressions for individual tensorial products of double tensors. It is worth noting that such expressions can only be derived at $j_1 = 1/2$ (j_2 is arbitrary).

The above relationships describe the behaviour of tensors (23.52) and their tensorial products. A further task is to express the operators corresponding to physical quantities in terms of these tensors. Specifically, two-particle operator (23.27) is expanded by going over to the second-quantization representation and coupling the quasispin ranks

$$G = \left[1 + P(1 \leftrightarrow 2)\sum_{i=1}^3 G_i\right], \qquad (23.62)$$

where

$$G_1 = -\left[b_{1122}^k c_{1212} + b_{1212}^k c_{1122}\right]; \qquad (23.63)$$

$$G_2 = -b_{1221}^k\left[c_{1221} + \frac{(-1)^{k+j_1+j_2}\sqrt{2}}{\sqrt{[k, j_1]}}\sum_K T^{(K0)}(j_1, j_1)\right]; \qquad (23.64)$$

$$G_3 = -\left[c_{2212}(b_{1222}^k + b_{2122}^k) + c_{2122}(b_{2212}^k + b_{2221}^k)\right]. \qquad (23.65)$$

We have used here the notation

$$c_{rstp} = \frac{1}{2}\sum_{KK'K''}\frac{1}{2k+1}\begin{bmatrix} K & K' & K'' \\ 0 & 0 & 0 \end{bmatrix}\left[T^{(Kk)}(j_r, j_s) \times T^{(K'k)}(j_t, j_p)\right]^{(K''0)}.$$

$$(23.66)$$

We have thus arrived at the expansion of the two-particle operator in terms of irreducible tensorial products having certain ranks in relation to the total and quasispin angular momenta of both subshells. This representation of two-particle operators can conveniently be used for the form of multi-configuration approximation in which the wave function of two subshells is constructed using the vectorial coupling scheme for quasispin angular momenta of individual subshells

$$|(j_1+j_2)^N\alpha_1 Q_1 J_1\alpha_2 Q_2 J_2 JM\rangle$$

$$= \sum_{M_{Q_1}M_{Q_2}} \begin{bmatrix} Q_1 & Q_2 & Q \\ M_{Q_1} & M_{Q_2} & M_Q \end{bmatrix} |j_1^{N_1} j_2^{N_2}\alpha_1 Q_1 J_1\alpha_2 Q_2 J_2 JM\rangle, \quad (23.67)$$

where $M_Q = -(j_1+j_2+1-N)/2$.

Thus, the use of wave function in the form (23.67) and operators in the form (23.62)–(23.66) makes it possible to separate the dependence of multi-configuration matrix elements on the total number of electrons using the Wigner–Eckart theorem, and to regard this form of superposition-of-configuration approximation itself as a one-configuration approximation in the space of total quasispin angular momentum.

23.3 Isospin basis for two $n_1l_1j^{N_1}n_2l_2j^{N_2}$ subshells

The isospin operator has been introduced for the non-relativistic configuration $n_1l^{N_1}n_2l^{N_2}$. Here we shall outline the uses of this concept in the relativistic case of the two subshells $n_1l_1j^{N_1}n_2l_2j^{N_2}$. Note that many of the relations in Chapter 18 can readily be adapted to the case of jj coupling if we substitute the $n_1l_1j^{N_1}n_2l_2j^{N_2}$ configuration for the $n_1l^{N_1}n_2l^{N_2}$ configuration, the tensor rank j for l and s, and the factor $2j+1$ for those of type $(2l+1)(2s+1)$. This accounts for the briefness of the presentation in this section, although we should still have in mind that the operators of physical quantities (Hamiltonian, operators of electronic transitions) in the relativistic approach have another form, therefore there is no way of drawing complete analogy between the LS and jj couplings.

We shall now introduce the electron creation operators $a_m^{(j)}$ and $b_m^{(j)}$ and the electron annihilation operators $\tilde{a}_m^{(j)} = (-1)^{j-m}a_{-m}^{(j)\dagger}$, $\tilde{b}_m^{(j)} = (-1)^{j-m}b_{-m}^{(j)\dagger}$ for the electrons in the subshells $n_1l_1j^{N_1}$ and $n_2l_2j^{N_2}$, respectively. They are irreducible tensors of rank $\tau = 1/2$

$$a_{\frac{1}{2}m}^{(\tau j)} = b_m^{(j)}; \qquad a_{-\frac{1}{2}m}^{(\tau j)} = a_m^{(j)}; \qquad \tilde{a}_{\frac{1}{2}m}^{(\tau j)} = \tilde{a}_m^{(j)}; \qquad \tilde{a}_{-\frac{1}{2}m}^{(\tau j)} = -\tilde{b}_m^{(j)} \quad (23.68)$$

in the space of isospin angular momentum

$$T_\rho^{(1)} = \frac{i}{2}\sqrt{[\tau, j]}\left[a^{(\tau j)} \times \tilde{a}^{(\tau j)}\right]_{\rho 0}^{(10)}, \qquad (23.69)$$

so that

$$\tilde{a}_{\eta m}^{(\tau j)} = (-1)^{\tau - \eta + j - m} a_{-\eta - m}^{(\tau j)\dagger}. \qquad (23.70)$$

In analogy with (18.18)–(18.20) we conclude that the operator $T_1^{(1)}$ annihilates an electron in the first subshell and creates it in the second (and the operator $T_{-1}^{(1)}$ vice versa), so that the total number of electrons in the subshells will remain unchanged ($N = N_1 + N_2 = N_1' + N_2'$). The operator $T_0^{(1)}$ has the form $i(\hat{N}_2 - \hat{N}_1)/2$, where $\hat{N}_1$ and $\hat{N}_2$ are the particle number operators for the first and second subshells, respectively.

Out of tensors (23.68) we can form irreducible operators (cf. (18.24)–(18.25))

$$U_{\Gamma\rho}^{(\gamma k)} = \left[a^{(\tau j)} \times \tilde{a}^{(\tau j)}\right]_{\Gamma\rho}^{(\gamma k)}; \qquad U_{\Gamma\rho}^{(\gamma k)\dagger} = (-1)^{\Gamma + \rho} U_{-\Gamma-\rho}^{(\gamma k)}, \qquad (23.71)$$

in terms of which (or in terms of their tensorial products) we can express operators corresponding to physical quantities. So the particle number operator and the total angular momentum operator $J^{(1)}$ will be

$$\hat{N} = \sqrt{[\tau, j]}\, U^{(00)}; \qquad (23.72)$$

$$J_\rho^{(1)} = i[2j(j+1)(2j+1)/3]^{1/2} U_{0q}^{(01)}. \qquad (23.73)$$

The many-electron wave function in the isospin basis $|(jj)^N \beta T J M_T M_J\rangle$ (where β denotes all additional quantum numbers required for the levels to be classified uniquely) will be an eigenfunction of a set of commuting operators $\hat{N}$, $\mathbf{T}^2$, $T_z = (\hat{N}_2 - \hat{N}_1)/2$, $\mathbf{J}^2$, J_z. Here, as in Chapter 18, T takes on the values $N/2$, $N/2 - 1\ldots$, $1/2(0)$. The last value $T = 1/2$ (0) means that T equals 1/2 or 0 depending on the parity of N.

The antisymmetric wave function for N electrons ($0 \le N \le 2(2j+1)$) in the isospin basis can be constructed out of conventional antisymmetric functions obtainable by vectorial coupling of momenta of individual shells, i.e.

$$\left|n_1 n_2 l_1 l_2 (jj)^N \beta T J M_T M_J\right\rangle = \sum_{\alpha_1 J_1 \alpha_2 J_2} \left|n_1 n_2 l_1 j^{N_1} l_2 j^{N_2} \alpha_1 J_1 \alpha_2 J_2 J M_J\right\rangle$$

$$\times \left(j^{N_1} j^{N_2} \alpha_1 J_1 \alpha_2 J_2 J | (jj)^N \beta T J\right), \qquad (23.74)$$

where the last factor is the matrix of transformation between the conventional and isospin bases, which, in turn, is represented in terms of CFP with N_2 detached electrons defined in the isospin basis (provided that

$T_1 = N_1/2$, $T_2 = N_2/2$)

$$\left(j^{N_1} j^{N_2} \alpha_1 J_1 \alpha_2 J_2 J | (jj)^N \beta T J M_T \right) = (-1)^{T-N/2}$$

$$\times \left[\frac{(2T+1)N!}{(N/2-T)!(N/2+T+1)!} \right]^{1/2}$$

$$\times \left(j^{N_1}(\alpha_1 T_1 J_1) j^{N_2}(\alpha_2 T_2 J_2) \| j^N \beta T J \right). \quad (23.75)$$

At $T = N/2$ these expressions become the conventional CFP of one subshell

$$(j^{N_1}(\alpha_1(T_1 = N_1/2)J_1) j^{N_2}(\alpha_2(T_2 = N_2/2)J_2) \| (jj)^N \beta (T = N/2)J)$$

$$= \left(j^{N_1}(\alpha_1 J_1) j^{N_2}(\alpha_2 J_2) \| j^N \beta J \right). \quad (23.76)$$

If we remember that isospin behaves as angular momentum in a certain additional space, we shall be able to apply the Wigner–Eckart theorem to the matrix elements of appropriate tensors in that space, and also the entire technique of the SU_2 group. So, having applied this theorem to a certain matrix element in both spaces, we obtain

$$\left(n_1 n_2 (jj)^N \beta T J M_T M_J | F_{\Gamma \rho}^{(\gamma k)} | n_1 n_2 (jj)^{N'} \beta' T' J' M_T' M_J' \right)$$

$$= \frac{(-1)^{2\gamma+2k}}{[T,J]^{1/2}} \begin{bmatrix} T' & \gamma & T \\ M_T' & \Gamma & M_T \end{bmatrix} \begin{bmatrix} J' & k & J \\ M_J' & \rho & M_J \end{bmatrix}$$

$$\times \left(n_1 n_2 (jj)^N \beta T J \| T^{(\gamma k)} \| n_1 n_2 (jj)^{N'} \beta' T' J' \right). \quad (23.77)$$

The $N \le 2j+1$ configuration is assumed to be partially filled and the $N > 2j+1$ configuration almost filled. Rotations of wave functions in isospin space will also be described by a formula of type (18.41). As noted in Chapter 18, using the isospin basis aggravates the problem of classifying repeating states. For jj coupling it emerges at $j \ge 3/2$. For $j = 1/2$ we have

$$\hat{T}^2 + \hat{J}^2 = 2 \left[U^{(11)} \times U^{(11)} \right]^{(00)} = \hat{N}(4 - \hat{N})/2, \quad (23.78)$$

i.e. for these configurations the isospin characteristic is extraneous. As j increases, the number of degenerate states rises. To classify them we can resort to groups of higher ranks, with the characteristics of irreducible representations used as additional quantum numbers. Let us consider the set of operators

$$U_{\Gamma \rho}^{(\gamma k)}, \quad V_{\Gamma \rho}^{(\gamma k)}, \quad \tilde{V}_{\Gamma \rho}^{(\gamma k)}, \quad (23.79)$$

where the first operator is given by (23.71) and

$$V_{\Gamma\rho}^{(\gamma k)} = \left[a^{(\tau j)} \times a^{(\tau j)}\right]_{\Gamma\rho}^{(\gamma k)};$$
(23.80)

$$\tilde{V}_{\Gamma\rho}^{(\gamma k)} = (-1)^{\Gamma+\rho} V_{-\Gamma-\rho}^{(\gamma k)\dagger} = \left[\tilde{a}^{(\tau j)} \times \tilde{a}^{(\tau j)}\right]_{\Gamma\rho}^{(\gamma k)}.$$
(23.81)

In (23.80) and (23.81) the rank sum $\gamma + k$ is an odd number, otherwise these operators are identically equal to zero. We shall separate sets of operators that are scalars in the space of total angular momentum but tensors in isospin space. If we go through a similar procedure for one subshell of equivalent electrons we shall end up with the quasispin classification of its states. It turns out that ten operators $U^{(00)}$, $U_{\rho 0}^{(10)}$, $V_{\rho 0}^{(10)}$, $\tilde{V}_{\rho 0}^{(10)}$ are generators of a group of five-dimensional quasispin, which can be easily verified by comparing their commutation relations with the standard commutation relations for generators of that group.

Here, for jj coupling, irreducible representations of Sp_4 group will be characterized by two parameters (v, t) where v is the quantum number of the five-dimensional quasispin, and t is the quantum number of the reduced isospin (at $N = v$, the quantum number t coincides with isospin T). Out of the remaining operators (23.79) we shall select a maximal complete (under commutation) subset of operators commuting with all the generators of Sp_4 group. These are operators $U_{0\rho}^{(0k)}$ (k is odd) that serve as generators of Sp_{2j+1} group and then we get direct product $Sp_4 \times Sp_{2j+1}$. For the Sp_4 group it is convenient to use the reduction scheme $Sp_4 \supset SU_2^T \times U_1$, where SU_2^T is the group of isospin angular momentum, and U_1 is the group whose generator is the operator of the particle number $\hat{N}$. This scheme enables states to be classified according to the particle number, the quantum numbers of isospin and its projection.

For Sp_{2j+1} a convenient way is the reduction scheme $Sp_{2j+1} \supset SU_2^J$ where SU_2^J is the group of total angular momentum that enables states to be classified according to the quantum numbers J, M_J. So in this case the states of the $n_1 l_1 j^{N_1} n_2 l_2 j^{N_2}$ configuration can be characterized by the set of quantum numbers $(vt)TJM_J$.

23.4   Relativistic energy of electron–electron interaction in isospin basis

Now we shall have to express operators for physical quantities in terms of irreducible tensors in the spaces of total angular momentum and quasispin. One-electron terms of relativistic energy operator (2.1) (formulas (2.2)–(2.4)) are expressed in terms of operators (23.69), (23.71)–(23.73) in a trivial way. With two-electron operators the procedure of deriving the pertinent relations is more complex. The relativistic counterpart of (18.50) looks like

$$\hat{G} = \sum_k \left\{ \frac{1}{2} \sum_{T_1 T_2 T} \begin{bmatrix} T_1 & T_2 & T \\ 0 & 0 & 0 \end{bmatrix} \left[U^{(T_1 k)} \times U^{(T_2 k)} \right]_{00}^{(T0)} \right.$$

$$\times \left[F_k(\lambda_2 \lambda_2, \lambda_2 \lambda_2) + (-1)^T F_k(\lambda_1 \lambda_1, \lambda_1 \lambda_1) + 2(-1)^{T_1} F_k(\lambda_1 \lambda_2, \lambda_1 \lambda_2) \right]$$

$$- 2 \sum_T \begin{bmatrix} 1 & 1 & T \\ -1 & 1 & 0 \end{bmatrix} \left[U^{(1k)} \times U^{(1k)} \right]_{00}^{(T0)} F_k(\lambda_1 \lambda_1, \lambda_2 \lambda_2)$$

$$- (-1)^k \sqrt{\frac{2k+1}{2(2j+1)}} \sum_T U_{00}^{(T0)} \left[F_k(\lambda_2 \lambda_2, \lambda_2 \lambda_2) + (-1)^T F_k(\lambda_1 \lambda_1, \lambda_1 \lambda_1) \right.$$

$$\left. + 2(-1)^T F_k(\lambda_1 \lambda_1, \lambda_2 \lambda_2) \right] + \sqrt{2} \sum_{T_2 T; M_T = \pm 1} (-1)^{(T_2 + T)(1 + M_T)/2}$$

$$\times \begin{bmatrix} 1 & T_2 & T \\ -1 & 0 & -1 \end{bmatrix} \left[U^{(1k)} \times U^{(T_2 k)} \right]_{M_T 0}^{(T0)} \left[F_k(\lambda_1 \lambda_2, \lambda_2 \lambda_2) \right.$$

$$\left. + (-1)^{T_2} F_k(\lambda_1 \lambda_1, \lambda_1 \lambda_2) \right] + (-1)^k 2 \sqrt{\frac{2k+1}{2j+1}} \left[U_{10}^{(10)} F_k(\lambda_1 \lambda_1, \lambda_1 \lambda_2) \right.$$

$$\left. - U_{-10}^{(10)} F_k(\lambda_1 \lambda_2, \lambda_2 \lambda_2) \right] + \sum_{M_T = \pm 2} \left[U^{(1k)} \times U^{(1k)} \right]_{M_T 0}^{(20)} F_k(\lambda_1 \lambda_1, \lambda_2 \lambda_2) \right\},$$

(23.82)

where $\lambda = nlj$, and the quantity $F_k(\lambda_1 \lambda_2, \lambda_3 \lambda_4)$ is related to the reduced matrix elements of the relativistic energy operator by

$$\left(n_1 l_1 j n_2 l_2 j J \| G^{(kk)} \| n_3 l_3 j n_4 l_4 j J \right)$$

$$= (-1)^{J+k+1} 2[k, J]^{1/2} \begin{Bmatrix} j & j & J \\ j & j & k \end{Bmatrix} F_k(\lambda_1 \lambda_2, \lambda_3 \lambda_4). \quad (23.83)$$

This expression holds both for electrostatic interaction energy operator (2.5) and for magnetic and retardation interactions (2.6) and (2.7).

Note that the first three terms of operator (23.82) in the braces, which contain only the summation over the isospin quantum numbers, describe the part of interaction that is diagonal in relation to the quantum numbers of a configuration (appropriate tensors have zero projection of the isospin operator, i.e. $M_T = 0$). The remaining three terms contain $M_T \neq 0$; their matrix elements are only non-zero when the wave functions have different electron distributions over the subshells. So, the matrix elements of operators with projections $M_T = \pm 1$ couple the $n_1 l_1 j^{N_1} n_2 l_2 j^{N_2}$ and $n_1 l_1 j^{N_1 \pm 1} n_2 l_2 j^{N_2 \mp 1}$ configurations, and those with $M_T = \pm 2$ couple the $n_1 l_1 j^{N_1} n_2 l_2 j^{N_2}$ and $n_1 l_1 j^{N_1 \pm 2} n_2 l_2 j^{N_2 \mp 2}$ configurations.

In consequence, the first group of terms in (23.82) corresponds to the interaction energy within a given relativistic configuration and the others

Table 23.2. The weights of usual (α_i) and isospin (β_i) bases for configuration $1s^2 2s^2 2p\left(\frac{1}{2}\right)^N \left(\frac{3}{2}\right)^{N_1} 3p\left(\frac{3}{2}\right)^{N_2} \left(J = \frac{3}{2}\right)$

	$N; N_1, N_2$	$J_1, J_1'; J_2, J_2'$	T, T'	α_1 α_2	β_1 β_2
Pb^{73+}	2;2,1	$0, 2; \frac{3}{2}, \frac{3}{2}$	$\frac{3}{2}, \frac{1}{2}$	0.887	0.784
				−0.462	0.621
Pb^{73+}	2;2,1			0.937	0.702
				−0.350	0.712
Pb^{73+}	2;1,2	$\frac{3}{2}; \frac{3}{2}; 2, 0$		0.931	0.999
				−0.366	0.046
W^{67+}	0;1,2			0.929	0.999
				0.360	−0.043
Xe^{47+}	0;1,2			0.916	1.000
				−0.400	−0.009

to the interaction between them. The heavy reliance on the tensorial properties of operators and wave functions in their respective spaces permits a unified approach to both diagonal and non-diagonal (in relation to the quantum numbers of the configuration) matrix elements and enables some of the correlation effects to be included.

Now we shall take some numerical examples to illustrate the applicability of the isospin basis for description of relativistic electronic configurations. We should take into account that jj coupling within a shell of equivalent electrons is only encountered when relativistic effects begin to dominate. So we should use the appropriate, e.g. relativistic Hartree–Fock, wave functions and include magnetic and retardation interactions. Some results of this kind are provided in Table 23.2. The Dirac–Hartree–Fock radial orbitals are used; in the first line only the part of the Hamiltonian corresponding to electrostatic interaction is taken into account, in the other cases the Hamiltonian in the Breit approximation is diagonalized. It is seen from the table that there are cases, notably for configurations of the type $N_2 = N_1 + 1$ and $n_2 = n_1 + 1$, where the isospin basis for ions under consideration is closer to reality than the conventional one, and then isospin is a fairly accurate quantum number. Consequently, the use of the isospin basis can shed new light on the nature of some classes of excited atoms and ions. However, further studies of this problem are highly desirable.

Part 6

Electric and Magnetic Multipole Transitions

24

General expressions for electric (Ek) and magnetic (Mk) multipole radiation quantities

24.1 Line and multiplet strengths

A number of ideas of the theory of electronic transitions were discussed in Chapter 4. In Part 6 we are going to consider this issue in more detail. Let us start with the definition of the main characteristics of electronic transitions, common for both electric and magnetic multipole radiation.

As we have seen while considering energy spectra, the energy levels of free atoms are always degenerate relative to the projections M of the total angular momentum J. Further we shall learn that the characteristics of spontaneous electronic transitions do not depend on them. Let us define the line strength of the electronic transition of any multipolarity k as the modulus of the relevant matrix element squared, i.e.

$$S(\alpha J, \alpha' J') = \sum_{MM'q} \left| \left(\alpha J M | O_q^{(k)} | \alpha' J' M' \right) \right|^2, \qquad (24.1)$$

where the operator $O_q^{(k)}$ describes either electric $Q_q^{(k)}$ or magnetic $M_q^{(k)}$ multipole radiation. Line strengths are symmetric with regard to the transposition of the initial and final states, namely,

$$S(\alpha J, \alpha' J') = S(\alpha' J', \alpha J). \qquad (24.2)$$

The total line strength does not depend on the coupling scheme, i.e.

$$S = \sum_{a_i a_f} \left| \left(a_i | O_q^{(k)} | a_f \right) \right|^2 = \sum_{b_i b_f} \left| \left(b_i | O_q^{(k)} | b_f \right) \right|^2, \qquad (24.3)$$

Equation (24.3) is also valid for intermediate coupling. Applying the Wigner–Eckart theorem (5.15) to (24.1) and performing the summation, we arrive at the result

$$S(\alpha J, \alpha' J') = \left| \left(\alpha J \| O^{(k)} \| \alpha' J' \right) \right|^2, \qquad (24.4)$$

293

showing that the line strength is equal to the square of the submatrix element of the electron transition operator. Suppose that LS coupling takes place in the configuration considered. Then the energy levels will be characterized by the αLSJ quantum numbers. A set of lines occurring due to the transitions between all levels of one term and those of the other one composes the multiplet. The total strength of all lines is called the multiplet strength

$$S(\alpha LS, \alpha' L'S') = \sum_{JJ'} S(\alpha LSJ, \alpha' L'S'J'). \tag{24.5}$$

For a pure LS coupling scheme, both the electric and magnetic multipole transitions are diagonal with regard to S and S'. The multiplet strength is also symmetric with respect to the transposition of the initial and final terms

$$S(\alpha LS, \alpha' L'S') = S(\alpha' L'S', \alpha LS). \tag{24.6}$$

Equations of the kind (24.5) and (24.6) hold for any vectorial coupling scheme, including an intermediate one. The analogue of Eq. (24.4) for (24.5) will be

$$S(\alpha LS, \alpha' L'S') = \left| \left(\alpha LS \| O^{(k)} \| \alpha' L'S' \right) \right|^2. \tag{24.7}$$

The ratio of the strength of one line in the multiplet to the strength of the whole multiplet is called the relative line strength $\bar{S}$ in the multiplet

$$\bar{S}(LSJ, L'SJ') = S(\alpha LSJ, \alpha' L'SJ')/S(\alpha LS, \alpha' L'S). \tag{24.8}$$

It may be written in the form

$$\bar{S}(LSJ, L'SJ') = \frac{(2J+1)(2J'+1)}{2S+1} \left\{ \begin{matrix} L & J & S \\ J' & L' & k \end{matrix} \right\}^2, \tag{24.9}$$

showing that the relative line strength in the multiplet depends only on quantum numbers L, L', S, J and J'. It follows from (24.8) that the relative line strengths are normalized to unity

$$\sum_{JJ'} \bar{S}(LSJ, L'SJ') = 1 \tag{24.10}$$

and are symmetric

$$\bar{S}(LSJ, L'SJ') = \bar{S}(L'SJ', LSJ). \tag{24.11}$$

Making use of the orthogonality condition for $6j$-coefficients (6.25), by (24.9), we establish the following sum rules for relative line strengths:

$$\sum_{J'} \bar{S}(LSJ, L'SJ') = g(J)/g(LS) = (2J+1)/[(2L+1)(2S+1)], \tag{24.12}$$

$$\sum_{J} \bar{S}(LSJ, L'SJ') = g(J')/g(L'S) = (2J'+1)/[(2L'+1)(2S+1)]. \tag{24.13}$$

The quantity $g(J)/g(LS)$ stands for the relative statistical weight of the J level in the LS term. Therefore, formulas (24.12) and (24.13) show that the sum of relative line strengths of the multiplet with given initial or final level is equal to the relative statistical weight of this level in the term.

In a number of specific cases, for fixed values of part of the parameters, it is possible to find simple algebraic expressions for the $6j$-coefficient in (24.9) and to establish certain regularities in the quantities of the relative and absolute line strengths. Unfortunately, as a rule they are approximate due to the deviations from the pure coupling scheme. For the same reasons the line strengths themselves must be calculated in the intermediate coupling scheme, by determining the submatrix elements of the appropriate operators in the pure coupling scheme with the exact phase multipliers, and only afterwards squaring them.

24.2 Oscillator strength, transition probability, lifetime and line intensity

Line and multiplet strengths are useful theoretical characteristics of electronic transitions, because they are symmetric, additive and do not depend on the energy parameters. However, they are far from the experimentally measured quantities. In this respect it is much more convenient to utilize the concepts of oscillator strengths and transition probabilities, already directly connected with the quantities measured experimentally (e.g. line intensities). Oscillator strength f^k of electric or magnetic electronic transition $\alpha J \rightarrow \alpha' J'$ of multipolarity k is defined as follows:

$$f^k(\alpha J, \alpha' J') = \frac{(2k+1)(k+1)m\omega^{2k-1}}{[(2k+1)!!]^2 k\hbar e^2 c^{2k-2}(2J+1)} S^k(\alpha J, \alpha' J'), \qquad (24.14)$$

where $\omega = \frac{1}{\hbar}(E_{\alpha J} - E_{\alpha' J'})$. Thus, the oscillator strength is a dimensionless quantity and is proportional to the line strength. The definition of the oscillator strength in the form (24.14) implies that the oscillator strength is positive for radiation. There follows from (24.14) and (24.2) the symmetry property

$$(2J+1)f^k(\alpha J, \alpha' J') = -(2J'+1)f^k(\alpha' J', \alpha J). \qquad (24.15)$$

Thus, the quantities gf (oscillator strengths multiplied by statistical weights $g = 2J + 1$) are more symmetric than f. The definition of oscillator strength (24.14) in terms of line strength is fairly general. It is valid both for electric and magnetic multipole radiation, and for any coupling scheme, including the intermediate one. The whole dependence of this quantity on coupling scheme is contained in the line strength.

In the system of atom plus electromagnetic field, there must be valid energy and angular momentum conservation laws. A free atom, being in an excited state, passes to another state having lower energy spontaneously radiating the photon, which takes away the energy $\hbar\omega = 2\pi\hbar\nu$ and certain angular momentum with regard to the radiation source.

The general definition of the electron transition probability is given by (4.1). More concrete expressions for the probabilities of electric and magnetic multipole transitions with regard to non-relativistic operators and wave functions are presented by formulas (4.10), (4.11) and (4.15). Their relativistic counterparts are defined by (4.3), (4.4) and (4.8). They all are expressed in terms of the squared matrix elements of the respective electron transition operators. There are also presented in Chapter 4 the expressions for electric dipole transition probabilities, when the corresponding operator accounts for the relativistic corrections of order α^2. If the wave functions are characterized by the quantum numbers LJ, $L'J'$, then the right sides of the formulas for transition probabilities must be divided by the multiplier $2J + 1$.

The excited state has a finite lifetime. The lifetime τ of any state of excited level αJ (their number is $2J + 1$) of a free atom is defined as

$$\tau_{\alpha J} = 1/\sum_{\alpha' J'} W(\alpha J, \alpha' J'), \tag{24.16}$$

where the summation is over all transitions into lower levels. In principle, these transitions may be either of the electric or magnetic type. However, as a rule, one type of transition and its multipolarity prevails. Due to the finiteness of the lifetime of the excited level, it has a certain natural width which is usually very small.

Estimations indicate that if the wavelength of the radiation is large when compared with the size of the source (this is the case for all atoms and the overwhelming majority of ions), then the probability for radiation of a photon of multipolarity k is a rapidly decreasing function of k. Therefore, in the expansions over multipoles of certain parity it is usually sufficient to take into account the first non-zero term. However, contributions of Ek- and $M(k-1)$-transitions may be of the same order.

Oscillator strength or transition probability is the individual characteristic of a separate atom or ion. However, in reality we usually have to deal with a large number of them, where, depending on the specific physical situation, various elementary processes of excitation, ionization, recombination, etc. may take place. Real spectral lines are characterized by the intensity of radiation, defined in the conditions of natural isotropic excitation as

$$I(\alpha J, \alpha' J') = N_{\alpha J}\hbar\omega W(\alpha J, \alpha' J'), \tag{24.17}$$

where $N_{\alpha J}$ is the population of the excited level αJ (number of atoms or ions in 1 cm^3, occupying the level αJ). It follows from (24.17) that the intensity of the line is proportional to the transition probability, its wavelength and the number of atoms radiating this spectral line. Thus

$$I \sim gW \sim gf \sim S. \tag{24.18}$$

Line intensities are additive

$$I = \sum_{kq} (I_q^{Ek} + I_q^{Mk}). \tag{24.19}$$

Unlike line or oscillator strengths and transition probabilities, line intensities are directly measured quantities.

Spontaneous transitions are not the only possible transitions. Electronic transitions may be also induced by, for example, an external radiation field. According to the detailed equilibrium principle, the rate of transitions from all states of the lower level $\alpha'J'$ into all states of the upper level αJ, caused by the absorption of photons from the radiation field, must be equal to the rates of spontaneous and induced transitions from the level αJ into $\alpha'J'$, i.e.

$$g_{\alpha'J'} B(\alpha'J', \alpha J) N_{\alpha'J'} \rho(\alpha J, \alpha'J')$$
$$= (g_{\alpha J} W(\alpha J, \alpha'J') + g_{\alpha J} B(\alpha J, \alpha'J') \rho(\alpha J, \alpha'J')) N_{\alpha J}, \tag{24.20}$$

where $\rho(\alpha J, \alpha'J')$ is the radiation energy density in the unit frequency interval. $B(\alpha'J', \alpha J)$ and $B(\alpha J, \alpha'J')$ are called the Einstein coefficients for absorption and induced radiation, respectively. Since ρ usually is small, the induced radiation may often be neglected. If $N_{\alpha'J'} \gg N_{\alpha J}$ then the absorption, including the self-absorption, may be strong. For $N_{\alpha J} \gg N_{\alpha'J'}$ we have the so-called inverse population of the levels, that is the necessary condition for the generation of laser radiation. In such a case the induced radiation plays the leading role.

24.3 'Exact' selection rules for electronic transitions

Selection rules for electronic transitions consist of the conditions of non-zero values of the corresponding matrix elements. These conditions may be presented in the form of the polygon rules: numerical values of the parameters (ranks of tensors and quantum numbers describing the atom or ion considered) must form a polygon having an integer perimeter. The degree of accuracy of various quantum numbers is different, therefore selection rules may also be divided into 'exact' and 'approximate'. However, this classification is rather conditional, which is why these words are written within quotation marks. Usually the parity π of the electronic configuration and the total momentum J are considered as the exact quantum numbers. If we call the characteristics of electronic configuration

K itself the quantum numbers, then in a single-configuration approach they may be treated as exact quantum numbers as well.

However, due to the admixture of weak interactions it may occur that the parity is no longer a completely exact quantum number. The same is true for J if we account for hyperfine interactions. Fortunately, due to the weakness of the above-mentioned interactions, the parity and total momentum are the most accurate quantum numbers. In many cases a single-configuration approximation describes fairly accurately atomic characteristics, then the configuration may also be treated as an exact quantum number. However, quite often one has to account for the admixtures (superposition) of other configurations.

Let us consider first the selection rules with respect to $K\pi J$. As already stated, the configuration unambiguously fixes the parity of the state under consideration, therefore we shall not denote it. In general, the energy level is denoted by the quantum numbers $K\alpha J$. Then for LS coupling the level will be $K\alpha LSJ$, and the term, $K\alpha LS$. Electronic transitions take place between two levels of one and the same configuration K or of two configurations K and K'. The concrete realization of these possibilities is defined by the appropriate selection rules for radiation. They differ for electric and magnetic multipole transitions. Therefore below we shall consider them separately, specifying the tensorial structure of the respective operators.

Operators of electric and magnetic multipole radiation (e.g. see their forms (4.5), (4.6), (4.9), (4.12), (4.13), (4.16) and (4.21)) are the one-electron quantities of type (13.22). One-configurational wave functions of two different physical states (in our case, two different electronic configurations) are orthogonal to each other. Therefore only those matrix elements of one-electron operators will be non-zero which combine the same configurations (then we have diagonal, with respect to configurations, matrix elements) or the configurations differing by the quantum numbers of one electron. Thus, in a single-configurational approach, electronic transitions are permitted between the levels of one and the same configuration ($K = K'$) or between the levels of configurations differing by the quantum numbers of only one electron.

In order to establish more detailed selection rules, we have to examine the tensorial structure of the respective operators. Let us start with selection rules for orbital momentum L. The operator of Ek-radiation (4.12) contains the tensor $C^{(k)}$, whereas that of Mk-radiation (4.16) contains tensorial products $\left[C^{(k-1)} \times L^{(1)}\right]^{(k)}$ and $\left[C^{(k-1)} \times S^{(1)}\right]^{(k)}$. Therefore, matrix elements of these operators contain one-electron submatrix elements $\left(l\|C^{(k)}\|l'\right)$, $\left(l\|\left[C^{(k-1)} \times L^{(1)}\right]^{(k)}\|l'\right)$ and $\left(l\|\left[C^{(k-1)} \times S^{(1)}\right]^{(k)}\|l'\right)$. The first submatrix element is non-zero at the triangular condition $\{ll'k\}$, and if $l+l'+k$ is even. The second and third submatrix elements are non-zero at $\{ll'k-1\}$, and $l+l'+k-1$ is even. Hence, the Ek-transitions are permitted between configurations for which the parity of Δl coincides with that of multipolarity k, whereas for Mk-transitions it must be opposite to that of k.

In the particular case of electric dipole radiation $\Delta l = 1$, i.e. $E1$-transitions are permitted between configurations of opposite parity. For $E2$-transitions $\Delta l = 0, \pm 2$ (excluding transitions $ns - n's$), i.e. they are allowed between levels of one and the same configuration or between configurations of the same parity. $M1$-transitions may take place only between levels of one and the same configuration. There are no restrictions on Δn for Ek-transitions. Selection rules for J and M follow from the Clebsch–Gordan coefficient

$$\begin{bmatrix} J' & k & J \\ M' & q & M \end{bmatrix}$$

occurring while applying the Wigner–Eckart theorem (5.15) to the appropriate matrix element of the electronic transition in question. Summarizing, the 'exact' selection rules for Ek- and Mk-transitions are as follows:

$$Ek : \begin{cases} \{ll'k\} \text{ (or } \Delta l = 0, \pm 1, \ldots, \pm k) \text{ and } l + l' + k \text{ is even,} \\ \{JJ'k\} \text{ (or } \Delta J = 0, \pm 1, \ldots, \pm k \text{ and } J + J' \geq k), \ M' + q = M. \end{cases}$$
$$(24.21)$$

$$Mk : \begin{cases} \{ll'k - 1\} \text{ (or } \Delta l = 0, \pm 1, \ldots, \pm(k-1)) \text{ and } l + l' + k \text{ is odd,} \\ \{JJ'k\} \text{ (or } \Delta J = 0, \pm 1, \ldots, \pm k \text{ and } J + J' \geq k), \ M' + q = M. \end{cases}$$
$$(24.22)$$

Let us notice that selection rules for M, q and M' are important only if we are interested in the polarization of the radiation, otherwise they may be neglected, because the other radiation characteristics do not depend on these parameters.

24.4 'Approximate' selection rules. Intermediate coupling. Intercombination lines

Let us consider 'approximate' selection rules for the case of LS coupling. Similar examples for other coupling schemes will be presented in the next two sections together with the corresponding expressions for the submatrix elements of the respective transition operators in terms of the coefficients of fractional parentage, $3nj$-coefficients and one-electron submatrix elements. The dependence of each submatrix element of Ek-

transition on J is contained in the $6j$-coefficient

$$\begin{Bmatrix} L & J & S \\ J' & L' & k \end{Bmatrix}.$$

For S and S' the conditions $\delta(S, S')$ occur. The above-mentioned $6j$-coefficient is non-zero if four triangle conditions $\{LSJ\}$, $\{L'SJ'\}$, $\{JJ'k\}$ and $\{LL'k\}$ are fulfilled. The first two describe LS coupling, the third is already known from the Wigner–Eckart theorem and only the last triangle gives us the selection rules for total orbital angular momenta L and L'. It turned out that they are identical to those for J and J'. Thus, having in mind that $\delta(S, S')$ is equivalent to $\{SS'0\}$, we establish the following selection rules for L and S:

$$Ek : \begin{cases} \{LL'k\} \text{ or } \Delta L = 0, \pm 1, \dots, \pm k, \ L + L' \geq k, \\ \{SS'0\} \text{ or } \Delta S = 0. \end{cases} \tag{24.23}$$

For the special case of electric dipole transitions we obtain from (24.23) $\Delta L = 0, \pm 1$ (transitions $S - S$ are forbidden) and $\Delta S = 0$.

The analogous selection rules for magnetic multipole (Mk) radiation (operator (4.16)) have the form

$$Mk : \begin{cases} \{LL'(k-1)\} \text{ or } \Delta L = 0, \pm 1, \dots, \pm(k-1) \text{ and } L + L' \geq k - 1, \\ \{SS'(k-1)\} \text{ or } \Delta S = 0, \pm 1, \dots, \pm(k-1) \text{ and } S + S' \geq k - 1. \end{cases} \tag{24.24}$$

Magnetic dipole transitions are permitted, if $K\alpha LS = K'\alpha'L'S'$, i.e. only transitions between levels of one and the same terms are allowed.

For the expressions of matrix elements of multipole transitions, containing the CFP, the selection rules of the kind $\Delta v = \pm 1$ may be established, however they are rather approximate.

A total set of selection rules consists of the sum of all selection rules both 'exact' and 'approximate'. Transition is forbidden if at least one selection rule is violated. The transitions may be forbidden to a different extent – by one, two, three, etc. violated conditions. If an electronic transition is between complex configurations, then, as we shall see in the next section, there may be a large number of selection rules. However, the majority of them, especially with regard to the quantum numbers of intermediate momenta, are rather approximate, even when a specific pure coupling scheme is valid. This is explained by the presence of interaction between the momenta.

All selection rules are contained in the expressions for matrix elements of the electron transition operators. Therefore, not knowing them cannot cause the occurrence of false transitions, because they will be automatically equal to zero. However, studies of the selection rules allow us in many cases to determine the transition type and multipolarity (Ek or Mk), and to estimate their relative intensities without carrying out accurate calculations. Comparing, for example, the selection rules with regard to π, J for $E2$- and $M1$-transitions, we learn that the transitions with $\Delta J = \pm 1$ inside one and the same configuration may have both electric and magnetic origin. Some conclusions on relative intensities of lines in a multiplet follow from analytical expressions for line strengths of electronic transitions. These expressions, as a rule, may be established for the simple cases $k = 1$ and 2. Unfortunately, these conclusions are rather approximate in the majority of cases, when actually some intermediate coupling takes place.

Electronic transitions are usually divided into two groups: permitted and forbidden. However, such grouping is rather relative. Sometimes $E1$-transitions are called permitted, whereas all the rest are considered as forbidden. More exact and general is classification of transitions satisfying the above-mentioned selection rules as permitted. Otherwise, if at least one selection rule is violated, the appropriate transition is called forbidden.

While diagonalizing the energy matrix and, thus, employing intermediate coupling, the selection rules for approximate quantum numbers change considerably. Actually they disappear; all transitions between any approximate quantum numbers, leading to the given J and J', become allowed. Occurrence of the so-called intercombination lines, caused by violation of the selection rule $\Delta S = 0$ while using the LS coupling, may serve as one example of this phenomenon. From a physical point of view it is explained by the existence of spin–orbit interactions, which lead to the occurrence of transitions between terms of different multiplicities.

Accounting for the deviations from pure coupling changes the line strengths of permitted transitions. As a rule these changes have the same order of magnitude as the strengths of forbidden lines. In the majority of cases, particularly when dealing with complex electronic configurations, carrying out the calculations in intermediate coupling becomes the only way to obtain reliable results. In a number of cases diagonalization of the energy matrix with respect to configurations has to be employed. This leads to the occurrence of many-electron transitions (see Chapter 29).

24.5 Sum rules

In the central field approximation, when radial wave functions not depending on term are usually employed, the line strengths of any transition may be represented as a product of one radial integral and of a number of $3nj$-coefficients, one-electron submatrix elements of standard operators ($C^{(k)}$ and/or $L^{(1)}$, $S^{(1)}$), CFP (if the number of electrons in open shells changes) and appropriate algebraic multipliers. It is usually assumed that the radial integral does not depend on the quantum numbers of the vec-

Table 24.1. Algebraic values of g_0

$K - K'$	g_0
$l_0^{N_0} l_1 - l_0^{N_0} l_2$	$g(l_0^{N_0})$
$l_0^{N_0} l_1^2 - l_0^{N_0} l_1 l_2$	$g(l_0^{N_0})(4l_1 + 1)$
$l_0^{N_0} l_1 l_2 - l_0^{N_0} l_1 l_3$	$g(l_0^{N_0})(4l_1 + 2)$
$l_0^{N_0} - l_0^{N_0-1} l_1$	$g(l_0^{N_0-1})[1 - (N_0 - 1)/(4l_0 + 2)]$
$l_0^{N_0} l_1 - l_0^{N_0-1} l_1^2$	$g(l_0^{N_0-1})(4l_1 + 1)[1 - (N_0 - 1)/(4l_0 + 2)]$
$l_0^{N_0} l_1 - l_0^{N_0-1} l_1 l_2$	$g(l_0^{N_0-1})(4l_1 + 2)[1 - (N_0 - 1)/(4l_0 + 2)]$

torial model. In many cases these formulas may be summed over some quantum numbers (of the type J, L, S, etc.). As a result we arrive at certain sum rules for line strengths. Let us examine them here for the cases when they do not depend on the particular type of transition. More specific sum rules will be presented further on in Part 6.

Formulas (24.5), (24.10), (24,12) and (24.13) may be considered as certain sum rules. Let us define the total strength of the transitions between all levels of two configurations K and K'

$$S(K, K') = \sum_{\alpha J, \alpha' J'} S(K\alpha J, K'\alpha' J'). \tag{24.25}$$

In LS coupling this may be expressed in terms of the multiplet strength (24.5), i.e.

$$S(K, K') = \sum_{\alpha LS, \alpha' L'S'} S(K\alpha LS, K'\alpha' L'S'). \tag{24.26}$$

This quantity does not depend on coupling scheme and may be shown to be proportional to the line strength $s(nl, n'l')$ of the 'jumping' electron, namely

$$S(K, K') = g_0 s(nl, n'l'), \qquad (nl \neq n'l'). \tag{24.27}$$

The algebraic values of the multiplier g_0 are given in Table 24.1 for the most practically needed cases of electronic transitions. The statistical weight of a shell is defined by formula (9.4).

The following sum rule can also be defined:

$$S(KLS, K'L'S) = \sum_{\alpha\alpha'} S(K\alpha LS, K'\alpha' L'S). \tag{24.28}$$

It is possible to obtain a number of the other sum rules for particular cases of electronic transitions, however as a rule all such sums are valid only approximately because the respective calculations must be performed in the intermediate coupling scheme where significant deviations from such sum rules can occur.

Certain sum rules are known for oscillator strengths and transition probabilities as well. So, we can define the oscillator strength and probability of the transition between terms, respectively:

$$f(\alpha LS, \alpha' L'S)$$
$$= \frac{1}{(2L+1)(2S+1)} \sum_{JJ'} (2J+1) f(\alpha LSJ, \alpha' L'SJ'), \qquad (24.29)$$

$$W(\alpha LS, \alpha' L'S)$$
$$= \frac{1}{(2L+1)(2S+1)} \sum_{JJ'} (2J+1) W(\alpha LSJ, \alpha' L'SJ'). \qquad (24.30)$$

It follows from the expressions for line strengths that $\sum_{J'} S(\alpha J, \alpha' J')$ is proportional to $(2J+1)$, whereas $\sum_{J'} W(\alpha J, \alpha' J')$ does not depend on J. Hence, the total probability of transitions from the level αJ to all levels within a given multiplet does not depend on J. The same is true for oscillator strengths. Therefore, if the populations N_1 and N_2 of the levels J_1 and J_2 follow the rule

$$N_1 : N_2 = (2J_1 + 1) : (2J_2 + 1) \qquad (24.31)$$

(the Boltzmann distribution with the condition $kT \gg \Delta E_{J_1 J_2}$), then the sum of intensities of all lines of the multiplet which have one and the same initial level is proportional to its statistical weight. Evidently, such a conclusion also holds for the transitions to one and the same final level.

Oscillator strengths obey the following general sum rule, which does not depend on coupling scheme ($\mathcal{N}$ is the total number of electrons in an atom):

$$\sum_{\alpha' J'} f(\alpha J, \alpha' J') = \mathcal{N}. \qquad (24.32)$$

The summation in (24.32) must be performed over all levels $\alpha' J'$ of the discrete and continuous spectrum. Moreover, transitions of all electrons must be accounted for. This sum rule is too general, because we are usually interested in the transitions of one outer electron. There are also certain sum rules (already approximate), for such transitions, e.g. (K_0 stands for closed shells):

$$\sum_{n'l'L'S'J'} f(K_0 nl^N \alpha LSJ, K_0 nl^{N-1} (\alpha_1 L_1 S_1) n'l'L'S'J') = N. \qquad (24.33)$$

For one electron above filled shells, we have

$$\sum_{n'l'j'} f(nlj, n'l'j') = 1. \qquad (24.34)$$

Thus, the sum rules reveal certain regularities in the behaviour of electronic transitions. Unfortunately, many of them depend on a particular model of coupling scheme and therefore hold only approximately. Nevertheless, their use may help to estimate the accuracy of the model employed.

25

Non-relativistic matrix elements of the Ek-transitions

25.1 Ek-transitions of electrons above the core configurations

Operators of electronic transitions, except the third form of the Ek-radiation operator for $k > 1$, may be represented as the sums of the appropriate one-electron quantities (see (13.20)). Their matrix elements for complex electronic configurations consist of the sums of products of the CFP, $3nj$-coefficients and one-electron submatrix elements. The many-electron part of the matrix element depends only on tensorial properties of the transition operator, whereas all pecularities of the particular operator are contained in its one-electron submatrix element.

Electronic transitions between complex configurations may be divided into two groups: transitions without and with participation of the core electrons. Let us denote the whole configuration K of an atom as the sum of two parts

$$K = K_0 + K_1. \tag{25.1}$$

Here K_0 stands for all closed and open shells whose electrons do not participate in the transition (passive shells). It is called the core configuration or simply the core. K_1 describes the electrons (usually one or two electrons above the core), one of which performs the transition. Typical patterns of the first sort of transitions are given by the following configurations:

$$l_1^{N_1} l_2 - l_1^{N_1} l_3, \tag{25.2}$$

$$l_1^{N_1} l_2^2 - l_1^{N_1} l_2 l_3, \tag{25.3}$$

$$l_1^{N_1} l_2 l_3 - l_1^{N_1} l_2 l_4. \tag{25.4}$$

Configuration $l_1^{N_1}$ represents the core in (25.2) and (25.3), whereas in (25.4) the core is $l_1^{N_1} l_2$. The momenta of outer electrons may be coupled with those of the core by all coupling schemes considered in Chapter 11. Therefore, the expressions of matrix elements of electronic transitions

305

between complex configurations in various coupling schemes must be available. We shall need the following formulas for one-electron submatrix elements of Ek-transition operators (4.12) and (4.13), respectively:

$$\left(n_2 l_2 \| Q^{(k)} \| n_1 l_1\right) = -\left(l_2 \| C^{(k)} \| l_1\right)\left(n_2 l_2 \| r^k \| n_1 l_1\right), \tag{25.5}$$

$$\left(n_2 l_2 \| Q'^{(k)} \| n_1 l_1\right) = -\left(l_2 \| C^{(k)} \| l_1\right)\left[k\left(n_2 l_2 \| r^{k-1}\frac{\partial}{\partial r} \| n_1 l_1\right)\right.$$
$$\left. -\frac{1}{2}\left[l_2(l_2 + 1) - l_1(l_1 + 1) - k(k - 1)\right]\left(n_2 l_2 \| r^{k-2} \| n_1 l_1\right)\right]. \tag{25.6}$$

Let us notice that for $n_2 l_2 = n_1 l_1$ and the same radial orbitals expression (25.6) equals zero. Therefore the radial orbitals, depending on the term quantum numbers, must be employed when calculating Ek-transitions using the second form (4.13).

The third form of the Ek-transition operator contains even two-particle terms [77], therefore its matrix elements are expressed in terms of the corresponding two-electron quantities. Unfortunately, these matrix elements are very cumbersome, and therefore are of little use. For the special case of transitions in shell l^N they were established in [171]. Submatrix elements of Ek-transitions with relativistic corrections (4.18)–(4.20) and (4.22) are considered in detail for one-electron and many-electron configurations in [172].

As we have seen, the line and oscillator strengths as well as the probabilities of electronic transitions between configurations of any type are expressed in terms of the submatrix elements of the appropriate operators. These submatrix elements may be found in a similar way as for the energy operator (see Part 5). Therefore, further on we shall consider only submatrix elements and present only final results. It is convenient to write the submatrix element of the non-relativistic operators of Ek-radiation (4.12) and (4.13) for transitions of the type (25.2), namely $l_1^{N_1} l_2 - l_1^{N_1} l_3$, as follows:

$$\left(l_1^{N_1} l_2 \alpha_1 L_1 S_1 l_2 s T_1 T_2 J \| A^{(k)} \| l_1^{N_1} l_3 \alpha_1' L_1' S_1' l_3 s U_1' U_2' J'\right)$$
$$= (-1)^{\varphi} \delta(\alpha_1 L_1 S_1, \alpha_1' L_1' S_1') Q(\alpha, T_1 T_2, U_1' U_2', JJ')\left(l_2 \| A^{(k)} \| l_3\right). \tag{25.7}$$

Here $(l_2 \| A^{(k)} \| l_3)$ represents quantities (25.5) and (25.6), $T_1 T_2$ and $U_1' U_2'$ describe the coupling scheme of the momenta of the outer electron with those of the core (formulas (11.2)–(11.5)) whereas α stands for all remaining quantum numbers, pointed out in the left side of formula (25.7) and omitted in its right side. Below we present expressions for φ and Q in the case of LS, LK, JK and JJ coupling schemes, including the cases when the coupling schemes in the initial and final configurations differ from each other (let us also recall that $[x, y, \ldots] = (2x+1)(2y+1)\ldots$):

$T_1 T_2 = LS, \quad U_1' U_2' = L'S'$:

$$\varphi = L_1 + S + l_3 + J',$$

$$Q = \delta(S, S')\sqrt{[L, L', J, J']} \begin{Bmatrix} L & J & S \\ J' & L' & k \end{Bmatrix} \begin{Bmatrix} l_2 & L & L_1 \\ L' & l_3 & k \end{Bmatrix}; \qquad (25.8)$$

$T_1 T_2 = LK, \quad U_1' U_2' = L'K'$:

$$\varphi = l_2 + L_1 + S_1 + s + K + K' + J',$$

$$Q = \sqrt{[L, L', K, K', J, J']} \begin{Bmatrix} K & J & s \\ J' & K' & k \end{Bmatrix} \begin{Bmatrix} L & K & S_1 \\ K' & L' & k \end{Bmatrix}$$

$$\times \begin{Bmatrix} l_2 & L & L_1 \\ L' & l_3 & k \end{Bmatrix}; \qquad (25.9)$$

$T_1 T_2 = J_1 K, \quad U_1' U_2' = J_1' K$:

$$\varphi = s + l_3 - J_1 + J',$$

$$Q = \delta(J_1, J_1')\sqrt{[K, K', J, J']} \begin{Bmatrix} K & J & s \\ J' & K' & k \end{Bmatrix} \begin{Bmatrix} l_2 & K & J_1 \\ K' & l_3 & k \end{Bmatrix}; \qquad (25.10)$$

$T_1 T_2 = J_1 j_2, \quad U_1' U_2' = J_1' j_3$:

$$\varphi = l_1 - s + J_1 + J,$$

$$Q = \delta(J_1, J_1')\sqrt{[j_2, j_3, J, J']} \begin{Bmatrix} j_2 & J & J_1 \\ J' & j_3 & k \end{Bmatrix} \begin{Bmatrix} l_2 & j_2 & s \\ j_3 & l_3 & k \end{Bmatrix}; \qquad (25.11)$$

$T_1 T_2 = LS, \quad U_1' U_2' = L'K'$:

$$\varphi = l_3 + L_1 + S_1' - S + s + L',$$

$$Q = \sqrt{[L, L', S, K', J, J']} \begin{Bmatrix} S & L' & J' \\ K' & s & S_1 \end{Bmatrix} \begin{Bmatrix} L & J & S \\ J' & L' & k \end{Bmatrix}$$

$$\times \begin{Bmatrix} l_2 & L & L_1 \\ L' & l_3 & k \end{Bmatrix}; \qquad (25.12)$$

$T_1 T_2 = LS, \quad U_1' U_2' = J_1' K$:

$$\varphi = l_2 + L_1 - S + J,$$

$$Q = \sqrt{[L, S, J_1', K', J, J']} \begin{bmatrix} L & k & S_1 & K' \\ S & L_1 & J' & l_3 \\ J & s & l_2 & J_1' \end{bmatrix}; \qquad (25.13)$$

$T_1 T_2 = LS, \quad U_1' U_2' = J_1' j_3$:

$$\varphi = 0,$$

$$Q = \sqrt{[L, S, J_1', j_3, J, J']} \begin{Bmatrix} l_3 & s & S & J \\ j_3 & S_1 & L & k \\ J' & J_1' & L_1 & l_2 \end{Bmatrix}; \qquad (25.14)$$

$\underline{T_1 T_2 = LS, \ U_1' U_2' = J_1' K':}$

$$\varphi = L + S_1 + s + J,$$

$$Q = \sqrt{[J_1', L, K, K', J, J']} \begin{Bmatrix} J_1' & l_2 & K \\ L & S_1 & L_1 \end{Bmatrix} \begin{Bmatrix} K' & J' & s \\ J & K & k \end{Bmatrix}$$

$$\times \begin{Bmatrix} l_3 & K' & J_1' \\ K & l_2 & k \end{Bmatrix};$$

(25.15)

$\underline{T_1 T_2 = LK, \ U_1' U_2' = J_1' j_3:}$

$$\varphi = l_2 + L + S_1 + J_1,$$

$$Q = \sqrt{[L, K, J_1', j_3, J, J']} \begin{Bmatrix} J_1' & l_2 & K \\ L & S_1 & L_1 \end{Bmatrix} \begin{Bmatrix} J & s & K \\ k & l_3 & l_2 \\ J' & j_3 & J_1' \end{Bmatrix}; \quad (25.16)$$

$\underline{T_1 T_2 = J_1 K, \ U_1' U_2' = J_1' j_3:}$

$$\varphi = 0,$$

$$Q = \delta(J_1, J_1') \sqrt{[K, j_3, J, J']} \begin{Bmatrix} j_3 & J_1 & J' \\ s & K & J \\ l_3 & l_2 & k \end{Bmatrix}.$$

(25.17)

The definitions of the $3nj$-coefficients ($6j$-, $9j$-, and $12j$-) may be taken from Chapter 6. Analogous expressions for electronic transitions of types (25.3) and (25.4) follow from formulas describing transitions in the case of three open shells. The most important of them will be presented later in this chapter.

25.2   Selection and sum rules

Conditions of non-zero values of the submatrix elements of the electron transition operators define the selection rules for radiation. The latter coincide with those non-zero conditions for the quantity Q. On the other hand, the selection rules for Q are defined by the conditions of polygons for $3nj$-coefficients, in terms of which they are expressed. The requirement (24.21) must also be kept in mind. The selection rules for transitions (25.8)–(25.17) are summarized in Table 25.1. In all cases the selection rules $\{JJ'k\}$, $\{l_2 l_3 k\}$ and $l_2 + l_3 + k$ is even number are valid. The table contains only those polygons which have the quantum numbers of both configurations, because only in such a case do these conditions serve as the selection rules for radiation. If a certain quantum number has no restrictions from this point of view, this means that it does not form a polygon with quantum numbers of the other configuration. Such quantities are placed in curly brackets.

The most complicated selection rules – the quadrangle conditions – occur for transitions between configurations where the energy levels of

Table 25.1. Selection rules for the transition $l_1^{N_1} l_2 \alpha_1 L_1 S_1 T_1 T_2 J - l_1^{N_1} l_3 \alpha_1' L_1' S_1' U_1' U_2' J'$

$LS - L'S'$	$LK - L'K'$	$J_1 K - J_1' K'$	$J_1 j_2 - J_1' j_3$	$LS - L'K'$
$\{L\ L'\ k\}$	$\{L\ L'\ k\}$	$\{K\ K'\ k\}$	$\{j_2\ j_3\ k\}$	$\{L\ L'\ k\}$
$\{S\ S'\ 0\}$	$\{K\ K'\ k\}$	$\{J_1\ J_1'\ 0\}$	$\{J_1\ J_1'\ 0\}$	$\{S\ L'\ J'\}$
$\{L_1\ L_1'\ 0\}$	$\{S_1\ S_1'\ 0\}$			$\{L_1\ L_1'\ 0\}$
	$\{L_1\ L_1'\ 0\}$			$\{K'\}$

$LS - J_1' K'$	$LS - J_1' j_3$	$LK - J_1' K'$	$LK - J_1' j_3$	$J_1 K - J_1' j_3$
$\{L\ S_1\ K'\ k\}$	$\{s\ l_2\ J_1'\ J\}$	$\{l_2\ K\ J_1'\}$	$\{l_2\ K\ J_1'\}$	$\{J_1\ J_1'\ 0\}$
$\{L_1\ l_3\ S\ J'\}$	$\{L_1\ l_3\ S\ J'\}$	$\{K\ K'\ k\}$	$\{L\}\ \{j_3\}$	$\{K\}\ \{j_3\}$
$\{s\ l_2\ J_1'\ J\}$	$\{L\}\ \{j_3\}$	$\{L\}$		

the first configuration are classified in LS coupling whereas those of the second one are classified with the $J_1' K'$ coupling scheme. These rules follow from the conditions of non-zero values of the appropriate $12j$-coefficients. However, as was emphasized in Chapter 24 and as we shall see once more in Chapter 30, all selection rules involving the intermediate quantum numbers are more or less approximate and, therefore, are of little use while performing calculations in the intermediate coupling scheme.

Employing the algebraic or graphical (see Chapter 8) methods of summing the $3nj$-coefficients we can establish a number of sum rules for electronic transitions. Let us discuss here the frequently considered transition (25.8), when LS coupling is used for both configurations. Adopting the shortened version of the submatrix element in the form $(N_1 1 \| A^{(k)} \| N_1 1)$, we arrive at the following sum rules for transitions (25.7) in LS coupling:

$$\sum_J \left(N_1 1 \| A^{(k)} \| N_1 1 \right)^2 = [L, J'] \left\{ \begin{matrix} l_2 & L & L_1 \\ L' & l_3 & k \end{matrix} \right\}^2 \left(l_2 \| A^{(k)} \| l_3 \right)^2, \qquad (25.18)$$

$$\sum_{JJ'} \left(N_1 1 \| A^{(k)} \| N_1 1 \right)^2 = [L, L', S] \left\{ \begin{matrix} l_2 & L & L_1 \\ L' & l_3 & k \end{matrix} \right\}^2 \left(l_2 \| A^{(k)} \| l_3 \right)^2 \qquad (25.19)$$

$$\sum_{LJ} \left(N_1 1 \| A^{(k)} \| N_1 1 \right)^2 = \frac{[J']}{[l_3]} \left(l_2 \| A^{(k)} \| l_3 \right)^2, \qquad (25.20)$$

$$\sum_{LJJ'} \left(N_1 1 \| A^{(k)} \| N_1 1 \right)^2 = \frac{[L', S]}{[l_3]} \left(l_2 \| A^{(k)} \| l_3 \right)^2. \qquad (25.21)$$

We have in mind that the condition $\delta(\alpha_1 L_1 S_1, \alpha_1' L_1' S_1')$ is included in (25.18)–(25.21). It is possible to establish many other sum rules with respect to L, L', etc., however, they are of little importance. Analogous sum rules may be easily found in a similar way for the transitions (25.9)–(25.17) as well.

25.3   Ek-transitions between levels of one and the same configuration

The submatrix element of non-relativistic Ek-transitions between the levels of a shell of equivalent electrons is as follows:

$$\left(nl^N \alpha LSJ \| A^{(k)} \| nl^N \alpha' L'S'J'\right) = \delta(S,S')(-1)^{L+J'+S+k}$$

$$\times \sqrt{\frac{[J,J']}{[S]}} \begin{Bmatrix} L & J & S \\ J' & L' & k \end{Bmatrix} \left(l^N \alpha LS \| U^{(k)} \| l^N \alpha' L'S'\right) \left(l \| A^{(k)} \| l\right), \quad (25.22)$$

whereas for two shells

$$\left(n_1 n_2 l_1^{N_1} l_2^{N_2} \alpha_1 L_1 S_1 \alpha_2 L_2 S_2 LSJ \| A^{(k)} \| n_1 n_2 l_1^{N_1} l_2^{N_2} \alpha'_1 L'_1 S'_1 \alpha'_2 L'_2 S'_2 L'S'J'\right)$$

$$= (-1)^{L_1+L_2+L+L'+S+J'} \delta(\alpha_2 L_2 S_2, \alpha'_2 L'_2 S'_2) \delta(S_1 S, S'_1 S')$$

$$\times \left(n_1 l_1 \| A^{(k)} \| n_1 l_1\right) \sqrt{[L,L',J,J']/[S]} \begin{Bmatrix} L_1 & L & L_2 \\ L' & L'_1 & k \end{Bmatrix} \begin{Bmatrix} L & J & S \\ J' & L' & k \end{Bmatrix}$$

$$\times \left(l_1^{N_1} \alpha_1 L_1 S_1 \| U^{(k)} \| l_1^{N_1} \alpha'_1 L'_1 S_1\right)$$

$$+ (-1)^{L_1+L'_2+S+J'} \delta(\alpha_1 L_1 S_1, \alpha'_1 L'_1 S'_1) \delta(S_2 S, S'_2 S') \left(n_2 l_2 \| A^{(k)} \| n_2 l_2\right)$$

$$\times \sqrt{[L,L',J,J']/[S_2]} \begin{Bmatrix} L_2 & L & L_1 \\ L' & L'_2 & k \end{Bmatrix} \begin{Bmatrix} L & J & S \\ J' & L' & k \end{Bmatrix}$$

$$\times \left(l_2^{N_2} \alpha_2 L_2 S_2 \| U^{(k)} \| l_2^{N_2} \alpha'_2 L'_2 S_2\right). \quad (25.23)$$

Actually, only transitions described by one term in (25.23), take place. Formulas (25.22) and (25.23) are valid for the first and second forms of the Ek-transition operator (formulas (4.12) and (4.13)). The corresponding one-electron submatrix elements are given by (25.5) and (25.6). Analogous expressions for the third form of the Ek-radiation operator are established in [77]. The appropriate selection and sum rules may be found in a similar way as was done for transitions between different configurations. It is interesting to mention that the non-zero conditions for submatrix elements for the operator U^k with regard to a seniority quantum number suggest new selection rules for the transitions in the shell of equivalent electrons: $v = v'$ at odd and $v = v'$, $v' \pm 2$ at even k values.

25.4   Ek-transitions with participation of core electrons

In such cases the transition quantities depend not only on resultant quantum numbers of the core, but also on the remaining quantum numbers of the configurations. A general case of such transitions may be described by the formula

$$n_1 l_1^{N_1} n_2 l_2^{N_2} - n_1 l_1^{N_1-1} n_2 l_2^{N_2+1}. \tag{25.24}$$

In (25.24) there may be other open shells, whose electrons do not participate in a transition. The most frequent special cases of (25.24) are the following transitions:

$$n_1 l_1^{N_1} - n_1 l_1^{N_1-1} n_2 l_2, \tag{25.25}$$

$$n_1 l_1^{N_1} n_2 l_2 - n_1 l_1^{N_1-1} n_2 l_2^2, \tag{25.26}$$

$$n_1 l_1^{N_1} n_2 l_2 - n_1 l_1^{N_1-1} n_2 l_2 n_3 l_3. \tag{25.27}$$

In order to find expressions for the submatrix elements of these transitions, we have to transform them, using the CFP with one detached electron, to the form in which the electrons of the principal shell ($l_1^{N_1-1}$ for (25.25)–(25.27)) do not participate in the transition under consideration.

A submatrix element of the Ek-transitions, caused by operators (4.12) or (4.13) and generally denoted as $A^{(k)}$ for the transitions of type (25.24), not specifying a coupling scheme between shells, may be written as follows:

$$
\begin{aligned}
&(n_1 n_2 l_1^{N_1} l_2^{N_2} \alpha_1 L_1 S_1 \alpha_2 L_2 S_2 T_1 T_2 J \| A^{(k)} \| \\
&\quad \times n_1 n_2 l_1^{N_1-1} l_2^{N_2+1} \alpha_1' L_1' S_1' \alpha_2' L_2' S_2' U_1' U_2' J') \\
&= (-1)^{N_2} \sqrt{N_1(N_2+1)} \left(l_1^{N_1} \alpha_1 L_1 S_1 \| l_1^{N_1-1}(\alpha_1' L_1' S_1') l_1 \right) \\
&\quad \times \left(l_2^{N_2}(\alpha_2 L_2 S_2) l_2 \| l_2^{N_2+1} \alpha_2' L_2' S_2' \right) \\
&\quad \times ((L_1' S_1' l_1) L_1 S_1, (L_2 S_2), T_1 T_2 J \| A^{(k)} \| \\
&\quad \times (L_1' S_1'), (L_2 S_2 l_2) L_2' S_2', U_1' U_2' J').
\end{aligned}
\tag{25.28}
$$

Specifying a coupling scheme between shells we arrive at a number of concrete formulas for the transitions studied. As in the case of (25.7), let us present the submatrix element in the right side of (25.28) as follows:

$$
\begin{aligned}
&\left((L_1' S_1' l_1) L_1 S_1, (L_2 S_2), T_1 T_2 J \| A^{(k)} \| (L_1' S_1'), (L_2 S_2 l_2) L_2' S_2', U_1' U_2' J' \right) \\
&= (-1)^\varphi Q(\alpha, T_1 T_2, U_1' U_2', J J') \left(l_1 \| A^{(k)} \| l_2 \right).
\end{aligned}
\tag{25.29}
$$

For LS coupling formula (25.29) turns into
$T_1 T_2 = LS$, $U_1' U_2' = L'S'$:

$$\varphi = l_1 + L_2 + L_2' + L + S_1' + S_2' - J',$$

$$Q = \delta(S, S') \sqrt{[L_1, S_1, L_2', S_2', L, L', J, J']}$$

$$\times \begin{Bmatrix} L & J & S \\ J' & L' & k \end{Bmatrix} \begin{Bmatrix} s & S_2 & S_2' \\ S & S_1' & S_1 \end{Bmatrix} \begin{Bmatrix} L_1' & L_1 & l_1 \\ L' & L & k \\ L_2' & L_2 & l_2 \end{Bmatrix}. \tag{25.30}$$

Appropriate expressions for the other coupling schemes, including different ones for the upper and lower configurations, are rather cumbersome and of little use, therefore they are not presented here. They were obtained in [173] and generalized to cover the case of relativistic corrections in [172]. We shall present here the more complete set of formulas for the useful special case of transitions (25.25). Then $N_2 = 0$, $L_2 = S_2 = 0$, $L_1 S_1 = LS$, $L_2' S_2' = l_2 s_2$ and the second coefficient of fractional parentage in (25.28) equals unity. Hence, formula (25.28) turns into

$$\left(n_1 l_1^{N_1} \alpha LSJ \| A^{(k)} \| n_1 n_2 l_1^{N_1-1} l_2 \alpha_1' L_1' S_1' l_2 s_2 U_1' U_2' J \right)$$
$$= \sqrt{N_1} \left(l_1^{N_1} \alpha LS \| l_1^{N_1-1} (\alpha_1' L_1' S_1') l \right)$$
$$\times \left((L_1' S_1' l_1) LSJ \| A^{(k)} \| (L_1' S_1') l_2 s_2, U_1' U_2' J \right). \tag{25.31}$$

For the first configuration in a non-relativistic approach only LS coupling must be used, whereas for the second one all four schemes LS, LK, $J_1 K$ and $J_1 j_2$ may occur. A submatrix element in the right side of (25.31), presented in form (25.29), becomes equal to
$T_1 T_2 = LS$, $U_1' U_2' = L'S'$:

$$\varphi = l_2 + L_1' - S - J',$$
$$Q = \delta(S, S') \sqrt{[L, L', J, J']} \left\{ \begin{matrix} l_1 & L & L_1' \\ L' & l_2 & k \end{matrix} \right\} \left\{ \begin{matrix} L & J & S \\ J' & L' & k \end{matrix} \right\}; \tag{25.32}$$

$T_1 T_2 = LS$, $U_1' U_2' = L'K'$:

$$\varphi = l_2 + L_1' + L' - S + S_1' + s,$$
$$Q = \sqrt{[L, L', S, K', J, J']} \left\{ \begin{matrix} l_1 & L & L_1' \\ L' & l_2 & k \end{matrix} \right\} \left\{ \begin{matrix} L & J & S \\ J' & L' & k \end{matrix} \right\}$$
$$\times \left\{ \begin{matrix} K' & L' & S_1' \\ S & s & J' \end{matrix} \right\}; \tag{25.33}$$

$T_1 T_2 = LS$, $U_1' U_2' = J_1' K'$:

$$\varphi = l_1 + L_1' - S + J,$$
$$Q = \sqrt{[L, S, J_1', K', J, J']} \begin{bmatrix} J' & l_2 & S & L_1' \\ s & J & J_1' & l_1 \\ K' & S_1' & k & L \end{bmatrix}; \tag{25.34}$$

$T_1 T_2 = LS$, $U_1' U_2' = J_1 j_2$:

$$\varphi = 0,$$
$$Q = \sqrt{[L, S, J_1, j_2, J, J']} \left\{ \begin{matrix} l_2 & l_1 & L_1' & J_1 \\ k & L & S_1' & \\ J' & J & S & s \end{matrix} \; j_2 \right\}. \tag{25.35}$$

Similar expressions for submatrix elements of the operators in question for transitions of the type (25.26) may be easily obtained from general formula (25.28) as special cases. Analogous equalities for the transition (25.27) already follow from the cases of configurations consisting of three open shells. We shall consider them only in LS coupling. Thus, the submatrix element of the operator under consideration for the transition

$$n_1 l_1^{N_1} n_2 l_2^{N_2} n_3 l_3^{N_3} - n_1 l_1^{N_1-1} n_2 l_2^{N_2+1} n_3 l_3^{N_3} \tag{25.36}$$

is equal to

$$\begin{aligned}
&\left(n_1 n_2 n_3 l_1^{N_1} l_2^{N_2} l_3^{N_3} \alpha_1 L_1 S_1 \alpha_2 L_2 S_2 (L_{12} S_{12}) \alpha_3 L_3 S_3 LSJ \| A^{(k)} \| \right. \\
&\times \left. n_1 n_2 n_3 l_1^{N_1-1} l_2^{N_2+1} l_3^{N_3} \alpha_1' L_1' S_1' \alpha_2' L_2' S_2' (L_{12}' S_{12}') \alpha_3' L_3' S_3' L'S'J' \right) \\
&= (-1)^{N_2 + l_2 + L_2 + L_3 + L_{12} - L_2' + S_1' + S_2' + S_{12} + L + L' + S + J'} \\
&\times \delta(S_{12}S, S_{12}'S') \delta(\alpha_3 L_3 S_3, \alpha_3' L_3' S_3') \\
&\times \sqrt{N_1(N_2+1)[L_1, L_2', S_1, S_2', L_{12}, L_{12}', L, L', J, J']} \\
&\times \left(l_1^{N_1} \alpha_1 L_1 S_1 \| l_1^{N_1-1} (\alpha_1' L_1' S_1') l_1 \right) \left(l_2^{N_2} (\alpha_2 L_2 S_2) l_2 \| l_2^{N_2+1} \alpha_2' L_2' S_2' \right) \\
&\times \left(n_1 l_1 \| A^{(k)} \| n_2 l_2 \right) \begin{Bmatrix} L & J & S \\ J' & L' & k \end{Bmatrix} \begin{Bmatrix} L_{12}' & k & L_{12} \\ L & L_3 & L' \end{Bmatrix} \\
&\times \begin{Bmatrix} S_1 & S_1' & s \\ S_2' & S_2 & S_{12} \end{Bmatrix} \begin{Bmatrix} L_1 & L_1' & l_1 \\ L_2 & L_2' & l_2 \\ L_{12} & L_{12}' & k \end{Bmatrix}.
\end{aligned} \tag{25.37}$$

If the operator connects the second and third shells, then

$$\begin{aligned}
&\left(n_1 n_2 n_3 l_1^{N_1} l_2^{N_2} l_3^{N_3} \alpha_1 L_1 S_1 \alpha_2 L_2 S_2 (L_{12} S_{12}) \alpha_3 L_3 S_3 LSJ \| A^{(k)} \| \right. \\
&\times \left. n_1 n_2 n_3 l_1^{N_1} l_2^{N_2-1} l_3^{N_3+1} \alpha_1' L_1' S_1' \alpha_2' L_2' S_2' (L_{12}' S_{12}') \alpha_3' L_3' S_3' L'S'J' \right) \\
&= (-1)^{N_3 + L_1 + L_3 + L_{12} + L_2' - L_3' + S_1' - S_3' + S_{12} + S_2' - S_{12}' + s + L + J'} \\
&\times \delta(S, S') \delta(\alpha_1 L_1 S_1, \alpha_1' L_1' S_1') \\
&\times \sqrt{N_2(N_3+1)[L_2, L_{12}, L_3', L_{12}', S_2, S_{12}', S_3', S_{12}, L, L', J, J']} \\
&\times \left(l_2^{N_2} \alpha_2 L_2 S_2 \| l_2^{N_2-1} (\alpha_2' L_2' S_2') l_2 \right) \left(l_3^{N_3} (\alpha_3 L_3 S_3) l_3 \| l_3^{N_3+1} \alpha_3' L_3' S_3' \right) \\
&\times \left(n_2 l_2 \| A^{(k)} \| n_3 l_3 \right) \begin{Bmatrix} L & J & S \\ J' & L' & k \end{Bmatrix} \begin{Bmatrix} L_2' & L_1 & L_{12}' \\ L_{12} & l_2 & L_2 \end{Bmatrix} \\
&\times \begin{Bmatrix} S_{12} & S_{12}' & s \\ S_3' & S_3 & S \end{Bmatrix} \begin{Bmatrix} S_2' & S_1 & S_{12}' \\ S_{12} & s & S_2 \end{Bmatrix} \begin{Bmatrix} L_{12} & L_{12}' & l_2 \\ L_3 & L_3' & l_3 \\ L & L' & k \end{Bmatrix}.
\end{aligned} \tag{25.38}$$

If the operator $A^{(k)}$ connects the first and third shells, then the expression we are seeking can be obtained from (25.38), interchanging in the latter the subscripts $1 \leftrightarrow 2$ and adding N_2 in the phase multiplier. By the way, all cases of electronic transitions (25.2)–(25.4) directly follow from the above-mentioned transitions between three-shell configurations.

In conclusion of this section let us briefly discuss the selection and sum rules for *Ek*-transitions with the participation of core electrons.

We shall not present here complete sets of selection rules for all transitions in question. For these transitions they become much more complex, e.g. already for transition (25.24), the pentagon conditions can occur. Interested readers may find them in [173]. Selection rules differ considerably for the various coupling schemes. They change if we account for the relativistic corrections to the non-relativistic *Ek*-transition operators. However, all selection rules involving intermediate momenta are more or less approximate. Due to the presence of interaction between coupled momenta a certain intermediate coupling scheme takes place, and these selection rules are violated.

Again, employing the algebraic or graphical methods of the summation of $3nj$-coefficients, we can establish a number of sum rules with respect to quantum numbers J, J', $T_1 T_2$ and/or $U_1' U_2'$. Further summing may be carried out by accounting for the orthonormality of the CFP. However, all these sum rules are only approximately valid, that is why we are not presenting them here.

Thus, the expressions for the submatrix elements of the *Ek*-transition operators presented in this chapter allow one to study electronic transitions between any sort of practically needed configurations.

26

Relativistic matrix elements of Ek-transitions

26.1 Ek-transitions of electrons above the core configurations

As we have seen in Chapter 4, relativistic operators of the Ek-transitions in the general case have several forms and are dependent on the gauge condition of the electromagnetic field potential. These forms are equivalent and do not depend on gauge for exact wave functions. Unfortunately, we are always dealing with the more or less approximate wave functions of many-electron systems, therefore we need general expressions for the appropriate matrix elements.

Making use of relativistic wave functions (2.15), expressing $\alpha^{(1)}$ in terms of $\sigma^{(1)}$ (formula 2.8), employing the properties of $3nj$- and Clebsch–Gordan coefficients, we arrive at the following expressions for the one-electron submatrix elements of relativistic Ek-transition operators (4.5) and (4.6)

$$\left\langle n_2 l_2 j_2 \| _r Q^{(k)} \| n_1 l_1 j_1 \right\rangle = (-1)^{(l_1 - k - l_2 - 1)/2 - j_2}$$

$$\times \sqrt{\frac{(2j_1 + 1)(2j_2 + 1)}{2k + 1}} \begin{bmatrix} j_1 & j_2 & k \\ 1/2 & -1/2 & 0 \end{bmatrix}$$

$$\times \left\{ (l_1 l_2 k) \left[A(\lambda_1 \lambda_2 \mid 1k) - \frac{2k+1}{k+1} B(\lambda_1 \lambda_2' \mid 1k) \right] \right.$$

$$\left. + (l_1' l_2' k) \left[A(\lambda_1' \lambda_2' \mid 1k) + \frac{2k+1}{k+1} B(\lambda_1' \lambda_2 \mid 1k) \right] \right\}, \tag{26.1}$$

$$\left\langle n_2 l_2 j_2 \| _v Q^{(k)} \| n_1 l_1 j_1 \right\rangle = (-1)^{(l_2 - l_1 + k + 1)/2 - j_2}$$

$$\times \sqrt{\frac{(2j_1 + 1)(2j_2 + 1)}{2k + 1}} \begin{bmatrix} j_1 & j_2 & k \\ 1/2 & -1/2 & 0 \end{bmatrix}$$

$$\times \left\{ \frac{1}{2} (-1)^{l_2 + j_2 + 1/2} \left[2j_2 + 1 + (-1)^{j_1 + j_2 + k} (2j_1 + 1) \right] \right.$$

315

$$\times \left[(l_1 l_2' k - 1) A(\lambda_1 \lambda_2' \mid 0k) + (l_1' l_2 k - 1) A(\lambda_1' \lambda_2 \mid 0k) \right]$$
$$+ k \left[(l_1 l_2' k - 1) B(\lambda_1 \lambda_2' \mid 0k - 1) + (l_1 l_2' k + 1) B(\lambda_1 \lambda_2' \mid 0k + 1) \right.$$
$$\left. - (l_1' l_2 k - 1) B(\lambda_1' \lambda_2 \mid 0k - 1) - (l_1' l_2 k + 1) B(\lambda_1' \lambda_2 \mid 0k + 1) \right] \bigg\}. \quad (26.2)$$

Here the radial integrals are defined as follows:

$$A(\lambda_1 \lambda_2 \mid \kappa k) = \int_0^\infty f(n_1 l_1 j_1 \mid r) f(n_2 l_2 j_2 \mid r) r^\kappa \Phi_k(z) r^2 dr, \qquad (26.3)$$

$$B(\lambda_1 \lambda_2 \mid \kappa k) = \int_0^\infty f(n_1 l_1 j_1 \mid r) f(n_2 l_2 j_2 \mid r) r^\kappa g_k(z) r^2 dr, \qquad (26.4)$$

where λ_i, λ_i' means $n_i l_i j_i$ and $n_i l_i' j_i$, respectively, and

$$\Phi_k(z) = g_{k-1}(z) - \frac{k}{k+1} g_{k+1}(z). \qquad (26.5)$$

Integral A may be expressed in terms of integrals B

$$A(\lambda_1 \lambda_2 \mid \kappa k) = B(\lambda_1 \lambda_2 \mid \kappa k - 1) - \frac{k}{k+1} B(\lambda_1 \lambda_2 \mid \kappa k + 1). \qquad (26.6)$$

If the integrals A or B contain λ_i', this means that we have to replace the corresponding function $f(n_i l_i j_i \mid r)$ by $g(n_i l_i' j_i \mid r)$.

Let us also mention that using a number of functional relations between the products of $3nj$-coefficients and submatrix elements $(l \| C^{(k)} \| l')$, the spin–angular parts of matrix elements (26.1) and (26.2) are transformed to a form, whose dependence on orbital quantum numbers (as was also in the case of matrix elements of the energy operator, see Chapters 19 and 20) is contained only in the phase multiplier. In some cases this mathematical procedure is rather complicated. Therefore, the use of the relativistic radial orbitals, expressed in terms of the generalized spherical functions (2.18), is much more efficient. In such a representation this final form of submatrix element of relativistic Ek-radiation operators follows straightforwardly [28].

Selection rules for Ek-radiation in the case of one-electron configurations consist of the triangle conditions, which the parameters $j_1 j_2$ and k in the Clebsch–Gordan coefficients must satisfy. The triangular condition $(l_1 l_2 k)$ with the extra requirement that $l_1 + l_2 + k$ be even takes care of the selection rule with respect to the parities of the configurations.

A submatrix element of the Ek-radiation with regard to antisymmetric wave functions for the relativistic analogue of transition (25.2) is as follows:

$$\left\langle n_1 n_2 l_1 j_1^{N_1} l_2 j_2 \alpha_1 J_1 j_2 J \| _{e,m} O^{(k)} \| n_1 n_3 l_1 j_1^{N_1} l_3 j_3 \alpha_1' J_1' j_3 J' \right\rangle$$
$$= (-1)^{j_3 + J_1 + J + k} \delta(\alpha_1 J_1, \alpha_1' J_1') \sqrt{(2J+1)(2J'+1)}$$
$$\times \begin{Bmatrix} j_2 & k & j_3 \\ J' & J_1 & J \end{Bmatrix} \left\langle n_2 l_2 j_2 \| _{e,m} O^{(k)} \| n_3 l_3 j_3 \right\rangle . \tag{26.7}$$

Quantities (26.1) and (26.2) may play the role of one-electron submatrix element in (26.7). In the next paragraph we shall see that formula (26.7) is valid for Mk-radiation as well, that is why the operator $O^{(k)}$ has the subscripts e, m. Selection rules for this transition follow from the four triangle conditions $\{j_2 k j_3\}$, $\{j_2 J_1 J\}$, $\{j_3 J_1 J'\}$ and $\{J J' k\}$ for $6j$-coefficient and selection rules for one-electron submatrix element, just discussed above.

Summing two $6j$-coefficients over J' and $J_1 J'$, we can establish a number of sum rules for (26.7), namely,

$$\sum_{J'} \left\langle N_1 1 \| _{e,m} O^{(k)} \| N_1 1 \right\rangle^2$$
$$= \delta(\alpha_1 J_1, \alpha_1' J_1') \frac{2J+1}{2j_2+1} \left\langle n_2 l_2 j_2 \| _{e,m} O^{(k)} \| n_3 l_3 j_3 \right\rangle^2 , \tag{26.8}$$

$$\sum_{JJ'} \left\langle N_1 1 \| _{e,m} O^{(k)} \| N_1 1 \right\rangle^2$$
$$= \delta(\alpha_1 J_1, \alpha_1' J_1')(2J_1 + 1) \left\langle n_2 l_2 j_2 \| _{e,m} O^{(k)} \| n_3 l_3 j_3 \right\rangle^2 . \tag{26.9}$$

The following formula holds for Ek- and Mk-transitions between the levels of one and the same configuration $n_1 l_1 j_1^{N_1} n_2 l_2 j_2^{N_2}$ $(N_1 + N_2 = N)$:

$$\left\langle n_1 n_2 l_1 j_1^{N_1} l_2 j_2^{N_2} \alpha_1 J_1 \alpha_2 J_2 J \| _{e,m} O_N^{(k)} \| n_1 n_2 l_1 j_1^{N_1} l_2 j_2^{N_2} \alpha_1' J_1' \alpha_2' J_2' J \right\rangle$$
$$= \delta(\alpha_2 J_2, \alpha_2' J_2')(-1)^{J_1 + J_2 + J' + k} \sqrt{[J, J']} \left\langle n_1 l_1 j_1 \| _{e,m} O^{(k)} \| n_1 l_1 j_1 \right\rangle$$
$$\times \left(j_1^{N_1} \alpha_1 J_1 \| T^k \| j_1^{N_1} \alpha_1' J_1' \right) \begin{Bmatrix} J_1 & J & J_2 \\ J' & J_1' & k \end{Bmatrix} + \delta(\alpha_1 J_1, \alpha_1' J_1')$$
$$\times (-1)^{J_1 + J_2' + J + k} \sqrt{[J, J']} \left\langle n_2 l_2 j_2 \| _{e,m} O^{(k)} \| n_1 l_1 j_1 \right\rangle$$
$$\times \left(j_2^{N_2} \alpha_2 J_2 \| T^k \| j_2^{N_2} \alpha_2' J_2' \right) \begin{Bmatrix} J_2 & J & J_1 \\ J' & J_2' & k \end{Bmatrix} . \tag{26.10}$$

Here submatrix elements of the operator T^k are defined in accordance with (7.5). As in the non-relativistic case (25.23), only transitions described by one term in (26.10) take place, depending on which subshell an electron is 'jumping' in. In this case the other summand must be considered as being equal to zero.

26.2   *Ek*-transitions with participation of core electrons

The submatrix element of the relativistic electronic transition from one subshell into the other

$$n_1 l_1 j_1^{N_1} n_2 l_2 j_2^{N_2} - n_1 l_1 j_1^{N_1-1} n_2 l_2 j_2^{N_2+1} \tag{26.11}$$

has the form

$$\left\langle n_1 n_2 l_1 j_1^{N_1} l_2 j_2^{N_2} \alpha_1 J_1 \alpha_2 J_2 J \|_{e,m} O^{(k)} \| n_1 n_2 l_1 j_1^{N_1-1} l_2 j_2^{N_2+1} \alpha_1' J_1' \alpha_2' J_2' J' \right\rangle$$

$$= (-1)^{N_2} \sqrt{N_1(N_2+1)} \left(j_1^{N_1} \alpha_1 J_1 \| j_1^{N_1-1}(\alpha_1' J_1') j_1 \right)$$

$$\times \left(j_2^{N_2}(\alpha_2 J_2) j_2 \| j_2^{N_2+1} \alpha_2' J_2' \right)$$

$$\times \left\langle J_1' j_1(J_1) J_2 J \|_{e,m} O^{(k)} \| J_1', J_2 j_2(J_2'), J' \right\rangle, \tag{26.12}$$

where the last multiplier is equal to

$$\left\langle J_1' j_1(J_1) J_2 J \|_{e,m} O^{(k)} \| J_1', J_2 j_2(J_2'), J' \right\rangle$$

$$= (-1)^{j_2+J_2-J_2'} \sqrt{[J_1, J_2', J, J']} \begin{Bmatrix} j_2 & k & j_1 \\ J_2 & J & J_1 \\ J_2' & J' & J_1' \end{Bmatrix}$$

$$\times \left\langle n_1 l_1 j_1 \|_{e,m} O^{(k)} \| n_2 l_2 j_2 \right\rangle. \tag{26.13}$$

If $N_2 = 0$ in (26.11), then we have the transition

$$n_1 l_1 j_1^{N_1} - n_1 l_1 j_1^{N_1-1} n_2 l_2 j_2, \tag{26.14}$$

for which (26.12) becomes

$$\left\langle n_1 l_1 j_1^{N_1} \alpha J \|_{e,m} O^{(k)} \| n_1 n_2 l_1 j_1^{N_1-1} l_2 j_2 \alpha_1' J_1' j_2 J' \right\rangle$$

$$= (-1)^{k+j_2+J_1'+J} \sqrt{N_1[J,J']} \left(j_1^{N_1} \alpha J \| j_1^{N_1-1}(\alpha_1' J_1') j_1 \right)$$

$$\times \begin{Bmatrix} j_1 & k & j_2 \\ J' & J_1' & J \end{Bmatrix} \left\langle n_1 l_1 j_1 \|_{e,m} O^{(k)} \| n_2 l_2 j_2 \right\rangle. \tag{26.15}$$

Much more general is the transition

$$n_1 l_1 j_1^{N_1} n_2 l_2 j_2^{N_2} n_3 l_3 j_3^{N_3} n_4 l_4 j_4^{N_4} - n_1 l_1 j_1^{N_1-1} n_2 l_2 j_2^{N_2} n_3 l_3 j_3^{N_3+1} n_4 l_4 j_4^{N_4}. \tag{26.16}$$

As mentioned in Chapter 6, four momenta (here the momenta of four subshells) may be coupled in two different ways: successive and pair. First we arrange, with the help of transformation matrices, the wave functions of four subshells so that the subshells on which the operator acts (here $j_1^{N_1}$ and $j_2^{N_2}$) are side by side and coupled together. We then express these transformation matrices in terms of $3nj$-coefficients, perform summations, and after rather tedious mathematical manipulations, arrive at the equality

$$\left\langle N_1 N_2 N_3 N_4 \|_{e,m}O_N^{(k)}\| N_1 - 1 N_2 N_3 + 1 N_4 \right\rangle_A$$
$$= (-1)^{N_2+N_3}\delta(\alpha_2 J_2 \alpha_4 J_4, \alpha_2' J_2' \alpha_4' J_4')\sqrt{N_1(N_3+1)}$$
$$\times \left(j_1^{N_1}\alpha_1 J_1 \| j_1^{N_1-1}(\alpha_1' J_1') j_1\right)\left(j_3^{N_3}(\alpha_3 J_3) j_3 \| j_3^{N_3+1}\alpha_3' J_3'\right)$$
$$\times \left\langle \cdots \|_{e,m}O^{(k)}\| \cdots \right\rangle_A. \tag{26.17}$$

We have used the shortened forms $\langle N_1 N_2 N_3 N_4|$ and $|N_1 - 1 N_2 N_3 + 1 N_4\rangle$ for the wave function of the configurations in (26.16) and have specified the form of coupling scheme A. Here $N = N_1 + N_3$ and the last multiplier in (26.17) for the pair and successive couplings of momenta is equal to

$$\left\langle J_1' j_1(J_1)J_2(J_{12}), J_3 J_4(J_{34}), J \|_{e,m}O^{(k)}\| J_1' J_2'(J_{12}'), J_3 j_3(J_3')J_4'(J_{34}'), J' \right\rangle$$
$$= (-1)^{j_1+J_1+J_2+J_4+J_3'+J_{12}'+J_{34}'}\sqrt{[J_1, J_{12}, J_{12}', J_3', J_{34}, J_{34}', J, J']}$$
$$\times \begin{Bmatrix} J_1' & j_1 & J_1 \\ J_{12} & J_2 & J_{12}' \end{Bmatrix}\begin{Bmatrix} J_3 & J_3' & j_3 \\ J_{34}' & J_{34} & J_4 \end{Bmatrix}\begin{Bmatrix} j_3 & k & j_1 \\ J_{34} & J & J_{12} \\ J_{34}' & J' & J_{12}' \end{Bmatrix}$$
$$\times \left\langle n_1 l_1 j_1 \|_{e,m}O^{(k)}\| n_3 l_3 j_3 \right\rangle, \tag{26.18}$$

and

$$\left\langle J_1' j_1(J_1)J_2(J_{12})J_3(J_{123})J_4 J \|_{e,m}O^{(k)}\| J_1' J_2'(J_{12}'), J_3 j_3(J_3'), (J_{123}')J_4'J' \right\rangle$$
$$= (-1)^{J_1+J_2+J_4+J_{12}+J_{123}'+J'+1}\delta(\alpha_2 J_2 \alpha_4 J_4, \alpha_2' J_2' \alpha_4' J_4')$$
$$\times \sqrt{[J_1, J_{12}, J_{123}, J_3', J_{12}', J_{123}', J, J']}\begin{Bmatrix} J_1' & j_1 & J_1 \\ J_{12} & J_2 & J_{12}' \end{Bmatrix}$$
$$\times \begin{Bmatrix} J' & J_4 & J_{123}' \\ J_{123} & k & J \end{Bmatrix}\begin{Bmatrix} J_{12} & J_3 & J_{123} \\ j_1 & j_3 & k \\ J_{12}' & J_3' & J_{123}' \end{Bmatrix}$$
$$\times \left\langle n_1 l_1 j_1 \|_{e,m}O^{(k)}\| n_3 l_3 j_3 \right\rangle, \tag{26.19}$$

respectively.

Employing the rules for interchanging the quantum numbers of separate subshells in (26.17), the appropriate formulas for other types of transitions may be easily established. Thus, while interchanging the first and second subshells in the right side of (26.17), the additional phase multiplier $(-1)^{J_1+J_2-J_{12}+J_1'+J_2'-J_{12}'+N_2}$ occurs. While interchanging the first and third subshells, the right side of (26.17) must be multiplied by the transformation matrix of the appropriate momenta and phase. Transition (26.11) follows from (26.17)–(25.19) if $N_2 = N_4 = 0$. Assuming $N_4 = 0$ we can establish an expression for the submatrix element of operator $_{e,m}O^{(k)}$ for transitions of the type

$$n_1 l_1 j_1^{N_1} n_2 l_2 j_2^{N_2} n_3 l_3 j_3^{N_3} - n_1 l_1 j_1^{N_1-1} n_2 l_2 j_2^{N_2} n_3 l_3 j_3^{N_3+1}. \tag{26.20}$$

The expression is $(N = N_1 + N_3)$:

$$\left\langle N_1 N_2 N_3 \| e,m O_N^{(k)} \| N_1 - 1 N_2 N_3 + 1 \right\rangle$$

$$= (-1)^{N_2+N_3+j_1+j_3+J_1+J_2+J_3-J_3'+J_{12}'} \delta(\alpha_2 J_2, \alpha_2' J_2')$$

$$\times \sqrt{N_1(N_3+1)[J_1, J_{12}, J_{12}', J_3', J, J']} \left(j_1^{N_1} \alpha_1 J_1 \| j_1^{N_1-1} (\alpha_1' J_1') j_1 \right)$$

$$\times \left(j_3^{N_3} (\alpha_3 J_3) j_3 \| j_3^{N_3+1} \alpha_3' J_3' \right) \begin{Bmatrix} J_1' & j_1 & J_1 \\ J_{12} & J_2 & J_{12}' \end{Bmatrix}$$

$$\times \begin{Bmatrix} j_1 & j_3 & k \\ J_{12} & J_3 & J \\ J_{12}' & J_3' & J' \end{Bmatrix} \left\langle n_1 l_1 j_1 \| e,m O^{(k)} \| n_3 l_3 j_3 \right\rangle. \tag{26.21}$$

Assuming $N_3 = 0$ we have from (26.20) the transition

$$n_1 l_1 j_1^{N_1} n_2 l_2 j_2^{N_2} - n_1 l_1 j_1^{N_1} n_2 l_2 j_2^{N_2-1} n_3 l_3 j_3, \tag{26.22}$$

for which

$$\left\langle N_1 N_2 0 \| e,m O^{(k)} \| N_1 N_2 - 1 1 \right\rangle = (-1)^{k-j_2+j_3+J_1+J_2'-J_{12}'}$$

$$\times \sqrt{N_2[J_2, J_{12}', J, J']} \delta(\alpha_1 J_1, \alpha_1' J_1') \left(j_2^{N_2} \alpha_2 J_2 \| j_2^{N_2-1} (\alpha_2' J_2') j_2 \right)$$

$$\times \begin{Bmatrix} J_2' & j_2 & J_2 \\ J & J_1 & J_{12}' \end{Bmatrix} \begin{Bmatrix} j_2 & J & J_{12}' \\ J' & j_3 & k \end{Bmatrix} \left\langle n_2 l_2 j_2 \| e,m O^{(k)} \| n_3 l_3 j_3 \right\rangle. \tag{26.23}$$

Transition (26.7) directly follows from (26.17) as a special case. Generally, having in mind the rules of the permutations of the quantum numbers of separate shells, we are in a position to deduce from the above formulas all expressions for other kinds of electronic transitions between complex configurations practically needed. Let us recall that the relativistic transition of type (26.16) corresponds to non-relativistic transitions (25.24), (25.26) and (25.27). The non-relativistic counterpart of (26.22) is transition (25.25), which is the simplest example of transitions with the participation of core electrons.

26.3   Selection and sum rules

Selection rules for electronic transitions with the participation of core electrons are similar to those for transitions when the core is left unchanged, with the exception of the selection rules following from the CFP with one detached electron. Then seniority quantum numbers v_i and v_i' of the subshells, between which the electron is 'jumping', must be changed by unity, i.e. $\Delta v_i = \pm 1$, $\Delta v_i' = \pm 1$.

Summing formula (26.12) squared over J and J' we find

$$\sum_{JJ'} \left\langle N_1 N_2 \|_{e,m} O_N^{(k)} \| N_1 - 1 N_2 + 1 \right\rangle^2 = N_1(N_2+1) \frac{[J_1, J_2']}{[j_1, j_2]}$$

$$\times \left(j_1^{N_1} \alpha_1 J_1 \| j_1^{N_1-1}(\alpha_1' J_1') j_1 \right)^2 \left(j_2^{N_2}(\alpha_2 J_2) j_2 \| j_2^{N_2+1} \alpha_2' J_2' \right)^2$$

$$\times \left\langle n_1 l_1 j_1 \|_{e,m} O^{(k)} \| n_2 l_2 j_2 \right\rangle^2. \tag{26.24}$$

Performing the further summation over $\alpha_1' J_1'$, accounting for the orthonormality properties of the CFP, we obtain

$$\sum_{\alpha_1' J_1' JJ'} \left\langle N_1 N_2 \|_{e,m} O_N^{(k)} \| N_1 - 1 N_2 + 1 \right\rangle^2 = N_1(N_2+1) \frac{[J_1, J_2']}{[j_1, j_2]}$$

$$\times \left(j_2^{N_2}(\alpha_2 J_2) j_2 \| j_2^{N_2+1} \alpha_2' J_2' \right)^2 \left\langle n_1 l_1 j_1 \|_{e,m} O^{(k)} \| n_2 l_2 j_2 \right\rangle^2. \tag{26.25}$$

And, finally, summing (26.25) over $\alpha_2' J_2'$ we arrive at the equality

$$\sum_{\alpha_1' J_1' \alpha_2' J_2' JJ'} \left\langle N_1 N_2 \|_{e,m} O_N^{(k)} \| N_1 - 1 N_2 + 1 \right\rangle^2$$

$$= N_1(2j_2 + 1 - N_2) \frac{[J_1, J_2]}{[j_1, j_2]} \left\langle n_1 l_1 j_1 \|_{e,m} O^{(k)} \| n_2 l_2 j_2 \right\rangle^2. \tag{26.26}$$

Assuming in (26.24)–(26.26) that $N_2 = 0$, we establish the appropriate sum rules for simpler transition (26.14).

Such selection rules may be found for the general case (26.16) as well. Unfortunately, they are rather cumbersome and therefore of little use. That is why we present below only two expressions of the kind (26.26) for pair A_1 and successive A_0 couplings of the momenta of separate subshells, namely,

$$\sum_{\alpha_1' J_1' \alpha_3' J_3' J_{12}' J_{34}' JJ'} \left\langle N_1 N_2 N_3 N_4 \|_{e,m} O^{(k)} \| N_1 - 1 N_2 N_3 + 1 N_4 \right\rangle_{A_1}^2$$

$$= \delta(\alpha_2 J_2 \alpha_4 J_4, \alpha_2' J_2' \alpha_4' J_4') N_1(2j_3 + 1 - N_3) \frac{[J_{12}, J_{34}]}{[j_1, j_3]}$$

$$\times \left\langle n_1 l_1 j_1 \|_{e,m} O^{(k)} \| n_3 l_3 j_3 \right\rangle^2, \tag{26.27}$$

$$\sum_{\alpha_1' J_1' J_{12}' J_{123}' \alpha_3' J_3' J_{123} JJ'} \left\langle N_1 N_2 N_3 N_4 \|_{e,m} O^{(k)} \| N_1 - 1 N_2 N_3 + 1 N_4 \right\rangle_{A_0}^2$$

$$= \delta(\alpha_2 J_2 \alpha_4 J_4, \alpha_2' J_2' \alpha_4' J_4') N_1(2j_3 + 1 - N_3) \frac{[J_3, J_4, J_{12}]}{[j_1, j_3]}$$

$$\times \left\langle n_1 l_1 j_1 \|_{e,m} O^{(k)} \| n_3 l_3 j_3 \right\rangle^2. \tag{26.28}$$

However, as was already emphasized for the non-relativistic transitions, all these sum rules are not rigorous due to interactions between the intermediate momenta and the presence of deviations from pure coupling schemes.

The expressions for *Ek*-transitions presented here allow one to perform calculations of the respective matrix elements in relativistic approximation for any type of practical configuration and any multipolarity of radiation.

27

Mk-transitions. Particular cases of $E2$- and $M1$-transitions

27.1 Relativistic and non-relativistic Mk-transitions

As was pointed out in Chapter 4, division of the radiation into electric and magnetic is connected with the existence of two types of multipoles, characterized by the parities $(-1)^k$ and $(-1)^{k+1}$, respectively. The first ones we have studied quite thoroughly in Chapters 24–26. Here let us consider in a similar way the Mk-transitions. Again, as we have seen in Chapter 4, the potential of the electromagnetic field in this case does not depend on gauge. Therefore only one relativistic expression (4.8) was established for the probability of Mk-radiation, described by the appropriate operator (4.9). The probability of non-relativistic Mk-transitions (in atomic units) is given by formula (4.15), whereas the corresponding non-relativistic operator has the form (4.16).

In order to be able to calculate oscillator strengths or transition probabilities of Mk-radiation, we need the corresponding expressions for the submatrix elements of appropriate operators (4.9) and (4.16). The one-electron submatrix element of relativistic operator (4.9) is equal to

$$\langle n_2 l_2 j_2 \| {}_m Q^{(k)} \| n_1 l_1 j_1 \rangle = (-1)^{j_2 + (l_2 + 1 + l_1 + k)/2} \sqrt{2j_1 + 1}$$

$$\times \begin{bmatrix} j_1 & k & j_2 \\ -1/2 & 1 & 1/2 \end{bmatrix} \{ (l_1' l_2 k) B(\lambda_1' \lambda_2 | 0k) + (l_1 l_2' k) B(\lambda_1 \lambda_2' | 0k) \} . \quad (27.1)$$

Here the radial integral is defined by (26.4). The selection rules for relativistic Mk-transitions between one-electron configurations directly follow from the non-zero conditions of submatrix element (27.1). They consist of triangular condition $\{j_1 k j_2\}$ and symbols (abc) at radial integrals, ensuring the even values of the perimeter of the corresponding triangle. It follows that relativistic $M1$-transitions may also occur between configurations for which $\Delta l = 2$ (not only $\Delta l = 0$, as is the case for the non-relativistic operator). The one-electron submatrix element of the non-relativistic operator of Mk-transitions is as follows:

$$\left(l_2 s_2 j_2 \|_m Q^{(k)}\| l_1 s_1 j_1\right) = i^{-k-1}(-1)^{-j_1} \frac{(l_1 l_2 k - 1)}{2(k+1)}$$

$$\times \sqrt{\frac{(2j_1 + 1)(2j_2 + 1)}{2k+1}} \left\{1 - 2k^2 + (-1)^{l_2 + 1/2 - j_2}\right.$$

$$\times \left.\left[2j_2 + 1 + (-1)^{j_1 + j_2 + k}(2j_1 + 1)\right]\right\}. \qquad (27.2)$$

It can be straightforwardly obtained while using generalized spherical functions (see Chapter 2). Selection rules for Mk-radiation, described by non-relativistic formulas as usual, follow from the non-zero conditions for the quantities in this expression. They were discussed in Chapter 24 (see formulas (24.22) and (24.24)).

As was shown in Chapter 26, the pecularities of relativistic operators of electronic transitions are confined in their one-electron submatrix elements. Therefore formulas (26.7), (26.10), (26.12), (26.13), (26.15), (26.17)–(26.19), (26.21) and (26.23) are equally applicable for both the relativistic Ek- and Mk-transitions between complex electronic configurations. This is denoted by the subscripts e, m at the operator $O^{(k)}$. The same holds for sum rules (26.8), (26.9), (26.24)–(26.28). Therefore, we have only to present the appropriate expressions for the submatrix elements of non-relativistic operator (4.16) of Mk-transition in general cases of complex electronic configurations. For Mk-transitions between the levels of a shell of equivalent electrons the following formula is valid:

$$\left(nl^N \alpha L S J \|_m Q^{(k)}\| nl^N \alpha' L' S' J'\right) = \frac{i}{c} k \sqrt{[J, J', k]} \left(l \| C^{(k-1)} \| l\right)$$

$$\times \left(nl|r^{k-1}|nl\right) \left\{(-1)^{J'+L+S+k} \delta(S, S') \frac{\sqrt{(2l+k+1)(2l-k+1)}}{2(k+1)\sqrt{[S,k]}}\right.$$

$$\times \left\{\begin{matrix} L & S & J \\ J' & k & L' \end{matrix}\right\} \left(l^N \alpha L S \| U^k \| l^N \alpha' L' S'\right)$$

$$+ \sqrt{\frac{2k-1}{k}} \left\{\begin{matrix} L' & S' & J' \\ k-1 & 1 & k \\ L & S & J \end{matrix}\right\} \left(l^N \alpha L S \| V^{k-1,1} \| l^N \alpha' L' S'\right)\right\}. \qquad (27.3)$$

The appropriate expression for Mk-transitions between the levels of one and the same two-shell configuration is given by the formula

$$(n_1 l_1^{N_1} n_2 l_2^{N_2} \alpha_1 L_1 S_1 \alpha_2 L_2 S_2 L S J \|_m Q^{(k)}\| \times n_1 l_1^{N_1} n_2 l_2^{N_2} \alpha'_1 L'_1 S'_1 \alpha'_2 L'_2 S'_2 L' S' J')$$

$$= \frac{i}{c} \sqrt{k[k-1,k,J,J',L]} \left[\sqrt{2S+1} \left\{ \begin{matrix} L' & S' & J' \\ k-1 & 1 & k \\ L & S & J \end{matrix} \right\} E \right.$$

$$\left. + (-1)^{J'+L+S} \left\{ \begin{matrix} L & J & S \\ J' & L' & k \end{matrix} \right\} M \right], \tag{27.4}$$

where

$$E = (-1)^{L_1+L_2+L'+k+S_1+S_2+S'} \delta(\alpha_2 L_2 S_2, \alpha'_2 L'_2 S'_2) \left(l_1 \| C^{(k-1)} \| l_1 \right)$$

$$\times \left(n_1 l_1 \| r^{k-1} \| n_1 l_1 \right) \sqrt{[L', S']} \left(l_1^{N_1} \alpha_1 L_1 S_1 \| V^{k-1,1} \| l_1^{N_1} \alpha'_1 L'_1 S'_1 \right)$$

$$\times \left\{ \begin{matrix} L'_1 & k-1 & L_1 \\ L & L_2 & L' \end{matrix} \right\} \left\{ \begin{matrix} S'_1 & 1 & S_1 \\ S & S_2 & S' \end{matrix} \right\} + (-1)^{L_2+L_1+L+k+S'_2+S_1+S}$$

$$\times \delta(\alpha_1 L_1 S_1, \alpha'_1 L'_1 S'_1) \left(n_2 l_2 \| r^{k-1} \| n_2 l_2 \right) \left(l_2 \| C^{(k-1)} \| l_2 \right)$$

$$\times \sqrt{[L', S']} \left(l_2^{N_2} \alpha_2 L_2 S_2 \| V^{k-1,1} \| l_2^{N_2} \alpha'_2 L'_2 S'_2 \right)$$

$$\times \left\{ \begin{matrix} L'_2 & k-1 & L_2 \\ L & L_1 & L' \end{matrix} \right\} \left\{ \begin{matrix} S'_2 & 1 & S_2 \\ S & S_1 & S' \end{matrix} \right\}, \tag{27.5}$$

$$M = \frac{1}{2(k+1)\sqrt{[k-1,k]}} \left\{ (-1)^{L_1+L_2+L'} \delta(S_1, S'_1) \delta(\alpha_2 L_2 S_2, \alpha'_2 L'_2 S'_2) \right.$$

$$\times \sqrt{(2L'+1)k(2l_1+k+1)(2l_1-k+1)} \sqrt{[S_1]} \left(l_1 \| C^{(k-1)} \| l_1 \right)$$

$$\times \left(n_1 l_1 \| r^{k-1} \| n_1 l_1 \right) \left(l_1^{N_1} \alpha_1 L_1 S_1 \| U^k \| l_1^{N_1} \alpha'_1 L'_1 S'_1 \right) \left\{ \begin{matrix} L'_1 & k & L_1 \\ L & L_2 & L' \end{matrix} \right\}$$

$$+ (-1)^{L_2+L_1+L} \delta(S_2, S'_2) \delta(\alpha_1 L_1 S_1, \alpha'_1 L'_1 S'_1)$$

$$\times \sqrt{[L'] k(2l_2+k+1)(2l_2-k+1)} \sqrt{[S_2]} \left(l_2 \| C^{(k-1)} \| l_2 \right)$$

$$\times \left(n_2 l_2 \| r^{k-1} \| n_2 l_2 \right) \left\{ \begin{matrix} L'_2 & k & L_2 \\ L & L_1 & L' \end{matrix} \right\}$$

$$\times \left(l_2^{N_2} \alpha_2 L_2 S_2 \| U^k \| l_2^{N_2} \alpha'_2 L'_2 S'_2 \right) \right\}. \tag{27.6}$$

As we have seen from the selection rules for non-relativistic magnetic dipole ($M1$) radiation, these transitions are permitted only between levels of one and the same configuration. However, this is not so for higher multipolarities ($k > 1$). Therefore we present here the appropriate formulas to cover the general cases needed in practice. So, it may be of use to have the following expression for the submatrix element of the operator of Mk-transitions between the levels of two different two-shell configurations:

$$
\begin{aligned}
\Big(n_1 & l_1^{N_1} n_2 l_2^{N_2} \alpha_1 L_1 S_1 \alpha_2 L_2 S_2 L S J \|_m Q^{(k)} \| \\
& \times n_1 l_1^{N_1-1} n_2 l_2^{N_2+1} \alpha_1' L_1' S_1' \alpha_2' L_2' S_2' L' S' J' \Big) \\
= & \frac{i}{c} (-1)^{N_2+L_2-L_2'+l_2} \left(l_1^{N_1} \alpha_1 L_1 S_1 \| l_1^{N_1-1} (\alpha_1' L_1' S_1') l_1 \right) \\
& \times \left(l_2^{N_2} (\alpha_2 L_2 S_2) l_2 \| l_2^{N_2+1} \alpha_2' L_2' S_2' \right) \left(n_1 l_1 \| r^{k-1} \| n_2 l_2 \right) \\
& \times \sqrt{ N_1 (N_2+1) k [k-1, L_1, S_1, L_2', S_2', L, L', J, J'] } \\
& \times \Big[(-1)^{L-J'-k+S_1'+S_2'} \delta(S,S') \left(l_1 \| \left[C^{(k-1)} \times L^{(1)} \right]^{(1)} \| l_2 \right) \\
& \times \begin{Bmatrix} L & S & J \\ J' & k & L' \end{Bmatrix} \begin{Bmatrix} S_2 & S_2' & 1/2 \\ S_1' & S_1 & S \end{Bmatrix} \begin{Bmatrix} L' & L_1' & L_2' \\ k & l_1 & l_2 \\ L & L_1 & L_2 \end{Bmatrix} \\
& + (-1)^{S_2-S_2'+1/2} \sqrt{[k,S,S']} \left(l_1 s \| \left[C^{(k-1)} \times S^{(1)} \right]^{(k)} \| l_2 s \right) \\
& \times \begin{Bmatrix} L' & S' & J' \\ k-1 & 1 & k \\ L & S & J \end{Bmatrix} \begin{Bmatrix} L' & L_1' & L_2' \\ k-1 & l_1 & l_2 \\ L & L_1 & L_2 \end{Bmatrix} \\
& \times \begin{Bmatrix} S' & S_1' & S_2' \\ 1 & 1/2 & 1/2 \\ S & S_1 & S_2 \end{Bmatrix} \Big].
\end{aligned}
\tag{27.7}
$$

The one-electron submatrix element of operator r^{k-1} stands for the radial integral of this quantity and the appropriate radial wave functions. Indeed, we can easily show that the right side of formula (27.7) equals zero for $k = 0$. Therefore, $M1$-transitions occur only between levels of one and the same configuration.

27.2 $E2$- and $M1$-transitions between levels of one and the same configuration

As has been emphasized in Chapter 24 the probabilities of Ek- and Mk-transitions are rapidly decreasing functions of k (see also Chapter 30). Therefore, we can usually restrict ourselves to examination of the radiation of the lowest multipolarity permitted by the appropriate selection rules. Between levels of one and the same configuration both $E2$- and $M1$-transitions are allowed, that is why we must consider them together.

Oscillator strengths of non-relativistic Ek- and Mk-transitions may be calculated by the general formula (24.14). Transition probabilities may be found for Ek-radiation using formula (4.10) or (4.11) and for Mk-

radiation by employing definition (4.15). Concrete formulas for E1- and E2-radiation will be presented in Chapter 30, whereas for M1-transitions we give them below:

$$f^{M1} = \frac{2}{3(2J+1)}\Delta E \left|(J\|_m Q^{(1)}\|J')\right|^2, \tag{27.8}$$

$$W^{M1} = \frac{214.2 \times 10^8}{2J+1}(\Delta E)^3 \left|(J\|_m Q^{(1)}\|J')\right|^2. \tag{27.9}$$

In these formulas the energy difference ΔE is measured in atomic units, transition probabilities are obtained in seconds and the submatrix element in the cases of one or two shells of equivalent electrons for LS coupling may be taken from (27.3) or (27.4). When calculating using intermediate coupling one has to bear in mind that the appropriate wave functions are of the form (11.10).

Let us study the relative role of E2- and M1-radiation for the example of such transitions inside the configurations $3d^3$ and $2p^3$. The selection rules for E2- and M1-transitions in LS coupling directly follow from (24.21)–(24.24), namely,

$$E2 : \Delta S = 0; \ \Delta L = 0, \pm 1, \pm 2 \quad (L + L' \geq 2);$$

$$\Delta J = 0, \pm 1, \pm 2 \quad (J + J' \geq 2); \tag{27.10}$$

$$M1 : \Delta S = \Delta L = 0; \quad \Delta J = \pm 1. \tag{27.11}$$

However, the quantum numbers L and S are not rigorous, due to the existence of the spin–orbit interaction between the respective momenta. Therefore, the above-mentioned selection rules hold only approximately. In intermediate coupling the selection rules with respect to L and S change and allow many more transitions. For example, the isolation of the condition $\Delta S = 0$ leads to the occurrence of the so-called intercombination E2- and M1-lines. For the configuration $3d^3$ in intermediate coupling, instead of (27.10) and (27.11) we obtain

$$E2 : \Delta S = 0, \pm 1; \quad \Delta L = 0, \pm 1, \ldots, \pm 4;$$

$$\Delta J = 0, \pm 1, \pm 2 \quad (J + J' \geq 2); \tag{27.12}$$

$$M1 : \Delta S = 0, \pm 1; \quad \Delta L = 0, \pm 1, \pm 2;$$

$$\Delta J = 0, \pm 1 \quad (J + J' \geq 1). \tag{27.13}$$

Formally in (27.11) there must be $\Delta J = 0, \pm 1$ $(J + J' \geq 1)$ instead of only $\Delta J = \pm 1$. However, for pure LS coupling and $\Delta J = 0$, we have the transition for which $\Delta E = 0$.

Selection rules (27.10) for E2- and (27.11) for M1-transitions permit in the configuration under consideration the existence of 63 and 30 transitions, respectively. When employing the wave functions of intermediate coupling with selection rules (27.12) and (27.13) valid, these rules allow 131

$E2$- and 89 $M1$-transitions. Of them, transitions with $\Delta J = \pm 2$ are purely of $E2$ nature (43 transitions), 88 transitions are conditioned by both the $E2$- and $M1$-transition operators, and only the transition ${}^2_3P_{1/2} - {}^4_3P_{1/2}$ is purely of a magnetic dipole type.

The effect of diagonalization of the energy matrix for such transitions is much more essential compared to $E1$-radiation (for details see Chapter 30). While calculating the oscillator strengths and transition probabilities we need the energy differences of the respective levels. For the permitted $M1$-radiation, transitions are allowed only between levels with the same L. Therefore ΔE is of the order of a term fine structure. While accounting for (27.13), $M1$-transitions occur between levels belonging to different terms and, as a result of the large ΔE in such cases, the corresponding values of the transition probabilities have the same or fairly often even larger order than for the transitions described by selection rules (27.11), even though their line strengths are much smaller. Therefore, when calculating the $M1$-transitions, accounting for the deviations from pure (in this case LS) coupling (diagonalization of the energy matrix) is obligatory. The same is true, although to a slightly lesser extent, for $E2$-radiation, because in such a case there are also transitions with $\Delta L = 0$.

For the majority of transitions in neutral and not highly ionized atoms, according to the quantities of the transition probabilities, one radiation sort ($E2$ or $M1$) predominates, and only in a few cases are the contributions of both operators of the same order. For $\Delta L = 0$ the $M1$-radiation has usually the larger probability compared to that of $E2$. For $\Delta L = \pm 1$ both sorts of radiation, as a rule, compete with each other, whereas for $\Delta L = \pm 2$ the $E2$-transitions predominate, because in this case they are already permitted in pure LS coupling. Usually the energy difference between levels, ΔE plays the decisive role. Therefore, in the overwhelming majority of cases the lines for which $\Delta L \neq 0$ are more intensive. For $\Delta S = 0$ the lines with $\Delta L \neq 0$ are more probable because usually they are permitted for $E2$ radiation already in pure LS coupling. However, we emphasize that all these regularities hold more or less approximately, they are more tendencies than rigorous rules.

In conclusion, let us briefly discuss the relative role of $E2$- and $M1$-transitions depending on the ionization degree and, thus, on the plasma temperature. Table 27.1 lists wavelengths $\lambda(\text{Å})$ and probabilities (in s^{-1}) of $E2$- and $M1$-transitions inside configuration $1s^2 2s^2 2p^3$, calculated in the intermediate coupling scheme, starting with the numerical Hartree–Fock radial functions and accounting for relativistic effects of order α^2 for the ions OII, MgVI, ArXII and FeXX.

The data of Table 27.1 illustrate a number of the more or less rigorous regularities in the behaviour of $E2$- and $M1$-transitions for various ions and various transitions in a given ion. The utilization of intermediate

Table 27.1. Wavelengths (Å) and probabilities (s^{-1}) of E2- and M1-transitions between the levels of configuration $1s^22s^22p^3$ for OII, MgVI, ArXII and FeXX

$LSJ - L'S'J'$		OII	MgVI	ArXII	FeXX
	λ	1983.3	1015.8	588.8	352.6
$^4S_{3/2} -^2 P_{1/2}$	$E2$	3.49-8*	5.06-5	3.16-2	8.32+0
	$M1$	2.97-2	5.92+0	5.72+2	3.24+4
	λ	> 100000	~100000	23590.7	1750.4
$^2P_{1/2} -^2 P_{3/2}$	$E2$	~0	~0	4.47-8	1.58-2
	$M1$	~ 10^{-10}	2.44-5	6.52-1	1.29+3
	λ	1983.2	1014.4	574.5	293.5
$^4S_{3/2} -^2 P_{3/2}$	$E2$	4.31-9	6.11-6	2.89-3	5.61-2
	$M1$	7.42-2	1.47+1	1.31+3	3.74+4
	λ	3307.3	1698.3	1012.5	706.3
$^4S_{3/2} -^2 D_{3/2}$	$E2$	3.03-5	1.62-3	4.84-2	6.99-1
	$M1$	7.79-6	4.11-2	6.08+1	1.28+4
	λ	> 100000	~100000	31806.1	2809.1
$^2D_{3/2} -^2 D_{5/2}$	$E2$	~0	~0	1.54-8	1.83-3
	$M1$	~ 10^{-11}	1.27-5	3.21-1	3.86+2
	λ	4954.9	2534.6	1472.2	940.1
$^2D_{5/2} -^2 P_{1/2}$	$E2$	4.14-1	8.40-1	1.50+0	2.47+0
	$M1$	0	0	0	0
	λ	3307.1	1695.3	981.3	564.4
$^4S_{3/2} -^2 D_{5/2}$	$E2$	4.71-5	2.53-3	8.65-2	2.92+0
	$M1$	1.72-7	9.20-4	1.64+0	1.00+3
	λ	4954.4	2527.8	1407.0	704.3
$^2D_{3/2} -^2 P_{1/2}$	$E2$	6.24-1	1.28+0	2.70+0	1.24+1
	$M1$	1.48-2	2.96+0	2.66+2	7.80+3
	λ	4953.8	2518.9	1327.8	502.3
$^2D_{3/2} -^2 P_{3/2}$	$E2$	3.12-1	6.44-1	1.57+0	1.59+1
	$M1$	2.38-2	4.78+0	5.07+2	4.55+4
	λ	4954.3	2525.6	1385.7	611.6
$^2P_{3/2} -^2 D_{5/2}$	$E2$	7.27-1	1.49+0	3.39+0	2.85+1
	$M1$	1.34-2	2.67+0	2.55+2	1.35+4

* The notation $a.bc \pm d$ in this table means $a.bc \cdot 10^{\pm d}$.

coupling has conditioned the occurrence of highly probable transitions with $\Delta L \neq 0$ and $\Delta S \neq 0$. Their high probability can be explained by the considerably larger ΔE values for transitions with $\Delta L \neq 0$ and $\Delta S \neq 0$ in comparison with transitions for which $\Delta L = \Delta S = 0$. In the cases when both $E2$ and $M1$ transitions are permitted, for not highly ionized atoms the $E2$-transitions usually prevail. However, with increase of ionization degree, $M1$-radiation begins to dominate. For example, for FeXX the probability of $M1$-transition exceeds the probability of $E2$-transition on average by three orders of magnitude and even more.

Besides, with the increase of ionization degree, the wavelengths of the usual $E1$-transitions move to the far ultraviolet or even X-ray spectral region. The lines of $E2$- and $M1$-transitions occur in the visible wavelength region. Therefore they become very useful for diagnostics of high temperature plasmas. On the other hand, such data give us a wealth of information on the relative role of separate intra-atomic interactions, the changes of the character of coupling scheme along the isoelectronic sequences, etc.

Part 7

Calculation of Energy Spectra and Electronic Transitions in the Case of Complex Configurations

28

Methods of determination
of radial orbitals

28.1 Non-relativistic numerical radial orbitals

While calculating matrix elements of various operators we learned that integration over angular variables led to the occurrence of multipliers consisting of $3nj$-coefficients, coefficients of fractional parentage as well as submatrix elements of operators of the kind U^k or V^{k1}. Their numerical values may be found using the appropriate formulas, tables or computer programs. Here let us discuss the problem of calculation of radial wave functions (orbitals) and radial integrals occurring in the expressions for the relevant matrix elements. Radial orbitals may be presented in numerical or analytical forms, and for many-electron systems they are more or less approximate. The Hartree–Fock method is the most universal and efficient way of finding the radial orbitals of many-electron atoms or ions. It is based on the central field approximation and the Rayleigh–Ritz variational principle. Let us build the functional of the system under consideration in the form

$$E(\psi) = (\psi|\hat{H}_0|\psi)/(\psi|\psi), \qquad (28.1)$$

where $(\psi|\psi)$ ensures normalization of the wave function of the system, and $\hat{H}_0$ is defined by formula (1.15). The Hartree–Fock equations in the single-configuration approximation follow from the many-electron Schrödinger equation while varying functional (28.1) with regard to the radial orbitals of the single-electron approach $P(nl|r)$, which are the same for a given shell of equivalent electrons and which guarantee the stationarity of the total energy E, written in the form

$$E = \sum_{nl} \left\{ NI(nl) + \sum_k f_k(l^N)F^k(nl,nl) + \frac{1}{2}\sum_{n'l' \neq nl} \right.$$

$$\left. \times \left[\sum_k f_k(l^N, l'^{N'})F^k(nl, n'l') + \sum_k g_k(l^N, l'^{N'})G^k(nl, n'l') \right] \right\}. \quad (28.2)$$

Here the radial integral $I(nl)$ and Slater integrals F^k and G^k are defined by (19.23), (19.31) and (20.10), respectively. Coefficients $f_k(l^N)$, $f_k(l^N, l'^{N'})$ and $g_k(l^N, l'^{N'})$ have the expressions (19.30), (20.14) and (20.15), respectively. The summation over nl runs through all shells of equivalent electrons, whereas the sum over $n'l' \neq nl$ accounts for the electrostatic interaction between shells. In general, these coefficients have quite complex expressions, they are different for each term. Therefore, we face the necessity to solve appropriate equations for each term. However, the calculations indicate that the dependence of radial orbitals on the term is weak, at least for configurations for which the phenomenon of the collapse of the orbit of the excited electron does not take place [174]. That is why simplified expressions for these coefficients are usually employed. Thus, for example, in [175] it is suggested to omit the part of these coefficients depending on the term. However, most frequently, coefficients averaged with regard to all terms of the configuration considered, are utilized:

$$\bar{X} = \sum_{LS}(2L+1)(2S+1)X(LS)/\left[\sum_{LS}(2L+1)(2S+1)\right]. \quad (28.3)$$

The values of f_k and g_k, averaged in this way, are equal

$$\bar{f}_0(l^N) = f_0(l^N) = N(N-1)/2, \quad (28.4)$$

$$\bar{f}_{k>0}(l^N) = -N(N-1)(l\|C^{(k)}\|l)^2/[(4l+2)(4l+1)], \quad (28.5)$$

$$\bar{f}_0(l^N, l'^{N'}) = f_0(l^N, l'^{N'}) = NN', \quad (28.6)$$

$$\bar{f}_{k>0}(l^N, l'^{N'}) = 0, \quad (28.7)$$

$$\bar{g}_k(l^N, l'^{N'}) = -NN'(l\|C^{(k)}\|l')^2/[2(2l+1)(2l'+1)]. \quad (28.8)$$

In the case of $N = 4l + 2$ these expressions lead to exact values of the interactions considered for closed shells.

The radial orbitals we are looking for are usually orthogonalized according to condition (10.1). Let us notice that the orthogonality requirement is not compulsory, there exists the method of solving the Hartree–Fock equations for non-orthogonal radial orbitals, but it is much more complex than the conventional one (see also the final section of this chapter). The orthonormality of the functions is secured by introduction into the varied functional terms containing integrals of the type (10.1), multiplied by the so-called Lagrange multipliers $\lambda_{nl,n'l}$. The orthogonality with regard to l is guaranteed by the properties of the respective spherical functions.

Variation of the expression, obtained in this way,

$$E' = E + \frac{1}{2} \sum_{nl,n'l} \lambda_{nl,n'l} \int_0^\infty P(nl|r)P(n'l|r)dr \qquad (28.9)$$

straightforwardly leads to the system of integro-differential equations, usually called the Hartree–Fock equations. For each nl-shell a separate equation must be written, namely,

$$\left[\frac{d^2}{dr^2} - 2Y(nl|r) - \frac{l(l+1)}{r^2} - \epsilon_{nl,nl} \right] P(nl|r)$$

$$- \chi(nl|r) - \sum_{n'\neq n} \epsilon_{nl,n'l} P(n'l|r) = 0. \qquad (28.10)$$

Here $Y(nl|r)$ and $\chi(nl|r)$ are defined in a.u. as follows:

$$Y(nl|r) = -\frac{Z}{r} + \frac{1}{r} \sum_{n'l'} \frac{1 + \delta(nl,n'l')}{N}$$

$$\times \sum_k f_k(l^N, l'^{N'}) Y_k(n'l', n'l'|r), \qquad (28.11)$$

$$\chi(nl|r) = \frac{2}{r} \sum_{n'l'\neq nl} \sum_k \frac{g_k(l^N, l'^{N'})}{N} Y_k(nl, n'l'|r) P(n'l'|r), \qquad (28.12)$$

where

$$Y_k(nl, n'l'|r) = r^{-k} \int_0^r r_1^k P(nl|r_1)P(n'l'|r_1)dr_1$$

$$+ r^{k+1} \int_r^\infty r_1^{-(k+1)} P(nl|r_1)P(n'l'|r_1)dr_1, \qquad (28.13)$$

and

$$\epsilon_{nl,n'l} = \lambda_{nl,n'l}/N. \qquad (28.14)$$

For $N \neq N'$

$$\lambda_{nl,n'l} = \frac{NN'}{N'-N} \int_0^\infty \left\{ 2 \left[Y(n'l|r) - Y(nl|r) \right] P(nl|r)P(n'l|r) \right.$$

$$\left. + \chi(n'l|r)P(nl|r) - \chi(nl|r)P(n'l|r) \right\}dr. \qquad (28.15)$$

The system of equations (28.10), each describing a shell of equivalent electrons $n_il_i^{N_i}$ with boundary conditions $P(n_il_i|0) = P(n_il_i|\infty) = 0$ is usually solved numerically by computer. The methods of solving the various modifications of such equations are described in detail in [16, 44, 45].

The Hartree–Fock approach is also called the self-consistent field method. Indeed, the potential of the field in which the electron nl is orbiting is also expressed in terms of the wave functions we are looking for. Therefore the procedure for determination of the radial orbitals must be coordinated with the process of finding the expression for the potential: starting with the initial form of the wave function, we find the expression for the potential needed to determine the more accurate wave function. We must continue this process until we reach the desired consistency between these quantities.

The main difficulty in solving the Hartree–Fock equation is caused by the non-local character of the potential in which an electron is orbiting. This causes, in turn, a complicated dependence of the potential, particularly of its exchange part, on the wave functions of electronic shells. There have been a number of attempts to replace it by a local potential, often having an analytical expression (e.g. universal Gaspar potential, Slater approximation for its exchange part, etc.). These forms of potential are usually employed to find wave functions when the requirements for their accuracy are not high or when they serve as the initial functions.

Let us consider a few examples. If we replace the non-local potential by a local one, then we arrive at the homogeneous equation for each shell $n_i l_i^{N_i}$:

$$\left[-\frac{d^2}{dr^2} + \frac{l_i(l_i+1)}{r^2} + V^i(r) \right] P(n_i l_i | r) = \epsilon_i P(n_i l_i | r), \qquad (28.16)$$

where $V^i(r)$ is some given function of the potential energy of the field in which the electron $n_i l_i$ is orbiting. It is obvious that the solution of a homogeneous equation is much simpler than that of an inhomogeneous one.

There exists a whole number of approximate expressions for $V^i(r)$ (see, for example [139]). The simplest, called the Thomas–Fermi potential, follows from the statistical model of an atom. Unfortunately, it leads to results of very low accuracy. More accurate is the Thomas–Fermi–Dirac model, in which an attempt is made to account for the exchange part of the potential energy of an electron in the framework of the free electron gas approach. Various forms of the parametric potential method are fairly widely utilized, particularly for multiply charged ions. Such potentials may look as follows [16]:

$$V(r) = -\frac{2}{r} \left\{ (N-1)e^{-\alpha_1 r} + a_2 r e^{-\alpha_3 r} + \ldots + a_{n-1} r^k e^{-\alpha_n r} + Z - N + 1 \right\}. \qquad (28.17)$$

Free parameters α_i and a_j may be found from experimental data (but in such a case the method becomes semi-empirical, see also Chapter 21), or from the more accurate theoretical calculations for the ground and a few excited configurations.

The simplest version of the self-consistent field approach is the Hartree method, in which the variational principle is applied to a non-symmetrized product of wave functions, and the orthogonality conditions for functions with different n are neglected. This leads to neglecting the exchange part of the potential, which causes errors in the results.

The so-called Hartree–Fock–Slater method is much more widely utilized, and is a hybrid of the Hartree and Thomas–Fermi–Dirac methods. In this method the direct part of the potential is calculated using the Hartree–Fock approach, whereas the exchange part is approximated by some statistical expression of the model of free electrons. The Slater potential is given by:

$$V^i(r) = -\frac{2Z}{r} + V_c(r) - \frac{3}{2}\left(\frac{24\rho}{\pi}\right)^{1/3} \quad Ry, \qquad (28.18)$$

where (q denotes the number of shells):

$$V_c(r) = \sum_{j=1}^{q} N_j \int_0^\infty \frac{2}{r_>} P_j^2(r_2)dr_2 \qquad (r_> = \max(r_2, r)), \qquad (28.19)$$

whereas the density of the charges of all electrons in an atom (in a.u.)

$$\rho(r) = \sum_{j=1}^{q} N_j \rho_j(r) = \frac{1}{4\pi r^2}\sum_{j=1}^{q} N_j P_j^2(r). \qquad (28.20)$$

The main advantage of this method is automatic orthogonality of all functions $P_i(r)$. Its principal disadvantage is an inaccurate account of the energy of self-interaction of the ith electron. We can avoid this by employing instead of (28.19) the Hartree-type term

$$V_H(r) = \sum_{j=1}^{q}(N_j - \delta(i,j)) \int_0^\infty \frac{2}{r_>} P_j^2(r_2)dr_2. \qquad (28.21)$$

This expression excludes self-interaction. There have been a number of attempts to include into the Hartree–Fock equations the main terms of relativistic and correlation effects, however without great success, because the appropriate equations become much more complex. For a large variety of atoms and ions both these effects are fairly small. Therefore, they can be easily accounted for as corrections in the framework of first-order perturbation theory. Having in mind the constantly growing possibilities of computers, the Hartree–Fock self-consistent field method in various approximations remains the main and most universal, as well as efficient, method to determine accurate wave functions of any atom or ion of the Periodical Table.

28.2   Relativistic Hartree–Fock radial orbitals

These (see Chapter 2) may be obtained utilizing the relativistic analogue of the Hartree–Fock method, normally called the Dirac–Hartree–Fock method [176–178]. The relevant equations may be found in an analogous manner to the non-relativistic case, therefore here we shall present only final results (in a.u.; let us recall that $\lambda = nlj$, $\lambda' = nl'j$):

$$\frac{dP(\lambda|r)}{r} = -\frac{\kappa}{r}P(\lambda|r) + \frac{1}{c}\left\{\left[2c^2 - \epsilon_{\lambda,\lambda} + \frac{Z}{r} - V(\lambda|r)\right]Q(\lambda'|r) - \chi(\lambda|r)\right\},$$
(28.22)

$$\frac{dQ(\lambda'|r)}{dr} = \frac{\kappa}{r}Q(\lambda'|r) + \frac{1}{c}\left\{\left[\epsilon_{\lambda,\lambda} - \frac{Z}{r} + V(\lambda|r)\right]P(\lambda|r) + \chi(\lambda'|r)\right\}.$$

In a relativistic case we have a system of two equations for each electron nlj due to the two-component character of wave function (2.15). Here κ is defined by (19.70) and $V(\lambda|r)$, $\chi(\lambda|r)$ and $\chi(\lambda'|r)$, respectively, given by

$$V(\lambda|r) = \frac{1}{r}\left\{\sum_{\lambda_i}[N_{\lambda_i} - \delta(\lambda, \lambda_i)]\, Y_0(\lambda_i, \lambda_i|r) + \sum_{k>0}\frac{2}{N_\lambda}f_k(j^N)Y_k(\lambda, \lambda|r)\right\},$$
(28.23)

$$\chi(\lambda|r) = \frac{1}{r}\sum_{\lambda_i \neq \lambda}\left\{\sum_k N_\lambda^{-1}g_k(j^N j_i^{N_i})Y_k(\lambda, \lambda_i|r) + \epsilon_{\lambda,\lambda_i}\delta(lj, l_ij_i)\right\}Q(\lambda'|r),$$ (28.24)

$$\chi(\lambda'|r) = \frac{1}{r}\sum_{\lambda_i \neq \lambda}\left\{\sum_k N_\lambda^{-1}g_k(j^N j_i^{N_i})Y_k(\lambda, \lambda_i|r) + \epsilon_{\lambda,\lambda_i}\delta(lj, l_ij_i)\right\}P(\lambda_i|r),$$
(28.25)

where

$$Y_k(\lambda_1, \lambda_2|r) = r^{-k}\int_0^r r_1^k\,[P(\lambda_1|r_1)P(\lambda_2|r_1) + Q(\lambda_1'|r_1)Q(\lambda_2'|r_1)]\, dr_1$$

$$+ r^{k+1}\int_r^\infty r_1^{-k-1}\,[P(\lambda_1|r_1)P(\lambda_2|r_1) + Q(\lambda_1'|r_1)Q(\lambda_2'|r_1)]\, dr_1. \quad (28.26)$$

Coefficients f_k and g_k are defined by (19.73) and (20.30), respectively, whereas N_{λ_i} denotes the number of electrons in subshell $n_il_ij_i$. Quantities ϵ, proportional to Lagrange multipliers, are in charge of orthonormality (2.17) of the radial functions.

The system of equations (28.22) must be solved accounting for the boundary conditions $P(nlj|0) = P(nlj|\infty) = 0$ for the large component of the relativistic wave function. While determining the wave function near the origin, the finiteness of the nuclear dimensions must be taken into consideration. Usually the nuclear charge is considered to be distributed uniformly inside the sphere. As a rule the calculations are performed for configurations averaged over the states corresponding to the given distribution of electrons in the subshells [179]. Then for an atom with a subshells we arrive at $2a$ equations. We also must have in mind that

$$\bar{f}_k(j_i^{N_i} j_m^{N_m}) = \delta(k,0) N_i N_m, \tag{28.27}$$

$$\bar{f}_k(j^N) = -\frac{N(N-1)}{4j} \begin{bmatrix} k & j & j \\ 0 & 1/2 & 1/2 \end{bmatrix}^2, \tag{28.28}$$

$$\bar{g}_k(j_i^{N_i} j_m^{N_m}) = -\frac{N_i N_m}{2j_m + 1} \begin{bmatrix} k & j_m & j_i \\ 0 & 1/2 & 1/2 \end{bmatrix}^2. \tag{28.29}$$

Equations (28.22) are valid for both ground and excited configurations of atoms and ions for the cases when all radial orbitals are varied, or some of them are 'frozen'.

28.3 Analytical radial orbitals

Usually the solutions of any version of the Hartree–Fock equations are presented in numerical form, producing the most accurate wave function of the approximation considered. Many details of their solution may be found in [45]. However, in many cases, especially for light atoms or ions, it is very common to have analytical radial orbitals, leading then to analytical expressions for matrix elements of physical operators. Unfortunately, as a rule they are slightly less accurate than numerical ones.

Analytical radial orbitals may be found from the variational principle starting with a given analytical form of the initial function. The accuracy of the final radial orbital obtained depends essentially on the type of initial function and on the number of parameters varied. With increase of the number of shells in an atom the number of varied parameters grows rapidly. This, in turn, complicates substantially the optimization procedure. For these reasons analytical radial orbitals are impractical for obtaining the energy spectra of middle and, particularly, heavy atoms.

The simplest analytical radial orbitals may be found by solving the radial Schrödinger equation for a one-electron hydrogen-like atom with arbitrary Z. They are usually called Coulomb functions and are expressed in terms of the confluent hypergeometric function $F(a; b; x)$, namely

$$P(nl|r) = \frac{(2Zr)^{l+1}}{n^{l+2}(2l+1)!}\sqrt{\frac{Z(n+l)!}{(n-l-1)!}}e^{-Zr/n}F(-n+l+1, 2l+2; 2Zr/n).$$

(28.30)

So-called Slater orbitals

$$P(nl|r) = A_{nl}r^{n^*}e^{-Z_{nl}r/n^*},$$

(28.31)

are widely used for many-electron atoms and in molecular calculations, when molecular orbitals are expressed in terms of the linear combinations of atomic ones. Slater orbitals are found when the effective potential of the central field is chosen in the form

$$V(r) = -Z_{nl}/r + n^*(n^* - 1)/2r^2.$$

(28.32)

In (28.31) and (28.32) Z_{nl} is the effective nuclear charge, dependent on n and l and defined by the equality $Z_{nl} = Z - S_{nl}$, where S_{nl} stands for screening constant, n^* denotes the effective principal quantum number, and A_{nl} is the normalizing factor.

There exist a number of attempts to generalize hydrogenic radial orbitals to cover the case of a many-electron atom. In [180] the so-called generalized analytical radial orbitals (Kupliauskis orbitals) were proposed:

$$P(nl|r) = A_{nl}\sum_{i=1}^{\max(2,n-l)}c_i^{nl}r^{\min(l+i,n)}e^{-\alpha_i^{nl}r}.$$

(28.33)

Here α_i^{nl} and c_i^{nl} for the cases $n = l + 1$ are found from the variational principle requiring the minimum of the non-relativistic energy, whereas c_i^{nl} ($n > l + 1$) – form the orthogonality conditions for wave functions. More complex, but more accurate, are the analytical approximations of numerical Hartree–Fock wave functions, presented as the sums of Slater type radial orbitals (28.31), namely

$$P(nl|r) = \sum_{i=1}^{M}N_ic_ir^{n_i}e^{-\alpha_i r},$$

(28.34)

where N_i stands for the normalizing factor of the separate Slater orbital and is equal to

$$N_i = \sqrt{(2\alpha_i)^{2n_i+1}/(2n_i)!};$$

(28.35)

M is an positive integer number. Optimal c_i and α_i values are found by solving the Hartree–Fock equations in an analytical form, i.e., varying the energy functional with respect to these parameters. Determining non-linear parameters α_i is rather difficult. Therefore quite often they are chosen to be the same for all shells with the given l values. An efficient method of the approximation of numerical Hartree–Fock radial functions by Slater and Gaussian type orbitals is described in [181].

The main advantage of analytical radial orbitals consists in the possibility to have analytical expressions for radial integrals and compact tables of numerical values of their parameters. There exist computer programs to find analytical radial orbitals in various approximations. Unfortunately, the difficulties of finding optimal values of their parameters grow very rapidly as the number of electrons increases. Therefore they are used only for light, or, to some extent, for middle atoms. Hence, numerical radial orbitals are much more universal and powerful.

The pecularities of solving the various versions of the Hartree–Fock equations are described in more detail in monographs [16, 45]. There are a number of widely used universal computer programs to solve the non-relativistic [16, 45] and relativistic [57, 182] versions of the Hartree–Fock equations, used separately or as a part of the more general complex, e.g. calculating energy spectra, etc. [183].

28.4 Collapse of the orbit of the excited electron

While determining matrix elements of any quantity, including physical ones, normally we have to calculate a number of radial integrals, adopting numerical or analytical radial wave functions. The accuracy requirements of the energy levels and wavelengths of electronic transitions are fairly high. Therefore, the appropriate radial integrals must be calculated with very high precision. The accuracy of both the theoretical and experimental values of oscillator strengths in many cases is of the order of 10% or more, therefore small differences in radial orbitals do not play a significant role. Much more important is accounting for relativistic and correlation effects.

Usually the dependence of the radial integrals of electronic transitions on spectral terms is neglected, and their value averaged with respect to these terms is employed. However, there are cases when the role of this dependence may increase substantially and may even become decisive. This is the case, when the radial integral itself becomes small and then it is rather sensitive even to insignificant improvements of wave function.

However, it is most important to account for the dependence of a radial wave function on terms for transitions between configurations, in one of which the phenomenon of collapse of the orbit of excited electron occurs. Its essence consists in rapid change of the mean distance of an electron from the nucleus as a result of the change of its state. This phenomenon is explained by the fact that the effective potential of the

Table 28.1. Influence of the localization of the excited 4*f*-electron in configuration $4d^9 4f LS$ on various atomic properties (integrals in a.u., ϵ_{4f} in eV)

Atom	LS	ϵ_{4f}	$G^1(4d,4f)$	$F^2(4d,4f)$	η_{4f}	$(4d\lvert r\rvert 4f)$
Xe I	^3P	-0.85	$1.04\cdot10^{-5}$	$3.37\cdot10^{-4}$	$1.27\cdot10^{-7}$	$-6.46\cdot10^{-3}$
	^3D	-0.85	$5.9\cdot10^{-6}$	$3.29\cdot10^{-4}$	$1.13\cdot10^{-7}$	$-5.32\cdot10^{-3}$
	^1P	-0.85	$1.2\cdot10^{-6}$	$3.27\cdot10^{-4}$	$1.03\cdot10^{-7}$	$-3.57\cdot10^{-3}$
	av.	-0.85	$5.5\cdot10^{-6}$	$3.28\cdot10^{-4}$	$1.12\cdot10^{-7}$	$-5.19\cdot10^{-3}$
Ba I	^3P	-15.22	$5.54\cdot10^{-1}$	$4.65\cdot10^{-1}$	$2.66\cdot10^{-3}$	$-8.12\cdot10^{-1}$
	^3D	-10.95	$5.20\cdot10^{-1}$	$4.39\cdot10^{-1}$	$2.43\cdot10^{-3}$	$-8.04\cdot10^{-1}$
	^1P	-0.86	$1.58\cdot10^{-5}$	$3.47\cdot10^{-4}$	$1.61\cdot10^{-7}$	$-7.38\cdot10^{-3}$
	av.	-10.39	$5.15\cdot10^{-1}$	$4.35\cdot10^{-1}$	$2.40\cdot10^{-3}$	$-8.02\cdot10^{-1}$

central symmetric field for a number of atoms has the form of two wells separated by a potential barrier. The wave function of the electron under consideration may be localized in the outward or inward well. With the increase of nuclear charge (sometimes only with the change of term in electronic configuration) the maximum of the electron wave function is displaced from the outward well towards the inward one, and as a result the mean distance of an electron from the nucleus changes drastically, i.e. the electron collapses. This phenomenon causes substantial non-monotonicities in the behaviour of numerous spectral characteristics of atoms and ions along isoelectronic sequences.

This effect is most pronounced for the excited configurations of type $nl^{4l+1}n(l+1)$, particularly for neutral atoms with a collapsing f-electron. Dependence of localization of the outer electron nl on the term diminishes with decreasing l. Table 28.1 lists the dependence on term LS and averaged, with respect to terms, numerical values of one-electron energies ϵ_{4f}, radial integrals $G^1(4d,4f)$ and $F^2(4d,4f)$ of the electrostatic interaction, fine structure constant η_{4f} as well as the integral of electric dipole transition $(4d\lvert r\rvert 4f)$ as an example of the influence of the collapse phenomenon on atomic properties, for neutral xenon and barium atoms.

The results of Table 28.1 show that for xenon, the 4*f*-electron is localized in the outward potential well far from the nucleus. Therefore, its radial orbital overlaps with those of inward electrons only a little. For this reason all integrals are very small and are of the same order for all LS values of the configuration considered. However, for barium we have a completely different picture, illustrating the collapse phenomenon. Only for the term ^1P does the electron remain not collapsed; for the rest of the terms the values of radial integrals have increased by 2–5 orders, evidencing the process of collapse. Changes of the radial integral of electric dipole transition by two orders for part of the terms leads to gigantic changes of the numerical values of oscillator strengths and transition probabilities and to the occurrence of various inconsistencies and inhomogeneities in their behaviour along isoelectronic sequences. Configurations with double well effective potential may be easily but fairly accurately described adopting the simplified version of the extended method (see Chapter 29), when the radial orbitals of a shell of equivalent electrons are described by two different radial functions [184]. Many aspects of the phenomenon of the collapse of the orbit of excited electrons are considered in detail in review paper [174].

When determining independently the wave functions of two configurations (e.g. ground and excited) of an atom or ion, the wave functions of each shell are usually slightly different. However, the methods of finding the matrix elements described in this book imply identity of the orbitals of the electronic core of both configurations. Otherwise the orthogonality of separate wave functions is violated, and then we should account for overlap integrals while calculating electronic transitions [185]. However, practical calculations show that as a rule the relevant corrections are small and, therefore, may be neglected.

Such wave functions are orthogonal if we use the so-called 'frozen core' approximation, in which the core wave functions are fixed and only those of excited electrons are recalculated. Such calculations are much simpler and the functions obtained in this way are very close to those obtained with full self-consistency.

Analogous to the case of energy spectra considered in Chapter 21, electronic transitions may also be studied on the basis of semi-empirical methods. However, the methods of determination of transition radial integrals from oscillator strengths or transition probabilities are inefficient due to low accuracy of the quantities mentioned. Therefore, when calculating transition properties, theoretical values of radial integrals are utilized. Quite often, especially for simple configurations, various model potentials, parameters of which can be found from various experimental data, are exploited, achieving in this way fairly high accuracy of wave functions and transition quantities.

For atoms with a few electrons (particularly for configurations consisting of one electron above the closed shells) the semi-empirical method of Bates and Damgaard is fairly efficient [186]. It is assumed that, while calculating electron transition integrals, the deviations of the potential of an atom or ion from its asymptotic form may be neglected (in other words, the parts of radial functions with large r give the main contribution to the integral) and then the exact equation for radial orbital is replaced by the approximate one

$$\frac{d^2R}{dr^2} + \left(\frac{2G}{r} - \frac{G^2}{n^{*2}} - \frac{l(l+1)}{r^2}\right) R = 0, \qquad (28.36)$$

where G is the effective charge, equal to 1 for the neutral atom, 2 for the first ion, etc.; $n^* = G/\sqrt{\epsilon_n}$, where ϵ_n stands for the energy measured. The solution of this equation is expressed in terms of the Witticker function

$$R = W_{n^*,l+1/2}\left(\frac{2G}{n^*}r\right). \qquad (28.37)$$

The radial integral of the transition with respect to these functions is presented in the form of two multipliers, tables for which, covering the transitions $s - p$, $p - d$ and $d - f$, may be found in [186].

However, again, as in the case of calculations of energy spectra, it is much more efficient to use universal computer programs to perform accurate theoretical or semi-empirical calculations of electron transition and other properties and quantities of atoms or ions, particularly if they may be combined with the relevant experimental studies [187–191].

28.5   Non-orthogonal radial orbitals

Let us briefly discuss the necessity to utilize the so-called non-orthogonal radial orbitals (NRO) in order to calculate atomic properties, mainly following the review paper [47]. The general theory of NRO was elaborated by A. Jucys and coworkers [192–194].

The use of NRO is necessary in a number of cases, particularly connected with the studies of processes involving configurations where the relaxation effects are important (excited configurations having states of the same symmetry in the lower-lying configurations, electronic transitions between the discrete states and to the continuum, when the relaxation effects are strong, etc.). In such cases the effective nuclear charge which is felt by the electron in the initial and final states changes considerably and, therefore, rearrangement of the radial orbitals of the wave functions of these states occurs. The radial orbitals of the initial and final state may differ significantly and then they do not fulfil mutually the orthogonality condition.

For the ground and excited configurations, which have no terms of the same symmetry in the lower-lying ones, the equivalency of orthogonal and NRO was demonstrated in general form [195]. However, the calculations show that for excited configurations having the terms of the same symmetry as in the lower-lying configurations, disregarding the orthogonality condition may lead to rather poor term energies [194]. Calculations of the fine structure of the terms ^3P and ^3D of the configuration $1s^22s^22p^3p$ for the carbon atom also illustrated the necessity to use NRO for such cases [194].

General expressions for the Hartree–Fock equations using NRO were obtained and studied in [196–198]. The use of NRO for the excited configurations $n_1l^Nn_2l$ essentially improved the energies of the terms existing also in the lower-lying configurations, for all configurations $1s2s\ ^1S$, $1s3s$ ^1S $(Z = 2 - 10)$ and $1s^22s^22p^N3p\ LS$ $(N = 1 - 6)$ [198].

The NRO approach is very efficient when accounting for correlation effects in the framework of the so-called extended method of calculation (see also Chapter 29) applied to electronic configurations having several shells of equivalent electrons with the same values of orbital quantum numbers. Its general theory is described in [199–204].

The utilization of NRO is also essential when calculating the electronic transitions between the bound states and to the continuum, if the radial orbitals of the initial and final states differ significantly [205, 206]. Accounting for the non-orthogonality terms is of particular importance when studying many-electron transitions.

Unfortunately, the mathematical expressions of physical quantities in the NRO method are rather complicated and, therefore, their practical use is tedious. However, with the advent of powerful computers this difficulty is easily overcome.

29

Correlation effects. Perturbation theory

29.1 Methods of accounting for correlation effects

Correlation effects have already been mentioned a number of times. Let us discuss them here in more detail. In the Hartree–Fock approach (to be more exact, in its single-configuration version, which we have been considering so far) it is assumed that each electron in an atom is orbiting independently in the central field of the nucleus and of the remaining electrons (one-particle approach). However, this is not exactly so. There exist effects, usually fairly small, of their correlated, consistent movement, causing the so-called correlation energy.

In the language of quantum numbers it may be explained in the following way. Only the parity of the configuration and total angular momentum are exact quantum numbers. From the parity conservation law it follows that the energy operator matrix elements connecting configurations of opposite parity vanish. Matrix elements of the one- and two-particle operators (the operators we are considering in this book) also vanish if they connect wave functions of configurations differing by the quantum numbers of more than one and two electrons, respectively. However, the non-diagonal with respect to configurations matrix elements of, for example, an electrostatic interaction, are in general non-zero. Hence, the configuration, considered as a quantum number, is not always an exact quantum number, and then a single-configuration approach is not sufficiently accurate. For this reason, while forming the energy matrix, non-diagonal, with respect to configurations, matrix elements must also be accounted for. Then the wave functions obtained after diagonalization of such a matrix will be as follows (cf. (11.10)):

$$\psi(K\alpha J) = \sum_{\bar{K}} a(\bar{K}J)\psi(\bar{K}\alpha J), \qquad (29.1)$$

in other words, the atom is described by a mixture of configurations of the

346

same parity. Usually the coefficient with $\bar{K} = K$ is close to unity, whereas the remaining coefficients are close to zero, i.e. the admixture of other configurations is small, therefore the single-configuration approximation is fairly accurate.

However, there are cases, especially for the configurations with d^N and f^N shells as well as for some excited configurations of multiply charged ions whose energy levels lie very close or even overlap, in which the conventional single-configuration approach is completely unfit (see also Chapter 31).

As a rule, only the calculations that account for correlation effects can correctly predict the existence of stable negative ions, representing very weakly bound electronic systems [207–216].

There are a number of methods that take into account correlation effects: perturbation theory, random phase approximation with exchange, method of incomplete separation of variables, so-called extended method of calculation, superposition-of-configurations, multi-configuration approach, etc. The first two methods have so far been applied only to comparatively simple systems (e.g. configurations having few electrons above the closed shells).

A comprehensive review of non-relativistic and relativistic versions of the random phase approximation may be found in [217]. Some applications are described in [218]. A survey of various aspects of the modern theory of many-electron atoms is presented in [219].

The essence of the method of incomplete separation of variables consists in introducing the interelectronic distances (usually only in an open shell) in an atom. Then its wave function

$$\psi(\{nl^N\})\alpha LS) = K(\mathbf{r}_1, \mathbf{r}_2, \ldots, \mathbf{r}_N)\psi(nl^N\alpha LS), \tag{29.2}$$

where symmetric multiplier

$$K(\mathbf{r}_1, \mathbf{r}_2, \ldots, \mathbf{r}_N) = \mu\left(1 + \mu_1 \sum_{j>i=1} r_{ij} + \mu_2 \sum_{i=1}^{N} r_i + \ldots\right). \tag{29.3}$$

Here μ ensures the normalization condition of the wave function. The use of the concept of irreducible tensorial operators in this approach [220] opens up the possibility of exploiting such a method for complex electronic configurations; however, so far it has been applied only to light atoms and ions.

Another possibility to account for correlation effects in a shell of equivalent electrons is to use the so-called extended method of calculation. In a conventional single-configuration approach, all electrons of shell l^N are described by one and the same radial orbital. In an extended method each electron in shell l^N has different radial functions, angular parts remaining the same. Therefore, all techniques of irreducible tensorial operators and coefficients of fractional parentage may be exploited here, too. The way of obtaining the expressions for matrix elements in an extended method from those of a conventional one, is described in [199]. This enables one to successfully apply such an approach to atoms and ions, having p-, d- and even f-electrons [200, 202, 204]. Unfortunately, in this way we account only for the radial part of correlation effects in a shell. However, in many cases, angular as well as intershell correlations are most important, therefore more universal methods are necessary. Among them the methods of superposition-of-configurations (configuration interaction) and multi-configuration approach are nowadays the most widely used. The first is defined by (29.1), recalling that the radial functions of principal and each admixed configuration are obtained independently, solving, for example, the Hartree–Fock equations for each of them. This approach can be quite easily applied to each atom or ion. Its main disadvantage lies in slow convergence of the procedure (the necessity to take into consideration a large number of configurations and, thus, to diagonalize energy matrices of very high orders).

This difficulty may be overcome by using the so-called simplified super-position-of-configurations method, based on averaging the energy differences between configurations and on the algebraic summation of the spin–angular parts of interconfigurational matrix elements [221]. Acceleration of convergence is achieved by utilization, for the description of admixed configurations, of transformed Hartree–Fock radial functions of the configuration considered. According to perturbation theory, the correction to the energy of term αLS of configuration K under investigation, obtained with the help of the admixed configuration K', may be written as follows:

$$\Delta E(K\alpha LS, K'\alpha'LS) = -\sum_{\alpha'} \frac{|(K\alpha LS|\hat{H}|K'\alpha'LS)|^2}{E(K'\alpha'LS) - E(K\alpha LS)}. \tag{29.4}$$

Here the summation α' is over all intermediate quantum numbers of configuration K'. If, as is often the case, the energy difference between the configurations is large compared to their own widths, then the energies in the denominator of (29.4) may be replaced by energies averaged with respect to all terms of the configurations, i.e.

$$E(K'\alpha'LS) - E(K\alpha LS) \approx \bar{E}(K') - \bar{E}(K) = \Delta\bar{E}(K', K). \tag{29.5}$$

Then we can analytically perform the summation in (29.4). The expressions obtained in this way may be interpreted as matrix elements of certain effective many-electron (two-, three-, four-electron) operators.

For further acceleration of the convergence of this method one may use the so-called transformed radial orbitals to describe the admixed configurations. For example, admixed configuration $P^T(n'l'|r)$ may be presented as

$$P^T(n'l'|r) = Nr^{\Delta l}P(nl|r), \quad \Delta l = l' - l, \tag{29.6}$$

$$P^T(n'l|r) = N'(A - r^2)P(nl|r), \tag{29.7}$$

where N and N' are normalizing factors; A ensures the orthogonality of the radial orbital. Calculations show that such an approach successfully combines simplicity, universality and high efficiency. The procedure converges very fast and, thus, allows one to admix a great number of configurations. All this is achieved without an increase of the order of the matrix to be diagonalized, simply adding the extra terms to matrix elements of the usual single-configuration approximation.

In a multi-configuration approach the wave function of an atom is also a linear combination (29.1) of the wave functions of separate configurations. The latter are built from one-electron orbitals φ_i^k, where k denotes the number of a function in the sequence of the above-mentioned orbitals of the ith configuration. Usually the number of one-electron functions equals the number of shells of equivalent electrons of each configuration. Let us assume that $i = 1$ corresponds to the principal (studied) configuration, whereas $i \neq 1$ are the admixed ones. Varying then the energy functional with respect to one-electron orbitals, we arrive at the following system of equations:

$$\frac{\partial E_{11}^F}{\partial \varphi_1^k} + 2\sum_{j=2}^m a_{1j}\frac{\partial E_{1j}^F}{\partial \varphi_1^k} = 0, \quad k = 1, 2, \ldots, b_1, \tag{29.8}$$

$$\frac{\partial E_{ii}^F}{\partial \varphi_i^k} + 2\sum_{i \neq j=1}^m \frac{a_{1j}}{a_{1i}}\frac{\partial E_{ij}^F}{\partial \varphi_i^k} = 0, \quad i = 2, \ldots, m; \quad k = 1, 2, \ldots, b_i. \tag{29.9}$$

Here $a_{ij} = a_j/a_i$, where a_i and a_j correspond to coefficients $a(\bar{K}\alpha J)$ in (29.1), b_i denotes the number of shells in the ith configuration, m is the number of configurations, whereas

$$E_{ij}^F = \int_0^\infty \psi_i H \psi_j d\tau + \sum_k \epsilon_{ij}^k \int_0^\infty \psi_i \psi_j d\tau. \tag{29.10}$$

Expressions (29.8) and (29.9) are the Hartree–Fock equations in a multi-configurational approximation [222], or the Hartree–Fock–Jucys equations. They must be solved together with the equations

$$\sum_j a_j(E_{ij}^F - \delta(i,j)E) = 0 \tag{29.11}$$

until full consistency is achieved. Calculations show that although the utilization of a multi-configuration approximation (solution of the Hartree–Fock–Jucys equations) is much more complicated than the superposition-of-configurations method, it leads to much faster convergence of the method itself and requires far fewer admixed configurations [223, 224].

Having in mind that usually the admixture of other configurations is small, equations (29.8) and (29.9) may be reduced to the form [225]

$$\frac{\partial E_{11}^F}{\partial \varphi_1^k} = 0, \ k = 1, 2, \ldots, b_1, \tag{29.12}$$

$$\frac{\partial E_{ii}^F}{\partial \varphi_i^k} + \frac{2}{a_{1i}} \frac{\partial E_{1i}^F}{\partial \varphi_i^k} = 0, \ i \neq 1; \quad k = 1, 2, \ldots, b_i. \tag{29.13}$$

In this approach, for the main configuration ($i = 1$) , the single-configuration Hartree–Fock equations, whereas for the admixed ones ($i \neq 1$), the two-configurational equations dealing only with the functions of admixed configuration, must be solved. From the practical point of view, the process of solution of these equations is much easier than that of equations (29.8) and (29.9), without losing too much accuracy in the results. Formulas (29.12) and (29.13) are called the simplified Hartree–Fock–Jucys equations. Pecularities of their solution are discussed in [226].

Adopting the concept of quasispin for two shells of equivalent electrons we arrive at some specific version of superposition of configurations (see (17.56)). However, here the weights of admixed configurations follow not from the diagonalization of the energy matrix, but are simply the usual Clebsch–Gordan coefficients. The same may be claimed for the isospin basis (18.36). Calculations show that there are cases when quasispin and isospin bases are closer to reality than the conventional ones, describing in this way part of the correlation effects (see also Chapters 17 and 18). The superposition-of-configurations method and the multi-configuration approach are used fairly widely to account for correlation effects in atoms and ions. Their relativistic versions are elaborated, too. For a large domain of atoms and ions, including fairly highly ionized ones, the combinations of a non-relativistic single-configuration method with the accounting for relativistic effects as corrections and for the correlation effects – by the method of superposition of configurations or in a multi-configuration approach (see also Chapter 31) – are of high accuracy and efficiency. The papers [18, 227–232] illustrate the typical accuracy of the energy levels, wavelengths, lifetimes of excited states as well as the quantities of allowed and forbidden electronic transitions, calculated within the framework of this approach. However, for such calculations one needs the non-diagonal (with respect to configurations) matrix elements of the operator of the electrostatic interaction considered below.

29.2 Matrix elements connecting different electronic configurations

Such general expressions for matrix elements of electrostatic interactions, covering the cases of three and four open shells, may be found in Chapter 25 of [14]. However, they are rather cumbersome and, therefore of little use for practical applications. Quite often sets of simpler formulas, adopted for particular cases of configurations, are employed. Below we shall present such expressions only for the simplest interconfigurational matrix elements occurring while improving the description of a shell of equivalent electrons (the appropriate formulas for the more complex cases may be found in [14]):

$$\left(n_1 l_1^{N_1} \alpha_1 L_1 S_1 | H | n_1 l_1^{N_1-1} \bar{\alpha}_1 \bar{L}_1 \bar{S}_1 n_2 l_2 L_1 S_1 \right) = (-1)^{L_1+S_1+1}(N_1-1)$$

$$\times \sqrt{\frac{N_1}{2}} [\bar{L}_1, \bar{S}_1] \sum_{\bar{\alpha}_1 \bar{L}_1 \bar{S}_1} (-1)^{\bar{L}_1+\bar{S}_1} \left(l_1^{N_1-1} \bar{\alpha}_1 \bar{L}_1 \bar{S}_1 \| l_1^{N_1-2} (\bar{\bar{\alpha}}_1 \bar{\bar{L}}_1 \bar{\bar{S}}_1) l_1 \right)$$

$$\times \sum_{L_0 S_0} \sqrt{[L_0, S_0]} \begin{Bmatrix} L_1 & l_1 & \bar{L}_1 \\ l_1 & L_1 & L_0 \end{Bmatrix} \begin{Bmatrix} \bar{S}_1 & s_1 & \bar{S}_1 \\ s_1 & S_1 & S_0 \end{Bmatrix}$$

$$\times \left(l_1^{N_1} \alpha_1 L_1 S_1 \| l_1^{N_1-2} (\bar{\bar{\alpha}}_1 \bar{\bar{L}}_1 \bar{\bar{S}}_1) l_1^2 (L_0 S_0) \right) \left(n_1 l_1^2 L_0 S_0 | h | n_1 l_1 n_2 l_2 L_0 S_0 \right)$$

$$+ \delta(l_1, l_2) \sqrt{N_1} \left(l_1^{N_1} \alpha_1 L_1 S_1 \| l_1^{N_1-1} (\bar{\alpha}_1 \bar{L}_1 \bar{S}_1) l_1 \right) A(n_1 l_1, n_2 l_2), \quad (29.14)$$

$$\left(n_1 l_1^{N_1} \alpha_1 L_1 S_1 | H | n_1 l_1^{N_1-2} \bar{\bar{\alpha}}_1 \bar{\bar{L}}_1 \bar{\bar{S}}_1 n_2 l_2^2 L_2 S_2 L_1 S_1 \right)$$

$$= \sqrt{N_1(N_1-1)/2} \left(l_1^{N_1} \alpha_1 L_1 S_1 \| l_1^{N_1-2} (\bar{\bar{\alpha}}_1 \bar{\bar{L}}_1 \bar{\bar{S}}_1) l_1^2 (L_2 S_2) \right)$$

$$\times \left(n_1 l_1^2 L_2 S_2 | h | n_2 l_2^2 L_2 S_2 \right), \quad (29.15)$$

$$\left(n_1 l_1^{N_1} \alpha_1 L_1 S_1 | H | n_1 l_1^{N_1-2} \bar{\bar{\alpha}}_1 \bar{\bar{L}}_1 \bar{\bar{S}}_1, n_2 l_2 n_3 l_3 L_{23} S_{23}, L_1 S_1 \right)$$

$$= \sqrt{N_1(N_1-1)/2} \left(l_1^{N_1} \alpha_1 L_1 S_1 \| l_1^{N_1-2} (\bar{\bar{\alpha}}_1 \bar{\bar{L}}_1 \bar{\bar{S}}_1) l_1^2 (L_{23} S_{23}) \right)$$

$$\times \left(n_1 l_1^2 L_{23} S_{23} | h | n_2 l_2 n_3 l_3 L_{23} S_{23} \right). \quad (29.16)$$

A two-electron matrix element of the operator of the electrostatic interaction energy is equal to

$$(n_1 l_1 n_2 l_2 L S | h | n_3 l_3 n_4 l_4 L S) = \sum_k \left[(-1)^{l_3+l_4+L} \left(l_1 \| C^{(k)} \| l_3 \right) \right.$$

$$\times \left(l_2 \| C^{(k)} \| l_4 \right) \begin{Bmatrix} l_1 & l_2 & L \\ l_4 & l_3 & k \end{Bmatrix} R^k(n_1 l_1 n_3 l_3, n_2 l_2 n_4 l_4)$$

$$+ (-1)^S \left(l_1 \| C^{(k)} \| l_4 \right) \left(l_2 \| C^{(k)} \| l_3 \right)$$

$$\left. \times \begin{Bmatrix} l_1 & l_2 & L \\ l_3 & l_4 & k \end{Bmatrix} R^k(n_1 l_1 n_4 l_4, n_2 l_2 n_3 l_3) \right]. \quad (29.17)$$

Here the most general case of radial integral in such a matrix element looks like

$$R^k(nln'l', n''l''n'''l''')$$

$$= \int\limits_0^\infty\int \frac{r_<^k}{r_>^{k+1}} P(nl|r_1) P(n'l'|r_1) P(n''l''|r_2) P(n'''l'''|r_2) dr_1 dr_2. \quad (29.18)$$

If two electrons are equivalent, then (29.17) becomes

$$\left(n_1 l_1^2 LS|h|n_3 l_3 n_4 l_4 LS\right) = (-1)^L \sqrt{2} \sum_k \left(l_1\|C^{(k)}\|l_3\right)\left(l_1\|C^{(k)}\|l_4\right)$$

$$\times \begin{Bmatrix} l_1 & l_1 & L \\ l_3 & l_4 & k \end{Bmatrix} R^k(n_1 l_1 n_3 l_3, n_1 l_1 n_4 l_4). \quad (29.19)$$

If a matrix element connects two pairs of equivalent electrons, then instead of (29.17) we have

$$\left(n_1 l_1^2 LS|h|n_2 l_2^2 LS\right) = (-1)^L \sum_k \left(l_1\|C^{(k)}\|l_2\right)^2 \begin{Bmatrix} l_1 & l_1 & L \\ l_2 & l_2 & k \end{Bmatrix} G^k(n_1 l_1, n_2 l_2),$$

$$(29.20)$$

where integral (29.18) has turned into the usual Slater integral of the exchange type (20.10).

The term $A(n_1 l_1, n_2 l_2)$ in (29.14) is equal to

$$A(n_1 l, n_2 l) = I(n_1 l, n_2 l) + \sum_{n_0 l_0}\left[4(2l_0 + 1)R^0(n_0 l_0 n_0 l_0, n_1 l n_2 l)\right.$$

$$\left. - \frac{2}{2l+1} \sum_k \left(l_0\|C^{(k)}\|l\right)^2 R^k(n_0 l_0 n_2 l, n_1 l n_0 l_0)\right]. \quad (29.21)$$

This term occurs for matrix elements differing only by a principal quantum number of one electron. It describes terms corresponding to kinetic energy and electrostatic interaction of an electron with a nuclear field (integral of the type (19.23)) as well as to the interaction with closed shells (summation over $n_0 l_0$).

In a relativistic case let us consider for the same purposes two subshells of equivalent electrons $n_1 l_1 j_1^{N_1} n_2 l_2 j_2^{N_2}$. General expressions do not depend on the explicit form of a two-electron interaction, therefore we shall not specify the operator type. Here, non-diagonal matrix elements of two sorts occur: $N_1 N_2 - N_1 - 1\, N_2 + 1$ and $N_1 N_2 - N_1 - 2\, N_2 + 2$. In the first case we have

$$\left\langle n_1 n_2 l_1 j_1^{N_1} l_2 j_2^{N_2} \alpha_1 J_1 \alpha_2 J_2 J\|H_r\|n_1 n_2 l_1 j_1^{N_1-1} l_2 j_2^{N_2+1} \alpha_1' J_1' \alpha_2' J_2' J\right\rangle$$

$$= (-1)^{N_2+1} N_2 \sqrt{N_1(N_2+1)} \left(j_1^{N_1} \alpha_1 J_1\|j_1^{N_1-1}(\alpha_1' J_1')j_1\right)$$

$$\times \sum_{\alpha_2'' J_2'' J_{20}} \left(j_2^{N_2}\alpha_2 J_2 \| j_2^{N_2-1}(\alpha_2'' J_2'')j_2\right) \left(j_2^{N_2-1}(\alpha_2'' J_2'')j_2^2(J_{20}) \| j_2^{N_2+1}\alpha_2' J_2'\right) M_1$$

$$+ (-1)^{N_2}(N_1-1)\sqrt{N_1(N_2+1)} \left(j_2^{N_2}(\alpha_2 J_2)j_2 \| j_2^{N_2+1}\alpha_2' J_2'\right)$$

$$\times \sum_{\alpha_1'' J_1'' J_{10}} \left(j_1^{N_1}\alpha_1 J_1 \| j_1^{N_1-2}(\alpha_1'' J_1'')j_2^2(J_{10})\right)$$

$$\times \left(j_1^{N_1-2}(\alpha_1'' J_1'')j_1 \| j_1^{N_1-1}\alpha_1' J_1'\right) M_2. \tag{29.22}$$

Here

$$M_1 = (-1)^{J_1''-J_2''-J_{20}+J} \sqrt{[J_1,J_2,J_2',J]} \begin{Bmatrix} J_1' & j_1 & J_1 \\ J_2 & J & J_2' \end{Bmatrix}$$

$$\times \begin{Bmatrix} J_2'' & j_2 & J_2 \\ j_1 & J_2' & J_{20} \end{Bmatrix} \left\langle n_1 n_2 l_1 j_1 l_2 j_2 J_{20} \| h_r \| n_2 l_2 j_2^2 J_{20} \right\rangle, \tag{29.23}$$

$$M_2 = (-1)^{j_1+j_2+J_1'+J_2'+J_1+J_1''+J} \sqrt{[J_1,J_1',J_2',J]} \begin{Bmatrix} J_1'' & j_1 & J_1' \\ j_2 & J_1 & J_{10} \end{Bmatrix}$$

$$\times \begin{Bmatrix} J_2 & j_2 & J_2' \\ J_1' & J & J_1 \end{Bmatrix} \left\langle n_1 l_1 j_1^2 J_{12} \| h_r \| n_1 n_2 l_1 j_1 l_2 j_2 J_{10} \right\rangle. \tag{29.24}$$

In the second one

$$\left\langle n_1 n_2 l_1 j_1^{N_1} l_2 j_2^{N_2}\alpha_1 J_1\alpha_2 J_2 J \| H_r \| n_1 n_2 l_1 j_1^{N_1-2} l_2 j_2^{N_2+2}\alpha_1' J_1'\alpha_2' J_2' J \right\rangle$$

$$= \frac{1}{2}\sqrt{N_1(N_1-1)(N_2+1)(N_2+2)} \sum_{J''} \left(j_1^{N_1}\alpha_1 J_1 \| j_1^{N_1-2}(\alpha_1' J_1')j_1^2(J'')\right)$$

$$\times \left(j_2^{N_2}(\alpha_2 J_2)j_2^2(J'') \| j_2^{N_2+2}\alpha_2' J_2'\right) M_3, \tag{29.25}$$

where

$$M_3 = (-1)^{J_1'+J_2'+J} \sqrt{[J_1,J_2',J]/[J'']} \begin{Bmatrix} J_1' & J'' & J_1 \\ J_2 & J & J_2' \end{Bmatrix}$$

$$\times \left\langle n_1 l_1 j_1^2 J'' \| h_r \| n_2 l_2 j_2^2 J'' \right\rangle. \tag{29.26}$$

Cases $N_1 N_2 - N_1 + 1\ N_2 - 1$ and $N_1 N_2 - N_1 + 2\ N_2 - 2$ follow from (29.22) and (29.25), respectively, after permutation of subshells, taking into account the phase multiplier and the subsequent transposition of indices 1 and 2. Operator H_r may be replaced by concrete operators, for example, of electrostatic or magnetic interactions. Let us notice also that non-diagonal (with respect to configurations) matrix elements of retarding interactions vanish. From the formulas presented we see that in a relativistic approach we face the necessity for the usual shell of equivalent electrons to take into account the non-diagonal (with respect to configurations) matrix elements.

29.3   Perturbation theory

In conclusion of this chapter let us discuss briefly the use of the stationary many-body perturbation theory of the effective Hamiltonian, described in Chapter 3, to calculate the energy levels of an oxygen isoelectronic sequence as an example. Using such an approach it is possible simultaneously to calculate the correlation corrections to all energy levels of complex $n = 2$ (configurations $1s^2 2s^2 2p^4$, $1s^2 2s 2p^5$ and $1s^2 2p^6$, they form a model space) in the same approximation [233]. The relativistic corrections were included in the Breit–Pauli approximation (see Chapter 1). The Hamiltonian is presented as a sum of zero-order term H_0 and perturbation H_p, i.e.

$$H = H_0 + H_p. \tag{29.27}$$

Here

$$H_0 = T + P + U, \tag{29.28}$$

where $T + P$ describes the kinetic and potential energy of the electrons with respect to the nucleus (see also (1.15)). U is the spherically symmetric potential defined by Morrison [234]. The perturbation is

$$H_p = H_C + H_{Br}, \tag{29.29}$$

where

$$H_C = V - U \tag{29.30}$$

is the residual Coulomb interaction (V describes Coulomb interaction between electrons) and $H_{Br} = W$ (see 1.17). The $E_i (i = 1, 2, \ldots d)$ eigenvalues of operator H are the solutions of the secular equation

$$H_{eff} \phi_i = E_i \phi_i, \tag{29.31}$$

where the effective operator H_{eff} acts only on the model space, and ϕ_i is a model function. Wave-operator Ω, acting on ϕ_i, reproduces eigenfunction ψ_i of operator H, i.e.

$$\psi_i = \Omega \phi_i. \tag{29.32}$$

Here a complete model space (P-space), defined on eigenfunctions of H_0, representing all possible distributions of electrons between open shells, is utilized. In our case the model space consists of the $1s^2 2s^2 2p^4$ and $1s^2 2p^6$ configurations for the even parity states and of the $1s^2 2s 2p^5$ configuration for odd parity states. The wave-operator may be written as

$$\Omega = \Omega_C + \Omega_{Br}, \tag{29.33}$$

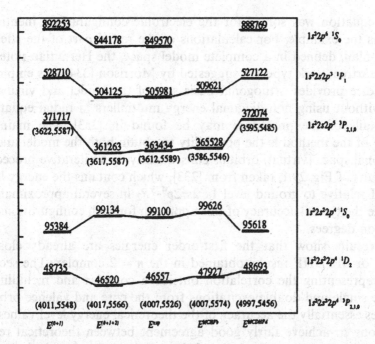

Fig. 29.1. Energy levels of Si VII relative to ground level $1s^2 2s^2 2p^4 \, ^3P_2$ in several approximations. $E^{(0+1)}$, $E^{(0+1+2)}$ — calculations in the first and second order [233]. E^{exp} — experimental data [235]. E^{MCHFb} — MCHF calculation results [229, 236]. E^{MCDHFd} — MCDHF calculation results [237]. In parentheses — energy values of the levels with $J = 1$ and $J = 0$, respectively.

where the indices C and Br denote parts of Ω which depend on H_C and H_{Br}, respectively. Then the first-order part of H_{eff} will be given by

$$H_{eff}^{(1)} = PH_pP = PH_CP + PH_{Br}P = H_C^{(1)} + H_{Br}^{(1)}. \qquad (29.34)$$

Here P is a projection operator ($P\psi_i = \phi_i$) onto the model space. We assume here that the residual interaction $H_C > H_{Br}$ and the second-order part of H_{eff} can be given by

$$H_{eff}^{(2)} = PH_C\Omega_C^{(1)}, \qquad (29.35)$$

i.e. we use condition $\Omega \approx \Omega_C^{(1)}$. It should be noticed that H_{eff} is a non-Hermitian operator. Fortunately, up to the third order

$$_hH_{eff} \approx \frac{1}{2}\left(H_{eff}^\dagger + H_{eff}\right) \qquad (29.36)$$

is a Hermitian operator. Then secular equation (29.31) becomes

$$\left(_hH_{eff}^{(0)} + {}_hH_{eff}^{(1)} + {}_hH_{eff}^{(2)}\right)\phi_i = {}_hH_{eff}^{(0+1+2)}\phi_i = E_i\phi_i. \qquad (29.37)$$

This equation was solved for the electronic configurations mentioned above as the example. For calculations of the radial part of the effective Hamiltonian, defined in a complete model space, the Hermitian potential of the Hartree–Fock type V^m suggested by Morrison [234] was employed. This choice provides orthogonal basis sets of the model and virtual orbitals without using non-diagonal energy multipliers in radial equations. The details of the procedure may be found in [233]. The main advantage of the method is the possibility to obtain both the model and the orthogonal space (virtual) orbitals consequently by an iterative procedure. The results of Fig. 29.1, taken from [233], which contains the energy levels of SiVII relative to ground level $1s^2 2s^2 2p^4 \, ^3P_2$ in several approximations, illustrate the typical accuracy of this approach for such configurations and ionization degrees.

The results show that the first-order energies are already close to MCHF or MCDHF results obtained in the $n = 2$ complex. The second-order, representing the correlation outside a complex and including all possible single and double excitations from the core and valence orbitals, improves essentially the accuracy of the theoretical energy level values and allows one to achieve fairly good agreement between theoretical results and experimental ones. Analysis of the root-mean-square deviations of calculations from experimental data also witnesses that such a version of perturbation theory leads to very accurate values of the energy levels considered. Therefore, it is very promising to apply such an approach to more complex atoms and ions.

There have been attempts to generalize the many-body perturbation theory to cover the relativistic regime in a rigorous and systematic manner [238–241]. Unfortunately, practical applications, so far, are only to simple atoms or ions.

30

The role of gauge dependence, relativistic and correlation effects in electronic transitions

30.1 Electronic transitions in intermediate coupling

As we have seen in Chapter 11, the energy levels of atoms and ions, depending on the relative role of various intra-atomic interactions, are classified with the quantum numbers of different coupling schemes (11.2)–(11.5) or their combinations. Therefore, when calculating electron transition quantities, the accuracy of the coupling scheme must be accounted for. The latter in some cases may be different for initial and final configurations. Then the selection rules for electronic transitions are also different. That is why in Part 6 we presented expressions for matrix elements of electric multipole (Ek) transitions for various coupling schemes.

In various pure coupling schemes the intensities of spectral lines may differ significantly. Some lines, permitted in one coupling scheme, are forbidden in others. Comparison of such theoretical results with the relevant experimental data may serve as an additional criterion of the validity of the coupling scheme used.

Studies of spectra of many-electron atoms and ions show that the presence of a pure coupling scheme is the exception rather than the rule. Therefore, their energy spectra must be calculated, as a rule, in intermediate coupling via diagonalization of the total energy matrix, starting with the coupling scheme assumed to be closest to reality. In such a case the electronic transitions must also be calculated in intermediate coupling.

The wave function in intermediate coupling is a linear combination of the relevant quantities of pure coupling (see (11.10)). Line strength in intermediate coupling is defined by (24.4), whose submatrix element equals

$$\left(\beta J \| O^{(k)} \| \beta' J'\right) = \sum_{i,j} a_i a_j \left(\psi_i(J) \| O^{(k)} \| \psi_j(J')\right). \tag{30.1}$$

Here a_i and a_j are the weights of the wave functions $\psi_i(J)$ and $\psi_j(J')$

of the pure coupling scheme, respectively, the summation over i and j in general runs over all possible states with given J and J' (e.g. in LS coupling they will be αLS and $\alpha'L'S'$). Explicit expressions for submatrix elements in the right side of (30.1) are presented in Chapters 25–27 for the majority of cases practically needed. Oscillator strengths and probabilities of the non-relativistic $E1$- and $E2$-transitions may be found from the following formulas, which are special cases of the expressions presented in Chapter 4:

$$f^{E1} = \frac{2\Delta E}{3(2J+1)} \left|\left(\beta J \|Q^{(1)}\| \beta' J'\right)\right|^2$$

$$= \frac{2}{3(2J+1)\Delta E} \left|\left(\beta J \|Q'^{(1)}\| \beta' J'\right)\right|^2, \tag{30.2}$$

$$W^{E1} = \frac{214.2 \cdot 10^8}{2J+1}(\Delta E)^3 \left|\left(\beta J \|Q^{(1)}\| \beta' J'\right)\right|^2$$

$$= \frac{214.2 \cdot 10^8}{2J+1}\Delta E \left|\left(\beta J \|Q'^{(1)}\| \beta' J'\right)\right|^2, \tag{30.3}$$

$$f^{E2} = \frac{177.5 \cdot 10^{-8}}{2J+1}(\Delta E)^3 \left|\left(\beta J \|Q^{(2)}\| \beta' J'\right)\right|^2$$

$$= \frac{177.5 \cdot 10^{-8}}{2J+1}\Delta E \left|\left(\beta J \|Q'^{(2)}\| \beta' J'\right)\right|^2, \tag{30.4}$$

$$W^{E2} = \frac{57030}{2J+1}(\Delta E)^5 \left|\left(\beta J \|Q^{(2)}\| \beta' J'\right)\right|^2$$

$$= \frac{57030}{2J+1}(\Delta E)^3 \left|\left(\beta J \|Q'^{(2)}\| \beta' J'\right)\right|^2. \tag{30.5}$$

Here $E(\beta J) - E(\beta'J') = \Delta E > 0$, the operators in general are defined by (4.12) and (4.13), ΔE and submatrix elements are presented in atomic units. Then W^{Ek} is in s^{-1}. Let us recall that the oscillator strengths are dimensionless quantities. The relevant expressions for magnetic dipole transitions are given by formulas (27.8) and (27.9), respectively.

As was shown in Chapters 24 and 27, the use of intermediate coupling essentially changes the selection rules with respect to approximate quantum numbers. Let us discuss a few examples of transitions between configurations, whose energy levels for neutral or a-few-times-ionized atoms are calculated in intermediate coupling, but classified with the help of LS coupling. Thus, for $E1$-transition $d^7p - d^8$, instead of selection rules described by (24.23) for $k = 1$, we have

$$\Delta L = 0, \pm 1, \pm 2, \pm 3; \qquad \Delta S = 0, \pm 1, \pm 2. \tag{30.6}$$

These selection rules allow 466 lines instead of the 184 permitted by (24.23). Thus, 282 lines must be attributed to forbidden radiation. Calculations show that line and oscillator strengths of many forbidden lines are of the same order of magnitude as those of allowed ones. The strengths of forbidden lines are of the same order as the changes of strengths of allowed lines after diagonalization of the energy matrix. The latter is caused by the degree of deviations from pure coupling and by the presence of quantum numbers, with respect to which the matrix elements of electrostatic interactions are not diagonal. Therefore, for configurations containing d^N electrons ($N = 3 - 7$) and, particularly, the f^N shell ($N = 3 - 11$), where the seniority quantum number v as well as other additional quantum numbers of this sort are employed, the classification of radiation as allowed and forbidden is meaningless. The use of intermediate coupling in such cases is a necessity.

These conclusions are supported by considering electronic transitions between the other complex configurations. Thus, in the case of the $d^3p - d^4$ transitions in intermediate coupling there are 1718 lines, of which only 637 belong to allowed transitions. In this case the distinction between allowed and forbidden transitions is even less pronounced. The cause of this is the presence, in configurations d^3p and d^4, of a large number of energy levels, whose characteristics differ from each other by only seniority quantum number. This leads to large non-diagonal terms in the energy matrix. As a result, large non-diagonal matrix elements also occur in the diagonalizing matrix, causing in this way the intense forbidden lines.

This phenomenon is particularly pronounced for $E1$-transitions of the sort $l^N l_1 - l^N l_2$. Indeed, for $d^3p - d^3s$ 2082 lines occur, among which only 215 transitions belong to transitions allowed in LS coupling. Hence, the number of transitions has increased by an order of magnitude. This fact is explained by breaking the selection rule $\delta(\alpha_0 L_0 S_0, \alpha'_0 L'_0 S'_0)$ with respect to the quantum numbers of the electronic core l^N. Then transitions between levels belonging to different parent terms become permitted. The energy separations for them as a rule are much larger than for levels with the same parent term, therefore the oscillator strengths of these forbidden transitions are quite large. The utilization of intermediate coupling in such cases plays an extremely important role.

Comparison of theoretical results obtained in the single-configuration approach and intermediate coupling, with experimental data indicates that oscillator strengths of allowed and forbidden transitions generally correspond to each other by the order of magnitude (for intense transitions between particular configurations also and quantitatively). The use of semi-empirical methods or improvement of the theory by accounting for correlation and relativistic effects leads to better agreement between calculated and measured data, but the main regularities remain unchanged. A similar picture occurs for electronic transitions of higher multipolarities (see Chapter 27).

30.2   Relativistic and correlation effects, gauge dependence

The influence of relativistic effects on the quantities of electronic transitions is manifested by relativistic changes of the energy level values and, hence, of wavelengths of electronic transitions, as well as by wave functions of intermediate coupling and by transition operators themselves. Line strengths calculated with non-relativistic electron transition operators, are influenced by relativistic effects only through intermediate coupling, because accounting for relativistic corrections to the energy operator changes the relative contributions of separate matrix elements in the energy matrix. This leads to changes in the weights of wave functions of the initial pure coupling scheme and, thus, to the changes of numerical values of line strengths. Additionally oscillator strengths and transition probabilities are altered due to the presence of certain degrees of the energy multiplier ΔE. These alterations may be particularly large in the presence of level crossings [242].

As was shown in Chapter 4 (formula (4.22)), relativistic corrections of the order α^2 to the intercombination $E1$-transitions in length form for accurate wave functions compensate each other. It follows from formulas (4.18)–(4.20) that for the velocity form of the $E1$-transition operator the relativistic corrections are of the order α^2 and may be presented in length, velocity and acceleration forms. Calculations of the $E1$-radiation for the Be isoelectronic sequence ($Z = 4 \div 92$) indicate that these relativistic corrections are more essential for faint intercombination lines of light neutral atoms. With an increase of ionization degree their contribution drops rapidly, because the relativistic corrections to the energy operator grow much faster than those for the transition operator. However, for complex many-electron atoms and ions these regularities may be quite different. Analogous calculations for neon and fluorine isoelectronic sequences testify to the complex character of the dependence of relativistic corrections for oscillator strengths on ionization degree and their remarkable role in fairly intense lines. However, again the decisive factor is the accounting for relativistic corrections to the energy. It is interesting, in this connection, to compare the wavelengths and oscillator strengths of $E1$-transitions calculated in the Hartree–Fock–Pauli (HFP) approximation (i.e. accounting for relativistic corrections of the order α^2 to the energy operator with respect to non-relativistic wave functions) with the relevant calculations in the Dirac–Hartree–Fock (DHF) approximation (relativistic energy operator, wave functions and transition operators). Table 30.1 lists such results for $E1$-transition $2p^4(^1D)3d\,^2S_{1/2} - 2p^5\,^2P_{1/2}$ in the fluorine isoelectronic sequence ($Z = 26 \div 100$), obtained in the DHF (λ_{DHF}, gf_K, $K = 0, -\sqrt{2}$) and in the HFP (λ_{HFP}, gf_r and gf_v) approximations.

Table 30.1. Oscillator strengths of the $E1$-transition $1s^2 2s^2 2p^4(^1D)3d$ $^2S_{1/2}$ − $1s^2 2s^2 2p^5$ $^2P_{1/2}$

Z	λ_{DHF}, Å	$gf_{K=0}$	$gf_{K=-\sqrt{2}}$	λ_{HFP}, Å	gf_r	gf_v
26	14.43	0.175	0.178	14.47	0.173	0.169
34	7.616	0.103	0.105	7.614	0.101	0.100
42	4.691	0.046	0.047	4.691	0.046	0.045
54	2.655	0.057	0.061	2.696	0.006	0.006
74	1.347	0.057	0.064	1.351	0.019	0.019
83	1.054	0.056	0.065	1.056	0.037	0.036
92	0.846	0.054	0.066	0.847	0.062	0.060
100	0.709	0.053	0.066	0.709	0.066	0.064

It follows from the table that with an increase of ionization degree the wavelength of the transition considered diminishes very rapidly and becomes of the order of 1 Å. For large ionization degrees, the classification of this transition in LS coupling is, of course, incorrect. Close values of wavelengths and oscillator strengths, obtained in both approaches, as well as the weak dependence of oscillator strengths on gauge condition in relativistic approximation and on the form of transition operator – in non-relativistic case – supports once more the conclusion that for both the energy spectra and electronic transitions in a large variety of atoms and ions, including very highly ionized atoms, the relativistic corrections to the energy are the most important. They are fairly accurately accounted for in the HFP approximation.

Let us consider in more detail the dependence of oscillator strengths of $E1$-transitions on gauge condition along the isoelectronic sequence and while accounting for correlation effects to a different extent. Figure 30.1 shows the dependence of oscillator strengths (gf) of the $E1$-transition $1s^2 2s2p$ $^1P_1 - 1s^2 2s^2$ $^1S_0$ on gauge K in the beryllium isoelectronic sequence in the DHF and HFP approximations. It follows that with an increase of ionization degree, the dependence of gf on K becomes less and less pronounced; the relevant curve (in the general case a parabola) approaches a straight line parallel to the K axis that indicates the growth of the accuracy of the approach used. Similar calculations for other cases show that the characteristic of the curves obtained are different not only for transitions between various types of configurations, but sometimes even for transitions between levels of the same configuration.

To demonstrate how the f-values change when a certain amount of correlation is taken into account, let us consider the $2s2p$ $^1P - 2p^2$ ^1S transition in Be-like OV. The dependence of oscillator strengths, obtained

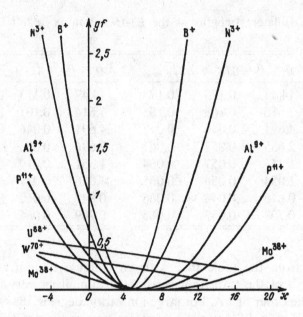

Fig. 30.1.   Dependence of gf of the $E1$-transition $1s^2 2s2p\ ^1P_1 - 1s^2 2s^2\ ^1S_0$ on gauge K in the Be isoelectronic sequence.

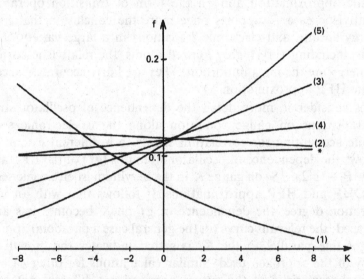

Fig. 30.2.   The dependence of the oscillator strength of the $2s2p\ ^1P - 2p^2\ ^1S$ transition in OV on the gauge condition K for various theoretical approches, HF, SC [243] and NCMET [244]. The curves refer to HF (1), SC without and with polarization and inner-shell effects, (2–4), NCMET (5). The experimental value $f = 0.103 \pm 0.007$ [246] is indicated with a star.

with Hartree–Fock (HF) superposition-of-configurations (SC) [243] and NCMET (non-closed shell many-electron theory) [244] methods, on the gauge condition K is shown in Fig. 30.2 [245]. The experimental value [246] is also indicated in the figure. The strongest dependence of the f-value on K is for the HF calculation, where $K_{f=0} = 13$. For the SC wave functions used by Hibbert [243], the gauge condition, for which f is zero, is very large, $K_{f=0} = 168$, and the parabola may be approximated by a straight line in the region $-10 < K < 10$. However, when core polarization is included in the SC calculation [243], the dependence of f on K is stronger and $K_{f=0} = -37$. When inner-shell correlations are also taken into account [243] the f-value is again almost independent of the gauge and $K_{f=0} = -165$. This is a well-known behaviour of f-values which oscillate as more and more correlation effects are included in the SC calculations. It is interesting that the f-value obtained by the sophisticated NCMET method [244] is rather strongly dependent on the gauge and $K_{f=0} = -19$ in this case. It can thus be seen that each of the theoretical methods gives different values of K which best reproduce the experimental result.

Thus, by employing general formulas for oscillator strengths, results can be obtained which are useful both in practical calculations and in studying the role of correlation effects on transition probabilities.

Let us also notice that slow variations of K with Z imply that the gauge condition K may be treated as a semi-empirical parameter in practical calculations to reproduce, with a chosen K, the accurate oscillator strength values for the whole isoelectronic sequence. Thus, dependence of transition quantities on K may serve as the criterion of the accuracy of wave functions used instead of the comparison of two forms of $E1$-transition operators. In particular, the relative quantities of the coefficients of the equation $f^{E1} = aK^2 + bK + c$ (the smaller the a value, the more exact the result), the position of the minimum of the parabola $K_f = 0$ (the larger the K value for which $f = 0$, the more exact is the approximation used, in the ideal case $f = 0$ for $K = \pm\infty$) may also help to estimate the accuracy of the method utilized.

30.3 Perturbation theory

In the previous chapter we have briefly discussed the use of the stationary many-body perturbation theory of the effective Hamiltonian, sketched in Chapter 3, to account for correlation effects. Here we shall continue such studies for electronic transitions on the example of the oxygen isoelectronic sequence. Having in mind (29.32) and the approximation $\Omega^{(1)} \approx \Omega_C^{(1)}$ the matrix element of the $E1$-transition operator O_{E1} between the initial state

Table 30.2.   Wavelengths λ (Å), oscillator strengths f and ratios of line intensities $I(^1P - {}^1D)/I(^1P - {}^1S)$ in various approximations: (a) for $1s^2 2s 2p^5\ {}^1P - 1s^2 2s^2 2p^4\ {}^1D$ transition; (b) for $1s^2 2s 2p^5\ {}^1P - 1s^2 2s^2 2p^4\ {}^1S$ transition

	Experiment Exp1	Exp2	NCMET	MBPT (0+1)	(0+1+2)'	(0+1+2)''	CI	MCDHF	MCHFb
	Ne III								
λ_a		379		349	381	381	372	351	374
λ_b		428		382	430	430	417	384	419
f_a	0.193	0.194	0.229	0.368	0.337	0.252	0.271	0.377	
f_b	0.052	0.045	0.078	0.146	0.129	0.076	0.110	0.150	
I_a/I_b	20.8	23.9	18.6	15.1	16.7	21.1		15.1	21.9
	Si VII								
λ_a		218		208	219	219	216	209	216
λ_b		246		231	247	247		232	244
f_a		0.149		0.238	0.227	0.187	0.195	0.241	
f_b		0.085		0.102	0.095	0.080	0.087	0.104	
I_a/I_b		12.2	15.4	14.4	15.3	14.9		14.3	15.2

i and the final one f is [247]

$$\langle\psi_f|O_{E1}|\psi_i\rangle = \langle\phi_f|O_{E1}|\phi_i\rangle + \left\langle\phi_f|\Omega_C^{(1)\dagger} O_{E1}|\phi_i\right\rangle + \left\langle\phi_f|O_{E1}\Omega_C^{(1)}|\phi_i\right\rangle. \quad (30.7)$$

The second and third terms on the right side of (30.7) represent the electron–electron corrections to the zero-order matrix element $\langle\phi_f|O_{E1}|\phi_i\rangle$ of the $E1$-transition operator. Relativistic corrections were included in the Breit–Pauli approximation and intermediate coupling was used to determine the transition matrix elements. Calculations show that the correlation terms in (30.7) change the values of the line strengths significantly. The results of such calculations are reported in Table 30.2 which uses the following notation: Exp1, experimental set-up at Giessen [248]; Exp2, experimental set-up at Bochum [248]; NCMET, theoretical results of Sinanoğlu [249]; CI, theoretical results of Baluja and Zeippen [250]; MCDHF, multi-configurational relativistic calculation results [237]; MCHFb, multi-configurational HF calculation results [236]; MBPT, calculation in the first-order (0+1) and the second-order of PT: $(0+1+2)'$, without correlation correction terms; $(0+1+2)''$, with correlation correction terms in transition operator [247].

It is evident from Table 30.2 that the oscillator strengths $gf''^{(0+1+2)}$, obtained using $E^{(0+1+2)}$ and the formula (30.7), are as a rule in a good agreement with the other calculations, in which the correlation effects are accounted for fairly accurately (for details see [247]). The calculated ratio of line intensities $I(1s^2 2s 2p^5\ {}^1P_1 - 1s^2 2s^2 2p^4\ {}^1D_2)/I(1s^2 2s 2p^5\ {}^1P_1 - 1s^2 2s^2 2p^4\ {}^1S_0)$ for NeIII is very close to the experimental data [248]. The calculations of lifetimes for the $1s^2 2s 2p^5\ {}^3P^0$ term also testify to the high accuracy of this approach. For NeIII $\tau''_{(0+1+2)} = 0.20$ ns, $\tau_{exp} =$

0.21 ± 0.01 ns, and for AlVI $\tau''_{(0+1+2)} = 0.101$ ns, $\tau_{exp} = 0.11 \pm 0.01$ ns, τ_{exp} are from [236]. More results of such kind may be found in [251].

Let us now proceed to $E2$- and $M1$-transitions. Formulas of the sort (30.7) hold for $E2$-transitions, too. The magnetic dipole transition operator O_{M1} has no radial part. Since the orthogonal radial orbitals are used in calculations, the operators $\Omega_C^{(1)\dagger} O_{M1}$ and $O_{M1}\Omega_C^{(1)}$ do not contribute to the matrix elements of the operator O_{M1}. Details of calculations and results of second-order MBPT calculations of energy spectra as well as electric quadrupole $(E2)$ and magnetic dipole $(M1)$ transitions in the oxygen isoelectronic sequence between the levels of the $1s^2 2s^2 2p^4$, $1s^2 2s 2p^5$ and $1s^2 2p^6$ configurations are reported in [233]. Here we shall only draw the main conclusions. It follows from the calculations that for the allowed transitions $(\Delta S = 0)$ correlation corrections have a minor effect: the values of $S_{E2}'^{(0+1+2)}$ are very slightly different from $S_{E2}^{(0+1)}$ whereas the $S_{E2}''^{(0+1+2)}$ values differ from $S_{E2}^{(0+1)}$ by no more than 7% in the case of the $^1S_0 - ^1D_2$ transition or 5% in the case of the $^3P_1 - ^3P_2$ transition for $10 \leq Z \leq 20$. A more significant effect of correlation corrections can be noticed for the allowed transition $^3P_0 - ^3P_2$ in the FeXIX ion.

Quite a different situation is found for the forbidden $(\Delta S \neq 0)$ transitions. Here the difference of the values of $S_{E2}'^{(0+1+2)}$ and $S_{E2}^{(0+1)}$ is in the range from 1% to 10% for the ions examined. Accounting for the $\Omega_C^{(1)\dagger} O_{E2}$ and $O_{E2}\Omega_C^{(1)}$ terms in the transition operator for the transitions with $J = 0$ ($^1S_0 - ^3P_2$, $^1D_2 - ^3P_0$ transitions) the calculated values of the line strengths $S_{E2}''^{(0+1+2)}$ are one order smaller in comparison with $S_{E2}'^{(0+1+2)}$ calculated without correlation correction terms. It should be noticed that the agreement of electric quadrupole line strengths and transition probabilities with those obtained by other accurate methods is fairly close.

As has already been mentioned, there are no correlation corrections to the magnetic dipole operator. Hence, the electron correlations manifest themselves only through the changing of the matrix elements of H_{eff}. For the $\Delta S = 0$ $M1$-transitions there is good agreement of line strengths as well as transition probabilities with other accurate methods. For intercombination transitions $(\Delta S \neq 0)$ this agreement is slightly worse, but the discrepancies of the line strength or transition probability values calculated in various approximations are in most cases not greater than a few per cent for $10 \leq Z \leq 20$.

Thus, the numerical procedure involving the treatment of both the first- and the second-order correlation energies in a complete finite basis demonstrates its high accuracy and flexibility in MBPT calculations. The MBPT version, described in Chapters 3 and 29, provides very accurate results, at least in the range of $Z = 10-26$. Accounting for the correlation energy due to the residual electrostatic interaction in the second-order of

MBPT significantly improves the electron transition wavelengths, line and oscillator strengths, transition probabilities as well as the lifetimes of excited levels. Therefore, it seems promising to generalize such an approach to cover the cases of more complex electronic configurations having several open shells, even with $n > 2$.

Two-electron transitions cause one more type of forbidden radiation. They become allowed when accounting, in some way, for correlation effects. The use of a superposition-of-configurations method or a multi-configurational approach is the easiest way to explain such transitions. For example, the transition $4d^9 5s^2 - 4d^{10} 5p$ in CdII is forbidden in the single-configuration approximation, because the operators of Ek- and Mk-transitions are one-electron operators, whereas the above mentioned configurations differ by quantum numbers of two electrons; for this reason these transitions are called two-electron transitions. They may be explained only when adopting more refined methods which take into consideration at least part of the correlation effects.

As was already mentioned, the configuration is not always an exact quantum number. Therefore, two-electron transitions may be interpreted as one-electron, if we expand the wave functions of both (or at least one) configurations between which the transition takes place, as a linear combination of certain single-configuration functions. First of all those configurations must be admixed, which, according to preliminary estimates, have the largest weights and have non-zero matrix elements of the operator of $E1$-transition.

Superposition of configurations already allows two-electron transitions in the pure coupling scheme. However, to obtain more accurate values, intermediate coupling must be utilized. Unfortunately, this leads to very rapid increase of the order of energy matrices to be diagonalized.

Because the above mentioned configurations, between which the transition takes place, are of opposite parity, the transition is of the $E1$ nature. To explain qualitatively its occurrence it is enough to assume the admixture of configuration $4d^9 5p^2$ to the principal one $4d^9 5s^2$ and then the two-electron transition becomes the one-electron $E1$-transition $4d^9 5p^2 - 4d^{10} 5p$. However, the oscillator strengths of such transitions are rather small ($f \sim 10^{-3} \div 10^{-4}$), and, therefore, they are extremely sensitive to the quality of accounting for correlation effects.

Another interesting example is presented by two-electron transition $s^2 - dp$, consisting of two transitions with $\Delta l = 1$ and $\Delta l = 2$. This may easily be explained, if we admix the configuration sp to dp (allowed $E1$-transition $s^2 - sp$ occurs) or the configuration p^2 to s^2 ($E1$-transition $p^2 - pd$). All admixed configurations must contain a term coinciding with the term of the principal configuration, because they are bound by electrostatic interaction, which is diagonal with respect to L and S.

Utilization of the multi-configurational approach ensures faster convergence than use of superposition-of-configurations, i.e. to achieve the same accuracy one needs to account for a much smaller number of admixed configurations. In a similar way, transitions having the participation of a greater number (e.g. three) of electrons may be explained. Many-electron transitions may also be described more or less successfully by employing other methods of accounting for correlation effects, including various versions of many-body perturbation theory.

Extremely accurate calculations of electronic transitions may contribute to explaining parity non-conserving effects in atoms [50, 252–254].

31

Peculiarities of the structure and spectra of highly ionized atoms

31.1 Peculiarities of multiply charged ions

Extensive studies of energy spectra and other characteristics of atoms and ions allow one to reveal general regularities in their structure and properties [255–257]. For example, by considering the lowest electronic configurations of neutral atoms, we can explain not only the structure of the Periodical Table of elements, but also the anomalies. The behaviour of the ionization energy of the outer electrons of an atom illustrates a shell structure of electronic configurations.

Multiply charged ions represent a peculiar world, essentially differing from the world of neutral atoms or their first ions. In this chapter we shall try to describe these peculiarities. Main attention will be paid to the need for a combination of theoretical and experimental studies of the spectra of highly ionized atoms, and to the role of relativistic and correlation effects for multiply charged ions, and to the regularities and irregularities of the behaviour of their spectral characteristics along isoelectronic sequences.

Recently intense beams of ultracold highly ionized atoms up to naked uranium (U^{92+}) have been obtained [258]. Single ions were confined in ion traps, ion crystals were observed, and the phase transition from ion cloud to crystal was noticed [259]. With increase of ionization degree the radiation of the ions moves to a shorter and shorter wavelength region, thus requiring special measurement techniques. For very highly ionized atoms (e.g. hydrogen-like bismuth [154]), transitions between the hyperfine structure components occur in the optical region. It is very useful, to know the wavelength domain of the radiation of such ions. The relevant calculations here are very helpful. Deviations from LS coupling due to the rapid growth of spin–orbit interactions along the isoelectronic sequence also increase very rapidly, therefore, other coupling schemes or intermediate coupling must be employed for identification and classification of the energy levels. This may be done only with the help of theoretical or semi-empirical calculations. Even the concept of the electronic configuration in its usual meaning is lost for very highly ionized atoms and then the totally relativistic approach must be utilized. The existing possibilities for performing improved theoretical calculations, accounting for relativistic and correlation effects, and their combination with semi-empirical methods allow us to successfully identify and classify, with the help of optimal sets of quantum numbers, the spectra of any atom or ion, including very highly ionized ones, obtained both in the laboratory or non-atmospheric astrophysical plasma.

An interesting pecularity of the spectra of highly ionized atoms is the occurrence of so-called satellite lines. They appear near resonance lines of the ion X^{n+} and belong to the ion $X^{(n-1)+}$, in which two electrons are excited. So, satellite lines of the resonant lines of H-like ions correspond to the transitions $2l2l' - 1s2l'$ in He-like ions, etc. In other words, the satellite transition $2s2p - 1s2s$ in He-like ion may be considered as the transition $2p - 1s$ in H-like ion in the presence of excited electron ('spectator') $2s$. Therefore, the wavelengths of the satellite lines of the ion $X^{(n-1)+}$ are close to those of the resonant lines of the ion X^{n+}. The occurrence of intense satellite lines in the spectra of highly ionized atoms may be explained by the rapid growth of the relevant transition probabilities (compared to that of autoionizing transitions) with increase of ionization degree. Considerations of their intensities with regard to that of resonant lines supplies us with an extremely efficient method for the diagnostics of the high-temperature plasma.

Some complex excited configurations possess a large number of levels. This leads to an abundance of close-lying lines, which, due to Doppler broadening cannot always be resolved. Thus, a new type of spectra, called 'quasicontinuous', appears for multiply charged ions. For the description of such spectra the statistical approach, discussed in the following chapter, seems to be very promising.

Practical calculations show that the Hartree–Fock–Pauli (HFP) approximation leads to fairly accurate values of the energy levels, wavelengths of electronic transitions, oscillator strengths and transition probabilities for a large variety of many-electron atoms and ions, including very highly ionized ones, particularly if this approach is combined with the accounting for correlation effects [18, 260–262]. For neutral or a-few-times-ionized atoms the correlation effects predominate over relativistic ones. For about ten-times-ionized atoms they are of the same order, whereas for higher ionization degrees the relativistic effects start to predominate. However, quite often, due to the particular structure of an ion and its configuration, the relative contributions of these two effects may differ from that mentioned above. A similar situation is valid for wavelengths of electronic

Table 31.1.   Wavelengths of the transition $2p^5 3d - 2p^6$ (in Å) for FeXVII and MoXXXIII

Ion	$LSJ - {}^1S_0$	PT	HFP	DHF	exp
	${}^3P_1 - {}^1S_0$	15.484	15.457	15.460	15.453
FeXVII	${}^3D_1-$	15.274	15.263	15.268	15.261
	${}^1P_1-$	15.023	14.996	14.004	15.012
	${}^3P_1 - {}^1S_0$	4.869	4.860	4.854	4.847
MoXXXIII	${}^3D_1-$	4.812	4.802	4.803	4.804
	${}^1P_1-$	4.643	4.637	4.630	4.630

transitions, however, here we have to distinguish two cases: transitions with the change and without the change of principal quantum number.

The binding energy E^{nl} of nl-electron may be presented in Ry as follows [16]:

$$E^{nl} = -(Z^*)^2/n^2 = -(Z - a)^2/n^2 = -(Z_c + b)^2/n^2, \qquad (31.1)$$

where $Z^* = Z - a = Z_c + b$, $b = Z - Z_c - a = (N-1) - a$. It is evident that the parameters a and b depend on nl. The quantity $b \to 0$ with increase of n and l. Then we find the following formula for the energy separation of two configurations with different n and l:

$$E^{n_2 l_2} - E^{n_1 l_1} = \frac{(Z - a_1)^2}{n_1^2} - \frac{(Z - a_2)^2}{n_2^2}. \qquad (31.2)$$

It follows from (31.2) that for large Z this energy separation (wavelength of the appropriate electronic transition) is proportional to Z^2. The analogous quantity for configurations with the same n (i.e. $\Delta n = 0$) is

$$E^{nl_2} - E^{nl_1} = \frac{(Z - a_1)^2 - (Z - a_2)^2}{n^2} = \frac{2(a_2 - a_1)}{n^2}\left[Z - \frac{a_1 + a_2}{2}\right], \qquad (31.3)$$

i.e. in this case it grows only linearly with Z and, therefore, rapidly decreases compared to the excitation energies for $\Delta n \neq 0$. It follows from (31.2) and (31.3) that for these two sorts of transitions the relative role of relativistic and correlation effects will be different; with an increase of Z the domains of the relevant transitions will move apart. For the case (31.2) they will move rapidly to the X-ray wavelengths region. Table 31.1 illustrates this statement. It also allows us to judge the accuracy of various methods (Z-expansion perturbation theory (PT) [263], Hartree–Fock–Pauli (HFP), Dirac–Hartree–Fock (DHF) and experiment [264]): all three methods, particularly the last two, allow us to obtain results which are very close to experimental ones. However, for the ion MoXXXIII the

Table 31.2. Wavelengths (in Å) of the $E1$-transition $1s^22s^22p^2 - 1s^22s2p^3$ in NiXXIII

$LSJ - L'S'J'$	HFP	SC	Exp.
$^3P_2 - ^3D_3$	131.37	128.26	128.43
$^3P_1 - ^3D_2$	129.40	126.45	126.35
$^3P_0 - ^3D_1$	114.98	112.11	111.84
$^3P_2 - ^3P_2$	108.50	106.37	105.44
$^3P_2 - ^3S_1$	92.90	91.33	91.79
$^3P_1 - ^3S_1$	88.33	86.94	87.60
$^1D_2 - ^1D_2$	103.25	102.02	102.02

DHF approach is the most accurate; only DHF may be utilized for even more highly ionized atoms, when the relativistic effects are already quite large and, therefore, cannot be accounted for as corrections.

Configurations, differing by the number of electrons in the shells with different l but the same n, are called degenerate. For them the formula (31.3) is valid. With an increase of Z in the isoelectronic sequence the energies of such configurations draw together, therefore, when accounting for correlation effects, these energetically closest configurations must be taken into consideration first. Generally, the main part of the correlation effects is accounted for by the configurations that are energetically closest to the one studied, as well as by configurations having the largest non-diagonal matrix elements of the electrostatic interaction. Indeed, the results of Table 31.2, in which the wavelengths of the $E1$-transition $1s^22s^22p^2 - 1s^22s2p^3$ in NiXXIII (calculated in HFP approximation and in the superposition-of-configuration method (SC), while admixing only one quasidegenerate configuration $1s^22p^4$ to $1s^22s^22p^2$) are presented, suggest that such a simple SC improves by an order of magnitude or more the results of the HFP approximation. The discrepancies with experimental data [265] are in the third or even fourth figure. Let us also emphasize that in all cases considered large deviations from pure LS coupling scheme are observed, therefore this classification of energy levels with its quantum numbers is rather conditional.

Thus, accounting for correlation effects only in the frameworks of the superposition of quasidegenerated configurations ensures essential improvement of the values of the energy levels and wavelengths of electronic transitions in highly ionized atoms. Further refinement of these data with the help of SC is inefficient due to slow convergence of this method. The multi-configurational approximation (MCA) looks much more promising in this respect. The use of MCA instead of SC enlarges significantly the

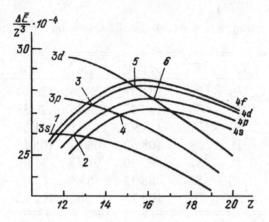

Fig. 31.1.   Dependence of the energy separations $1s^2 2s^2 2p^3 4l - 1s^2 2s^2 2p^4$ and $1s^2 2s 2p^4 3l - 1s^2 2s^2 2p^4$, divided by Z^3, on Z.

contribution of some admixed configurations. This contribution may become comparable to that of quasidegenerated configurations. MCA may increase by an order of magnitude the correction of separate admixed configurations to the energy in comparison with those obtained by SC.

31.2   Isoelectronic sequences

Studies of isoelectronic sequences are very useful and fruitful in theoretical, semi-empirical and experimental investigations of atomic spectra. They reveal important regularities and pecularities in the behaviour of various physical characteristics of many-electron atoms and ions, help to identify and classify their energy levels with an optimal coupling scheme, find initial values of semi-empirical parameters, etc.

Let us mention a few peculiarities of the spectra of some groups of highly ionized atoms. For certain Z values in an isoelectronic sequence very strong interaction between the electrons of some special sorts of configurations, e.g. $n_0 l_0^{4l_0+2} n_1 l_1^{N_1} n_2 l_2$ ($n_0 \leq n_1 < n_2$) and $n_0 l_0^{4l_0+1} n_1 l_1^{N_1} n_0' l_0' n_2' l_2'$ ($n_2' < n_2$, $n_0' > n_0$) may occur. These configurations are not quasidegenerated. The energies of these configurations depend on Z in a different way, therefore, it may happen that with an increase of ionization degree the lower-lying configuration may become the upper-lying one and vice versa. In such a case we observe the intersection of the energy levels of relevant ions, and then the single-configuration approach is completely unfit [266]. Figure 31.1 illustrates this conclusion. It shows the dependence of the energy separations between configurations $1s^2 2s^2 2p^3 4l$ and $1s^2 2s^2 2p^4$ and between $1s^2 2s 2p^4 3l$ and $1s^2 2s^2 2p^4$, divided by Z^3, on Z

along the oxygen isoelectronic sequence. Numbers 1–6 denote six points of intersection of the energies of these configurations which are of the same parity. In this domain the usual concept of configuration is not valid, because both these configurations have practically equal weights in the multiconfigurational wave function. Determination of such regularities allows one to predict in advance the ionization degrees and the sorts of excited configurations whose theoretical or semi-empirical study requires the multiconfigurational approach.

Let us discuss briefly the structure of multiply charged ions. It follows from calculations that the mean distance of the electronic shells from the nucleus rapidly diminishes along the isoelectronic sequence with an increase of ionization degree, i.e. the ion shrinks swiftly under the action of growing (due to less screening) nuclear charge field. It is interesting to notice that for high ionization stages the $2p$-shell becomes closer to the nucleus than $2s$; for very large Z values the splitting of electronic shells into subshells occurs, e.g. $2p^N$ shell becomes $2p_-^{N_1} 2p_+^{N_2}$.

Consideration of the behaviour of the total electronic charge distribution leads to similar conclusions. With growth of the ionization degree the maximum of the electronic charge distribution rapidly moves towards the nucleus and becomes significantly narrower, i.e. electronic shells are pressed to the nucleus and they themselves become thinner and thinner. In contrast to the neutral atom, which in fact is a comparatively weakly bound system, the highly ionized atom is a compact, strongly bound structure. Changing the electronic configuration of such an ion via excitation of the outer electrons does not alter significantly the distribution of charge density.

Typical peculiarities of multiply charged ions are reflected in their energy spectra, obeying the changes of the relative role of various intra-atomic interactions. For neutral and not highly ionized atoms, LS coupling is typical, the levels are grouped in multiplets and the fine structure of terms indeed is 'fine'. The level separations as a rule are much smaller compared to term separations. For very highly ionized atoms, all these features are lost, including the usual concept of electronic configuration. Instead we have practically pure jj coupling even for a shell of equivalent electrons, fine structure is not 'fine', and the term notion cannot be applied to classify the energy levels. These levels are grouped according to relativistic analogues of the configuration. For example, the energy levels of the configuration pd will make four subgroups according to four subconfigurations p_-d_-, p_-d_+, p_+d_- and p_+d_+. Such systems must be treated relativistically (usually in Dirac–Hartree–Fock approach) [18, 267, 268].

In a similar way the peculiarities of electronic transitions may be investigated along isoelectronic sequences. Line and oscillator strengths as well as transition probabilities usually change fairly smoothly along them. As was mentioned earlier, when analysing the validity of various coupling schemes, due to different selection rules and deviations from pure coupling, the quantities of electronic transitions may vary. So, for example, calculations of the $E1$-transitions $2p3s - 2p^2$ in the carbon isoelectronic sequence show that the probability of intercombination transition $^1D_2 - ^3P_2$, forbidden in LS coupling, with the increase of Z grows rapidly and for $Z \geq 35$ is almost equal to that of the allowed transition $^3P_1 - ^3P_0$. The probabilities of two other intercombination transitions initially also grow, reach a maximum, then start to decrease. Such behaviour of transition probabilities is explained by the changes of relative contributions of various interactions along the isoelectronic sequence and, thus, of coupling schemes, causing the occurrence of other selection rules.

However, in a number of cases various sorts of inhomogeneities and rapid changes for electron transition quantities may be noticed. One of the possible reasons for such phenomena is connected with the collapse of the orbit of the excited electron, already considered in Chapter 28. Let us discuss here another source of peculiarities, caused by cancellation (compensation) or interference effects in the sums, with which we usually have to deal when performing calculations of various spectral characteristics of atoms and ions. The necessity to fulfil the calculations in intermediate coupling (wave function (11.10)) leads to the use of the relevant expression for submatrix elements of the transition operator (30.1), being the sum, the terms of which may have different signs. Therefore, situations may occur, when the final result is the difference of two almost equal large numbers. Then small errors in initial data may lead to large discrepancies in final results.

The accuracy of oscillator strengths and transition probabilities in the best cases, depending on the method employed, approaches 10 or even a few per cent. However, in many cases, particularly for neutral or weakly ionized atoms having complex electronic configurations, discrepancies may be much larger (about 50–100% or even more). The largest errors, as a rule, are caused by cancellation effects.

Cowan [16] introduced the so-called 'cancellation factor'

$$F = \left[\frac{|\sum_{ij} a(iJ)b(jJ')(iJ\|O^{(k)}\|jJ')|}{\sum_{ij} |a(iJ)b(jJ')(iJ\|O^{(k)}\|jJ')|} \right]^2. \tag{31.4}$$

Calculations indicate that about 1% of the transitions between complex configurations has values of $F < 10^{-4}$. Computed line strengths may have large percentage errors for $F \leq 0.1$. Because of such interference of separate terms (so-called angular effects) some spectral lines disappear for certain ions. Figure 31.2 illustrates this statement for the example of the transition $2p^53d - 2p^6$ for the neon isoelectronic sequence. Indeed,

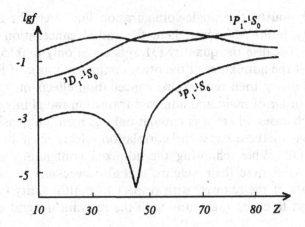

Fig. 31.2. Dependence of oscillator strengths of transition $2p^5 3d - 2p^6$ on Z in neon isoelectronic sequence.

the transitions $^1P_1 - ^1S_0$ and $^3D_1 - ^1S_0$ behave 'normally', whereas the transition $^3P_1 - ^1S_0$ 'disappears temporarily' for a number of ions in the domain $Z \approx 45$.

Another similar phenomenon is caused by radial effects. It may be estimated by the ratio of the appropriate transition radial integrals. For example, in the case of the length form of the $E1$-transition it is

$$F' = \int\limits_0^\infty rP(nl|r)P(n'l'|r)dr \Big/ \int\limits_0^\infty r\Big|P(nl|r)P(n'l'|r)\Big|dr. \qquad (31.5)$$

Indeed, radial orbitals have positive and negative parts, if $n > l + 1$ or $n' > l' + 1$. This may cause the cancellation of separate parts of the radial integrals and, thus, of line strengths. Sometimes this compensation may reach almost 100%. Its main peculiarity is that such cancellation excludes all transitions between given configurations, not separate lines, as was the case for angular compensation. For example, all transitions between the configurations $4p^5 6d$ and $4p^6$ in MoVII are extremely weak.

The third sort of cancellation effect is caused by superposition of configurations. Cowan [16] illustrates this case on the example of transitions $3p5p - 3p4s$ in SiI. The radial integral of this transition is obtained to be $(5p|r|4s) = 0.70$. Admixture of $5p$ to $4p$ is very small, but the integral $(4p|r|4s) = 7.41$, i.e. it is an order of magnitude larger than the main integral. Thus, admixture of only 1% of the configuration $4p$ to $5p$, due to multiplication by a large integral, gives the value of the relevant matrix element, proportional to ± 0.7 and 0.7, and then, depending on phase conditions for concrete transitions, the line strength is equal to zero or is four times larger than the quantity found in the single-configuration approach. Obviously, the single-configuration line strengths in this case are absolutely unfit. Moreover, here the radial cancellation effects are observed, too, because the quantity (31.5) is equal only to 0.15.

Of course, if the admixture of the other configurations is of the order of 10% or even more, then remarkable cancellation effects may occur even for the same order of main and admixed transition radial integrals. There are many such cases where it is enough only to mention quasidegenerate configurations. In these cases the correlation effects must be accounted for without fail. When choosing the admixed configurations, it is not enough only to analyse their weights, it is also necessary to calculate the radial integrals of the allowed with respect to multipolarity (usually $E1$) transitions and to check their ratio with the relevant integral of the main transition.

Let us also recall that the total strength of all transitions between the configurations considered is constant, therefore the cancellation effects in one or in several lines are compensated by increased values of other lines. Cancellation effects due to superposition of configurations lead to 'pumping' of the quantities of line strengths from one transition array to the other.

Starting with Z dependences of wavelengths (31.2) and (31.3), we can find the following estimates for transition quantities. For the $E1$-transition with $\Delta n \neq 0$ ($\Delta E \sim Z_c^2$, where $Z_c = Z - N + 1$)

$$S \sim Z_c^{-2}, \quad f^{E1} \sim \text{constant}, \quad W^{E1} \sim Z_c^4, \tag{31.6}$$

whereas for $\Delta n = 0$ ($\Delta E \sim Z_c$)

$$S \sim Z_c^{-2}, \quad f^{E1} \sim Z_c^{-1}, \quad W^{E1} \sim Z_c. \tag{31.7}$$

Thus, as was the case for wavelengths (formulas (31.2) and (31.3)), the oscillator strengths and transition probabilities behave differently for the transitions with $\Delta n = 0$ and $\Delta n \neq 0$. However, these regularities are valid only approximately. Especially large deviations may occur for the cases where correlation or relativistic effects are large.

In conclusion, the contemporary state of the theory and computational methods gives the real possibility of calculating the spectral characteristics of separate atoms and ions and their isoelectronic sequences, of studying the regularities and irregularities in them, and of modelling real physical objects and processes. Practical utilization of the methods described and their realization in the form of universal computer program packages gives the possibility of remarkable progress in the applications of theoretical results for the interpretation and classification of the spectra of atoms and ions, including multiply charged ions, obtained in both the laboratory and astrophysical plasmas. It also contributes significantly to revealing the peculiarities of their structure and properties, such as the changes of coupling schemes, selection rules for electronic transitions, etc. [269–271]. The accuracy of calculation achieved allows one to determine with high precision the wavelength regions in which the spectral lines of one or another ion will occur, to identify their energy levels, and to successfully apply theoretical data for plasma diagnostics and other purposes.

31.3 Astrophysical applications

There exist 'producers' and 'consumers' of atomic data. Astronomers as a rule are among the latter [272, 273, 319, 320]. In 1983 the Atomic Spectroscopy Group at the University of Lund organized a conference at Lund, whose purpose was to bring together 'producers' and 'consumers' of spectroscopic data to discuss the needs for data and the possibilities of meeting the demand, and to establish an international dialogue between scientists who used basic atomic spectroscopic data and scientists whose research produced such data. Other conferences of this sort continued this dialogue [274–277]. In this way the users review their present and future needs whereas the providers review their laboratory capabilities, new developments in data measurement and improved theoretical capabilities. The needs are not only for applications, such as tokamak plasmas, X-ray lasers and astrophysical models, but also in fundamental theory, for which the inclusion of the Breit interaction, the Lamb shift, and other quantum-electrodynamical corrections remain untested for very highly ionized atoms.

In stellar astronomy, spectroscopic studies are indirectly revealing the structure deep down in stars where one could not see otherwise. An internal structure shows up in the emergent radiation spectrum both kinematically and through abundance anomalies. The solution of the solar corona line problem in 1942 may serve as a typical example of astrophysical applications of atomic spectroscopy [256].

There are numerous needs for precise atomic data, particularly in the ultraviolet region, in heavy and highly ionized systems. These data include energy levels, wavelengths of electronic transitions, their oscillator strengths and transition probabilities, lifetimes of excited states, line shapes, etc. [278].

Astrophysicists enumerate a large number of present problems in astrophysics, in which atomic data are used or are needed. They ask about what is known and what can be measured or calculated at the moment and with what precision. Some groups successfully combine the production and use of atomic data [279].

Over the past decades we have witnessed the explosive growth of astrophysical spectroscopic observations in both the ultraviolet and infrared bands. With the launch of the Hubble Space Telescope, the precision and resolution of such data have reached remarkable levels, giving one the sense that the body of atomic data currently to be found in the literature lags far behind what is needed to adequately interpret the observations. Similarly, high temperature laboratory experiments in plasma physics, e.g. fusion energy and X-ray lasers, are demanding larger quantities of atomic data over a wide range of ionization states. Fortunately, the experimental and computational techniques of atomic physics have kept pace. One may cite, for example, the extraordinary precision inherent in recent laboratory work with laser-induced fluorescence spectroscopy and with Fourier transform spectrometers, and for data of highly ionized atoms, with ion traps and tokamak plasmas. The major challenge is to nurture and support expanding activity in those subdisciplines of atomic physics that apply such modern techniques to the production of extensive volumes of atomic data, and to reinvigorate such 'old fashioned' subjects as the term analysis of new, more accurate laboratory spectra [276, 280–283].

A contact between data users and providers has grown even more critical today with the prolific acquisition of data from space-based astrophysical observatories, with high temperature laboratory experiments in fusion energy and X-ray lasers, and with the demands for higher accuracy and large quantities of data from neutral to highly ionized atoms. The means of delivery of these data with new light sources such as the tokamak, the electron beam ion trap, laser generated plasmas, and inductively coupled plasmas have advanced considerably [277].

The advent of large-scale computing capabilities has been an important development for the atomic physics community, giving to it the opportunities to embark on comprehensive data generation projects that were unthinkable until recently. Indeed, several huge data projects have not only generated vast amounts of new data, but have also in many cases far surpassed the earlier data in their accuracy (see, for example, the Opacity Project [284]). Opacity is a quantity which determines the transport of radiation through matter, and is important for many problems in physics and astronomy. To calculate opacities one requires atomic data for a large number of processes involving the absorption and scattering of radiation. Stellar opacities are of particular interest in astrophysics because of their importance for the theories of stellar structure and stellar pulsations.

Taking advantage of advances in computational atomic and plasma physics and of the availability of powerful supercomputers, a collaborative effort – the international Opacity Project – has been made to compute accurate atomic data required for opacity calculations. The work includes computation of energy levels, oscillator strengths, photoionization cross-sections and parameters for pressure broadening of spectral lines. Several different methods of computation have been used to calculate the same data thereby providing a check on the validity of the data [284–286].

The dialogue between users and providers of atomic data is a two-way conversation, with atomic physicists beginning to view astrophysical and laboratory plasmas as unique sources of new information about the structure of complex atomic species. A number of monographs on theoretical atomic spectroscopy cowritten by theoreticians and astrophysicists and dedicated to astrophysicists also contribute to better mutual understanding [18, 320].

Utilization of data obtained from various plasma sources (e.g. beam-foil, tokamak and laser-produced plasma [287]) enabled the identification with high accuracy of the lines of highly ionized atoms in solar spectra. A special commision No14 on Atomic and Molecular Data of the International Astronomical Union coordinates the activity on systematization of spectroscopic data, informs the astrophysics community on new developments and provides assessments and recommendations. It also provides reports which highlight these new developments and list all important recent literature references on atomic spectra and wavelength standards, energy level analyses, line classifications, compilations of laboratory data, databases and bibliographies.

Among the publications let us mention an extensive review of atomic spectroscopic data for astrophysics [288], compilation of data for 2031 resonance lines of spectra of hydrogen through germanium [289], new compilations of wavelengths with energy level classifications for all spectra of magnesium, aluminium and sulphur [290–292], tables of energy levels and multiplets for hydrogen, carbon, nitrogen and oxygen spectra [293], an extensive collection of atomic data [294]. A biannual series of bulletins on atomic and molecular data for fusion includes references for recent papers with data on atomic energy levels and wavelengths [295].

The Data Center on Atomic Transition Probabilities at the National Institute of Standards and Technology, formerly the National Bureau of Standards, in Gaithersburg, Maryland 20899, USA is continuing its critical data compilation and is also engaged in developing a comprehensive numerical and bibliographical database.

32

Global methods in the theory
of many-electron atoms

32.1   Mean characteristics of atomic spectra

Complex atomic spectra may be studied by statistical (global) methods, particularly if we are interested only in general spectral properties and regularities. The atomic spectrum as a distribution of energy levels or spectral lines can be described by its statistical moments, which define its centre of gravity, width, asymmetry and other mean characteristics [296]. A number of principal moments make it possible to approximately describe the complex spectra in a model of unresolved arrays [297], to reveal statistical properties of spectra [296, 298], to analyse the relative influence of various interactions on the structure of the spectrum, to study changes of spectrum in atomic or isoelectronic sequences or due to configuration mixing [296–300]. The main problem is to find explicit expressions for such averaged characteristics, without detailed calculations of spectra by the methods described in previous chapters. The general theory of this approach is presented in the monograph [296] and in the papers [301, 302]. Below we shall briefly sketch this approach, mainly following [296].

The kth-order initial moment α_k for a set of levels or spectral line energies $E_1, E_2, \ldots, E_q$ is

$$\alpha_k = \sum_{i=1}^{q} E_i^k p(E_i). \tag{32.1}$$

Here $p(E_i)$ stands for the probability of finding this level with energy E_i in the spectrum and it equals the statistical weight of the level $q_i = 2J_i + 1$, divided by the total statistical weight of the spectrum, i.e.

$$p(E_i) = g_i/g = (2J_i + 1)/g. \tag{32.2}$$

For the spectrum corresponding to transitions between energy levels (transition array), the probability of the ith line can be expressed in terms

380

of the ratio of the intensity of this line I_i to the total intensity of the spectrum, namely,

$$p(E_i) = I_i/I. \tag{32.3}$$

For continuous E_i values the summation in (32.1) must be replaced by integration. This summation may be performed over the whole atomic spectrum; however, from a practical point of view the most important contributions are from the moments of spectra corresponding to the levels of one configuration or to transitions between the configurations chosen. Here we shall restrict ourselves only to the studies of such moments. If strong mixing of configurations is observed, then it is necessary to find the moments for all complexes of these configurations.

The first-order initial moment represents the average energy of the spectrum $\bar{E}$

$$\alpha_1 = \bar{E} = \sum_i E_i p(E_i). \tag{32.4}$$

It is convenient to use the centred moments μ_k defined with respect to the average energy, i.e.

$$\mu_k = \sum_i (E_i - \bar{E})^k p(E_i). \tag{32.5}$$

The first-order centred moment is equal to zero. The second one describes the variance of spectrum

$$\mu_2 = \sigma^2 = \sum_i (E_i - \bar{E})^2 p(E_i), \tag{32.6}$$

where $\sigma = \sqrt{\sigma^2}$ is the root-mean-square deviation of the spectrum from $\bar{E}$ and defines the width of the spectrum.

Instead of the third and the fourth centred moments it is convenient to use the following dimensionless quantities:

$$\kappa_1 = \frac{\mu_3}{(\sigma^2)^{3/2}} = \frac{\mu_3}{\mu_2^{3/2}}; \qquad \kappa_2 = \frac{\mu_4}{(\sigma^2)^2} - 3 = \frac{\mu_4}{\mu_2^2} - 3. \tag{32.7}$$

The skewness κ_1 characterizes the asymmetry of the spectrum, whereas the excess κ_2 describes deviation of the levels or line density from the normal (Gaussian) distribution density.

Central moments of the continuous spectrum may be defined by the formula

$$\mu_k = \int_{-\infty}^{\infty} (E - \bar{E})^k \rho(E) dE, \tag{32.8}$$

where $\rho(E)$ is the probability density, normalized to unity

$$\int_{-\infty}^{\infty} \rho(E)dE = 1. \tag{32.9}$$

The concept of probability density may also be utilized to describe the discrete spectrum or, particularly, to solve the inverse problem of the approximate reconstruction of the spectrum from its envelope line using a certain number of its lower moments. The density function of the energy levels of the discrete spectrum may have the form

$$\rho(E) = \sum_i \delta(E - E_i)p(E_i), \tag{32.10}$$

where $\delta(E - E_i)$ stands for the Dirac delta-function. Efficient exploitation of the method of statistical spectral characteristics is possible only when having explicit expressions for the above-mentioned moments. Let us consider the approximations and conditions for which such expressions may be deduced. In order to be able to perform the summations in (32.1) and (32.5) over many-electron quantum numbers, we have to assume that radial orbitals do not depend on these quantum numbers. Such an assumption is well grounded for the majority of physical phenomena (but, for example, not for the collapse of the orbit of excited electron). It holds even better for average quantities, because the latter depend only weakly on the approximation chosen.

For spectra corresponding to transitions from excited levels, line intensities depend on the mode of production of the spectra, therefore, in such cases the general expressions for moments cannot be found. These moments become purely atomic quantities if the excited states of the electronic configuration considered are equally populated (level populations are proportional to their statistical weights). This is close to physical conditions in high temperature plasmas, in arcs and sparks, also when levels are populated by the cascade of elementary processes or even by one process obeying non-strict selection rules. The distribution of oscillator strengths is also excitation-independent. In all these cases spectral moments become purely atomic quantities. If, for local thermodynamic equilibrium, the Boltzmann factor can be expanded in a series of powers $(\Delta E/kT)^n$ (this means the condition $\Delta E \ll kT$), then the spectral moments are also expanded in a series of purely atomic moments.

Atomic spectral moments can be expressed in terms of the averages of the products of the relevant operators. Let $O_1, O_2, \ldots, O_k$ be the operators of interactions in definite shells or the operators of electronic transitions between definite shells in the second quantization form. The average of their product with respect to the configuration K is expressed by the sum of products of their matrix elements

$$(O_1 O_2 \ldots O_k)^K = \frac{1}{g(K)} \sum_{\gamma_1 \gamma_2 \ldots \gamma_k} (\gamma_1|O_1|\gamma_2)(\gamma_2|O_2|\gamma_3) \ldots$$
$$\times (\gamma_{k-1}|O_{k-1}|\gamma_2)(\gamma_k|O_k|\gamma_1), \qquad (32.11)$$

where $g(K)$ is the statistical weight of the configuration K.

In practical calculations, it is convenient to use traceless operators with excluded average energy. Non-diagonal matrix elements of such operators change only their phases while replacing electrons by vacancies in all shells in which they are acting (radial functions are supposed to be frozen). This leads to the invariance (up to phase multiplier) of the average (32.11) under the substitutions $N_i \rightarrow 4l_i + 2 - N_i$ in all active shells. Averages and moments as mean quantities do not depend on the coupling scheme.

Let us use further on the central moments. The kth central moment of the energy level spectrum of configuration K in intermediate coupling may be presented in the form

$$\mu_k(K) = \frac{1}{g(K)} \sum_{\gamma} \left[(K\gamma|H|K\gamma) - \bar{E}(K) \right]^k . \qquad (32.12)$$

In the general case K may denote a single configuration or the superposition of a number of configurations. In the latter case the summation γ runs over many-electron quantum numbers of all these configurations. Excluding the average energy $\bar{E}(K)$ from the Hamiltonian H and transforming the wave functions to some pure coupling, taking into consideration (32.11) we find

$$\mu_k(K) = \frac{1}{g(K)} \sum_{\gamma} (K\gamma|\mathscr{H}|K\gamma)^k = \overbrace{\left(\mathscr{H} \ldots \mathscr{H} \right)}^{k \text{ operators}}_{K}, \qquad (32.13)$$

where

$$\mathscr{H} = H - \bar{E}(K). \qquad (32.14)$$

The expressions for moments of the emission spectrum, corresponding to the radiative transitions between the configurations K and K', take a simple form, if the condition

$$E(K\gamma) - E(K'\gamma') \gg \Delta E(K) + \Delta E(K') \quad \text{for all } \gamma, \gamma' \qquad (32.15)$$

is fulfilled, where $\Delta E(K_i)$ is the width of the energy spectrum of the configuration K_i. Then the central moment for radiation reads

$$\mu_k^r(K - K') = \left\{ \sum_{\gamma\gamma'} \left[(K\gamma|H|K\gamma) - (K'\gamma'|H|K'\gamma') \right. \right.$$

$$\left. \left. - \bar{E}(K - K') \right]^k S(K\gamma, K'\gamma') \right\} / S(K, K'). \quad (32.16)$$

Here $S(K, K')$ is the total line strength of the transition $K - K'$ and $\bar{E}(K - K')$ is defined by (32.43).

Let us emphasize that in single-configurational approach the terms of the Hamiltonian describing kinetic and potential energies of the electrons as well as one-electron relativistic corrections, contribute only to average energy and, therefore, are not contained in $\mathcal{H}$, which in the non-relativistic approximation consists only of the operators of electrostatic interaction $\mathcal{H}^e$ and the one-electron part of the spin–orbit interaction $\mathcal{H}^{so}$, i.e.

$$\mathcal{H} = \mathcal{H}^e + \mathcal{H}^{so}. \quad (32.17)$$

32.2 Average, variance, asymmetry, excess of a spectrum

The average energy of an atom or ion in a given configuration, being the first initial moment of the energy spectrum, is widely used as one of the main characteristics of the system considered. It may be defined as follows:

$$\alpha_1 = \bar{E}(K) = \sum_i N_i I(n_i l_i) + \sum_i \bar{E}^e(n_i l_i^{N_i}) + \sum_{i<j} \bar{E}_{ij}^e(n_i l_i^{N_i}, n_j l_j^{N_j}). \quad (32.18)$$

Here $I(n_i l_i)$ is given by (19.23), whereas $\bar{E}^e(n_i l_i^{N_i})$ and $\bar{E}_{ij}^e(n_i l_i^{N_i}, n_j l_j^{N_j})$ are the average energies of electrostatic interactions in the ith shell as well as between the ith and jth shells. The latter have the following definitions:

$$\bar{E}^e(nl^N) = \frac{N(N-1)}{2} F^0(nl, nl)$$

$$- \frac{N(N-1)}{(4l+2)(4l+1)} \sum_{k>0}^{2l} \left(l\|C^{(k)}\|l \right)^2 F^k(nl, nl), \quad (32.19)$$

$$\bar{E}_{12}^e(n_1 l_1^{N_1} n_2 l_2^{N_2}) = N_1 N_2 F^0(n_1 l_1 n_2 l_2)$$

$$- \frac{N_1 N_2}{2(2l_1+1)(2l_2+1)} \sum_{k=|l_1-l_2|}^{l_1+l_2} \left(l_1\|C^{(k)}\|l_2 \right)^2 G^k(n_1 l_1, n_2 l_2).$$

$$(32.20)$$

Radial integrals $F^k(n_1 l_1, n_2 l_2)$ and $G^k(n_1 l_1, n_2 l_2)$ are defined by (19.31) and (20.10), respectively. Let us recall that the average value of spin–orbit interaction equals zero.

The variance of spectrum consists of the contributions of spin–orbit and electrostatic interactions

$$\mu_2(K) = \sigma^2(K) = \sigma^2_{so}(K) + \sigma^2_e(K). \tag{32.21}$$

The spin–orbit contribution simply equals the sum of the relevant interactions in each shell of equivalent electrons

$$\sigma^2_{so}(K) = \sum_i \sigma^2_{so}(n_i l_i^{N_i}), \tag{32.22}$$

where (η_{nl} is defined by (19.59))

$$\sigma^2_{so}(nl^N) = N(4l + 2 - N)\frac{l(l+1)}{4(4l+1)}\eta^2_{nl}. \tag{32.23}$$

The electrostatic contribution consists of two sums, corresponding to the relevant interactions inside each shell and between them

$$\sigma^2_e(K) = \sum_i \sigma^2_e(n_i l_i^{N_i}) + \sum_{i<j} \sigma^2_e(n_i l_i^{N_i} n_j l_j^{N_j}). \tag{32.24}$$

They have the following definitions (principal quantum numbers are skipped):

$$\sigma^2_e(l^N) = \frac{N(N-1)(4l+2-N)(4l+1-N)}{(4l+2)(4l+1)4l(4l-1)}$$

$$\times \sum_{k>0}\sum_{k'>0} \left(l\|C^{(k)}\|l\right)^2 \left(l\|C^{(k')}\|l\right)^2$$

$$\times \left[\frac{2\delta(k,k')}{2k+1} - \left\{\begin{matrix} l & l & k \\ l & l & k' \end{matrix}\right\} - \frac{1}{(2l+1)(4l+1)}\right]$$

$$\times F^k(l,l)F^{k'}(l,l), \tag{32.25}$$

$$\sigma^2_e(l_1^{N_1} l_2^{N_2}) = \frac{N_1(4l_1+2-N_1)N_2(4l_2+2-N_2)}{(4l_1+2)(4l_1+1)(4l_2+2)(4l_2+1)}$$

$$\times \left\{\sum_{k>0}\sum_{k'>0} \frac{4\delta(k,k')}{2k+1} \left(l_1\|C^{(k)}\|l_1\right)\left(l_1\|C^{(k')}\|l_1\right)\left(l_2\|C^{(k)}\|l_2\right)\right.$$

$$\times \left(l_2\|C^{(k')}\|l_2\right) F^k(l_1,l_2)F^{k'}(l_1,l_2)$$

$$+ \sum_{kk'} \left[\frac{4\delta(k,k')}{2k+1} - \frac{1}{(2l_1+1)(2l_2+1)}\right]$$

$$\times (-1)^{k+k'} \left(l_1\|C^{(k)}\|l_2\right)^2 \left(l_1\|C^{(k')}\|l_2\right)^2 G^k(l_1,l_2)G^{k'}(l_1,l_2)$$

$$-4\sum_{k>0}\sum_{k'} \left(l_1\|C^{(k)}\|l_1\right)\left(l_2\|C^{(k)}\|l_2\right)\left(l_1\|C^{(k')}\|l_2\right)^2$$

$$\times \left.\left\{\begin{matrix} l_1 & l_1 & k \\ l_2 & l_2 & k' \end{matrix}\right\} F^k(l_1,l_2)G^{k'}(l_1,l_2)\right\}. \tag{32.26}$$

The appropriate expressions for the higher moments are much more complicated and are not presented here. Some of them (μ_3 and μ_4) may be found in the monograph [296] or deduced making use of the general methods described there.

Let us discuss now some aspects of practical applications of this method again following [296]. Explicit expressions for the lowest moments of the spectrum may be employed not only for an approximate description of the distribution function without carrying out the detailed calculations of separate levels, but also for studies of general properties of spectra, for evaluation of coupling schemes and the contributions of various interactions to the main spectral characteristics, for estimation of the accuracy of the approximation used, etc. Let us illustrate these statements using the example of the spectra of atoms and ions with one open shell.

For configurations with one electron or hole above closed shells the moments of the spectrum are conditioned in the single-configuration approach by only one-electron spin–orbit interaction, and then the dimensionless quantities skewness κ_1 and excess κ_2 (formulas (32.7)) as well as σ^2/η_l^2 ($l \neq 0$) are functions only of orbital quantum number and, therefore, do not depend on the accuracy of radial orbitals, namely,

$$\kappa_1(l) = -\kappa_1(l^{4l+1}) = -\sqrt{l(l+1)}, \tag{32.27}$$

$$\kappa_2(l) = \kappa_2(l^{4l+1}) = -\frac{2l^2 + 2l - 1}{l(l+1)}, \tag{32.28}$$

$$\sigma^2(l)/\eta_l^2 = \sigma^2(l^{4l+1})/\eta_l^2 = \frac{1}{4}l(l+1). \tag{32.29}$$

Their numerical values have been calculated to high accuracy (4–6 figures) and agree with experimental data for various shells of heavy elements [303]. For smaller n the discrepancies increase due to neglected correlation effects. For particular l values the expressions for moments depend only on radial integrals. Thus, for p^2 we have

$$\mu_2(p^2) = \sigma^2(p^2) = 0.0288(F^2)^2 + 0.8\eta_p^2, \tag{32.30}$$

$$\mu_3(p^2) = 0.006912(F^2)^3 + 0.072F^2\eta_p^2 - 0.2\eta_p^3, \tag{32.31}$$

$$\mu_4(p^2) = 0.00373248(F^2)^4 + 0.12672(F^2)\eta_p^2$$
$$-0.144F^2\eta_p^3 + 1.5\eta_p^4. \tag{32.32}$$

The coefficients contain different numbers of figures, because they are accurate values. The relevant expressions for the p^4 shell follow from (32.30)–(32.32) if we change the signs of the coefficients η_p with odd powers. For the p^3 shell the similar expressions are

$$\mu_2(p^3) = \sigma^2(p^3) = 0.0432(F^2)^2 + 0.9\eta_p^2, \tag{32.33}$$

$$\mu_3(p^3) = -0.005184(F^2)^3 + 0.108F^2\eta_p^2, \tag{32.34}$$

$$\mu_4(p^3) = 0.00435456(F^2)^4 + 0.13824(F^2)^2\eta_p^2 + 2.025\eta_p^4. \quad (32.35)$$

It is interesting to notice that the numerical coefficients at F^2 rapidly decrease with growth of the power of this integral. For atoms with small Z values $\eta_p \ll F^2$, therefore the moments $\mu_k(l^N)$ are mainly conditioned by the non-spherical part of the electrostatic interaction. In the case of neutral atoms the spin–orbit contribution together with the mixed term exceeds the electrostatic contribution to the variance and excess of the atomic spectrum of the configuration np^N only for $n \geq 6$.

The variance may be adopted to evaluate the accuracy of the coupling scheme used. Indeed, its formula (32.21) consists of two terms, corresponding to spin–orbit and electrostatic interactions, therefore the quantity σ_{so}/σ_e may also serve as a measure to estimate the coupling scheme. It is preferred over the ratio $\eta_{nl}/F^2(nl, nl)$, because the coupling scheme depends not only on radial integrals, but also on spin–angular parts of them. Let us also mention that σ characterizes the width of the spectrum ΔE, i.e. the energy separation between upper and lower levels.

Asymmetry of the spectrum, defined by skewness κ_1 in (32.7), is positive, if the majority of levels are in the lower part of the spectrum, and negative, if the opposite is true. The excess κ_2 of the spectra of the majority of atoms with one open shell for $1 < N < 4l + 1$ is also positive; hence, the density of the levels, while approaching the average energy, grows more rapidly in comparison to the case of normal distribution of the levels. The results of the Hartree–Fock calculations of σ^2, κ_1 and κ_2 for isoelectronic sequences with an open nd^N shell indicate that σ grows almost linearly with Z, whereas κ_1 and κ_2 decrease with increase of Z, more rapidly for larger n.

32.3 Configuration mixing

Mean characteristics of spectra, being averaged quantities, must weakly depend on the errors in calculation of particular levels. However, perturbations of many levels, especially by energetically close-lying configurations, may essentially change these characteristics. In the case of strong mixing of configurations, the mean characteristics of the whole complex of configurations must be considered. The appropriate moments may also be defined by formula (32.13) under the assumption that now the configuration is

$$K = K_1 + K_2 + \ldots + K_p \quad (32.36)$$

and the summation is performed over the states of the whole complex, whereas $g(K)$ and $\bar{E}(K)$ correspond to the statistical weight and the average energy of that complex

$$g(K) = \sum_{i=1}^{p} g(K_i), \qquad \bar{E}(K) = \frac{1}{g(K)} \sum_{i=1}^{p} g(K_i)\bar{E}(K_i). \tag{32.37}$$

The total moment of a complex of configurations may be presented as a sum of two terms: $\mu_k^0(K)$ representing the moment of the total spectrum in single-configuration approximation, and the second, $\Delta\mu_k^{sc}(K)$, describing the correction due to superposition of configurations, i.e.

$$\mu_k(K) = \mu_k^0(K) + \Delta\mu_k^{sc}(K). \tag{32.38}$$

For estimation of the degree of configuration mixing as well as for evaluation of the configurations, which must be accounted for first, the special mean characteristic – the strength of configuration mixing – can be adopted. It is defined as follows:

$$T(K, K') = \sum_{\gamma\gamma'} (K\gamma|H|K'\gamma') / \sigma^2(K, K'), \tag{32.39}$$

where

$$\sigma^2(K, K') = \sum_{\gamma\gamma'} [(K\gamma|H|K\gamma) - (K'\gamma'|H|K'\gamma')]^2$$
$$\times (K\gamma|H|K'\gamma')^2 / \sum_{\gamma\gamma'} (K\gamma|H|K'\gamma')^2 \tag{32.40}$$

is the variance of distances between mixing levels, with accounting for the quantities of matrix elements binding these levels. The levels of two configurations are more mixed for larger appropriate interconfigurational matrix elements and for energetically closer interacting levels.

The effect of mixing of a given configuration with many energetically distant configurations may be accounted for as corrections to the energy levels in second-order perturbation theory

$$\Delta E(K\gamma) = \sum_{K' \neq K} \sum_{\gamma'} (K\gamma|H|K'\gamma')^2 / [E(K\gamma) - E(K'\gamma')]. \tag{32.41}$$

If $E(K\gamma) - E(K'\gamma')$ is much larger than the width $\Delta E(K')$ of the spectrum of configuration K', then the correction to the energy of the level $K\gamma$ due to its interaction with the levels of configuration K' is equal to

$$\Delta E(K\gamma, K') = \bar{E}(K\gamma, K')^{-1} (K\gamma|HH|K\gamma), \tag{32.42}$$

where HH is the product of two Hamiltonians H in the second-quantized form. The main shortcoming of this method, where corrections are determined in the basis of single-configuration orbitals, is its slow convergence with increase in the number of admixed configurations K'. It may be improved by using transformed radial orbitals of the type (29.6).

Expressions for a number of main moments of the spectrum may be utilized to develop a new version of the semi-empirical method. Evaluation of the statistical characteristics of spectra with the help of their moments is also useful for studying various statistical peculiarities of the distribution of atomic levels, deviations from normal distribution law, etc. Such a statistical approach is also efficient when considering separate groups of levels in a spectrum (e.g. averaging the energy levels with respect to all quantum numbers but spin), when studying natural widths or lifetimes of excited levels, etc.

32.4 Statistical characteristics of radiation spectra

Global (statistical) characteristics of the radiation spectrum are also represented by its initial (32.1) or centred (32.5) moments. If the energy of each transition between two configurations K and K' obeys the condition (32.15), then the kth centred moment of the radiation spectrum, corresponding to the electronic transitions between these configurations, is defined by (32.16), where the average energy of the spectrum is given by

$$\bar{E}(K - K') = \left\{ \sum_{\gamma\gamma'} [(K\gamma|H|K\gamma) - (K'\gamma'|H|K'\gamma')] \, S(K\gamma, K'\gamma') \right\} \Big/ \sum_{\gamma\gamma'} S(K\gamma, K'\gamma'). \quad (32.43)$$

It is convenient to employ the shift of mean energy instead of the average energy of the radiation spectrum (32.43), excluding from $\bar{E}(K - K')$ the term equal to the difference of two large numbers (energies of final and initial configurations), i.e.

$$\delta\bar{E}(K - K') = \bar{E}(K - K') - [\bar{E}(K) - \bar{E}(K')]. \quad (32.44)$$

Hamiltonian H in the non-relativistic single-configuration approximation consists of the operators of electrostatic and spin–orbit interactions, however, the contribution of the latter to the shift vanishes. If in the radiation spectrum there are clearly pronounced groups of lines, then it is convenient to consider the statistical characteristics not of the whole spectrum, but of these particular groups. Mean energy and its shift for a group of transitions are defined by the same general formulas (32.43) and (32.44), but here K and K' already stand for the subconfigurations producing the appropriate transition array. Concrete expressions for the main moments of radiation spectra may be found in [302, 304]. They are useful for studying general regularities in radiation spectra as well as for approximate description and interpretation of complex spectra.

Various levels of initial and final configurations take different parts in radiative transitions. Therefore, it is convenient to introduce the concepts of emissive and receptive zones, which characterize the participation of the configurations in particular transitions [298]. The zones as the weighted distributions of the level energies (with the weight of each level equal to the total strength of all lines originating from it) may be expressed in terms of their moments. The initial moments of emissive (*em*) and receptive (*rec*) zones are defined, respectively, as

$$\alpha_k^{em}(K) = \sum_{\gamma\gamma'} (K\gamma|H|K\gamma)^k \, S(K\gamma,K'\gamma')/S(K,K'), \qquad (32.45)$$

$$\alpha_k^{rec}(K) = \sum_{\gamma\gamma'} (K'\gamma'|H|K'\gamma')^k \, S(K\gamma,K'\gamma')/S(K,K'). \qquad (32.46)$$

The moments of zones can be expressed in terms of the averages of products of energy and electron transition operators. Formulas of the kind (32.45) and (32.46) can also be used when K and K' represent a number of mixing configurations.

Having explicit formulas for a number of first moments we can approximately restore the envelope line of the radiation spectrum without its detailed calculations. If lines in the spectrum have one symmetric maximum, then its envelope line is approximated by a normal function whose reconstruction requires only the mean energy and variance of the spectrum. Such an approach is useful for the case of complex spectra consisting of many lines, which, due to low resolutions as well as Doppler and collisional broadening or large natural width, form continuous or quasi-continuous bands. Studies of variation of these statistical characteristics along isoelectronic sequences give a wealth of information on intra-atomic interactions.

Calculated values of the average energy shift, mean deviation and skewness agree qualitatively with the relevant quantities of more complete experimental spectra for a number of isoelectronic or isoionic sequences [302]. The shift and skewness are more sensitive to the quality of the initial data. Probably, the dispersion of global characteristic values of experimental spectra can serve as an indicator of their incompleteness. Analysis of contributions of various interactions to global characteristics reveals the role of various interactions in the formation of the main features of a spectrum. In particular, analysis shows that the narrowing of the emission spectrum is mainly conditioned by the interactions within open shells; the interaction between open shells contributes to this effect only when configurations with two energetically separated groups of levels are involved. The narrowing of spectrum becomes more pronounced for complex spectra, corresponding to transitions between configurations with several open shells.

If the spectral lines are broadened but have the same profile, the moments of the spectrum can be expressed in terms of the moments of the non-broadened spectrum and the moments of a separate line. When the moments of the separate line do not exist (for natural line shape), the global characteristics of the spectrum can be found simulating its structure by Monte-Carlo or distribution function methods and then convoluting each calculated line.

33
Peculiarities of configurations with vacancies in inner shells

33.1   Autoionization

Methods of theoretical studies of energy spectra and electron transitions, described in previous chapters, are also applicable to configurations with vacancies in inner shells. However, such types of configurations possess a number of peculiarities, mainly connected with problems of ensuring the orthogonality of the appropriate wave functions, with the changes of relative role played by relativistic and correlation effects. These peculiarities are manifest in the energy and radiation spectra of such systems. As a rule we shall only touch the main aspects of this problem. More detailed information may be found in the monograph by Karazija [305]. From a physical point of view, processes involving inner electronic shells of free atoms and ions lead to X-ray and electronic (photoelectron and Auger) spectra. Appropriate theory may be used to interpret the spectra of gases and vapours as well as such spectra of solid state, containing atoms with open d- and f-shells and corresponding to transitions in inner shells. Comparison of such spectra of free atoms with the spectra of the same atoms in molecules or solid states allows one to elucidate the effects of free atoms and chemical surroundings, e.g. of the crystalline field.

The single-configuration approach remains the main method used, allowing one to describe not only qualitatively, but in many cases quantitatively, many features of various such spectra and the regularities in them. However, having in mind that relativistic effects grow very rapidly with increase of nuclear charge and with shrinking electronic shells, accounting for relativistic effects for this kind of phenomena is of paramount importance. Again, they may be taken into consideration as corrections in the Hartree–Fock–Pauli approximation or, starting with a relativistic approach, in the Dirac–Hartree–Fock approximation. To achieve better quantitative agreement we have to combine accounting for both relativistic and correlation effects. For highly excited autoionizing states the correlation effects caused by the interactions of these states with the continuous spectrum are also significant. Therefore, the formula (29.1) describing the atomic wave function as the superposition of configurations, generally speaking, must contain integration over continuous spectrum, i.e.

$$\varphi_E = a\psi + \int b_{E'}\psi_{E'}dE', \qquad (33.1)$$

where ψ is defined by (29.1), a and $b_{E'}$ are the expansion coefficients, depending also on E. For ground as well as for not highly excited configurations the influence of continuum usually is small and therefore it may be neglected. That is why the approach (29.1) is fairly accurate for energy spectra and electron transition quantities of a large variety of electronic configurations of complex atoms and ions. However, it is not true for autoionizing states lying in the continuum with respect to the other ionization limit and corresponding to the so-called quasistationary states of an atom. For example, level $1s2s^22p^6\ ^2S$ belongs to the discrete spectrum of this configuration, but it lies in the continua of $1s^22s^02p^6\epsilon l\ ^2S$, $1s^22s2p^5\epsilon l\ ^2S$ and $1s^22s^22p^4\epsilon l\ ^2S$.

The levels of configurations with two excited electrons are autoionizing too (e.g. $2s2p\ ^1P$ of He is in continuum of $1s\epsilon p\ ^1P$), as well as configurations with a vacancy in an inner shell. As a consequence of mixing states of discrete and continuous spectra, the possibility of radiationless decay of excited states occurs. Such a process is called autoionization, or for decay of the state with an inner vacancy, Auger transition. The degree of the above mentioned mixing is determined by the value of interconfigurational matrix element of the energy operator H, and practically, by electrostatic interaction between electrons. Hence, the parity and total angular momentum conservation laws are valid for the autoionization process. The probability W_a of radiationless decay of an autoionizing state (per unit of time) is defined by the formula

$$W_a = \frac{2\pi}{\hbar}\ |(\varphi_E|H|\psi)|^2. \qquad (33.2)$$

Generally, when studying autoionizing levels, we have to take into consideration both (radiative and radiationless) channels of their decay. The total natural width of the autoionizing level will be the sum of its autoionizing and radiative widths.

Configurations with vacancies in inner shells possess a number of peculiarities: their states are autoionizing; they are short-lived; relativistic effects are essential for them; their energy spectrum has particular characteristics. Short lifetimes of excited states lead to large widths of relevant spectral lines.

In X-ray and electron spectroscopy the sets of shells with given $n = 1, 2, 3, 4, 5, 6, \ldots$ are often denoted by capital letters $K, L, M, N, O, P, \ldots$, whereas the subshells with given nlj are numbered as

$$
\begin{array}{ccccccccc}
1s_{1/2} & 2s_{1/2} & 2p_{1/2} & 2p_{3/2} & 3s_{1/2} & 3p_{1/2} & 3p_{3/2} & 3d_{3/2} & 3d_{5/2} & \ldots \\
K & L_1 & L_2 & L_3 & M_1 & M_2 & M_3 & M_4 & M_5 & \ldots
\end{array}
$$
(33.3)

A N-electron vacancy (hole) in a shell may be denoted as $nl^{-N} \equiv nl^{4l+2-N}$ (see also Chapters 9, 13 and 16). As we have seen in the second-quantization representation, symmetry between electrons and vacancies has deep meaning. Indeed, the electron annihilation operator at the same time is the vacancy creation operator and vice versa; instead of particle representation hole (quasiparticle) representation may be used, etc. It is interesting to notice that the shift of energy of an electron A due to creation of a vacancy B^{-1} is approximately (usually with the accuracy of a few per cent) equal to the shift of the energy of an electron B due to creation of a vacancy A^{-1}, i.e.

$$
\Delta E_A(B^{-1}) \approx \Delta E_B(A^{-1}). \tag{33.4}
$$

Unfortunately, vacancies cannot be treated as quasiparticles for the cases of strong configuration mixing, including mixing with continua.

33.2   Production of excited states in atoms

If the time period for creation of excited states and relaxation of electronic shells is much less than the lifetime of an excited state and if interference effects between various processes binding the same initial and final states may be neglected, then the creation and decay of excited states may be treated separately as a two-step process. Such an approach is widely used in X-ray and electron spectroscopy.

Excited states, the decay of which produces X-ray and Auger spectra, are created by the interaction of atoms with X-rays, electrons, ions as well as the result of nuclear processes. An important characteristic of an atom is the population of its excited state, defined as the number of atoms in a given state per unit volume. Usually relative populations are employed (e.g. with respect to population of ground state). If total population is normalized to unity then we have the probabilities for atoms to occupy various states. X-ray and Auger spectra normally are registered in a stationary regime, where the system is in dynamic equilibrium, i.e. when the number of atoms ΔN_i^{in}, occupying per unit time the state i, equals the number of atoms ΔN_i^{out}, leaving it, namely,

$$
\Delta N_i^{in} = \Delta N_i^{out}. \tag{33.5}
$$

Population of the excited level αJ is

$$N(\alpha J) = \sum_{\alpha' J'} N(\alpha' J') \sum_t W^t(\alpha' J' \to \alpha J)/W(\alpha J)$$

$$= \sum_{\alpha' J'} N(\alpha' J') \sum_t K^t(\alpha' J' \to \alpha J), \tag{33.6}$$

where $N(\alpha' J')$ is the population of level $\alpha' J'$, $W^t(\alpha' J' \to \alpha J)$ refers to the probability of the transition from $\alpha' J'$ to αJ via process t, $W(\alpha J)$ defines the decay rate (in atomic units it is equal to total width of level αJ) and $K^t(\alpha' J' \to \alpha J)$ is called the branching ratio. The summation in (33.6) is performed over all processes t and initial states $\alpha' J'$. Populations of excited levels are conditioned by the processes of excitation and ionization of atoms by particles as well as by radiative and radiationless transitions during spontaneous de-excitation of atoms. In order to estimate them, one needs the cross-section and probabilities of occurrence of various processes, including many radiative and Auger transitions, which are in charge of natural widths of relevant lines. Usually the excitation and ionization of an atom by electrons is considered in a non-relativistic approximation. In ion–atom collisions the electronic shells are strongly perturbed, producing vacancies in various shells as a consequence of multistep ionization and excitation of an atom as well as electron capture by ion. A theoretical description of these processes is very difficult and rather approximate to date. The so-called 'sudden-perturbation approximation' of an atom may be fairly successfully utilized for calculations of many-electron transition probabilities while interpreting X-ray and electronic spectra [306].

A new R-matrix approach for calculating cross-sections and rate coefficients for electron-impact excitation of complex atoms and ions is reviewed in [307]. It is found that accurate electron scattering calculations involving complex targets, such as the astrophysically important low ionization stages of iron-peak elements, are possible within this method.

33.3 X-ray absorption spectra

Absorption of X-rays of the energies corresponding to binding energies of electrons in atoms by electronic shells as a rule leads to photoexcitation of an atom and to the photoeffect. The microscopic quantity describing absorption of X-rays is the so-called effective absorption cross-section σ_ω, characterizing the absorption of X-rays of frequency ω by a single atom, and is defined as

$$\sigma_\omega = n_\omega/j_\omega, \tag{33.7}$$

where n_ω refers to the number of quanta absorbed by an atom per unit time and j_ω is the density of a beam of incident photons.

The dependence of photoabsorption cross-section on many-electron quantum numbers (sets of quantum numbers of a chain of electronic shells) is mainly determined by the submatrix element of the transition operator. Their non-relativistic and relativistic expressions for the most widely considered configurations are presented in Part 6. When exciting an atom by X-rays the main type of transitions are as follows:

$$n_1 l_1^{4l_1+2} n_2 l_2^{N_2} \rightarrow n_1 l_1^{4l_1+1} n_2 l_2^{N_2+1}. \tag{33.8}$$

The highest probabilities are for transitions between configurations with $n_1 = n_2$ and $l_2 = l_1 + 1$. In the final state the coupling, close to LS, holds for neighbouring shells; then the matrix element of electric multipole transition is defined by formulas (25.28), (25.30). Similar expressions for other coupling schemes may be easily found starting with the data of Part 6 and Chapter 12.

In non-relativistic dipole approximation the photoionization of closed electronic shell is described by the formula

$$\sigma_{nlj}^{ion}(\omega) = \sigma^{ion}\left(nl^{4l+2} \rightarrow nl^{4l+1} j\epsilon\right)$$

$$= \frac{4}{3}\pi^2 \alpha \frac{2j+1}{2l+1} \hbar\omega \left[l\left(nl|r|\epsilon l-1\right)^2 + (l+1)\left(nl|r|\epsilon l+1\right)^2\right]. \tag{33.9}$$

Here $\hbar\omega = I_{nlj} + \epsilon$, where ϵ refers to photoelectron energy and I_{nlj} is the binding energy of the nlj-electron in the atom. Performing the summation in (33.9) over j and neglecting the dependence of ϵ on j, we arrive at the following expression for total photoionization cross-section of the closed shell:

$$\sigma_{nl}^{ion}(\omega) = \sigma^{ion}(nl^{4l+2} \rightarrow nl^{4l+1}\epsilon) = \sum_j \sigma_{nlj}^{ion}(\omega)$$

$$\approx \frac{8}{3}\pi^2 \alpha \hbar\omega \left[l\left(nl|r|\epsilon l-1\right)^2 + (l+1)\left(nl|r|\epsilon l+1\right)^2\right]. \tag{33.10}$$

Such expressions can be easily generalized to cover the case of the electric multipole transition operator with an unspecified value of the gauge condition K of electromagnetic field potential (4.10) or (4.11).

Calculations show that cross-sections obtained in the Hartree–Fock approximation utilizing length and velocity forms of the appropriate operator, may essentially differ from each other for transitions between neighbouring outer shells, particularly with the same n. However, they are usually close to each other in the case of photoionization or excitation from an inner shell whose wave function is almost orthogonal with the relevant function of the outer open shell. In dipole approximation an electron from a shell l^N may be excited to $l' = l \pm 1$, but the channel $l \rightarrow l + 1$ prevails. For configurations $n_1 l_1^{N_1} n_2 l_2^{N_2}$ an important role is played by exchange electrostatic interaction (for $n_1 = n_2$ it becomes even decisive).

Unfortunately, the single-configuration Hartree–Fock approach quite often fails to explain even the main qualitative features of photo-absorption spectra. Only improved methods, which take into consideration correlations not only between atomic electrons but also with the photoelectron, allow one to achieve fairly good agreement of theoretical results with experimental data.

Correlation effects in the initial state of an atom as well as in the final state of an ion (photoionization) or an atom (excitation) may be effectively accounted for in the multi-configuration approximation. It is necessary to emphasize that the admixed configurations, which are connected with normal or admixed configuration of the other state by fairly large matrix elements of the transition operator, essentially contribute to the cross-section, although they may be insignificant to the transition energy. The method of strong coupling represents another way of mixing the different final states of a system consisting of an ion and an electron.

A number of versions of perturbation theory have been widely used to study photoabsorption spectra. Their main advantage is the possibility of accounting for correlation effects in initial and final states without dealing directly with improved wave functions, as well as convenience in dealing with diagrams. Unfortunately, there are also many disadvantages of this approach, mainly connected with the extremely rapid growth of computer time with increase in the number of terms taken into account, as well as slow convergence and problems with estimating the contributions of various diagrams. However, the success of the version of perturbation theory described in Chapter 3 for calculating energy levels and transitions between discrete states (Part 7), gives the hope that such a version will also be very efficient for studies of photoabsorption processes.

33.4 Photoelectron spectra

These may be generated by irradiating an atom with a beam of monochromatic X-rays or ultraviolet rays. X-ray and electron spectroscopy is one of the main methods used for studying the structure of atomic electronic shells, particularly inner ones, as well as the role of relativistic and correlation effects. A wealth of such information may also be obtained from the studies of angular distribution of photoelectrons. It is interesting to notice that with increase of the energy of X-rays the dipole approximation fails to correctly describe the angular distribution of electrons.

Recording the electrons emitted from an atom in a certain direction during a photoionization process allows one to obtain two main types of spectra: (a) distribution of photoelectrons with respect to their energies as a result of the action of a monochromatic beam of X- or ultraviolet rays of a given energy; (b) distribution of intensities of photoelectrons while ionizing a given shell as a function of photon energy. If in the initial state only the ground level of an atom is populated, then the relative intensities of the lines of the photoelectron spectrum are proportional to the cross-section of photoionization from this level to various final states, summed over the total momentum of the system ion plus photoelectron and its projection, as well as over photoelectron quantum numbers. Photoelectron energy ϵ, according to the energy conservation law, is given by

$$\epsilon = \hbar\omega - \left[E^i(\gamma' J') - E^a(\gamma J) \right], \qquad (33.11)$$

where E^i and E^a are the energies of an ion and atom, respectively, whereas ω refers to the incident radiation energy.

Correlations between electrons in initial and final states lead not only to the shift of photoelectron lines, changes of their forms and redistribution of their intensities, but also to the occurrence of so-called satellite lines, corresponding to photoionization with the excitation of the other electron. Correlation effects in photoelectron spectra are caused by mixing of configurations separately in the initial state, in the final ion and in the final state of continuum.

If an atom, before interacting with a photon, was in its ground state whereas an ion was created in the state with the lowest (for given vacancy ζ^{-1}) energy, then the quantity $E^i(\zeta^{-1}) - E^a$ in (33.11) is an electron binding energy I_ζ in the atom. Then (33.11) may be rewritten as

$$\epsilon = \hbar\omega - I_\zeta. \qquad (33.12)$$

X-ray and electron spectroscopy is a direct and very accurate method of determination of binding energies of inner electrons in free atoms, whereas emission and Auger spectra allow one to find only differences of binding energies of different electrons. Sometimes the equivalent notion of ionization energy of a shell or subshell is used instead of binding energy. Ionization energy is the energy required to remove one electron with zero kinetic energy from an atom or ion. Correlation effects are of relatively little importance for the energies of inner electrons [305].

33.5   Characteristic X-ray spectra

These occur as a result of spontaneous radiation of atoms with vacancies in inner electronic shells. Their nature reflects the structure of energy levels of excited atoms. As we have seen in Chapter 31, the radiation of highly ionized atoms is in the X-ray region, but their spectra are similar to optical ones. X-ray characteristic spectra of atoms with one inner vacancy consist of so-called diagram lines. The characteristic radiation spectrum of a given element is a set of lines of all possible X-ray transitions.

Lines, corresponding to different transitions from initial states with vacancy in the shells with the same n, compose a series of spectra, e.g. K-, L-, M-series etc. Main diagram lines correspond to electric dipole ($E1$) transitions between shells with different n. The lines of $E2$-transitions also belong to diagram lines. Selection rules of $E1$-radiation as well as the one-particle character of the energy levels of atoms with closed shells and one inner vacancy cause, as a rule, a doublet nature of the spectra, similar to optical spectra of alkaline elements. X-ray spectra are even simpler than optical spectra because their series consist of small numbers of lines, smaller than the number of shells in an atom. The main lines of the X-ray radiation spectrum, corresponding to transitions in inner shells, preserve their character also for the case of an atom with open outer shells, because the outer shells hardly influence the properties of inner shells.

Levels with the same n, l and $j = l \pm 1/2$ are called spin-doublet levels. Two lines of radiation spectra are also called spin-doublet if they correspond to one and the same initial or final level, whereas the second level of the transition considered represents a spin-doublet (e.g. $K_{\alpha 1}$ and $K_{\alpha 2}$ lines).

Radiation spectrum lines, which cannot be explained by the model of one vacancy, historically were also called satellite lines. There are cases, particularly in the long-wavelengths region, when the intensities of satellite lines become comparable to those of diagram lines, or even exceed them. Then the division of X-ray radiation into diagram and satellite lines becomes senseless.

It is interesting to underline that in the X-ray radiation region the ratio W_{E2}/W_{E1} increases by a few orders of magnitude compared to that in the optical region, but still remains $\ll 1$.

The presence of an outer open shell in an atom, even if this shell does not participate in the transitions under consideration, influences the X-ray radiation spectrum. Interaction of the vacancy with the open shell, particularly in the final state when the vacancy is not in a deep shell, splits the levels of the core. Depending on level widths and relative strength of various intra-atomic interactions, this multiplet splitting leads to broadening of diagram lines, their asymmetry, the occurrence of satellites, or splitting of the spectrum into large numbers of lines.

Radiative transitions with the participation of electrons from deep inner shells depend essentially on relativistic effects. This requires us to utilize relativistic operators and wave functions. Correlation effects are also significant but to a lesser extent. Superposition of configurations allows, for example, the so-called two-electron one-photon transitions, leading to the occurrence of groups of intense satellite lines. The complex satellite structure of the X-ray spectra of free atoms, even noble gases with all closed shells in the ground state, is conditioned not only by interelectronic correlations, but also by various radiationless and radiative processes, influencing the populations of excited levels. Thus, accurate mathematical modelling of the spectra requires consideration of the whole cascade of processes taking place in the presence of the interaction of an atom with an incident particle and of the subsequent rearrangement of electronic shells of the atom.

33.6 Auger spectra

These are produced by autoionization transitions from highly excited atoms with an inner vacancy. In many cases it is the main process of spontaneous de-excitation of atoms with a vacancy. Let us recall that the wave function of the autoionizing state (33.1) is the superposition of wave functions of discrete and continuous spectra. Mixing of discrete state with continuum is conditioned by the matrix element of the Hamiltonian (actually, of electrostatic interaction between electrons) with respect to these functions. One electron fills in the vacancy, whereas the energy (in the form of a virtual photon) of its transition is transferred by the above mentioned interaction to the other electron, which leaves the atom as a free Auger electron. Its energy ϵ_A equals the difference in the energies of the ion in initial and final states:

$$\epsilon_A = E(a^{-1}) - E(b^{-1}c^{-1}), \tag{33.13}$$

where a^{-1}, b^{-1} and c^{-1} are any vacancies, for which the Auger transition is energetically permitted, i.e. $\epsilon_A > 0$.

If there is a quasidiscrete state γ of an atom interacting with the continuum in which this state is, then the probability of its radiationless decay per unit time into an ion in the state γ' and a free electron with energy ϵ equals

$$W_A(\gamma \rightarrow \gamma'\epsilon\gamma'') = \frac{2\pi}{\hbar} |(\gamma'\epsilon\gamma''|H|\gamma)|^2, \tag{33.14}$$

where γ'' refers to the remaining quantum numbers of an electron and the whole system, whereas $\epsilon = E(\gamma) - E(\gamma')$. In practical calculations usually only the Coulomb interaction is accounted for in Hamiltonian H. Relativistic effects are important for Auger transitions, particularly for those involving deep inner shells [305].

Correlation effects in Auger spectra may be classified similarly to those in the case of photoelectron spectra. Strong correlation effects occur in

Auger spectra, involving electrons of outer open shells. Specific for Auger spectra is the interaction of the electronic configuration with two vacancies in final state of an ion, first of all, with quasidegenerated configurations, particularly if they are energetically close.

Configuration mixing may open new additional channels for Auger decay of an excited state, which must be taken into account when choosing the admixed configurations. For Auger transitions in inner shells the correlation and relativistic effects mix together. They must be accounted for simultaneously in the framework of the relativistic multiconfiguration approach or perturbation theory. More general is a united description of creation and decay of excited states, which includes both Auger and radiative decay channels.

33.7 Width and shape of spectral lines. Fluorescence yield

These features of lines of various spectra (X-ray, emission, photoelectron, Auger) are determined by the same reason, therefore they are discussed together. Let us briefly consider various factors of line broadening, as well as the dependence of natural line width and fluorescence yield, characterizing the relative role of radiative and Auger decay of a state with vacancy, on nuclear charge, and on one- and many-electron quantum numbers.

Population $N_k(t)$ of excited level k decreases with time, obeying an exponential law

$$N_k(t) = N_k(0)e^{-\sum_i W_{ki}t}, \qquad (33.15)$$

where $N_k(0)$ is the initial population of the level, W_{ki} refers to total probability of transitions to level i. Lifetime τ_k of excited level k is the time during which its population decreases by e times, i.e.

$$\tau_k = 1/\sum_i W_{ki}. \qquad (33.16)$$

Linewidth Γ_k is expressed in terms of W_{ki}, namely,

$$\Gamma_k = \hbar \sum_i W_{ki}. \qquad (33.17)$$

The width of the spectral line equals the sum of the widths of initial and final levels. Due to the short lifetime of highly excited states with an inner vacancy, their widths, conditioned by spontaneous transitions, are very broad. The other reasons for broadening of X-ray and electronic lines (apparatus distortions, Doppler and collisional broadenings) usually lead to small corrections to natural linewidth.

The lifetime of a separate atom in its ground state is infinite, therefore the natural width of the ground level equals zero. Typical lifetimes of excited states with an inner vacancy are of the order $10^{-14} - 10^{-16}$ s, giving a natural width $0.1 - 10$ eV. The closer the vacancy is to the nucleus, the more possibilities there are to occupy this vacancy and then the broader the level becomes. That is why $\Gamma_K > \Gamma_L > \Gamma_M$. Generally, the total linewidth Γ is the sum of radiative (Γ_r) and Auger (Γ_A) widths, i.e.

$$\Gamma = \Gamma_r + \Gamma_A, \tag{33.18}$$

where both Γ_r and Γ_A are expressed in terms of the sums of probabilities of relevant transitions (33.17).

Fluorescence yield ω_f is a quantity characterizing the ability of an atom with an inner vacancy to de-excite radiatively:

$$\omega_f = W_r/W = W_r/(W_r + W_A), \tag{33.19}$$

where W_r and W_A are total probabilities for spontaneous decay of an excited state by means of radiative or Auger transitions, respectively. Total probabilities do not depend on projections of momenta, therefore, fluorescence yields coincide for state and level. It follows from (33.17) and (33.19) that ω_f may be expressed also in terms of radiative Γ_r and Auger Γ_A linewidths, namely,

$$\omega_f = \Gamma_r/(\Gamma_r + \Gamma_A). \tag{33.20}$$

The yield of Auger electrons may also be used

$$\omega_A = \Gamma_A/(\Gamma_r + \Gamma_A), \tag{33.21}$$

with the condition

$$\omega_f + \omega_A = 1. \tag{33.22}$$

It is useful to know the dependence of natural linewidths and fluorescence yields on atomic characteristics. Considering Z-dependencies of the appropriate matrix elements, we can show that the radiative linewidth is proportional

$$\Gamma_r \sim (Z - \delta)^4, \tag{33.23}$$

where δ equals approximately the screening constant for the initial state of the jumping electron. Auger width of the level practically does not depend on Z, i.e.

$$\Gamma_A \approx \text{const.} \tag{33.24}$$

Fluorescence yield for $Z \gg \delta$ may be approximated by the formula

$$\omega = Z^4/(A + Z^4), \tag{33.25}$$

where A is a constant, determined from experimental data. For $Z \rightarrow \infty$, ω approaches 1.

Effective nuclear charge grows with increase of ionization and excitation of an atom, therefore, the fluorescence yield in ions and, to a less extent, in excited atoms, tends to increase. Radiative and Auger linewidths, as well as fluorescence yields, depend on relativistic and correlation effects to less extent in comparison with the probabilities of separate transitions, because in sums the individual corrections partly compensate each other. Therefore, calculations of radiative widths in the Hartree–Fock–Pauli approach lead to reasonably accurate results.

In conclusion, there is a large variety of processes in atoms connected with the vacancies in inner electronic shells. New fairly accurate measurements of various characteristics of such objects and processes continue to occur, e.g. [308–314]. Extensive bibliographies, data collections and centres, review articles and books on this subject matter are presented in [315–318]. Their theoretical study is a challenge worthy of the efforts of many theoreticians.

Epilogue

It took a very long time to write this book, especially to bring it to a relatively consistent and complete form. The journey of the reader to these final pages was also not easy and straightforward. What are your feelings after getting through the jungle of more than 1300 many-storeyed formulas? Perhaps, twofold. At first – relief and satisfaction: it is all over now, I made it! But secondly – are all these formulas correct? The answer is not so simple. I tried to do my utmost to be able to answer 'yes': compared with the original papers, deduced some of them again, checked numerically, looked for special cases, symmetry properties, etc. But I cannot assert that absolutely all signs, phases, indices, etc. are correct. Therefore, if you intend to do some serious research starting with one or other formula from the book, it is worthwhile carrying out additional checks, making use of one of the above mentioned methods.

Not all aspects of the theory are dealt with in equal depth. Some are just mentioned, some even omitted. For example, the method of effective (equivalent) operators deserves mentioning [321]. It allows one to take into account the main part of relativistic effects but at the same time to preserve the LS coupling used for classification of the energy spectra of the atoms or ions considered.

The symmetry of a number of atomic quantities (wave functions, matrix elements, $3nj$-coefficients etc.) with respect to certain substitution groups or simply substitutions like $l \to -l-1$, $L \to -L-1$, $S \to -S-1$, $j \to -j-1$, $N \to 4l+2-N$, $v \to 4l+4-v$ leads to new expressions or helps to check already existing formulas or algebraic tables [322, 323]. Some expressions are invariant under such transformations. For example, Eq. (5.40) is invariant with respect to substitutions $S \to -S-1$ and $v \to 4l+4-v$. Clebsch–Gordan coefficients in Table 7.2 are invariant under transformation $j \to -j-1$. However, applying this substitution to the coefficient in Table 7.2, we obtain the algebraic value of the other coefficient.

Atomic spectroscopy continues to be a rapidly expanding field of physics. Several large-scale sophisticated atomic structure calculations have recently been undertaken, and they have yielded very extensive data sets that are expected to be of good quality [324]. Applications of these new databases to various domains of physics and neighbouring branches of science and technology, astrophysical observations and modelling included, hold great promise. A number of new interesting results have been obtained when studying the interactions of atoms and ions with photons, electrons and solids [325]. Various atoms and their ionization degrees are of interest for fusion problems [326].

The physics of highly charged ions [327–329] is developing rapidly. Let us mention only the creation of longer and longer living single 'hollow' atoms [330], and the first observation of magnetic octupole decay corresponding to $n = 3 - 4$ spectra of nickel-like Th^{62+} and U^{64+} [331].

The European Physical Society has emphasized the increasing role of physics and its education, and has also expressed concern about public understanding of the role of physics [332].

What's next? Is this the end? Of course, not! It may be the beginning! The beginning of a new approach. This book certainly is not the last one on atomic spectroscopy. Some things in it already seem slightly obsolete. Indeed, it is possible to formulate a more general method to find expressions for non-relativistic and relativistic matrix elements of any one- and two-electron operator for an arbitrary number of shells in an atomic configuration, requiring neither coefficients of fractional parentage nor unit tensors. It is based on combination of the second-quantization method in the coupled tensorial form, angular momentum theory in three spaces (orbital, spin and quasispin) and a generalized graphical technique [333]. All matrix elements (diagonal and non-diagonal with respect to configurations) will differ then only by the values of the projections of the quasispin momenta of separate shells and will be expressed in terms of completely reduced (in all three spaces) matrix elements of the following second-quantization operator and its irreducible tensorial products:

$$a_{m_q}^{(q\lambda)},$$

$$\left[a_{m_{q1}}^{(q\lambda)} \times a_{m_{q2}}^{(q\lambda)} \right]^{(\Gamma_1)},$$

$$\left[a_{m_{q1}}^{(q\lambda)} \times \left[a_{m_{q2}}^{(q\lambda)} \times a_{m_{q3}}^{(q\lambda)} \right]^{(\Gamma_2)} \right]^{(\Gamma_3)},$$

$$\left[\left[a_{m_{q1}}^{(q\lambda)} \times a_{m_{q2}}^{(q\lambda)}\right]^{(\Gamma_1)} \times a_{m_{q3}}^{(q\lambda)}\right]^{(\Gamma_4)},$$

$$\left[\left[a_{m_{q1}}^{(q\lambda)} \times a_{m_{q2}}^{(q\lambda)}\right]^{(\Gamma_1)} \times \left[a_{m_{q3}}^{(q\lambda)} \times a_{m_{q4}}^{(q\lambda)}\right]^{(\Gamma_5)}\right]^{(\Gamma_6)},$$

where λ and Γ_i ($i = 1 - 6$) represent ls and L_iS_i (in LS coupling) or j and J_i (in jj coupling), respectively. Thus, both diagonal and non-diagonal matrix elements are treated uniformly in this methodology. Who will write a monograph on theoretical atomic spectroscopy based on such an approach?

References

[1] N. Bohr. *The Theory of Spectra and Atomic Calculation*, Cambridge University Press, Cambridge, 1922.

[2] E. U. Condon and G. H. Shortley. *The Theory of Atomic Spectra*, Cambridge University Press, Cambridge, 1935.

[3] A. R. Edmonds. *Angular Momentum in Quantum Mechanics*, Princeton University Press, Princeton, 1957.

[4] M. E. Rose. *Elementary Theory of Angular Momentum*, John Wiley and Sons, New York, 1957.

[5] U. Fano and G. Racah. *Irreducible Tensorial Sets*, Academic Press, New York, 1959.

[6] J. C. Slater. *Quantum Theory of Atomic Structure* (Vol. I and II), McGraw-Hill, New York, 1960.

[7] H. A. Bethe and E. C. Salpeter. *Quantum Theory of One- and Two-Electron Atoms*, Academic, New York, 1978.

[8] D. M. Brink and G. R. Satchler. *Angular Momentum*, Clarendon Press, Oxford, 1962.

[9] A. P. Jucys, I. B. Levinson and V. V. Vanagas. *Mathematical Apparatus of the Angular Momentum Theory*, Vilnius, 1960 (in Russian). English translations: Israel Program for Scientific Translations, Jerusalem 1962; Gordon and Breach, New York, 1963.

[10] B. R. Judd. *Operator Techniques in Atomic Spectroscopy*, McGraw-Hill, New York, 1963.

[11] A. P. Jucys and A. A. Bandzaitis. *The Theory of Angular Momentum in Quantum Mechanics*, Mintis Publishers, Vilnius, 1965 (in Russian, second extended edition in 1977).

[12] B. R. Judd. *Second Quantization and Atomic Spectroscopy*, Johns Hopkins University Press, Baltimore, 1967.

[13] E. ElBaz and B. Castel. *Graphical Methods of Spin Algebras in*

Atomic, Nuclear and Particle Physics, Marcel Dekker, New York, 1972.

[14] A. P. Jucys and A. J. Savukynas. *Mathematical Foundations of the Atomic Theory*, Mintis Publishers, Vilnius, 1973 (in Russian).

[15] I. I. Sobel'man. *An Introduction to the Theory of Atomic Spectra*, Nauka, Moscow, 1977 (in Russian, English Translation: Pergamon Press, Oxford, 1972).

[16] R. D. Cowan. *The Theory of Atomic Structure and Spectra*, University of California Press, Berkeley, 1981.

[17] I. Lindgren and J. Morrison. *Atomic Many-Body Theory* (Springer Series in Chemical Physics, **13**) Springer-Verlag, Berlin, 1982.

[18] A. A. Nikitin and Z. B. Rudzikas. *Foundations of the Theory of the Spectra of Atoms and Ions*, Nauka, Moscow, 1983 (in Russian).

[19] B. E. J. Pagel *et al.*, *Mon. Notic. Roy. Astron. Soc.*, **255**, 325–345 (1992).

[20] B. R. Judd, *Repts Progr. Phys.*, **48**, 907–954 (1985).

[21] G. Racah, *Phys. Rev.*, **61**, 186–197 (1941).

[22] G. Racah, *Phys. Rev.*, **62**, 438–462 (1942).

[23] G. Racah, *Phys. Rev.*, **63**, 367–382 (1943).

[24] G. Racah, *Phys. Rev.*, **76**, 1352–1365 (1949).

[25] A. Jucys. *Theory of Many-Electron Atoms* (Selected Papers), Mokslas Publishers, Vilnius, 1978 (in Russian).

[26] G. Gaigalas, J. Kaniauskas and Z. Rudzikas, *Lietuvos Fizikos Rinkinys**, **25**, No 6, 3–13 (1985).

[27] Z. B. Rudzikas, *Comments Atom. Mol. Phys.*, **26**, No 5, 269–286 (1991).

[28] Z. B. Rudzikas and J. M. Kaniauskas, *Int. J. Quantum Chem.*, **10**, 837–852 (1976).

[29] J. M. Kaniauskas and Z. B. Rudzikas, *Lietuvos Fizikos Rinkinys*, **15**, 693–706 (1975).

[30] V. Č. Šimonis, J. M. Kaniauskas and Z. B. Rudzikas, *Int. J. Quantum Chem.*, **25**, 57–62 (1984).

[31] Z. B. Rudzikas and J. M. Kaniauskas. *Quasispin and Isospin in the Theory of an Atom*, Mokslas Publishers, Vilnius, 1984 (in Russian).

[32] J. Kaniauskas, V. Šimonis and Z. Rudzikas, *J. Phys. B: Atom. Mol. Phys.*, **20**, 3267–3281 (1987).

[33] B. G. Wybourne. *Spectroscopic Properties of the Rare Earths*, John Wiley and Sons, New York, 1965.

[34] B. R. Judd, *Physica Scripta*, **T26**, 29–33 (1989).

[35] B. R. Judd, *J. Alloys and Compounds*, **180**, 105–110 (1992).

---

* English translation by Allerton Press, Inc.: *Soviet Physics-Collection*, since 1993: *Lithuanian Physics Journal*.

[36] B. R. Judd and G. M. S. Lister, *J. Phys. B: Atom. Mol. Opt. Phys.*, **25**, 577–602 (1992).

[37] B. R. Judd and G. M. S. Lister, *J. Phys. A: Math. Gen.*, **25**, 2615–2630 (1992).

[38] B. R. Judd and G. M. S. Lister, *J. Phys. B: Atom. Mol. Opt. Phys.*, **26**, 193–203 (1993).

[39] B. R. Judd, G. M. S. Lister and N. Vaeck, *J. Phys. A: Math. Gen.*, **26**, 4991–5005 (1993).

[40] B. G. Wybourne, *J. Phys. B: Atom. Mol. Opt. Phys.*, **25**, 1683–1696 (1992).

[41] B. R. Judd and G. M. S. Lister, *J. Phys. B: Atom. Mol. Opt. Phys.*, **26**, 3177–3188 (1993).

[42] M. Bentley, B. R. Judd and G. M. S. Lister, *J. Alloys and Compounds*, **180**, 165–169 (1992).

[43] B. R. Judd and G. M. S. Lister, *J. Physique II France*, **2**, 853–863 (1992).

[44] D. R. Hartree. *The Calculation of Atomic Structures*, John Wiley and Sons, New York, 1957.

[45] C. Froese Fischer. *The Hartree–Fock Method for Atoms (A Numerical Approach)*, John Wiley and Sons, New York, 1977.

[46] E. Clementi (ed.). *Modern Techniques in Computational Chemistry: MOTECC-90*, ESCOM Science Publishers B. V., Leiden, 1990.

[47] A. Kupliauskienė, *Lietvos Fizikos Žurnalas*, **35**, 113–121 (1995).

[48] M.-J. Vilkas, G. Gaigalas and G. Merkelis, *Lietuvos Fizikos Rinkinys*, 142–164 (1991).

[49] G. Feldman, T. Fulton and J. Ingham, *Ann. Phys.*, **219**, 1–41 (1992).

[50] I. Lindgren, I. Martinson and R. Schuch (eds). *Heavy-Ion Spectroscopy and QED Effects in Atomic Systems* (Proceedings of Nobel Symposium 85, Saltsjöbaden, Sweden, June 29 – July 3, 1992), *Physica Scripta*, **T46** (1993).

[51] G. Merkelis, J. Kaniauskas and Z. Rudzikas, *Lietuvos Fizikos Rinkinys*, **25**, No 5, 21–30 (1985).

[52] G. Merkelis, G. Gaigalas and Z. Rudzikas, *Lietuvos Fizikos Rinkinys*, **25**, No 6, 14–31 (1985).

[53] V. B. Berestetskii, E. M. Lifshitz and L. P. Pitaevskii. *Relativistic Quantum Theory*, Vol. 4 of Course of Theoretical Physics, Pergamon Press, Oxford, 1971.

[54] I. P. Grant, *Adv. Phys.*, **19**, 747–811 (1970).

[55] P. Pyykkö, *Adv. Quant. Chem.*, **11**, 353–409 (1979).

[56] I. P. Grant *et al.*, *Comput. Phys. Commun.*, **21**, 207–232 (1980).

[57] K. G. Dyall *et al.*, *Comput. Phys. Commun.*, **55**, 425–456 (1989).

[58] A. I. Akhiezer and V. B. Berestetskii. *Quantum Electrodynamics*,

Interscience Monographs and Texts in Physics and Astronomy, Vol. 11, Interscience Publishers, New York, 1965.

[59] I. S. Kičkin and Z. B. Rudzikas, *Lietuvos Fizikos Rinkinys*, **14**, 19–30, 31–44, 45–56 (1974).

[60] V. I. Sivcev *et al.*, *Lietuvos Fizikos Rinkinys*, **14**, 189–205 (1974).

[61] J. M. Kaniauskas, I. S. Kičkin and Z. B. Rudzikas, *Lietuvos Fizikos Rinkinys*, **14**, 463–475 (1974).

[62] I. S. Kičkin, J. M. Kaniauskas and Z. B. Rudzikas, *Lietuvos Fizikos Rinkinys*, **14**, 727–739 (1974).

[63] A. A. Slepcov *et al.*, *Lietuvos Fizikos Rinkinys*, **15**, 5–20 (1975).

[64] I. S. Kičkin *et al.*, *Lietuvos Fizikos Rinkinys*, **15**, 539–558 (1975).

[65] H. P. Kelly, *Adv. Chem. Phys.*, **14**, 129–190 (1969).

[66] U. I. Safronova and V. S. Senashenko. *Theory of the Spectra of Multiply Charged Ions.* Energoizdat, Moscow, 1984 (in Russian).

[67] H. Feshbach, *Ann. Phys.*, **19**, 287–313 (1962).

[68] P. O. Löwdin, *Phys. Rev.*, **A139**, 357–372 (1965).

[69] B. R. Judd, *Adv. Chem. Phys.*, **14**, 91–127 (1969).

[70] P. G. H. Sandars, *Adv. Chem. Phys.*, **14**, 365–419 (1969).

[71] I. Lindgren, *J. Phys. B: Atom. Mol. Opt. Phys.*, **23**, 1085–1093 (1990).

[72] I. Lindgren, *J. Phys. B: Atom. Mol. Opt. Phys.*, **24**, 1143–1159 (1991).

[73] I. Lindgren, *Physica Scripta*, **T34**, 36–46 (1991).

[74] I. Lindgren, Many-body problems in atomic physics. In T. L. Ainsworth *et al.* (eds). *Recent Progress in Many-Body Theories*, Vol. 3, Plenum Press, New York, 1992.

[75] O. Sinanoğlu, *Adv. Chem. Phys.*, **6**, 315–412 (1964).

[76] O. Sinanoğlu, *Adv. Chem. Phys.*, **14**, 237–282 (1969).

[77] J. M. Kaniauskas, G. V. Merkelis and Z. B. Rudzikas, *Lietuvos Fizikos Rinkinys*, **16**, 795–806 (1976).

[78] T. Brage and C. Froese Fischer, *Physica Scripta*, **45**, 43–48 (1992).

[79] A. Dalgarno and H. R. Sadeghpour, *Phys. Rev.*, **A46**, R3591–R3593 (1992).

[80] I. P. Grant, *J. Phys. B.: Atom. Mol. Phys.*, **7**, 1458–1475 (1974).

[81] I. P. Grant and A. F. Starace, *J. Phys. B: Atom. Mol. Phys.*, **8**, 1999–2000 (1975).

[82] Z. B. Rudzikas and J. M. Kaniauskas, *Lietuvos Fizikos Rinkinys*, **13**, 849–859 (1973).

[83] G. W. F. Drake, *J. Phys. B.: Atom. Mol. Phys.*, **9**, No 7, L169–L171 (1976).

[84] J. Burgdörfer and F. Meyer, *Physica Scripta*, **T46**, 225–230 (1993).

[85] V. Špakauskas, J. Kaniauskas and Z. Rudzikas, *Lietuvos Fizikos Rinkinys*, **17**, 563–574 (1977).

[86] J. Kaniauskas and Z. Rudzikas, *Lietuvos Fizikos Rinkinys*, **13**, 657–666 (1973).

[87] R. Karazija *et al.* *Tables for the Calculation of Matrix Elements of Atomic Quantities*, Computing Centre of the USSR Academy of Sciences, Moscow, 1967, second edition 1972 (English translation by E. K. Wilip, ANL-Trans-563 (National Technical Information Service, Springfield, VA, 1968)).

[88] D. Varshalovich, A. Moskalev and V. Khersonskii. *Quantum Theory of Angular Momentum*, World Scientific Publishers, Singapore, 1988.

[89] J. Kaniauskas and Z. Rudzikas, *Lietuvos Fizikos Rinkinys*, **13**, No 2, 191–205 (1973).

[90] Z. B. Rudzikas, *Lietuvos Fizikos Rinkinys*, **10**, 861–871 (1970).

[91] V. V. Špakauskas, J. M. Kaniauskas and Z. B. Rudzikas, *Lietuvos Fizikos Rinkinys*, **18**, 293–315 (1978).

[92] V. V. Špakauskas, I. S. Kičkin and Z. B. Rudzikas, *Lietuvos Fizikos Rinkinys*, **16**, 201–215 (1976).

[93] Z. B. Rudzikas and J. V. Čiplys, *Physica Scripta*, **T26**, 21–28 (1989).

[94] I. S. Kičkin *et al.*, *Lietuvos Fizikos Rinkinys*, **16**, 217–229 (1976).

[95] P. A. M. Dirac. *The Principles of Quantum Mechanics*, Clarendon Press, Oxford, 1930 (first edition), 1935 (second edition), 1947 (third edition).

[96] R. D. Lawson and M. H. MacFarlane, *Nucl. Phys.*, **66**, 80–96 (1965).

[97] C. W. Nielson and G. F. Koster. *Spectroscopic Coefficients for the p^n, d^n and f^n Configurations*, Mass. Inst. Technol., Cambridge, MA, 1963.

[98] B. F. Beiman. *Lectures on Applications of the Group Theory to Nuclear Spectroscopy*, Fizmatgiz, Moscow, 1961 (in Russian).

[99] B. R. Judd. Group theory in atomic spectroscopy. In E. M. Loebl (ed.) *Group Theory and Its Applications*, Academic Press, New York, 1968.

[100] B. G. Wybourne. *Symmetry Principles and Atomic Spectroscopy*, J. Wiley and Sons, New York, 1970.

[101] B. Wybourne. *Classical Groups for Physicists*. J. Wiley and Sons, New York, 1973.

[102] I. S. Kičkin and Z. B. Rudzikas, *Lietuvos Fizikos Rinkinys*, **11**, 757–768 (1971).

[103] Z. Rudzikas and V. Špakauskas, *Lietuvos Fizikos Rinkinys*, **13**, 181–190 (1973).

[104] B. Judd. Problems of the theory of atomic spectra. In B. Judd and B. Wybourne. *Theory of Complex Atomic Spectra*, Mír, Moscow, 1973 (in Russian).

[105] I. S. Kičkin and Z. B. Rudzikas, *Lietuvos Fizikos Rinkinys*, **11**, 743–755 (1971).

[106] K. G. Dyall and I. P. Grant, *J. Phys. B: Atom. Mol. Phys.*, **15**, L371–L373 (1982).

[107] C. Schwartz and A. de-Shalit, *Phys. Rev.*, **94**, 1257–1266 (1954).

[108] J. Kaniauskas, V. Šimonis and Z. Rudzikas, *Lietuvos Fizikos Rinkinys*, **19**, 637–648 (1979).

[109] P. J. Redmond, *Proc. Roy. Soc.*, **A222**, 84–94 (1954).

[110] I. Kičkin, J. Kaniauskas and Z. Rudzikas, *Lietuvos Fizikos Rinkinys*, **15**, 183–196 (1975).

[111] V. Šimonis, J. Kaniauskas and Z. Rudzikas, *Lietuvos Fizikos Rinkinys*, **21**, No 5, 15–26 (1981).

[112] S. Meshkov, *Phys. Rev.*, **91**, 871–876 (1953).

[113] S. Feneuille, *J. Physique*, **28**, 61–66 (1967).

[114] S. Feneuille, *J. Physique*, **28**, 315–327 (1967).

[115] P. H. Butler and B. G. Wybourne, *J. Math. Phys.*, **11**, 2512–2518 (1970).

[116] J. C. Morrison, *J. Math. Phys.*, **10**, 1431–1437 (1969).

[117] A. K. Kerman, R. D. Lawson and M. H. MacFarlane, *Phys. Rev.*, **124**, 162–167 (1961).

[118] O. Sinanoğlu and D. R. Herrick, *J. Chem. Phys.*, **62**, 886–892 (1975).

[119] D. R. Herrick and O. Sinanoğlu, *Phys. Rev.*, **A11**, 97–110 (1975).

[120] S. J. Nikitin and V. N. Ostrovsky, *J. Phys., B: Atom. Mol. Phys.*, **9**, 3141–3147 (1976).

[121] J. Kaniauskas, V. Špakauskas and Z. Rudzikas, *Lietuvos Fizikos Rinkinys*, **19**, 21–31 (1979).

[122] W. Heisenberg, *Z. Phys.*, **77**, 1–11 (1932).

[123] V. Šimonis, J. Kaniauskas and Z. Rudzikas, *Lietuvos Fizikos Rinkinys*, **22**, No 4, 3–15 (1982).

[124] P. Bogdanovič, R. Karazija and J. Boruta, *Lietuvos Fizikos Rinkinys*, **20**, No 2, 15–24 (1980).

[125] M. Moshinsky and C. J. Quesne, *J. Math. Phys.*, **11**, 1631–1639 (1970).

[126] V. I. Sivcev, I. S. Kičkin and Z. B. Rudzikas. *Matrix Elements of Relativistic Energy Operator for Four Subshells of Equivalent Electrons* (No 371–76 Dep); *Non-Diagonal with Respect to Configurations Matrix Elements of Energy Operator for Four Subshells of Equivalent Electrons* (No 2181–Dep76). Allunion Institute of Scientific and Technical Information, Moscow, 1976.

[127] K. Rajnak and B. G. Wybourne, *Phys. Rev.*, **132**, 280–290 (1963).

[128] B. G. Wybourne, *J. Chem. Phys.*, **40**, 1456–1457 (1964).

[129] J. Schrijver and P. E. Norman, *Physica*, **64**, 269–277 (1973).

[130] A. A. Ramonas and K. K. Ušpalis, *Lietuvos Fizikos Rinkinys*, **15**, 737–748 (1975).

[131] A. Kancerevičius *et al.*, *Lietuvos Fizikos Rinkinys*, **31**, 251–261 (1991).

[132] I. Martinson, *Repts Progr. Phys.*, **52**, 157–225 (1989).

[133] L. J. Curtis and I. Martinson, *Comments Atom. Mol. Phys.*, **24**, 213–233 (1990).

[134] C. Froese Fischer, J. E. Hansen and B. R. Judd, *J. Opt. Soc. Amer.*, **B5**, 2446–2451 (1988).

[135] G. J. van net Hof *et al.*, *Can. J. Phys.*, **72**, 193–205 (1994).

[136] L. J. Curtis, Z. B. Rudzikas and D. G. Ellis, *Phys. Rev.*, **A44**, 776–779 (1991).

[137] S. Fraga, J. Karwowski and K. M. S. Saxena. *Atomic Energy Levels*, Elsevier Scientific, Amsterdam, 1979.

[138] S. Fraga, K. M. S. Saxena and J. Karwowski. *Handbook of Atomic Data*, Elsevier Scientific, Amsterdam, 1976, 1979 (first and second printings).

[139] J. Migdałek. *Model-Potential Methods in Atomic Structure Calculations*, Zeszyty Naukove Universytetu Jagiellonskiego, MXV, Prace Fizyczne, Zeszyt 32, Krakow, 1990.

[140] M. Cornille *et al.*, *J. Phys. B: Atom. Mol. Phys.*, **20**, L623–L626 (1987).

[141] B. C. Fawcett and H. E. Mason, *Atom. Data Nucl. Data Tables*, **47**, 17–31 (1991).

[142] B. C. Fawcett, *Atom. Data Nucl. Data Tables*, **47**, 319–361 (1991).

[143] B. C. Fawcett and M. Wilson, *Atom. Data Nucl. Data Tables*, **47**, 241–317 (1991).

[144] K. L. Bell *et al.*, *Physica Scripta*, **50**, 343–353 (1994).

[145] W. Dankwort, J. Ferch and H. Gebauer, *Z. Phys.*, **267**, 229–237 (1974).

[146] J. Vizbaraitė, J. Kaniauskas and Z. Rudzikas, *Lietuvos Fizikos Rinkinys*, **15**, 707–719 (1975).

[147] I. Kičkin, V. Sivcev and Z. Rudzikas, *Lietuvos Fizikos Rinkinys*, **16**, 37–48 (1976).

[148] R. Huber *et al.*, *Phys. Lett.*, **55A**, No 3, 141–142 (1975).

[149] I. Lindgren and A. Rosen, *Case Studies in Atomic Physics*, **4**, No 3, 93–196, No 4, 197–298 (1974).

[150] S. Garpman *et al.*, *Phys. Rev.*, **A11**, 758–781 (1975).

[151] J. Carlsson, P.Jönsson and Ch. Froese Fischer, *Phys. Rev.*, **A46**, No 5, 2420–2425 (1992).

[152] P. Jönsson and Ch. Froese Fischer, *Phys. Rev.*, **A50**, 3080–3088 (1994).

[153] J. Blase *et al.*, *J. Opt. Soc. Amer.*, **B11**, 1897–1929 (1994).

[154] G. Huber. Hydrogen-like bismuth. In *Abstracts of the 26th EGAS Conference*, page I-6, Barcelona, 12-15 July 1994.

[155] R. Neumann, F. Träger and G. zu Putlitz. Laser-microwave spectroscopy. In H. J. Beyer and H. Kleinpoppen (eds). *Progress in Atomic Spectroscopy*, part D, Plenum, New York, 1987.

414   References

[156] J. Vizbaraitė, K. Eriksonas and J. Boruta, *Lietuvos Fizikos Rinkinys*, **19**, 487–493 (1979).

[157] A.-M. Mårtensson-Pendrill *et al.*, *Phys. Rev.*, **A45**, 4675–4681 (1992).

[158] A.-M. Mårtensson-Pendrill and A. Ynnermann, *J. Phys. B: Atom. Mol. Opt. Phys.*, **25**, L551–L559 (1992).

[159] C. Froese Fischer *et al.*, *Comput. Phys. Commun.*, **74**, 415–431 (1993).

[160] P. J. Mohr, *Ann. Phys.*, **88**, 52–87 (1974).

[161] P. J. Mohr, *Phys. Rev. Lett.*, **34**, 1050–1052 (1975).

[162] K. T. Cheng, W. R. Johnson and J. Sapirstein, *Phys. Rev. Lett.*, **66**, 2960–2963 (1991).

[163] W. R. Johnson and G. Soff, *Atom. Data Nucl. Data Tables*, **33**, 405–446 (1985).

[164] H. Gould, *Physica Scripta*, **T46**, 61–64 (1993).

[165] P. von Brentano *et al.*, *Physica Scripta*, **T46**, 162–166 (1993).

[166] P. J. Mohr, *Physica Scripta*, **T46**, 44–51 (1993).

[167] E. Savičius, J. Kaniauskas and Z. Rudzikas, *Lietuvos Fizikos Rinkinys*, **19**, 747–756 (1979).

[168] J. Kaniauskas and Z. Rudzikas, *J. Phys. B: Atom. Mol. Phys.*, **13**, 3521–3533 (1980).

[169] Z. Rudzikas *et al.*, *Lietuvos Fizikos Rinkinys*, **4**, 5–12 (1964).

[170] L. Bréhamet, *Il Nuovo Cimento*, **108** B, No 9, 1003–1015 (1993).

[171] J. Kaniauskas, G. Merkelis and Z. Rudzikas, *Lietuvos Fizikos Rinkinys*, **19**, 475–485 (1979).

[172] J. Kaniauskas and Z. Rudzikas, *Lietuvos Fizikos Rinkinys*, **16**, 491–503 (1976).

[173] J. Vizbaraitė *et al.*, *Lietuvos Fizikos Rinkinys*, **1**, 21–32 (1961).

[174] R. Karazija, *Uspekhi Fizitcheskich Nauk*, **135**, 79–115 (1981) (in Russian).

[175] A. Jucys and J. Vizbaraitė, *Lietuvos MA darbai*, **B1(24)**, 65–73 (1961) (in Russian).

[176] I. P. Grant. In G. L. Malli (ed.), *Relativistic Effects in Atoms, Molecules and Solids*, 55–72, 73–88, 89–100, 101–114, Plenum, New York, 1983.

[177] F. A. Parpia and I. P. Grant, *Information Quarterly*, No 30, 39–40 (1989).

[178] S. P. Goldman and A. Dalgarno, *Phys. Rev. Lett.*, **57**, 408–411 (1986).

[179] I. Kičkin *et al.*, *Lietuvos Fizikos Rinkinys*, **18**, 165–177 (1978).

[180] Z. Kupliauskis, A. Matulaitytė and A. Jucys, *Lietuvos Fizikos Rinkinys*, **11**, 557–563 (1971).

[181] A. Kancerevičius, *Lietuvos Fizikos Rinkinys*, **22**, No 1, 29–38 (1982).

[182] H. M. Quiney, I. P. Grant and S. Wilson, *J. Phys. B: Atom. Mol. Phys.*, **22**, L15–L19 (1989); **23**, L271–L278 (1990).

[183] C. Froese-Fischer, *Comput. Phys. Commun.*, **64**, 369–398 (1991).

[184] A. Kancerevičius and A. Karosienė, *Lietuvos Fizikos Rinkinys*, **29**, 159–168 (1989).

[185] A. Hibbert, C. Froese Fischer and M. R. Godefroid, *Comput. Phys. Commun.*, **51**, 285–293 (1988).

[186] D. R. Bates and A. Damgaard, *Philos. Trans.*, **242**, 101–122 (1949).

[187] J. Carlsson *et al.*, *Z. Phys. D: Atoms, Molecules, Clusters*, **3**, 345–351 (1986).

[188] J. O. Gaardsted *et al.*, *Physica Scripta*, **42**, 543–550 (1990).

[189] S. Mannervik *et al.*, *Phys. Rev.*, **A35**, 3136–3138 (1987).

[190] I. Martinson (ed.). *Frontiers in Atomic Structure Calculations and Spectroscopy* (Proceedings of a workshop held in Lund, Sweden, May 24–25, 1988), *Physica Scripta*, **RS15** (1989).

[191] D. Lecler and J. Margerie (eds). *Proceedings of the 25th EGAS Conference of the European Group for Atomic Spectroscopy* (University of Caen, France, July 13–16, 1993), *Physica Scripta*, **T51** (1994).

[192] A. P. Jucys, A. A. Bandzaitis and J. J. Grudzinskas, *Int. J. Quantum Chem.*, **3**, No 5, 913–930 (1969).

[193] A. P. Jucys and J. J. Grudzinskas, *Int. J. Quantum Chem.*, **6**, 445–464 (1972).

[194] A. P. Jucys, K. V. Sabas and Z. J. Kupliauskis, *Int. J. Quantum Chem.*, **9**, No 4, 721–741, 743–754 (1975).

[195] J. V. Čiplys and V. M. Lazauskas, *Lietuvos Fizikos Rinkinys*, **17**, 371–379 (1976).

[196] V. M. Lazauskas, Z. J. Kupliauskis and A. P. Jucys, *Lietuvos Fizikos Rinkinys*, **14**, 431–441 (1974).

[197] V. M. Lazauskas and Z. J. Kupliauskis, *Lietuvos Fizikos Rinkinys*, **15**, 731–735 (1975).

[198] A. J. Kancerevičius, *Lietuvos Fizikos Rinkinys*, **32**, 315–324 (1992).

[199] A. P. Jucys, *Int. J. Quantum. Chem.*, **1**, 311–319 (1967).

[200] A. P. Jucys, V. A. Kaminskas and V. J. Kaveckis, *Int. J. Quantum Chem.*, **2**, 405–411 (1968).

[201] A. P. Jucys and E. P. Našlėnas, *Int. J. Quantum Chem.*, **3**, 931–943 (1969).

[202] A. P. Jucys and V. J. Stasiukaitis, *Int. J. Quantum Chem.*, **4**, 333–335 (1970).

[203] A. P. Jucys, E. P. Našlėnas and P. S. Žvirblis, *Int. J. Quantum Chem.*, **6**, 465–472 (1972).

[204] J. J. Vizbaraitė, V. J. Stasiukaitis and T. D. Strockytė, *Lietuvos Fizikos Rinkinys*, **12**, 745–751 (1972).

[205] C. Nicolaides and D. Beck, *Can. J. Phys.*, **53**, 1224–1228 (1975).

[206] A. J. Kancerevičius and L. L. Kuzmickytė, *Lietuvos Fizikos Rinkinys*, **28**, 148–157 (1988).

[207] C. Froese Fischer, *Physica Scripta*, **40**, 25–27 (1989).

[208] C. Froese Fischer, *Phys. Rev.*, **A39**, 963–970 (1989).

[209] S. Salomonson and P. Oster, *Phys. Rev.*, **A41**, 4670–4681 (1990).

[210] C. Froese Fischer and D. Chen, *J. Mol. Struct.*, **199**, 61–73 (1989).

[211] T. Brage and C. Froese Fischer, *Phys. Rev.*, **A44**, 72–79 (1991).

[212] C. Froese Fischer and J. Hansen, *Phys. Rev.*, **A44**, 1559–1564 (1991).

[213] G. Miecznik, T. Brage and C. Froese Fischer, *J. Phys. B: Atom. Mol. Opt. Phys.*, **25**, 4217–4228 (1992).

[214] E. J. Knystautas, *Phys. Rev. Lett.*, **69**, 2635–2637 (1992).

[215] C. A. Nicolaides and G. Aspromallis, *J. Mol. Struct.*, **199**, 283–302 (1989).

[216] C. Froese Fischer, *Phys. Rev.*, **A41**, 3481–3488 (1990).

[217] W. R. Johnson, *Adv. Atom. Molec. Physics*, **25**, 375–391 (1988).

[218] W. R. Johnson and K. T. Cheng, *Phys. Rev.*, **A46**, 2952–2954 (1992).

[219] I. P. Grant, *Comput. Phys. Commun.*, **84**, 59–77 (1994).

[220] K. Ušpalis and A. Ramonas, *Lietuvos Fizikos Rinkinys*, **10**, 341–353 (1970).

[221] P. Bogdanovič and G. Žukauskas, *Lietuvos Fizikos Rinkinys*, **23**, No 3, 18–33 (1983).

[222] A. Jucys, *J. Exp. Theor. Phys. (ZETP)*, **23**, No 2(8), 129–139 (1952) (in Russian).

[223] C. Froese Fischer, *J. Phys. B: Atom. Mol. Opt. Phys.*, **26**, 855–862 (1993).

[224] T. Brage and C. Froese Fischer, *Physica Scripta*, **T47**, 18–28 (1993).

[225] A. Jucys *et al.*, *Optics and Spectroscopy*, **12**, 157–162 (1962) (in Russian).

[226] A. Kancerevičius, *Lietuvos Fizikos Rinkinys*, **25**, No 3, 12–20 (1985).

[227] C. Froese Fischer, *J. Phys. B: Atom. Mol. Phys.*, **16**, 157–165 (1983).

[228] M. Godefroid and C. Froese Fischer, *J. Phys. B: Atom. Mol. Phys.*, **17**, 681–692 (1984).

[229] C. Froese Fischer and H. P. Saha, *J. Phys. B: Atom. Mol. Phys.*, **17**, 943–952 (1984).

[230] C. Froese Fischer and H. P. Saha, *Physica Scripta*, **32**, 181–194 (1985).

[231] C. Froese Fischer and M. Godefroid, *J. Phys. B: Atom. Mol. Phys.*, **19**, 137–148 (1986).

[232] C. Froese Fischer, *Physica Scripta*, **49**, 51–61, 323–330 (1994).

[233] G. Gaigalas *et al.*, *Physica Scripta*, **49**, 135–147 (1994).

[234] J. Morrison, *Phys. Rev.*, **A6**, 643–650 (1972).

[235] B. C. Fawcett, *Atom. Data Nucl. Data Tables*, **16**, 135–164 (1975).

[236] C. Froese Fischer and H. P. Saha, *Phys. Rev.*, **A28**, 3169–3178 (1983).

[237] K. T. Cheng, Y.-K. Kim and J. P. Desclaux, *Atom. Data Nucl. Data Tables*, **24**, 111–189 (1979).

[238] I. Lindgren, *Nucl. Instrum. Meth. Phys. Res.*, **B31**, 102–114 (1988).

[239] W. R. Johnson, S. A. Blundell and J. Sapirstein, *Phys. Rev.*, **A42**, 1087–1095 (1990).

[240] C. Guet and W. R. Johnson, *Phys. Rev.*, **A44**, 1531–1535 (1991).

[241] W. R. Johnson and J. Sapirstein, *Phys. Rev.*, **A46**, R2197–R2200 (1992).

[242] C. Froese Fischer, *Phys. Rev.*, **A22**, 551–556 (1980).

[243] A. Hibbert, *J. Phys. B: Atom. Mol. Phys.*, **7**, 1417–1434 (1974).

[244] C. A. Nicolaides, D. R. Beck and O. Sinanoğlu, *J. Phys. B: Atom. Mol. Phys.*, **6**, 62–70 (1973).

[245] Z. B. Rudzikas, M. Szulkin and I. Martinson, *Physica Scripta*, **T8**, 141–144 (1984),

[246] L. Engström *et al.*, *Physica Scripta*, **24**, 551–556 (1981).

[247] M.-J. Vilkas *et al.*, *Physica Scripta*, **49**, 592–600 (1994).

[248] H.-J. Flaig and K.-H. Schartner, *Physica Scripta*, **31**, 255–261 (1985).

[249] O. Sinanoğlu, *Nucl. Instrum. Meth.*, **110**, 193–208 (1973).

[250] K. L. Baluja and C. J. Zeippen, *J. Phys. B: Atom. Mol. Opt. Phys.*, **21**, 15–24 (1988).

[251] G. Merkelis *et al.*, *Physica Scripta*, **51**, 233–251 (1995).

[252] S. A. Blundell, W. R. Johnson and J. Sapirstein, *Phys. Rev. Lett.*, **65**, 1411–1414 (1990).

[253] A.-M. Mårtensson-Pendrill. In S. Wilson (ed.). *Methods in Computational Chemistry, Vol.5: Atomic and Molecular Properties*, pp. 99–156, Plenum Press, New York, 1992.

[254] E. P. Plummer and I. P. Grant. *J. Phys. B: Atom. Mol. Phys.*, **18**, L315–L320 (1985).

[255] B. Edlen. *Atomic Spectra*. In S. Flugge (ed.). *Handbuch der Physik*, **27**, 80–220, Springer, Berlin (1964).

[256] B. Edlen, *Z. Astrophys.*, **22**, 30–64 (1942).

[257] B. Edlen. *Energy Level Structure of Highly Ionized Atoms*. In H. J. Beyer and H. Kleinpoppen (eds). *Progress in Atomic Spectroscopy*, Part D, pp. 271–293, Plenum, New York, 1987.

[258] P. H. Mokler *et al.*, *Physica Scripta*, **T51**, 28–38 (1994).

[259] H. Walther, *Physica Scripta*, **T51**, 11–19 (1994).

[260] J. Karwowski, J. Styszynski and W. H. E. Schwarz, *J. Phys. B: Atom. Mol. Opt. Phys.*, **24**, 4877–4886 (1991).

[261] Y.-K. Kim, W. C. Martin and A. W. Weiss, *J. Opt. Soc. Amer.*, **B5**, 2215–2224 (1988).

[262] S. Fritzsche and I. P. Grant, *Physica Scripta*, **50**, 473–480 (1994).

[263] U. I. Safronova and Z. B. Rudzikas, *J. Phys. B: Atom. Mol. Phys.*, **10**, 7–18 (1977).

[264] V. A. Boiko, S. A. Pikuz and A. Ya. Faenov, *Preprint FIAN*, No 20, Moscow, 1976.

[265] C. Breton *et al.*, *J. Opt. Soc. Amer.*, **69**, 1652–1658 (1979).

[266] M. Bogdanovičienė, *Lietuvos Fizikos Rinkinys*, **20**, No 5, 59–64 (1980).

[267] S. Wilson, I. P. Grant and B. L. Gyorffy (eds). *The Effects of Relativity in Atoms, Molecules and the Solid State*, Plenum Press, New York, 1991.

[268] I. P. Grant, *Adv. Atom. Molec. Opt. Physics*, **32**, 169–186 (1993).

[269] Z. B. Rudzikas, *Nucl. Instrum. Meth.*, **202**, 289–297 (1982).

[270] Z. B. Rudzikas, *Int. J. Quantum Chem.*, **25**, 47–55 (1984).

[271] S. S. Churilov *et al.*, *Physica Scripta*, **50**, 463–468 (1994).

[272] A. Dalgarno and D. Leyzer (eds). *Spectroscopy of Astrophysical Plasmas*, Cambridge University Press, London, 1987.

[273] P. L. Smith and W. L. Wiese (eds). *Atomic and Molecular Data for Space Astronomy. Needs, Analysis and Availability* (A Collection of Papers Presented at the Joint Commission Meeting III of the 21st IAU General Assembly Held in Buenos Aires, Argentina, 23 July – 1 August 1991), Springer-Verlag, Berlin, 1992.

[274] L. J. Curtis and L. S. Anderson (eds). *Proceedings of Colloquium on Atomic Spectra and Oscillator Strengths for Astrophysics and Fusion Research* (Toledo, Ohio, USA, August 11–13, 1986), *Physica Scripta*, **RS5** (1987).

[275] J. E. Hansen (ed.). *Atomic Spectra and Oscillator Strengths for Astrophysics and Fusion Research* (Invited and Contributed Papers of the Colloquium, Amsterdam, August 28–31, 1989), North-Holland, Amsterdam, 1990.

[276] D. S. Leckrone and J. Sugar (eds). *The 4th International Colloquium on Atomic Spectra and Oscillator Strengths for Astrophysical and Laboratory Plasmas*, Gaithersburg, MD, USA, September 14–17, 1992 (Invited Papers), *Physica Scripta*, **T47** (1993).

[277] J. Sugar, D. Leckrone (eds). *The 4th International Colloquium on Atomic Spectra and Oscillator Strengths for Astrophysical and Laboratory Plasmas*, Gaithersburg, Maryland, September 14–17, 1992 (Poster Papers), NIST Special Publication 850, Gaithersburg, MD 20899, 1993.

[278] Z. Rudzikas and P. Bogdanovich, *Baltic Astronomy*, **3**, 131–141 (1994).

[279] V. Escalante, A. Sternberg and A. Dalgarno, *Astrophys. J.*, **375**, 630–634 (1991).

[280] E. J. Knystautas, *Comments Atom. Mol. Phys.*, **12**, 239–246 (1983).

[281] I. Martinson, *J. Opt. Soc. Amer.*, **B5**, 2159–2172 (1988).

[282] G. O'Sullivan, *Comments Atom. Mol. Phys.*, **28**, 143–178 (1992).

[283] H. G. Berry, R. Dunford and L. Young (eds). *Atomic Spectroscopy and Highly Ionised Atoms* (Proceedings of the Symposium held in Lisle, IL, USA, 16–21 August 1987), North-Holland Physics Publishing, Amsterdam, 1988.

[284] M. J. Seaton (ed.). *The Opacity Project*, Vol. 1, Institute of Physics Publishing, London, 1994.

[285] C. J. Zeippen, *Abstracts of the 26th EGAS Conference* (Barcelona, 12–15 July 1994), Universitat Autonoma de Barcelona, I–5, 1994.

[286] R. K. Janev, L. P. Presniakov and V. P. Shevelko. *Physics of Highly Charged Ions*, Springer-Verlag, Berlin, 1985.

[287] C. Jupen, R. C. Isler and E. Träbert, *Mon. Notic. Roy. Astron. Soc.*, **264**, 627–635 (1993).

[288] W. C. Martin. *Lecture Notes in Physics*, **407**, 121–147, 1992. Chapter 8 in P. L. Smith and W. L. Wiese (eds). *Atomic and Molecular Data for Space Astronomy*, Springer-Verlag, Berlin, 1992.

[289] D. C. Morton, *Astrophys. J. Suppl. Ser.*, **77**, 119–202 (1991).

[290] V. Kaufman and W. C. Martin, *J. Phys. and Chem. Ref. Data*, **20**, 83–152 (1991).

[291] V. Kaufman and W. C. Martin, *J. Phys. and Chem. Ref. Data*, **20**, 775–858 (1991).

[292] V. Kaufman and W. C. Martin, *J. Phys. and Chem. Ref. Data*, **22**, 279–375 (1993).

[293] C. E. Moore. *Tables of Spectra of Hydrogen, Carbon, Nitrogen and Oxygen Atoms and Ion.*, CRC Press, Boca Raton, FL, 1993.

[294] C. S. Jeffery (ed.). *Daresbury Laboratory Newsletter on Analysis of Astronomical Spectra*, No 18. Dept. of Physics and Astron., Univ. of St. Andrews, Fife, Scotland, 1993.

[295] J. Batero (ed.). *International Bulletin on Atomic and Molecular Data for Fusion*, No 42–45, International Atomic Energy Agency, Vienna, 1992.

[296] R. Karazija. *Sums of Atomic Quantities and Mean Characteristics of Spectra*, Mokslas, Vilnius, 1991.

[297] J. Bauche, C. Bauche-Arnoult and M. Klapisch, *Adv. Atom. Molec. Physics*, **23**, 131–195 (1988).

[298] J. Bauche *et al.*, *Phys. Rev.* **A28**, 829–835 (1983).

[299] J. Karwowski and M. Bancewicz, *J. Phys. A: Math. Gen.*, **20**, 6309–6320 (1987).

[300] M. Bancewicz and J. Karwowski, *Phys. Rev.*, **A44**, 3054–3059 (1991).

[301] S. Kučas and R. Karazija, *Physica Scripta*, **47**, 754–764 (1993).

[302] S. Kučas *et al.*, *Physica Scripta*, **51**, 566–577; **52**, 639–648 (1995).

[303] C. E. Moore. *Atomic Energy Levels*, NBS, Washington, **I**, 1949; **II**, 1952, **III**, 1958.

[304] R. Karazija, *Acta Phys. Hung.*, **70(4)**, 367–379 (1991).

[305] R. Karazija. *Introduction to the Theory of X-Ray and Electronic Spectra of Free Atoms.* Mokslas, Vilnius, 1987. English translation by Plenum Publishing Corporation, New York, 1996.

[306] T. A. Carlson and C. W. Nestor, *Phys. Rev.*, **A8**, 2887–2894 (1973).

[307] P. G. Burke, V. M. Burke and K. M. Dunseath, *J. Phys. B: Atom. Mol. Opt. Phys.*, **27**, 5341–5373 (1994).

[308] M. Elango *et al.*, *Phys. Rev.*, **B47**, 11736–11748 (1993).

[309] R. Ruus *et al.*, *Phys. Rev.*, **B49**, 14836–14844 (1994).

[310] D. G. Ellis, I. Martinson and M. Westerlind, *Physica Scripta*, **49**, 561–564 (1994).

[311] D. J. Beideck *et al.*, *Phys. Rev.*, **A47**, 884–885 (1993).

[312] D. J. Beideck *et al.*, *J. Opt. Soc. Amer.*, **B10**, 977–981 (1993).

[313] E. Knystautas, N. Schlader and S. L. Bliman, *Nucl. Instrum. Meth. in Phys. Res.*, **B56/57**, 309–312 (1991).

[314] J. Reader *et al.*, *J. Opt. Soc. Amer.*, **B11**, 1930–1934 (1994).

[315] J. P. Connerade. *Adv. Atom. Molec. Opt. Phys.*, **29**, 325–367 (1992).

[316] H. P. Summers, *Adv. Atom. Molec. Opt. Phys.*, **33**, 275–319 (1994).

[317] J. W. Gallagher, *Adv. Atom. Molec. Opt. Phys.*, **33**, 373–389 (1994).

[318] E. W. McDaniel and E. J. Mansky, *Adv. Atom. Molec. Opt. Phys.*, **33**, 390–465 (1994).

[319] I. B. Levinson and A. A. Nikitin. *Handbook for Theoretical Computation of Line Intensities in Atomic Spectra* (translated from Russian), Israel Program for Scientific Translations, Jerusalem, 1965.

[320] Z. B. Rudzikas, A. A. Nikitin and A. F. Kholtygin. *Theoretical Atomic Spectroscopy. Handbook for Astronomers and Physicists*, Leningrad University Press, Leningrad, 1990 (in Russian).

[321] V. Sivcev *et al.*, *Lietuvos Fizikos Rinkinys*, **17**, 285–293 (1977).

[322] S. I. Ališauskas, Z. B. Rudzikas and A. P. Jucys, *Doklady Akademii Nauk SSSR*, **172**, No 1, 58–60 (1967) (in Russian).

[323] A. P. Jucys, Z. B. Rudzikas and A. J. Savukynas, *Int. J. Quantum Chem.*, **3**, 1001–1012 (1969).

[324] S. J. Adelman and W. L. Wiese (eds). *Astrophysical Applications of Powerful New Databases*, Astronomical Society of the Pacific Conference Series Volume 78, Book Crafters, Inc., San Francisco, 1995.

[325] L. J. Dubé *et al.* (eds). *The Physics of Electronic and Atomic Collisions* (XIX International Conference, Whistler, Canada July–August 1995), American Institute of Physics, Woodbury, New York, 1995.

[326] J. Botero and J. A. Stephens (eds). *International Bulletin on Atomic and Molecular Data for Fusion*, No 50–51, International Atomic Energy Agency, Vienna, 1996.

[327] Y. Awaya (ed.). *Eighth International Conference on the Physics of*

Highly Charged Ions (Omiya, Saitama, Japan, September 23–26, 1996), Book of Abstracts, RIKEN, Wako, Saitama, 1996.

[328] W. -Ü. L. Tchang-Brillet, J. -F. Wyart and C. Zeippen (eds). *The 5th International Colloquium on Atomic Spectra and Oscillator Strengths for Astrophysical and Laboratory Plasmas*, Meudon, France, August 28–31, 1995, *Physica Scripta*, **T65** (1996).

[329] D. Liesen (ed.). *Physics with Multiply Charged Ions*, NATO ASI Series, Series B, Physics Vol. 348., Plenum Press, New York, 1995.

[330] Y. Yamazaki *et al. J. Phys. Soc. Japan*, **65**, No 5, 1199–1202 (1996).

[331] P. Beiersdorfer *et al., Phys. Rev. Lett.* **67**, 2272–2275 (1991).

[332] H. Walther (ed.). *EPS10, Trends in Physics*, Abstracts of Contributed Papers of 10th General Conference of the European Physical Society, Sevilla, September 9–13, 1996.

[333] G. Gaigalas and Z. Rudzikas, *J. Phys. B: At. Mol. Opt. Phys.*, **29**, 3303–3318 (1996).

Index